AF457854

Finite Elements in Action

Finite Elements in Action

Modeling Quantum Mechanics and Electrodynamics in Nanoscale Systems

L. Ramdas Ram-Mohan

Worcester Polytechnic Institute

OXFORD
UNIVERSITY PRESS

Great Clarendon Street, Oxford, OX2 6DP,
United Kingdom

Oxford University Press is a department of the University of Oxford.
It furthers the University's objective of excellence in research, scholarship,
and education by publishing worldwide. Oxford is a registered trade mark of
Oxford University Press in the UK and in certain other countries.

© L. Ramdas Ram-Mohan 2026

The moral rights of the author have been asserted.

All rights reserved. No part of this publication may be reproduced, stored in a retrieval system, transmitted, used for text and data mining, or used for training artificial intelligence, in any form or by any means, without the prior permission in writing of Oxford University Press, or as expressly permitted by law, by licence or under terms agreed with the appropriate reprographics rights organization. Enquiries concerning reproduction outside the scope of the above should be sent to the Rights Department, Oxford University Press, at the address above.

You must not circulate this work in any other form
and you must impose this same condition on any acquirer.

Links to third party websites are provided by Oxford in good faith and
for information only. Oxford disclaims any responsibility for the materials
contained in any third party website referenced in this work.

Published in the United States of America by Oxford University Press
198 Madison Avenue, New York, NY 10016, United States of America

British Library Cataloguing in Publication Data
Data available

Library of Congress Control Number: 2025940071

ISBN 9780199563487

DOI: 10.1093/oso/9780199563487.001.0001

Printed and bound by
CPI Group (UK) Ltd, Croydon, CR0 4YY

The manufacturer's authorized representative in the EU for product safety is
Oxford University Press España S.A. of Parque Empresarial San Fernando de Henares,
Avenida de Castilla, 2 – 28830 Madrid (www.oup.es/en or product.safety@oup.com).
OUP España S.A. also acts as importer into Spain of products made by the manufacturer.

Preface

The central focus of this book is the elucidation of the interplay between the principle of stationary action and Schrödinger's equation, and its solution using the finite element method (FEM), in physical systems whose dimensions are on the order of nanometers. The treatment of the dynamics of electrons in such systems deserves a quantum mechanical description. Typical applications at the nanoscale also require the modeling of electrodynamic fields. For instance, nanoscale semiconductor laser design requires the interplay between electrons and photons to be modeled simultaneously. The coupling of electrons and photons in such systems leads us to consider the FEM for solving problems in Maxwell's electrodynamics along similar lines. In this book we explore the issues of developing variational methods and their implementation for several physical examples in the framework of the FEM and treat issues that are very common in modeling nanoscale systems.

The intended audience: The material presented in this book is aimed at senior undergraduates, graduate students, and those interested in computational modeling of the physical properties of nanoscale systems. The range of applications of the FEM with concrete examples presented here should be of interest to researchers specializing in many areas of physics, chemistry, material science, and other disciplines engaged in nanoscience. At the same time, this is not a cook-book with recipes for computer programs – other resources exist for that purpose. I have successfully engaged undergraduate students, including promising first year students, in exposing them to variational methods, the principle of stationary action, and their intimate connection with the FEM. I have guided them in writing the computer programs needed to work with 1D and 2D problems and produce results in a very short time.

Each chapter begins with a summary of the topics discussed in it. The subject material treated in every chapter with specific examples provides enough background for the definition of the problem under consideration, and results are given for typical problems in the area. References at the end of each chapter provide further guidance to the reader. Some of the chapters end with suggested problems. While the variational approach is emphasized throughout, I also show how Galerkin's approach to FEM is employed, and its equivalence to the principle of stationary action in cases where the differential equation being considered is derivable from the action. Most chapters are as complete as possible and can be read independently.

Why use the FEM for modeling nanoscale systems? In typical textbook examples, the differential equations that are considered for physical quantities are readily solved by using the methods of classical applied mathematics, such as the method of separation of variables. In simple domains with high symmetry, one can indeed invoke such methods with success. The boundary conditions are hopefully simple enough that analytical methods can provide solutions. However, in situations with complex geometry for the physical domain with the solutions satisfying complicated boundary conditions, we are inexorably moved toward numerical methods of solution. Amongst the general methods of solving differential equations, the FEM is by now well established and appreciated as a flexible and powerful method for several reasons. This power of the FEM arises from the fact that it may be thought of as being the discretization of the action integral; it is therefore applicable to a wide variety of problems. It transcends the geometrical constraints of the physical domain, and it is capable of incorporating essentially any boundary conditions that we are typically faced with in actual applications. The implementation of various boundary conditions is then of central interest in obtaining results appropriate to a particular physical problem. I devote several sections in various chapters displaying methods of doing so in the FEM.

It is this interplay between physics, mathematics, computation, and practical applications that makes the field so exciting, and the method so essential a component in the cycle of conception-optimized design-fabrication-testing-manufacture of devices based on the application of quantum mechanics and electrodynamics. The hope and the ambition is to significantly reduce the time to fruition of novel ideas into practical applications and at the same time to create a methodology that allows the very first prototype application that is constructed to be successful. In this sense, the combination of FEM and quantum mechanics has led to the concept of "wavefunction engineering," a paradigm which has evolved over the past two decades to maturity. Today, the development of new photonic, optoelectronic, and spintronic devices is gaining considerable momentum, and it is our contention that the FEM provides the best framework for their modeling and simulation.

Since the publication of my earlier book[†] there have been a very large number of research publications in the area of finite element modeling of quantum mechanical systems. Here, we will consider a small number of topics which should permit the readers to generalize the approach to problems of interest to them.

The modeling of electromagnetic waveguides, antennas, photonic crystals, and meta-materials require the solution of Maxwell's equations in complex geometries. The present-day trend has been to use what are known as vector finite elements, elements which ensure tangential continuity of the fields across their periphery as required by the curl operators in the field equations. This approach has been well documented in published literature and in textbooks. In this book, I indicate that there are inherent limitations to this method and put forward a new finite element that provides higher accuracy and at the same time allows the treatment of problems

[†] *Finite Element and Boundary Element Applications to Quantum Mechanics*, (Oxford University Press, 2002).

that couple quantum mechanics and electrodynamics. Results based on this approach are shown in some detail so that a researcher new to this area may gain expertise quickly.

The FEM and variational methods are a useful framework for thinking about other applications as well, and examples of these are relegated to the final chapters. The author believes that further applications of FEM to quantum mechanics and electrodynamics, and eventually to quantum field theory should provide a very fertile ground for research. The inclusion of differential geometry and its implementation into FEM is becoming prevalent and popular in recent research. This book is being presented in the hope that this will come true.

Acknowledgments: Several colleagues have influenced the presentation of the material in this book, and I thank them. I mention in particular John Albrecht, Stuart Solin, Anant Ramdas, and Keon-Ho Yoo who have interacted with me on some of the research problems discussed here. Shanshan Rodriguez and Leo Rodriguez showed tremendous excitement and interest in "geometrizing" the EMR phenomenon leading to our collaboration on the general relativistic consequences of the theory. I have interacted with Padmanabhan Aravind over several decades on issues in quantum theory, and this has helped in clarifying ideas.

My former and current students Sathwik Bharadwaj, Chris Boucher, Muping Chen, A. Debnath, Alexi Girgis, Andrei Ilyashenko, Paul Kassebaum, Zehao Li, Jonathan Moussa, Siddhant Pandey, Chris Pierce, Lisa Pugsley, Dzung Pham, and YuChen Wang have enthusiastically worked with me on many of the topics included in this book and have helped shape the presentation of the material. Arya Retheeshan has been working on issues in the phenomena of accidental degeneracy, a topic of long-term interest for me.

Patrick O'Mullan has been very helpful in solving LaTeX issues imaginatively, and I thank him for his help with the placement of tables, figures, and formatting.

I am particularly appreciative of the editorial support provided by Oxford University Press, specially by Dr. Sonke Adlung.

I thank my family for support and patience during the course of my writing this book.

L. R. Ram-Mohan
lrram@wpi.edu

Contents

Part II Quantum Scattering

Part V Electrodynamics

Part I

The Action Integral in Quantum Mechanics

1 Schrödinger's equation and the action

In this chapter:

- We begin by stating Schrödinger's equation for the wavefunction of a particle and describing its properties. It will be assumed that the reader has exposure to quantum theory at the level of an introductory course. The properties of the Schrödinger's wavefunction are defined.
- The boundary conditions for the wavefunction and its derivative that are used for solving Schrödinger's equation are discussed.
- We introduce the principle of stationary action for Schrödinger's equation, a principle that synthesizes much of physics through a single, central concept. The conserved probability current is derived directly from the Schrödinger Lagrangian.

1.1 Schrödinger's equation

The new physical phenomena that were being discovered at the end of the 19th century were such that "classical" concepts in physics were inadequate in explaining them in terms of the theoretical framework of that time. The particle description for the motion of objects and the continuous energy of physical systems, including electromagnetic radiation, that were usually invoked in the description of physical systems seemed to require a dramatic alteration. The idea of the "quantum" nature of electromagnetic radiation was put forward by Planck in 1900 to explain the frequency spectrum of black body radiation. The energy E of each quantum of radiation of frequency ν was postulated to be given by $E = h\nu$, where h is now known as Planck's constant. [1] The resulting energy spectrum agreed with the experimental results on the emission spectrum of a hot object over several orders of magnitude. The idea of energy quantization for electromagnetic radiation, in the form of photons having energy packets of magnitude $h\nu$, was used very effectively by Einstein in 1905 to explain the photoelectric effect.

Finite Elements in Action. L. Ramdas Ram-Mohan, Oxford University Press. © L. Ramdas Ram-Mohan (2026).
DOI: 10.1093/oso/9780199563487.003.0001

Other experiments on the nature of electromagnetic radiation further established that it has the dual nature of behaving as a wave displaying interference and diffraction phenomena, and also as a particle in the emission, absorption, and scattering of electrons in atoms. In 1924, this duality property in turn was extended to physical particles as a theoretical idea by de Broglie who postulated that *any* particle with momentum p has a wave nature associated with it, with a wavelength related to its momentum given by $\lambda = h/p$. This was analogous to the photon whose momentum is related to its wavelength in the same manner. As we go from the macroscopic to the nanoscale regime, the need for the wave description of particles becomes critical, and a wave equation for particles is called for. In 1926, Schrödinger provided a heuristic derivation of the wave equation obeyed by the wavefunction of a particle.†

Schrödinger's wave equation for a particle of mass m moving in a potential $V(\mathbf{r})$ is given by

$$-\frac{\hbar^2}{2m}\nabla^2\Psi(\mathbf{r},t) + V(\mathbf{r})\Psi(\mathbf{r},t) = i\hbar\frac{\partial}{\partial t}\Psi(\mathbf{r},t), \tag{1.1}$$

where $\hbar$ is Planck's constant divided by 2π. [1] The wavefunction of the particle, $\Psi(\mathbf{r},t)$, is a probability amplitude [2–4] such that the probability of finding the particle in a volume element d^3r located at $\mathbf{r}$ is given by $|\Psi(\mathbf{r},t)|^2\, d^3r$. Assuming we have one particle in the physical domain Ω, the probability of finding the particle somewhere in Ω is unity. The wavefunction $\Psi(\mathbf{r},t)$ then satisfies the normalization condition

$$\int_\Omega d^3r \mid \Psi(\mathbf{r},t) \mid^2 = 1. \tag{1.2}$$

Schrödinger's equation is constructed by considering the classical relation for the energy E of the particle

$$E = \frac{p^2}{2m} + V(\mathbf{r}), \tag{1.3}$$

in which the energy and momentum of the particle are represented by differential operators acting on the wavefunction with

$$\begin{aligned} E &\equiv i\hbar\left(\frac{\partial}{\partial t}\right), \\ \mathbf{p} &\equiv -i\hbar\nabla. \end{aligned} \tag{1.4}$$

This correspondence between the derivative operators and E and $\mathbf{p}$ is most directly understood by considering a plane-wave solution for the wavefunction to represent a

†Readers unfamiliar with the equation and the physical implications of quantum mechanics are encouraged to consult any of the several books on the subject that are referenced at the end of this chapter.

free particle moving along the x-direction, and noting that the derivatives acting on $\Psi(x,t) = A\exp(ik_x x - i\omega t)$ lead to

$$
\begin{aligned}
i\hbar\frac{\partial}{\partial t}\Psi(x,t) &= \hbar\omega\,\Psi(x,t) = E\,\Psi(x,t), \\
\hat{p}_x\Psi(x,t) &= -i\hbar\frac{\partial}{\partial x}\Psi(x,t) = \hbar k_x\Psi(x,t).
\end{aligned}
\tag{1.5}
$$

We will frequently use the relations $p_x = \hbar k_x$ and $E = \hbar\omega$, where the wavevector $k_x = 2\pi/\lambda$ and the angular frequency $\omega = 2\pi\nu$ of the wave define its propagation wavelength λ and frequency ν.

1.2 Boundary conditions

We are interested in solving Eq. (1.1) in physical systems of nanometer dimensions with boundary conditions specified by either (i) the particle being confined to the physical domain, or (ii) undergoing a scattering by a potential $V(\mathbf{r})$. With the present advances in technology, we can make structures with different materials and shapes in which these phenomena can be studied. For example, using molecular beam epitaxy, it is possible to grow layered semiconductor structures of GaAs and AlGaAs, atomic layer by layer, in which the layers are in the 2–100 nm range. The composite structure can be tailored to confine electrons in various ways that are determined by the choices in the composition and the layer thickness of the materials. In such systems, the boundary conditions at interfaces are that (a) the wavefunction and (b) the probability current must be continuous across the interfaces, a condition that follows from the conservation of probability for the particle. The probability current density is defined by [5, 6]

$$
\mathbf{j}(\mathbf{r},t) = \frac{\hbar}{2mi}\left[\Psi^*(\mathbf{r},t)\nabla\Psi(\mathbf{r},t) - \left(\nabla\Psi^*(\mathbf{r},t)\right)\Psi(\mathbf{r},t)\right]. \tag{1.6}
$$

We note that the conservation of probability, or the conservation of the number of particles, corresponds to a gauge symmetry, analogous to charge conservation and the gauge symmetry in electrodynamics (see Chap. 18 on the gauge in electrodynamics and its role in quantum mechanics). Later in this chapter, we will exploit this symmetry in order to derive the conserved current directly by a variational method.

If the particle is confined in a given region, the boundary conditions at the periphery of the physical domain would correspond to the wavefunction vanishing at the boundary. This gives rise to bound states, as in the text-book example of a simple rectangular infinite quantum well in one dimension.

In scattering problems such as, electrons incident on a finite potential energy barrier, it is assumed that we have an open system with a given incident flux of electrons that is scattered outwards or is reflected back. We shall then make use of scattering boundary conditions; these identify scattering as a mixed boundary value problem.

In the following, we will present the Schrödinger action from which we can derive Schrödinger's equation and also the current density using variational methods. The finite element method [7–10] is then used to evaluate the action and generate the simultaneous equations that correspond to the discretized Schrödinger's equation. As will become evident in the following chapters, the need for such computational methods arises when we consider potential functions that are not amenable to an analytic treatment and when complex boundary conditions are imposed.

1.3 The action integral and the variational method

The principle of stationary action pervades all of physics and provides a succinct and unifying approach to the description of the underlying dynamics in the physical systems. The action is an integral over time of the quantity called the Lagrangian which can be formulated for physical problems using simple rules and symmetry arguments. We give three examples of this method in the following.

The Lagrangian, L, in classical particle dynamics is the kinetic energy T minus the potential energy V of the particle. Thus, in classical mechanics the action is

$$A = \int dt\, L[q(t), \dot{q}(t)] = \int dt \left(\frac{1}{2} m\dot{q}^2 - V(q) \right), \tag{1.7}$$

where $q(t)$ is a generalized coordinate that is dependent on time, $V(q)$ is the potential energy of the particle of mass m, and square brackets are used around the arguments of L to indicate that it is a functional of $q(t)$ and $\dot{q}(t)$. For a mass attached to a spring, we have a simple harmonic oscillator for which the potential energy is $V(q) = \frac{1}{2}\kappa q^2$, where κ is the spring constant.

The equation of motion is derived by varying the action with respect to the coordinate $q(t)$, by means of a functional variation, and finding its stationary value. In other words, we consider the effect on the action of *changing the function* $q(t)$ *itself* over the entire physical range.[‡] Hamilton's principle states that the action is stationary for small functional variations about that function $q_0(t)$ for which this stationary value holds. Then Hamilton's dynamical principle requires that $q_0(t)$ is the physically allowed solution.

Consider the motion of a particle from an initial fixed coordinate $q(t_a)$ to a final given fixed coordinate $q(t_b)$. The principle of stationary action demands that

$$\begin{aligned} \delta_q A = 0 = \int dt \left[\left(\frac{\delta}{\delta \dot{q}} L[q, \dot{q}] \right) \delta\dot{q} + \left(\frac{\delta}{\delta q} L[q, \dot{q}] \right) \delta q \right] = 0 \\ = (m\dot{q}\delta q) \Big|_{t_a}^{t_b} + \int dt\, \delta q \left(-m\ddot{q} - \frac{\delta V(q)}{\delta q} \right), \end{aligned} \tag{1.8}$$

where an integration by parts has been performed on $\delta\dot{q}$ to obtain the second line of Eq. (1.8). The functional variation and the derivative with respect to time are

[‡]A simple visual analogy of altering the function is provided by a graphics equalizer on an audio system where we adjust the amplitude of the sound output by sliding controls which allows the listener to selectively raise or suppress the amplitude in different frequency windows. Here the shape of the function is altered at every point such as to make the action integral stationary under small changes in the function.

interchangeable. The variation of the function $q(t)$ is zero at the two end values t_a and t_b *since it is specified there and held fixed at the end points.* Hence, $\delta q(t)$ is zero at the end points, leading to the first term being zero. (Note that this is a Dirichlet boundary condition (BC) at two points. Other boundary conditions, such as the Neumann BC and the Cauchy BC will be treated in Chap. 2.)

In the above, we are assuming that the Lagrangian is not explicitly dependent on time. We obtain

$$-\int dt\, \delta q \left(m\ddot{q} + \frac{\delta V(q)}{\delta q} \right) = 0, \tag{1.9}$$

from the principle of stationary action. Since δq is arbitrary within the time interval $t_a < t < t_b$, we obtain the result that the integrand of the above equation, Eq. (1.9), must be zero. This gives us the Euler-Lagrange equation of motion for the particle

$$m\ddot{q} + \frac{\delta V(q)}{\delta q} = 0. \tag{1.10}$$

The solution $q_0(t)$ satisfies this equation of motion. Note that Eq. (1.10) is Newton's second law, $F = -\partial V(q)/\partial q = ma$, where $a = \ddot{q}$ is the acceleration.

It is well appreciated in classical mechanics that the three laws of Newton can be derived from a single idea: Hamilton's principle of stationary action. [11, 12] The vector relations representing the three laws of Newton are replaced by a scalar Lagrangian which is more directly constructed. Because of the symmetries enjoyed by the Lagrangian, we can do so without needing to guess at the details of internal forces.

Schrödinger's equation and its action:

When we consider fields, we allow the Lagrangian L to be the integral over the physical domain of a Lagrangian density $\mathcal{L}[\psi, \partial\psi, \dot{\psi}]$ that is a functional of the fields and their spatial and time derivatives under consideration. The construction of the Lagrangian is usually straightforward. The action is then given by the integral over the Lagrangian with respect to time. With the action at hand, given in terms of fields, the differential equations governing their behavior can be derived using the variational principle applied with field variables.

Here, let us construct the Schrödinger action from the 1D Schrödinger's equation, *thereby reversing the usual procedure of obtaining the differential equation from the principle of stationary action employing a variation of the action with respect to the fields.*

For simplicity, consider the case of stationary states, in which the wavefunction of a quantum state with energy $E = \hbar\omega$ is represented by

$$\Psi(x,t) = \psi(x)e^{-i\omega t}. \tag{1.11}$$

Substitution into Eq. (1.1) leads to the time-independent equation

$$-\frac{\hbar^2}{2m}\frac{d^2}{dx^2}\psi(x) + V(x)\psi(x) = E\psi(x). \tag{1.12}$$

Let us suppose that the boundary conditions on the wavefunction at the ends of the physical region $[x_a, x_b]$ are

$$\psi(x_a) = 0; \quad \psi(x_b) = 0. \tag{1.13}$$

In quantum mechanics, the wavefunction is a complex function in general, [4–6] and to construct the action integral we first multiply Eq. (1.12) by the functional differential $\delta\psi^*(x,t)$ and integrate over space to obtain

$$\int_{x_a}^{x_b} dx\, \delta\psi^*(x) \left[-\frac{\hbar^2}{2m}\frac{d^2}{dx^2}\psi(x) + V(x)\psi(x) - E\psi(x) \right] = 0. \tag{1.14}$$

The kinetic energy term is integrated by parts to write

$$\int_{x_a}^{x_b} dx \left[\delta_{\psi^*}\left(\frac{d}{dx}\psi^*(x)\,\frac{\hbar^2}{2m}\,\frac{d}{dx}\psi(x) \right) + \delta\psi^*(x)\,V(x)\psi(x) \right.$$
$$\left. - E\delta\psi^*(x))\,\psi(x) \right] - \left[\delta\psi^*(x)\frac{\hbar^2}{2m}\,\frac{d}{dx}\psi(x) \right]\Bigg|_{x_a}^{x_b} = 0. \tag{1.15}$$

As seen in the previous example, the functional differential δ_{ψ^*} and the derivative d/dx are interchangeable. Since $\psi(x)$ at x_a and at x_b are held fixed, the variations $\delta\psi^*(x_a) = 0 = \delta\psi^*(x_b)$, so that the contributions to the equation at the end points (what we can call the "surface terms," in anticipation of discussions in higher dimensions) vanish. We then have

$$\delta_{\psi^*} \int_{x_a}^{x_b} dx \left[\left(\frac{d}{dx}\psi^*(x)\,\frac{\hbar^2}{2m}\,\frac{d}{dx}\psi(x) \right) + \psi^*(x)(V(x) - E)\psi(x) \right] = 0. \tag{1.16}$$

An integral of Eq. (1.16) over the time interval $[t_a, t_b]$ is performed, in this case rather trivially since the above equation is not dependent on time. This leads to just a factor of time, T.

We can directly identify Eq. (1.16) as the result from the principle of stationary action, $\delta_{\psi^*} A[\psi, \psi^*] = 0$. As a consequence, we obtain

$$A[\psi,\psi^*] = \int_{t_a}^{t_b} dt \int_{x_a}^{x_b} dx \left[\left(\frac{d}{dx}\psi^*(x)\,\frac{\hbar^2}{2m}\,\frac{d}{dx}\psi(x) \right) + \psi^*(x)(V(x) - E)\psi(x) \right], \tag{1.17}$$

an expression in which we have not included a term that is not a functional of ψ, which may be thought of as the constant of integration in a functional integration. The above Lagrangian is appropriate for the case where we consider only stationary states.

We now return to the issue of conservation of probability. It can be shown [2–4] that the probability current density $\mathbf{j}(\mathbf{r},t)$ and the particle density $\rho(\mathbf{r},t) = \Psi^*(\mathbf{r},t)\Psi(\mathbf{r},t)$ satisfy the equation of continuity based on conservation of probability,

$$\nabla \mathbf{j}(\mathbf{r},t) + \frac{\partial\rho(\mathbf{r},t)}{\partial t} = 0. \tag{1.18}$$

Consider the current and density distributions to be localized in some region of space, this being a reasonable assumption for physical systems. We now integrate Eq. (1.18) over all space. Using Gauss's theorem in vector analysis, the volume integral over the divergence of the probability current can be expressed as a surface integral

$$\int d^3r\, \nabla \cdot \mathbf{j}(\mathbf{r},t) = \int ds\, \hat{\mathbf{n}} \cdot \mathbf{j}(\mathbf{r},t), \tag{1.19}$$

with $\hat{\mathbf{n}}$ being normal to the surface. The surface integral vanishes in the limit of a large radius for the surface on which the current can be considered to be zero; we have assumed that the physical distribution of particles and currents is over a finite volume. Hence, the time variation of the integral over the charge density is zero. In other words,

$$\frac{d}{dt}\int d^3r\, \rho(\mathbf{r},t) = \frac{d}{dt}N = 0, \tag{1.20}$$

where N is the number of particles in the physical system; here, we have no change in the number of particles with time. We are not allowing for particle creation or annihilation, which would require further detailed considerations of the issue of particle conservation. The conservation of particles gives rise to a gauge symmetry which states that the dynamics is invariant under the transformation of the wavefunction[§]

$$\Psi(x,t) \rightarrow e^{i\Lambda(x,t)}\Psi(x,t). \tag{1.21}$$

From Eq. (1.17), we note that the Lagrangian density for 1D for steady state problems is given by

$$\mathcal{L}[\psi^*,\psi] = \left\{\left(\frac{d}{dx}\psi^*(x)\right)\frac{\hbar^2}{2m}\left(\frac{d}{dx}\psi(x)\right)\right\} + \psi^*(x)\Big(V(x) - E\Big)\psi(x). \tag{1.22}$$

Under the gauge transformation $\psi(x) \rightarrow \exp(\Lambda(x))\psi(x)$, the new Lagrangian density is given by

$$\begin{aligned}
\mathcal{L}[\psi^*,\psi,\Lambda] &= \frac{d}{dx}\left(e^{-i\Lambda(x)}\psi^*(x)\right)\frac{\hbar^2}{2m}\frac{d}{dx}\left(e^{i\Lambda(x)}\psi(x)\right) + \psi^*(x)[V(x) - E]\psi(x) \\
&= \left(\frac{\hbar^2}{2m}\right)\Big[\psi'^*(x)\psi'(x) - i\Lambda'(x)\psi^*(x)\psi'(x) + i\Lambda'(x)\psi'^*(x)\psi(x) \\
&\qquad + \Lambda'(x)^2\,\psi^*(x)\psi(x)\Big] + \psi^*(x)[V(x) - E]\psi(x).
\end{aligned} \tag{1.23}$$

Under the gauge transformation of the particle number gauge, the Lagrangian density transforms as [13]

$$\mathcal{L} \rightarrow \mathcal{L} + j(x)\,[\hbar\,\Lambda'(x)] + \frac{\hbar^2}{2m}\Lambda'(x)^2\,\psi^*(x)\psi(x). \tag{1.24}$$

[§] We use the 1D example here for illustrative purposes.

Only the terms in the kinetic energy part of the Lagrangian density contribute to the current in general. Following Gell-Mann and Levy, [14] we define the conserved current as

$$j(x) = \lim_{\Lambda' \to 0} \left(\frac{1}{\hbar} \frac{\delta \mathcal{L}[\Lambda, \psi]}{\delta \Lambda'(x)} \right) = \frac{\hbar}{2mi} \left(\psi^*(x)\psi'(x) - \psi'^*(x)\psi(x) \right), \tag{1.25}$$

as mentioned earlier, in Eq. (1.6). This is one of the simplest examples of gauge variation to obtain a conserved current. The variational procedure based on the Lagrangian provides an unambiguous manner of defining the conserved current even when we have a multi-component wavefunction, or when the terms in the Schrödinger's equation have matrix coefficients, and also when we have an inhomogeneous or a composite material. This is particularly useful in the context of modeling layered quantum semiconductor structures, for example, where we have to include a multi-band description of the electronic states in the individual layers. Furthermore, this method of defining the conserved current comes into its own when finite element method is used in numerical simulations for the electronic band structure of composite materials. We shall elaborate on these applications in following chapters. For a derivation of the probability current directly from Schrödinger's equation, the reader is referred to the usual textbooks on quantum mechanics mentioned at the end of this chapter.

Poisson's equation and the variational method

As a third example, let us consider the Poisson equation satisfied by the electrostatic potential $\phi(\mathbf{r})$ in the presence of a charge distribution $\rho(\mathbf{r})$:

$$\nabla^2 \phi(\mathbf{r}) = -\rho(\mathbf{r})/\varepsilon_0. \tag{1.26}$$

Here, ε_0 is the permittivity of free space. [15] The open domain solution of this differential equation is readily derived [16, 17] since we know the potential of a point charge, and we can treat the charges constituting the charge density to be made up of a large number of point charges. We have

$$\phi(\mathbf{r}) = \frac{1}{4\pi\varepsilon_0} \int d^3r \, \frac{\rho(\mathbf{r})}{|\mathbf{r} - \mathbf{r}'|}. \tag{1.27}$$

It is more difficult to obtain the solution to the Poisson's equation with the boundaries at finite distances from the charge distribution. [18–21] The usual approach is to first obtain the Green's function for the inhomogeneous equation, Eq. (1.26), with the specific boundary conditions at hand, and then evaluate the potential. [22] In 2D and 3D geometries, the solutions can be obtained using this method when the physical region has some symmetry so that the method of separation of variables can be employed to advantage in determining the Green's function. In the following, we consider Poisson's equation in one dimension as an example for setting up the action integral.

Let us begin with the differential equation, Eq. (1.26), reduced to one dimension

$$\frac{d}{dx}\varepsilon_0\frac{d\phi(x)}{dx}+\rho(x)=0, \tag{1.28}$$

with the physical domain corresponding to the interval $x_a \leq x \leq x_b$. Let the function $\phi(x)$ satisfy the boundary conditions

$$\begin{aligned}\phi(x_a) &= V_a,\\ \phi(x_b) &= V_b.\end{aligned} \tag{1.29}$$

We wish to derive the action integral from Eq. (1.28). Since this is a time-independent problem, we need to consider constructing the Lagrangian L for this problem. Here, L is a function of ϕ and its derivative with respect to the coordinate, and no time derivative term is present in it. To construct L from the differential equation and the boundary conditions, we multiply Eq. (1.28) by $\delta\phi$, the incremental change in the functional form of $\phi(x)$. Note that the function itself is varied to obtain $\phi(x)+\delta\phi(x)$. Here the meaning of the functional variation of a given functional $h(\phi)$ is that $\delta_\phi\, h(\phi) = (\delta h(\phi)/\delta\phi)\delta\phi$.

Performing an integration over the physical domain, we obtain

$$\int_{x_a}^{x_b} dx\,\delta\phi\left[\frac{d}{dx}\varepsilon_0\frac{d\phi(x)}{dx}+\rho(x)\right]=0. \tag{1.30}$$

An integration by parts leads to

$$\int_{x_a}^{x_b} dx\left[-\left(\delta\frac{d\phi}{dx}\right)\cdot\varepsilon_0\cdot\frac{d\phi}{dx}+\delta\phi\,\rho(x)\right]+\left[(\delta\phi)\,\varepsilon_0\frac{d\phi}{dx}\right]_{x_a}^{x_b}=0. \tag{1.31}$$

As in the earlier examples, we can freely interchange the functional differential δ and the derivative d/dx. Since $\phi(x_a)$ and $\phi(x_b)$ are fixed by the boundary conditions, we have no variation of the function at the end points, so that $\delta\phi=0$ at x_a and at x_b. The end-point contributions vanish, and we have

$$\int_{x_a}^{x_b} dx\,\delta\left[-\frac{1}{2}\left(\frac{d\phi(x)}{dx}\varepsilon_0\frac{d\phi(x)}{dx}\right)+\phi(x)\rho(x)\right]=0. \tag{1.32}$$

Hence,

$$\delta_\phi L=\delta_\phi\int_{x_a}^{x_b} dx\left[-\frac{1}{2}\left(\frac{d\phi(x)}{dx}\right)\varepsilon_0\left(\frac{d\phi(x)}{dx}\right)+\phi\rho(x)\right]=0. \tag{1.33}$$

Note the factor of 1/2 before the first term which appears when we pull out the functional differential outside the square brackets.¶ We can recognize Eq. (1.28) as arising from the principle of stationary action, or the Lagrangian in this case of no time dependence. Thus,

$$A = T \times L = T \int_{x_a}^{x_b} dx \left[-\frac{1}{2} \left(\frac{d\phi(x)}{dx} \right) \varepsilon_0 \left(\frac{d\phi(x)}{dx} \right) + \phi\rho(x) \right]. \tag{1.34}$$

We have thus constructed the action

$$A = TL = \int_0^T dt \int_{x_a}^{x_b} dx \left[-\frac{1}{2} \left(\frac{d\phi}{dx} \right) \varepsilon_0 \left(\frac{d\phi}{dx} \right) + \phi \frac{\rho(x)}{\varepsilon_0} \right], \tag{1.35}$$

from which we can derive the "equation of motion" – our differential equation for the field $\phi(x)$ by reversing the order of the above steps. A constant of integration is dropped here as not being relevant to the physics and to the variational method. [11, 12, 23–25]

1.4 Final comments

This chapter introduced the Schrödinger's equation and simple properties of the wavefunction. The boundary conditions for the wavefunction at interfaces between two regions are the continuity of the wavefunction and the continuity of the probability current $\mathbf{j}(\mathbf{r}, t)$ defined in Eq. (1.6) across the interface. For bound states, the wavefunction, and hence the current, vanishes at the boundaries. For scattering problems, the incoming probability current for particles incident on the physical region must be conserved after scattering. We are content in this chapter with this limited introduction to quantum mechanics.

The principle of stationary action was introduced, and we have considered three standard examples of setting up the action and deriving the equation of motion. The steps involved in deriving the action appropriate for a given differential equation were displayed through the examples. The motivation for setting up the action and the remarkable advantages of using it in a finite element framework will become apparent in the following chapters. Here, we mention that the finite element approach is to discretize the action integral. The fields appearing in the action are represented in each element by a polynomial representation in terms of a few parameters in each element. The spatial integrals can then be evaluated in each finite element. The continuity across the finite elements is ensured while putting together the full discretized action integral. The result is that the action is simplified to a bilinear function of the interpolation parameters; the functional variational principle is then reduced to a variation of the parameters. The principle of stationary action is then reduced to a set of simultaneous equations for the parameters that are then determined by solving the equations. More on these concepts is presented in the following chapters.

¶This factor of 1/2 did not occur in the Schrödinger Lagrangian since the variation was performed using δ_{ψ^*} and because the wavefunction is complex in quantum mechanics.

1.5 Problems

1. The Helmholtz equation in 1D is given by $f''(x) + k^2 f(x) = 0$, where the function $f(x)$ is defined over the physical region $a \leq x \leq b$. Construct (i) the Lagrangian density, (ii) the Lagrangian, and (iii) the action for this differential equation. The boundary conditions are given by $f(a) = 0 = f(b)$.
2. Consider the 1D Helmholtz equation: $f''(x) + k^2 f(x) = 0$, where the function $f(x)$ is defined over the physical region $a \leq x \leq b$. Suppose now that the boundary conditions are $f'(a) = \mathcal{G}$ and $f(b) = \mathcal{H}$. Construct (i) the Lagrangian density, (ii) the Lagrangian, and (iii) the action for this differential equation.
3. The wavefunction for a quantum mechanical simple harmonic oscillator satisfies the differential equation

$$-\frac{\hbar^2}{2m}\frac{d^2\psi(x)}{dx^2} + \frac{1}{2}m\omega^2 x^2 \psi(x) = E\psi(x), \tag{1.36}$$

with the boundary conditions $\psi(\pm\infty) = 0$.

(a) Show that the equation reduces to

$$\frac{d^2\psi(\tilde{x})}{d\tilde{x}^2} + (\epsilon - \tilde{x}^2)\psi(\tilde{x}) = 0, \tag{1.37}$$

when the coordinate is rescaled using $x = \tilde{x}\ell_0$, with $\ell_0^2 = \hbar/(m\omega)$. Here the dimensionless quantity ϵ is the energy E divided by the factor $(\hbar\omega/2)$.

(b) Construct the Lagrangian density and the Lagrangian in the scaled variables. Write down the action for this problem.

References

[1] We will use the SI or MKS system of units, in which Planck's constant is $h = 6.626\,068\,96 \times 10^{-34}$ J s, and $\hbar = h/2\pi = 1.054\,571\,628 \times 10^{-34}$ J s. It is also useful to "tame" these numbers by multiplying them by the velocity of light c and converting Joules into electron-Volts (eV). Thus $\hbar_c = 197.326\,9631$ eV nm.

[2] D. Bohm, *Quantum Theory* (Prentice Hall, Englewood Cliffs, NJ, 1951).

[3] J. L. Powell and B. Crasemann, *Quantum Mechanics* (Addison-Wesley, New York, 1961).

[4] D. J. Griffiths, *Introduction to Quantum Mechanics*, 2nd edition (Benjamin Cummings, NY, 2004).

[5] L. I. Schiff, *Quantum Mechanics*, 3rd edition (McGraw-Hill, New York, 1968).

[6] L. D. Landau and E. M. Lifshitz, *Quantum Mechanics – Nonrelativistic Theory* (Pergamon, London, 1965).

[7] K.-J. Bathe and E. Wilson, *Numerical Methods in Finite Element Analysis* (Prentice Hall, Englewood Cliffs, NJ, 1976); K.-J. Bathe, *Finite Element Procedures in Engineering Analysis* (Prentice Hall, Englewood Cliffs, NJ, 1982); K.-J. Bathe, *Finite Element Procedures* (Prentice Hall, Englewood Cliffs, NJ, 1996).

[8] O. C. Zienkiewicz and Y. K. Cheung, *Finite Element Methods in Structural and Continuum Mechanics* (McGraw-Hill, New York, 1967); O. C. Zienkiewicz, *The Finite Element Method* (McGraw-Hill, New York, 1977).
[9] T. J. R. Hughes, *The Finite Element Method* (Prentice Hall, Englewood Cliffs, NJ, 1987).
[10] L. R. Ram-Mohan, *Finite Element and Boundary Element Applications to Quantum Mechanics* (Oxford University Press, Oxford, UK, 2002).
[11] L. D. Landau, and E. M. Lifshitz, *Mechanics*, Volume 1 of *Course of Theoretical Physics*, 3rd edition (Pergamon Press, Oxford, England, 1976).
[12] H. Goldstein, C. P. Poole, and J. L. Safco, *Classical Mechanics*, 3rd edition (Addison-Wesley, New York, 2001).
[13] J. Schwinger, Phys. Rev. **81**, 914–937 (1951); "The theory of quantized fields, I". Phys. Rev. **91**, 713–727 (1953); "The theory of quantized fields II."
[14] M. Gell-Mann and M. Levy, Il Nuovo Cimento **16**, 705–726 (1960); "The axial vector current in beta decay."
[15] In the SI or MKS units the potential is measured in Volts, the charge density ρ is given in Coulombs/m^3, and the permittivity is given by $\varepsilon_0 = 8.854\,187\,817 \times 10^{-12}$ Farad/m.
[16] D. J. Griffiths, *Introduction to Electrodynamics*, 3rd edition (Prentice-Hall, Upper Saddle River, NJ, 1999).
[17] J. R. Reitz, F. J. Milford, and R. W. Christy, *Foundations of Electromagnetic Theory*, 4th edition (Addison-Wesley, Reading, MA, 1992).
[18] R. Courant and D. Hilbert, *Methods of Mathematical Physics* (Interscience Publishers, NY, 1953).
[19] P. M. Morse and H. Feshbach, *Methods of Theoretical Physics* (McGraw-Hill Book Co., NY, 1953).
[20] L. A. Pipes and L. R. Harvill, *Applied Mathematics for Engineers and Physicists*, 3rd edition (McGraw-Hill Book Co., NY, 1971).
[21] D. A. McQuarrie, *Mathematical Methods for Scientists and Engineers* (University Science Books, Sausalito, CA, 2003).
[22] J. D. Jackson, *Classical Electrodynamics*, 3rd edition (John Wiley & Sons, New York, 1999).
[23] C. Lanczos, *The Variational Principles of Mechanics* (Dover Publications, NY, 1986).
[24] W. Yourgrau and S. Mandelstam, *Variational Principles in Dynamics and Quantum Theory* (Dover, New York, 1968).
[25] R. Weinstock, *Calculus of Variations* (Dover, New York, 1974).

2

Action, FEM, and BCs

In this chapter:

- We develop a unique symbolic way of setting up the discretization of the action integral and its evaluation using the method of finite elements. We first consider the Poisson's equation in order to introduce the concepts before working with the Schrödinger's equation.
- Without explicitly evaluating finite element matrices, we can discuss the form of element matrices, and the assembly of element matrices into the global matrix with its occupancy pattern for the physical problem. The procedure of applying BCs within the finite element method (FEM) can then be described in detail without an explicit evaluation of the element/global matrix elements. This critical chapter provides an introduction to the implementation of the FEM and ways of imposing BCs. With derivative BCs, we show how to define a modified action that is consistent with the BCs and that is amenable to an FEM treatment.
- We show how Dirichlet and Neumann BCs are imposed in FEM in the action integral formulation as well as in the Galerkin approach in some detail.
- We consider the example of 1D scattering to illustrate the application of mixed or Cauchy BCs for Schrödinger's equation. Eigenvalue problems with Schrödinger's equation are deferred to a later chapter. The 1D scattering example illustrates the method of modifying the action to include the mixed BCs. The scattering problem is then seen to be amenable to a variational description, so that this powerful method can be brought to bear to improve solutions in a systematic manner.

Note that experience with 1D FEM is crucial for success in setting up and solving 2D and 3D problems using the FEM. It is for this reason that this chapter has a somewhat lengthy description of the procedures in the calculations. The reader is encouraged to solve problems at the end of the chapter.

Finite Elements in Action. L. Ramdas Ram-Mohan, Oxford University Press. © L. Ramdas Ram-Mohan (2026).
DOI: 10.1093/oso/9780199563487.003.0002

2.1 Action and FEM

In Chap. 1, we introduced the action integral for classical particle dynamics, and showed how we can derive the action integral for the Schrödinger's and Poisson's equations. By first using the Poisson's equation in the following, we will illustrate the method of finite elements and the application of BCs. In this chapter, we show the advantages of the action formulation over the usual procedure of solving the differential equation directly.

The action A is a number and has the units of energy $\times$ time. It is the integral over time of the Lagrangian, L; this is, in turn, expressed as an integral over space of the Lagrangian density, $\mathcal{L}$, which is expressed as a function of fields. For the Poisson problem, as noted in Chap. 1, the action is

$$\begin{aligned} A &= \int dt\, L = \int dt \int dx\, \mathcal{L}[\phi(x),\, d\phi(x)/dx] \\ &= T \int_{x_a}^{x_b} dx \left[-\frac{1}{2} \left(\frac{d\phi(x)}{dx}\, \varepsilon_0\, \frac{d\phi(x)}{dx} \right) + \phi(x)\rho(x) \right]. \end{aligned} \tag{2.1}$$

The factor of $1/2$ in the first term arises because we are considering real fields.† The charge distribution density $\rho(x)$ is a given function that is the source of the electric potential ϕ. Here the Lagrangian has no explicit time dependence, and the potential is assumed to be independent of time. Hence, the Lagrangian also does not include any time derivative of the potential, $d\phi/dt$. The time integral in the above equation is then trivial, and we could define our variational principle to apply directly to the Lagrangian. Let the BCs on the potential be $\phi(x_a) = V_a$, at $x = x_a$, and $\phi(x_b) = V_b$, at $x = x_b$.

In the FEM, we directly evaluate the action with what we can consider as variational guesses for the form of the function $\phi(x)$. We remind the reader that in the usual physics textbook applications of the variational method, a global form is typically assumed for $\phi(x)$ over the entire physical domain. For example, $\phi(x)$ could be a sum over functions of the form $a_n\, x'^n/x^{n+1}$ for $x > x'$, or for the quantum harmonic oscillator, the choice $h_n(x)\, \exp(-x^2/2)$ with h_n being a polynomial, for instance. The spatial dependence of the Lagrangian density is then integrated out. The action is thus reduced to a simple bilinear function of the parameters a_n that occur in the assumed form of $\phi(x)$. These parameters are varied to minimize the action and are determined through the minimization condition. Once the parameters are obtained in this variational sense, we have the desired solution consistent with the assumed form of the function, with the parameters that are consistent with the principle of stationary action. This in essence has been the traditional approach. The limitation of this approach is that the assumed global functions may not have the freedom or flexibility in the parameterization to reach the stationary value or an acceptable minimum of the action.

†With the Schrödinger fields being complex, we did not have this factor in the Schrödinger action in Chap. 1; the variation of the action was performed with the combination of the real and imaginary components of $\psi^*(x)$ simultaneously.

Here, we allow for more flexibility in the choice of $\phi(x)$ by discretizing the integral into smaller regions called *elements.* [1–4] The functions are then chosen in terms of suitable parameters within each element *iel.* We write

$$A/T = L \cong \sum_{iel} L^{(iel)}. \tag{2.2}$$

Let us suppose that the physical domain of integration is $x_a \le x \le x_b$ and that we have split this region into "*nelem*" elements [5] of equal length as a matter of convenience. In a given element with index[‡] *iel*, ($iel = 0, 1, \ldots, nelem - 1$), the coordinate x ranges over $[x_{(iel)}, x_{(iel+1)}]$. Here x is the global coordinate.

It is usual to define a "standard element" with a local coordinate ξ ranging over $[-1, 1]$. This choice for the local element is made because of the eventual use of numerical integration methods to evaluate L, and we return to this issue in the following. Any function given in the element *iel* at the global coordinate x is now mapped to a function in the range $[-1, 1]$ of the standard element. The relation between global and local coordinates is linear and given by

$$x = \frac{x_{(iel+1)} + x_{(iel)}}{2} + \frac{x_{(iel+1)} - x_{(iel)}}{2}\xi, \tag{2.3}$$

and conversely,

$$\xi = \left(x - \frac{x_{(iel+1)} + x_{(iel)}}{2}\right) \Big/ \left(\frac{x_{(iel+1)} - x_{(iel)}}{2}\right). \tag{2.4}$$

We express the field $\phi(x)$ in terms of local interpolation polynomials $N_i(\xi)$, called *shape functions*, multiplied by as-yet unknown coefficients ϕ_i called nodal variables. [6] The polynomials N_i are defined to be unity at special points ξ_i called *nodes* and satisfy the condition

$$N_i(\xi_j) = \delta_{ij}; \quad (i, j = 0, 1, \ldots, nodelm - 1). \tag{2.5}$$

These are called Lagrange interpolation polynomials. Let there be *nodelm* nodes in each element. They are labeled with index *inode*, (= 0, 1, ..., *nodelm* − 1). Since the mapping Eq. (2.3) between x and ξ is linear, we can write

$$\phi(x) = \sum_{i=0}^{nodelm} \phi_i \, N_i(\xi(x)). \tag{2.6}$$

[‡]Unless otherwise stated, the index in arrays is chosen to start with *zero* rather than unity; this is usual for programming in the *C*-language. We call such an array to be a 0-based array, as opposed to a 1-based array.

The shape function N_i has been chosen to be unity at the node i, while it is zero at other nodes j, $(j \neq i)$. These are the properties of Lagrange polynomials, and the nth-degree Lagrange polynomial is given by

$$N_i(\xi) = \prod_{\substack{j=0 \\ j \neq i}}^{nodelm-1} \frac{(\xi - \xi_j)}{(\xi_i - \xi_j)}. \tag{2.7}$$

If the nodal values ϕ_i were known, then we would reproduce $\phi(x)$ exactly at the nodes using Eq. (2.6), and $\phi(x)$ would be represented approximately between nodes by Eq. (2.6) as a sum of shape functions modulated by the nodal values. Here the *nodelm* polynomials N_i are of degree $(nodelm - 1)$.

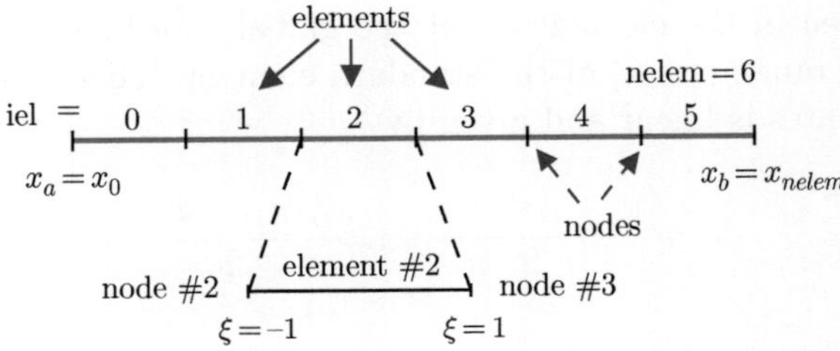

Figure 2.1 The physical region is divided into elements as shown. The nodes at the boundaries of the elements are indicated. Each of the physical elements is mapped on to the standard element over $[-1, 1]$, as shown for one of the elements. The array indices for elements and nodes are 0-based.

The reader is encouraged to work with linear interpolation polynomials in each element to work out the above for an explicit example. In this case, an element with a node at each end at $\xi_i = \mp 1$ allows us to write these polynomials as

$$N_1(\xi) = \frac{(1 - \xi)}{2}; \quad N_2(\xi) = \frac{(1 + \xi)}{2}, \tag{2.8}$$

where we see that $N_i(\xi_j) = \delta_{ij}$.

The shape functions are postulated to be valid only within the given element and are zero outside it. In each of the elements we can use the same set of interpolation polynomials since each element is mapped onto the same standard element. Of the *nodelm* nodes, two are placed at the beginning and at the end of each element. The internal nodes are typically distributed in an equidistant manner. The end node of an element coincides with the beginning node of the next element so that the value of the solution $\phi(x)$ is continuous across the element boundary. This follows from the element continuity, since the elements will share the same nodal value of $\phi^{(iel)}_{nodelm-1} = \phi^{(iel+1)}_0$. The Lagrange interpolation polynomials are said to provide function continuity, and this is denoted by $\mathcal{C}_{(0)}$-continuity.

2.2 Implementing Gauss quadrature over elements

Let us return to the issue of using the range $[-1, 1]$ for the standard element. It is usual to perform the integration in Eq. (2.1) by numerical methods using Gauss-Legendre quadrature. In this integration scheme, the properties of Legendre polynomials are used to write any integral I in the form

$$\begin{aligned} I &= \int_a^b f(x)dx = \int_{-1}^{+1} d\xi \left(\frac{dx}{d\xi}\right) f(x(\xi)) \\ &\cong \sum_{ig}^{ngaus} \left(\frac{b-a}{2}\right) f(x(\xi_{ig}))\, w_{ig}, \end{aligned} \tag{2.9}$$

where the Jacobian of transformation from $[a, b]$ to $[-1, 1]$ is given very simply by $|J| = dx/d\xi = (b-a)/2$. In the last step above, the integral is approximated by evaluating the function at $ngaus$ special points ξ_{ig}, called Gauss points, and multiplied by the corresponding weights w_{ig}. The values of ξ_{ig} and w_{ig} are predetermined and obtained from a look-up table stored on the computer. The order of the Gauss quadrature is given by the number of points $ngaus$ used in the calculation of the integral.[§]

To get another perspective on the formula, Eq. (2.9), we can give an analogous calculation by noting that if the area under the curve is discretized into rectangles, the area is given by the sum of the product of the value of the function at the mid-point of the rectangles (the height) and the width of the rectangles. The special feature of the Gauss method is that it approximates $f(x(\xi))$ by a polynomial of degree $p = 2 \times ngaus - 1$ but evaluates the function only at $ngaus$ points. Thus, if $f(x)$ were a polynomial of degree p, the method provides an exact (double precision) value for the integral with $ngaus$ evaluations of the function. [7] Gauss quadrature is very much more accurate than the rectangular discretization of the area under the curve for the same number of function evaluations. In simple terms, in the rectangular breakup of the area under the curve, the function is treated as a constant in each rectangle, whereas the Gauss method fits the function to a high order polynomial and obtains the area under the curve by integrating the polynomial. The Gauss points and weights are stored in double precision in order to get ~ 15 digit accuracy for a polynomial integrand of order p. A typical look-up table could contain points and weights for say $ngaus$ = 2, 4, 8, 12, 16 and 20; this would suffice for most applications.

2.3 Linear interpolation and element matrix calculations

Let us for the moment suppose that linear interpolation suffices to represent $\phi(x)$ in each element. In Fig. 2.1, the physical range is divided into $nelem = 6$ elements labeled by iel, ($iel = 0, 1, \ldots, nelem - 1$). The ($nelem + 1$) nodes range over the values

[§]The Gauss quadrature method had to await the development of the computer in order to become widely used, even though it was recognized that it allows integrals to be evaluated very accurately and efficiently.

(0, 1, ..., 6). Then $\phi(x)$ in the element *iel* is given in terms of its values at the two ends of the element,

$$\phi(x) = \phi_{(iel)} \frac{x_{(iel+1)} - x}{x_{(iel+1)} - x_{(iel)}} + \phi_{(iel+1)} \frac{x - x_{(iel)}}{x_{(iel+1)} - x_{(iel)}}. \tag{2.10}$$

Here $\phi_{(iel)}$ and ϕ_{iel+1} are called the *nodal values* of $\phi(x)$ at the *nodes* located at $x_{(iel-1)}$ and $x_{(iel)}$. Their polynomial coefficients are the linear Lagrange interpolation polynomials. [6] Expressed in terms of the local coordinate variable, Eq. (2.10) is

$$\phi(x) = \phi_{(iel)} \left(\frac{1-\xi}{2}\right) + \phi_{(iel+1)} \left(\frac{1+\xi}{2}\right). \tag{2.11}$$

Note that the shape functions

$$N_0(\xi) = \frac{1-\xi}{2}; \quad N_1(\xi) = \frac{1+\xi}{2}, \tag{2.12}$$

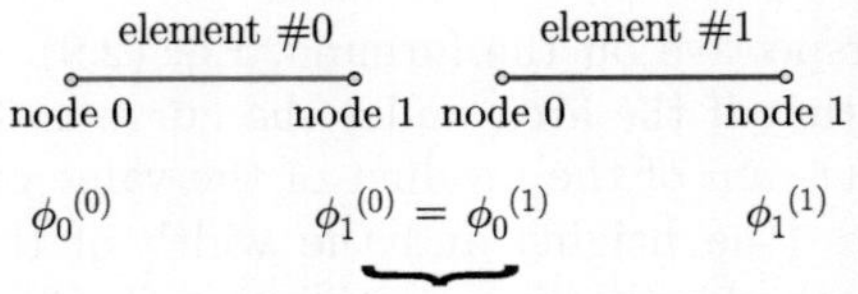

Figure 2.2 Two adjacent elements with their nodes and their nodal variables are shown.

have the required property: $N_0 = 1$ at $\xi = -1$, and it is zero at the other node at $\xi = 1$, while $N_1 = 0$ at $\xi = -1$ and equals unity at $\xi = +1$. We have the Lagrangian in the element *iel* given by

$$L^{(iel)} = \int_{x_{(iel)}}^{x_{(iel+1)}} dx \left[-\frac{1}{2}\phi'(x)\varepsilon_0\phi'(x) + \phi(x)\rho(x) \right]. \tag{2.13}$$

The interpolated form of the field $\phi(x)$ is substituted into Eq. (2.13). The derivative of $\phi(x)$ is expressed as

$$\begin{aligned} \frac{d\phi(x)}{dx} &= \left(\phi_{(iel)} \frac{d}{d\xi} N_0(\xi) + \phi_{(iel+1)} \frac{d}{d\xi} N_1(\xi) \right) \frac{d\xi}{dx} \\ &= \left(\phi_{(iel)} \left(-\frac{1}{2}\right) + \phi_{(iel+1)} \left(\frac{1}{2}\right) \right) \frac{2}{(x_{(iel+1)} - x_{(iel)})}. \end{aligned} \tag{2.14}$$

This substitution allows us to integrate the spatial dependence of the fields in the Lagrangian, leaving the unknown nodal coefficients ϕ_i in L. Note that in the integral, the change of variables from x to ξ requires the Jacobian of transformation

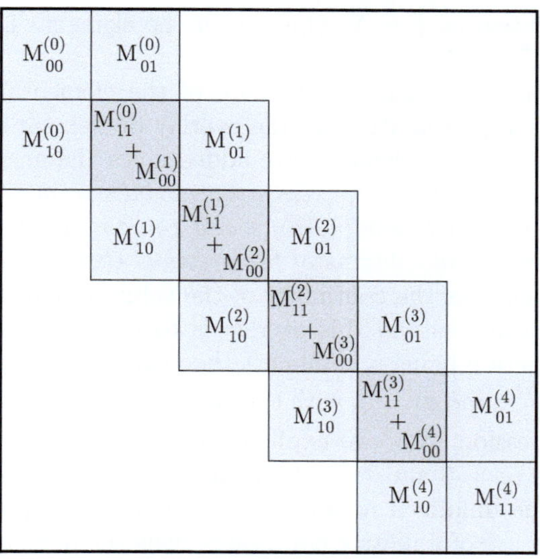

Figure 2.3 In a 1D problem with linear elements, the global matrix representing a part of the action integral is composed of the element matrices overlaid as shown. This overlay ensures inter-element continuity and corresponds to the action being a sum over the action discretized into elemental actions.

$dx/d\xi = (x_{(iel+1)} - x_{(iel)})/2$. In the present instance, we have for the first term in the integral

$$\begin{aligned} L_1^{(iel)} &= -\frac{1}{2}\int_{-1}^{1} d\xi \, \frac{dx}{d\xi}\,[\phi(x)\varepsilon_0\phi'(x)] \quad (x_{iel} \leq x \leq x_{iel+1}) \\ &= -\frac{1}{2}\frac{\varepsilon_0}{(x_{(iel+1)} - x_{(iel)})}\left[\phi_{iel}^2 - 2\phi_{iel}\phi_{iel+1} + \phi_{iel+1}^2\right], \end{aligned} \tag{2.15}$$

and the second term is

$$\begin{aligned} L_2^{(iel)} &= \phi_{iel}\cdot\int_{-1}^{1} d\xi \frac{(x_{(iel+1)} - x_{(iel)})}{2}\,[N_0(\xi)\rho(x(\xi)] \\ &+ \phi_{iel+1}\cdot\int_{-1}^{1} d\xi \frac{(x_{(iel+1)} - x_{(iel)})}{2}\,[N_1(\xi)\rho(x(\xi))]\,. \end{aligned} \tag{2.16}$$

We can write the resulting $L^{(iel)}$ in a matrix form:

$$\begin{aligned} L^{(iel)} &= -\frac{1}{2}\{\phi_{(iel)}, \phi_{(iel+1)}\}\begin{bmatrix} M_{(iel),(iel)} & M_{(iel),(iel+1)} \\ M_{(iel+1),(iel)} & M_{(iel+1),(iel+1)} \end{bmatrix}\begin{pmatrix} \phi_{(iel)} \\ \phi_{(iel+1)} \end{pmatrix} \\ &+ \{\phi_{(iel)}, \phi_{(iel+1)}\}\begin{pmatrix} R_{(iel)} \\ R_{(iel+1)} \end{pmatrix}, \end{aligned} \tag{2.17}$$

with R_i being the integrals $\int dx\, N_i(\xi)\rho(x)$ over the element. The matrix M is called the element matrix.¶

The total action corresponds to the sum of the elemental actions, and we can construct its matrix representation by the overlay of element matrices into a global matrix. The overlaid matrix elements are added since the action integral sums over the elemental actions. *Furthermore, since the nodal variables are common between adjacent elements, the corresponding rows and columns of the element matrix have to be added to the rows and columns of the previous element that is inserted into the global matrix.* This ensures the continuity of the solution across.

A simple way to understand this overlay is to write out the action as a sum of two elements. The action being a number is then expressed as a bilinear expansion in $\phi_0^{(iel=0)}, \phi_1^{(iel=0)}, \phi_0^{(iel=1)}$, and $\phi_1^{(iel=1)}$, with the integrals from the two elements contributing to the full expression. Now we explicitly equate $\phi_1^{(iel=0)} = \phi_0^{(iel=1)}$ for continuity across the elements $iel = 0$ and $iel = 1$. Relabeling the nodal values in a continuous manner, we have the unknown nodal values as a three-component array ϕ_0, ϕ_1, ϕ_2, and we see that the use of a matrix notation reveals the overlay pattern as described above, and for more elements as in Fig. 2.3.

For linear elements, the matrix occupancy pattern for the global matrix can be determined very easily. This helps us proceed with the following section without having to actually evaluate the matrices. Writing N for *nelem* for compactness, we have

$$L = -\frac{1}{2}\{\phi_0, \phi_1, \ldots, \phi_N\}\begin{bmatrix} M_{00} & M_{01} & & & & 0 \\ M_{10} & M_{11} & M_{12} & & & 0 \\ 0 & M_{21} & M_{22} & & & \vdots \\ \vdots & & & \ddots & & 0 \\ 0 & \cdots & \cdots & 0 & M_{N-1,N-1} & M_{N-1,N} \\ 0 & \cdots & \cdots & 0 & M_{N,N-1} & M_{N,N} \end{bmatrix}\begin{pmatrix} \phi_0 \\ \phi_1 \\ \phi_2 \\ \vdots \\ \vdots \\ \phi_N \end{pmatrix}$$

$$+ \{\phi_0, \phi_1, \ldots, \phi_N\}\begin{pmatrix} R_0 \\ R_1 \\ \vdots \\ R_N \end{pmatrix}$$

$$= -\frac{1}{2}\phi_\alpha M_{\alpha,\beta}\phi_\beta + \phi_\alpha R_\alpha; \quad (\alpha, \beta = 0, 1, \ldots, N). \tag{2.18}$$

The matrix element M_{11}, for example, is composed of the sum of the matrix elements from the first and the second element matrices,

$$M_{11} = M_{11}^{(iel=0)} + M_{00}^{(iel=1)}.$$

The matrix M is seen to be tridiagonal with linear elements (i.e. elements with two nodes and linear shape functions.) At this point, the Lagrangian is a bilinear function

¶The subscripts on M, the row and column indices of the matrix, are sometimes separated by a comma for clarity.

of the nodal variables, ϕ_i. We now invoke the variational principle to set $\delta_{\phi_\alpha} A = 0$. With no time dependence in the present problem, we have $\delta_{\phi_\alpha} A \equiv 0 = T\delta_{\phi_\alpha} L$. From Eq. (2.18), this nodal variational procedure leads to

$$\delta_{\phi_\alpha} L = -M_{\alpha\beta}\phi_\beta + R_\alpha = 0.$$

The discretized equation of motion is seen to be of the form

$$\begin{bmatrix} M_{00} & M_{01} & & & \\ M_{10} & M_{11} & & & \\ & \ddots & \ddots & \ddots & \\ & & \ddots & M_{N-1,N-1} & M_{N-1,N} \\ & & & M_{N,N-1} & M_{N,N} \end{bmatrix} \begin{pmatrix} \phi_0 \\ \phi_1 \\ \vdots \\ \phi_{N-1} \\ \phi_N \end{pmatrix} = \begin{pmatrix} R_0 \\ R_1 \\ \vdots \\ R_{N-1} \\ R_N \end{pmatrix}, \tag{2.19}$$

before the BCs are implemented.

2.3.1 Applying Dirichlet BCs to the action

We have yet to apply the BCs associated with the problem at hand. Here ϕ_0 is the solution value at x_a and is given by the BC to be V_a. Similarly, $\phi_N = V_b$, according to the BC at x_b. In the $N+1$ set of equations (Eqs. 2.19), we have terms involving $\phi_0 = V_a$ and $\phi_N = V_b$ which are known, and hence they should be moved to the right side of the simultaneous equations, Eq. (2.19), with a change of sign; we are then left with $N+1$ equations for $N-1$ variables!

After the rearrangement of the terms in the first and the last column, we have the equations

$$\begin{bmatrix} 0 & M_{01} & & & & \\ 0 & M_{11} & M_{12} & & & \\ & M_{21} & \ddots & \ddots & & \\ & & \ddots & \ddots & \ddots & \\ & & & \ddots & M_{N-1,N-1} & 0 \\ & & & & M_{N,N-1} & 0 \end{bmatrix} \begin{bmatrix} \phi_0 \\ \phi_1 \\ \vdots \\ \vdots \\ \vdots \\ \phi_N \end{bmatrix} = \begin{bmatrix} R_0 - M_{00}V_a \\ R_1 - M_{10}V_a \\ R_2 \\ \vdots \\ R_{N-2} \\ R_{N-1} - M_{N-1,N}V_b \\ R_N - M_{N,N}V_b \end{bmatrix}. \tag{2.20}$$

Without the implementation of BCs, the global matrix M is not invertible as can be verified by constructing a global matrix with a few elements. With numerical evaluation, it is found that the first equation is linearly dependent on the others so that a consistent solution to the simultaneous equations requires that we drop the first equation. A similar consideration holds for the last equation. We therefore delete the entries in the first and the last rows in the matrix M in Eq. (2.20). The corresponding columns are already free of any entries since the first and last terms in all the equations have been moved to the right side. We now have $N-1$ equations for $N-1$ unknowns, so that we may proceed with their solution.

In order to maintain the size of the matrix M, we can enter $M_{00} = 1$ and $M_{NN} = 1$ along the diagonal and supply the known values of the variables $\phi_0 = V_a$ and $\phi_N = V_b$ on the right side. The first and last equations then automatically provide these values to the end points, and we have no need to change the matrix dimension in mid-stream, as it were. If the reader moves their arm vertically and horizontally to erase the entries in the first column and the first row and enters unity in the diagonal element, they will appreciate that we are applying "benediction" to the matrix M. The solution of these simultaneous equations for the nodal values of ϕ is performed using standard matrix solvers contained in LAPACK, a matrix analysis package that is available freely on the web. [8]

To summarize, in order to apply benediction [9] to the matrix, we start with Eq. (2.19) and multiply column "0" of the matrix M by V_a and move the resulting numbers with a change of sign to the right-hand side. We do the same to the last column, multiplying it by V_b and moving it to the right side additively with a change of sign. Now we set the rows "0" and "N" to zero, and insert $M_{00} = 1$, and $M_{N,N} = 1$. This keeps the matrix size the same and leaves the known ϕ_0 and ϕ_N in place in the vector of unknowns. Next we set $R_0 \equiv V_a$, and $R_N \equiv V_b$. These set of changes lead to:

$$\begin{bmatrix} 1 & 0 & 0 & & & \\ 0 & M_{11} & M_{12} & & & \vdots \\ & M_{21} & M_{22} & \ddots & & \vdots \\ \vdots & \ddots & \ddots & M_{N-2,N-2} & M_{N-2,N-1} & \vdots \\ & & & M_{N-1,N-2} & M_{N-1,N-1} & 0 \\ & & & 0 & 0 & 1 \end{bmatrix} \begin{bmatrix} \phi_0 \\ \phi_1 \\ \vdots \\ \vdots \\ \phi_{N-1} \\ \phi_N \end{bmatrix} = \begin{bmatrix} V_a \\ R_1 - M_{10} V_a \\ R_2 \\ \vdots \\ R_{N-2} \\ R_{N-1} - M_{N-1,N} V_b \\ V_b \end{bmatrix}. \tag{2.21}$$

By defining two $1 \otimes 1$ subspaces, we ensure that the values of ϕ_0 and ϕ_N match the BCs, and the $N-1$ variables $\phi_1, \ldots, \phi_{(N-1)}$ solve to their appropriate values on inverting the modified M-matrix.

2.4 Galerkin's method and Dirichlet BCs

We have seen that for linear interpolation, there are two shape functions in each element associated with the beginning and end points of each of the 1D elements. In Galerkin's approach to solving the differential equation

$$\frac{d}{dx}\left(\varepsilon_0 \frac{d}{dx}\phi(x)\right) + \rho(x) = 0, \tag{2.22}$$

the left side of Eq. (2.22) is projected onto the shape functions: we multiply the left side of the differential equation by $N_i^{(iel)}(\xi)$ and integrate over the physical domain. We write

$$\int_{x_a}^{x_b} N_i^{(iel)}(\xi)\left[\frac{d}{dx}\left(\varepsilon_0\frac{d}{dx}\phi(x)\right)+\rho(x)\right]dx=0. \tag{2.23}$$

The potential $\phi(x)$ is also expressed in terms of the interpolation functions and their corresponding nodal variables ϕ_j. Projecting the differential equation on to the space of "basis functions," and demanding that each of these projections be zero, leads to constraints on the nodal variables. This is a "weak constraint" in the sense that we are not demanding that the differential equation be satisfied at every point. We are enforcing only the projection of the left side of the equation on basis functions to have null values. Hence, it is an approximate constraint, and it is called the "weak formulation" of the solution. An integration by parts leads to

$$\int_{x_a}^{x_b} dx\left[-\frac{dN_i(\xi)}{dx}\varepsilon_0\frac{dN_j(\xi)}{dx}\right]\phi_j+[N_i(\xi)\varepsilon_0\phi'(x)]_{x_a}^{x_b}$$
$$+\int_{x_a}^{x_b} N_i(\xi)\rho(x)dx=0. \tag{2.24}$$

Here, the shape functions take on their forms in each element. The first and last integrals in the above equation are the same as the ones obtained earlier in the action integral formulation (see Eq. (2.19)). The terms arising from integration by parts are labeled as "surface terms" for convenience, a label that is more natural in the 2D and 3D discussions to follow. The surface terms from each element cancel pairwise with those arising from neighboring elements leaving us with just the terms at the two end points x_a and x_b. We write

$$-M_{i,j}\phi_j+R_i+\left[N_i(\xi)\frac{dN_j(\xi)}{dx}\phi_j\right]_{x_a}^{x_b}=0. \tag{2.25}$$

Here R_i is the projection of the source term ρ in the differential equation. In arriving at this equation, we recognize that there are two shape functions for each linear element. One might think that we have $2\times nelem$ independent equations. However, the nodal variables at the nodes common to two adjacent finite elements have to be the same for the continuity of the solution. Here this constraint can be implemented by a simple overlay of the "element matrices" as we had done in the action integral formulation.

It is also useful to think in terms of the nodal point of view as opposed to the element point of view. For each node other than the first or the last, we can combine $N_1^{(iel)}(\xi)$ with $N_0^{(iel+1)}(\xi)$ to construct a different basis. In the case of linear elements, these are called "hat-functions," as is natural given their shape. In the nodal basis, it is very clear that we will end up with $nelem+1$ equations for the $nelem+1$ nodal values ϕ_j.

Let us investigate the form of the surface terms. For the last element terminating at x_b, we have $N_0 = (1-\xi)/2$ and $N_1 = (1+\xi)/2$, and

$$N_i(\xi(x)) \times \frac{d}{dx} N_j(\xi(x))\phi_j \Big|_{x=x_b} = \\ (1 \times \delta_{i,N}) \times \frac{2}{(x_N - x_{N-1})} \times \left(-\frac{1}{2}\phi_{N-1} + \frac{1}{2}\phi_N\right). \tag{2.26}$$

Here the shape function associated with the last node, evaluated at the last node itself, has the value 1. Similarly, at $x = x_a$ we have

$$N_i(\xi(x)) \times \frac{d}{dx} N_j(\xi(x))\phi_j \Big|_{x=x_a} = \\ (1 \times \delta_{i,0}) \times \frac{2}{(x_1 - x_0)} \times \left(-\frac{1}{2}\phi_0 + \frac{1}{2}\phi_1\right), \tag{2.27}$$

leading to

$$-M_{i,j}\phi_j + R_i + \delta_{i,N}\left[\beta_{N-1}\delta_{j,N-1}\phi_j + \beta_N \delta_{j,N}\phi_j\right] \\ - \delta_{i,0}\left[\alpha_0\delta_{j,0}\phi_j + \alpha_1\delta_{j,1}\phi_j\right] = 0. \tag{2.28}$$

Here α_0, α_1 and β_{N-1}, β_N are just numbers obtained by simplifying the above Eqs. (2.26, 2.27). On rearranging terms, we have

$$\begin{bmatrix} M_{00}+\alpha_0 & M_{0,1}+\alpha_1 & 0 & & \\ M_{10} & M_{11} & M_{12} & & \\ 0 & M_{21} & M_{22} & & \\ & & \ddots & \ddots & \\ & & & M_{N-1,N-1} & M_{N-1,N} \\ & & & M_{N,N-1}-\beta_{N-1} & M_{N,N}-\beta_N \end{bmatrix} \begin{bmatrix} \phi_0 \\ \phi_1 \\ \phi_2 \\ \vdots \\ \phi_{N-1} \\ \phi_N \end{bmatrix} \\ = \begin{bmatrix} R_0 \\ R_1 \\ R_2 \\ \vdots \\ R_{N-1} \\ R_N \end{bmatrix}. \tag{2.29}$$

By applying benediction, we see that the resulting equation

$$\begin{bmatrix} 1 & 0 & 0 & \cdots & & & \\ 0 & M_{11} & M_{12} & & & & \\ 0 & M_{21} & M_{22} & \ddots & & & \\ & & \ddots & \ddots & \ddots & & \\ & & \ddots & \ddots & \ddots & & \\ & & & \ddots & M_{N-1,N-1} & 0 \\ & & & \cdots & 0 & 1 \end{bmatrix} \begin{bmatrix} \phi_0 \\ \phi_1 \\ \phi_2 \\ \vdots \\ \vdots \\ \phi_{N-1} \\ \phi_N \end{bmatrix} = \begin{bmatrix} V_a \\ R_1 - M_{10}V_a \\ R_2 \\ \vdots \\ \vdots \\ R_{N-1} - M_{N-1,N}V_b \\ V_b \end{bmatrix}. \tag{2.30}$$

is identical with Eq. (2.21).

2.5 Applying Neumann BCs

2.5.1 The action integral and Neumann BCs

Let us consider Neumann BCs at both ends, x_a and x_b, for our example differential equation $\phi'(x)\varepsilon_0\phi'(x) + \rho(x) = 0$. The solution in this case will be unique only up to a constant since the solution function has no reference potential value allocated to it at some point. Typically, Neumann BCs are applied at one end, with the other end having a Dirichlet BC. We can, however, have physical situations where the value of $\phi'(x)$ is defined at the ends, supplemented by the value of ϕ specified somewhere in the physical domain. For the moment, let us suppose that the BCs are

$$\phi'(x_a) = \mathcal{F}_a; \quad \phi'(x_b) = \mathcal{F}_b. \tag{2.31}$$

We have to modify the action itself when Neumann or derivative BCs are present. We have to add additional terms to the action in this case in such a manner that the surface terms generated on integration by parts in deriving the equation of motion are eliminated. Otherwise, the condition $\delta_\phi A = 0$ does not allow us to draw any concrete conclusions about the differential equation. Let us first see the consequences of *not* modifying the action by considering Eq. (2.1),

$$A_0 = T\left[\int_{x_a}^{x_b} dx\left(-\frac{1}{2}\phi'(x)\varepsilon_0\phi'(x) + \phi(x)\rho(x)\right)\right]. \tag{2.32}$$

We have labeled this action as A_0, to distinguish it for this discussion. The application of the principle of stationary action leads to the requirement that $\delta_\phi A_0 = 0$, and we have

$$\delta_\phi A_0/T = 0 = \int_{x_a}^{x_b} dx\Big[-\delta_\phi(\phi'(x))\,[\varepsilon_0\phi'(x)] + \delta\phi(x)\,\rho(x)\Big]. \tag{2.33}$$

An integration by parts for the first term leads to

$$\delta_\phi A_0/T = 0 = \int_{x_a}^{x_b} dx\left[\delta\phi(x)\left(\frac{d}{dx}(\varepsilon_0\phi'(x)) + \rho(x)\right)\right] - \delta\phi(x)\,\varepsilon_0\phi'(x)\Big|_{x_a}^{x_b}. \tag{2.34}$$

Note that if we had Dirichlet BCs, then $\delta\phi$ would be zero at the two end points x_a and x_b; since ϕ is specified at these points, it cannot be varied. This eliminates the surface terms for the Dirichlet BCs. However, in the Neumann BC, these surface terms do not vanish, with $\phi'(x)$ having well-defined values at the two ends and $\delta\phi$ being arbitrary there. Under these conditions, the integral in Eq. (2.34) is not zero because these surface terms do not vanish. So even though $\delta\phi$ is arbitrary, we cannot conclude that the integrand is zero to obtain the differential equation. We see that the motivation to add correction terms to the above action is to ensure that it is compatible with the differential equation.

We add two terms to A_0 so that the new action is

$$A = A_0 + \phi(x_b)\mathcal{F}_b - \phi(x_a)\mathcal{F}_a. \tag{2.35}$$

The principle of stationary action requires that $\delta_\phi A = 0$. We perform this variation, and after an integration by parts, we obtain

$$\begin{aligned}\delta_\phi A/T = 0 = \int_{x_a}^{x_b} dx\,\delta\phi\left(\frac{d}{dx}[\varepsilon_0\phi'] + \rho(x)\right) - [\delta\phi\,\varepsilon_0\phi']\Big|_{x_a}^{x_b} \\ +\ \delta\phi(x_b)\mathcal{F}_b - \delta\phi\,(x_a)\,\mathcal{F}_a.\end{aligned} \tag{2.36}$$

With Neumann BCs, $\delta\phi$ is not fixed at x_a, x_b so that all the above terms appear in the variation of the action.

We proceed by choosing the arbitrary variation $\delta\phi$ in a suitable manner; we make the following three choices:

1. We choose $\delta\phi$ to be arbitrary in the range $x_a < x < x_b$, with $\delta\phi = 0$ at x_a and at x_b. The "surface terms" in Eq. (2.36) then vanish. For $\delta_\phi A$ to vanish as required by the principle of stationary action, we must now have the argument of the integral to vanish. Since $\delta\phi$ is arbitrary, we must have

$$\frac{d}{dx}[\varepsilon_0\phi'(x)] + \rho(x) = 0. \tag{2.37}$$

   We have thus derived the equation of motion from our action, as anticipated.
2. Next, we select $\delta\phi = 0$, except at $x = x_a$. Recall that $\delta\phi$ is arbitrary so that we are free to choose it in this manner. The first integral of Eq. (2.36) is absent, and Eq. (2.36) reduces to

$$\delta\phi(x_a)\varepsilon_0\phi'(x_a) - \delta\phi(x_a)\mathcal{F}_a = 0. \tag{2.38}$$

   With $\delta\phi(x_a)$ arbitrary, we obtain $\varepsilon_0\phi'(x_a) = \mathcal{F}_a$, at $x = x_a$. In other words, we have built in the Neumann BC into the action itself in Eq. (2.35).
3. Finally, we select $\delta\phi = 0$, except at $x = x_b$. Following the steps used earlier, we have

$$\delta\phi(x_b)\,(-\varepsilon_0\phi'(x_b) + \mathcal{F}_b) = 0,$$

   and with $\delta\phi(x_b)$ arbitrary, we arrive at $\varepsilon_0\phi'(x_b) = \mathcal{F}_b$. The modified action contains all the constraints of the derivative BCs.

The reader is directed to a very lucid discussion of derivative BCs and variational methods paralleling the above treatment in Courant and Hilbert [10].

2.5.2 FEM and the action with Neumann BC

We now proceed to apply the FEM to directly discretize the action, Eq. (2.35). Using linear interpolation, we obtain

$$L = A/T = -\frac{1}{2}\phi_\alpha M_{\alpha\beta}\phi_\beta + \phi_\alpha R_\alpha + \phi_\alpha \delta_{\alpha,N}\mathcal{F}_b - \phi_\alpha \delta_{\alpha,0}\mathcal{F}_a. \tag{2.39}$$

The discretized form of the variational principle corresponds to varying A, or equivalently L here, with respect to ϕ_α. Following the steps used in Sec. 2.4, we obtain the equation

$$-M_{\alpha\beta}\phi_\beta + R_\alpha + \delta_{\alpha,N}\mathcal{F}_b - \delta_{\alpha,0}\mathcal{F}_a = 0. \tag{2.40}$$

We write Eq. (2.40) in matrix form

$$\begin{bmatrix} M_{00} & M_{01} & & & & \\ M_{10} & M_{11} & M_{12} & & & \\ & M_{21} & M_{22} & \ddots & & \\ & & & \ddots & \ddots & M_{N-1,N} \\ & & & & M_{N,N-1} & M_{N,N} \end{bmatrix} \begin{bmatrix} \phi_0 \\ \phi_1 \\ \phi_2 \\ \vdots \\ \phi_{N-1} \\ \phi_N \end{bmatrix} = \begin{bmatrix} R_0 - \mathcal{F}_a \\ R_1 \\ R_2 \\ \vdots \\ R_{N-1} \\ R_N + \mathcal{F}_b \end{bmatrix}. \tag{2.41}$$

This set of linear equations is solved to determine the unknown coefficients $\{\phi_0, \ldots, \phi_N\}$.

The following comments are in order:

- If $\rho(x)$ is not zero we will naturally have a "driving term" even if $\phi'(x_a) = 0 = \phi'(x_b)$.
- All the ϕ_α are unknown in this case, and so there is no application of benediction here. However, it is usual to assign a reference value to one of the ϕ_α as mentioned earlier. For example, in the interior, we could set $\phi_{N/2} = 0$, as a reference value for the potential of zero at that node. Naturally, we will have to apply benediction to the matrix for the row and column of index $N/2$, in this case.
- With Lagrange interpolation, $\mathcal{F}_a, \mathcal{F}_b$ are treated as known quantities and are placed on the right side of the set of equations as above. We now need to ensure that the slopes at the two ends are explicitly given by $\mathcal{F}_a, \mathcal{F}_b$. Since $\phi(x)$ is known in its Lagrange interpolated form in the first and the last elements, for linear elements, we write

$$\begin{aligned} \mathcal{F}_a &= \varepsilon_0 \left(\phi_0 \frac{dN_1(\xi)}{d\xi} + \phi_1 \frac{dN_2}{d\xi} \right) \frac{d\xi}{dx} \\ &= \frac{1}{2}\varepsilon_0 \left(\frac{d\xi}{dx} \right) (-\phi_0 + \phi_1) = \varepsilon_0 \frac{\phi_1 - \phi_0}{x_1 - x_0}; \end{aligned} \tag{2.42}$$

$$\mathcal{F}_b = \varepsilon_0 \left(\phi_{N-1} \frac{dN_1}{d\xi} + \phi_N \frac{dN_2}{d\xi} \right) \frac{d\xi}{dx}$$
$$= \frac{1}{2}\varepsilon_0 \left(\frac{d\xi}{dx}\right)(-\phi_{N-1} + \phi_N) = \varepsilon_0 \frac{\phi_N - \phi_{N-1}}{x_N - x_{N-1}}. \quad (2.43)$$

Note that the derivative involves both the nodes in the linear element. This feature, that all the nodal values participate in the definition of the derivative at one end of the element, is true for higher order Lagrange elements as well. The above equations provide relations between the two nodal values and the derivative at each end, respectively. We could then replace ϕ_0 and ϕ_N in favor of $\{\mathcal{F}_a, \phi_1\}$ and $\{\phi_{N-1}, \mathcal{F}_b\}$, respectively.

2.5.3 Galerkin's method and the Neumann BC

When we employ Galerkin's scheme for solving the differential equation, we project $d/dx(\varepsilon_0\phi'(x)) + \rho(x) = 0$ onto the space of shape functions N_α and find

$$\int dx(-N'_\alpha \varepsilon_0 N'_\beta \phi_\beta + N_\alpha \rho(x)) + [N_\alpha \phi']\Big|_{x_a}^{x_b} = 0.$$

This can be written in the matrix form

$$-M_{\alpha\beta}\phi_\beta + R_\alpha + [\delta_{\alpha,N}\mathcal{F}_b - \delta_{a,0}\mathcal{F}_a] = 0,$$

or

$$\begin{bmatrix} M_{00} & M_{01} & & & \\ M_{10} & M_{11} & \ddots & & \\ & \ddots & \ddots & \ddots & \\ & & \ddots & \ddots & M_{N-1,N} \\ & & & M_{N,N-a} & M_{N,N} \end{bmatrix} \begin{bmatrix} \phi_0 \\ \phi_1 \\ \vdots \\ \phi_{N-1} \\ \phi_N \end{bmatrix} = \begin{bmatrix} R_0 - \mathcal{F}_a \\ R_1 \\ \vdots \\ R_{N-1} \\ R_N - \mathcal{F}_b \end{bmatrix}. \quad (2.44)$$

Equation (2.44) is the same as Eq. (2.41).

2.6 Hermite interpolation and BCs

So far, we have made use of Lagrange interpolation polynomials in the finite elements. The earlier discussion was limited to linear interpolation. We can go to polynomials of higher degree by increasing the number of nodes in each element and defining the corresponding interpolation polynomials. This is discussed in Appendix A. The accuracy of the calculations can be improved by increasing the degree of the interpolation polynomials in each finite element; this is called *p*-convergence.

We can improve the accuracy of the solution *substantially* if we make use of Hermite interpolation polynomials[‖] for the interpolation of functions in each element. These

[‖] The Hermite interpolation polynomials should not be confused with Hermite functions which are solutions to Hermite's differential equation. The Hermite functions are associated with the wavefunction of the quantum harmonic oscillator.

polynomials are invoked when we have not only the function value but also the first derivative of the function at nodes. This is referred to as having two degrees of freedom (f_i and f_i') at each node for $f(x)$. The function is said to have $\mathcal{C}_{(1)}$-continuity within the element. With nodes at element boundaries, Hermite interpolation functions provide derivative continuity across element boundaries. If we represent a general function $f(x)$ through interpolation in an element with two nodes located at the ends of the element, the Hermite interpolation polynomial representation requires that the values of the function and also its derivative be known at the nodes. Consider the element ranging over $x_a \le \xi \le x_b$ with node 0 located at x_a and node 1 located at x_b. Let

$$
\begin{aligned}
f(x_a) &= f_0; \quad \frac{df(x_a)}{dx} = f_0', \\
f(x_b) &= f_1; \quad \frac{df(x_b)}{dx} = f_1'.
\end{aligned} \tag{2.45}
$$

We then have the interpolated form

$$
f(\xi) = \Big(f_0 N_0(\xi) + f_1 N_1(\xi) \Big) + \Big(f_0' \overline{N}_0(\xi) + f_1' \overline{N}_1(\xi) \Big) \times \frac{dx}{d\xi}. \tag{2.46}
$$

Here the shape functions N_i and $\overline{N}_j$ are such that $f(\xi)$ takes on the value f_1, f_2 at the two nodes, and at the same time f_1', f_2' are the values of the derivative of $f(x)$ at the two nodes. The derivative degree of freedom leads to a larger element matrix, and we have doubled the size of the element matrix with the inclusion of this additional degree of freedom at each node. The additional factor of the Jacobian of transformation, obtained from the element in global coordinates to the standard element on which the shape functions are defined, is necessary to ensure that the derivative of $f(x)$ with respect to the global coordinate x has the correct value of $f'(x)$. For further details, the reader is referred to Appendix A.

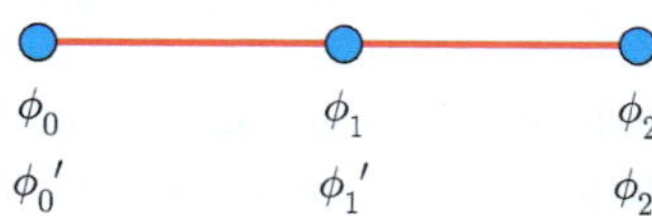

Figure 2.4 The nodal degrees of freedom at the three nodes are shown for $\mathcal{C}_{(\infty)}$ or Hermite interpolation in one element. The polynomials are of degree five.

Note that element matrices have dimension *nodelm* for Lagrange elements and *nodelm* $\times$ *ndof* for Hermite elements. Here *nodelm* is the number of nodes in each element and *ndof* is the number of degrees of freedom at each of the nodes. The degree of the interpolation polynomials is *nodelm* -1 for Lagrange shape functions and is *nodelm* $\times$ *ndof* -1 for Hermite shape functions. Thus, for example, quadratic shape functions are associated with Lagrange elements with three nodes and one degree of freedom at each node. An element with three nodes and *ndof*$=2$ requires quintic Hermite shape functions.

In order to simplify the following discussion about Hermite shape functions, consider just a single quintic Hermite element over the entire physical region. We wish to illustrate the application of BCs in this case. Now the node numbers in the element are given by $inode = 0, 1, 2$. The nodal variables are shown in Fig. 2.4 over one element with 3 nodes, and the shape functions are explicitly given in Appendix B.

2.6.1 Hermite interpolation and action: Dirichlet BC

In the case of Dirichlet BC, we begin with the action

$$A = TL = T\int_{x_a}^{x_b}\left[-\frac{1}{2}\left(\phi'(x)\varepsilon_0\phi'(x)\right) + \phi(x)\rho(x)\right]dx. \tag{2.47}$$

The substitution of Hermite shape functions for the field $\phi(x)$ in Eq. (2.47) allows us once again to integrate away the spatial dependence. We write the interpolation with care here. If the range of the element, $[x_a, x_b]$, does not coincide with $[1, -1]$, we must include scale factors as shown below in order that a differentiation with respect to x reproduces the derivative of $\phi(x)$ at the nodes, ϕ_i'. We can guarantee this if we scale the shape functions $\overline{N}_i$ by a factor of $dx/d\xi$. This does not affect the value of the *function* at the nodes. We write

$$\begin{aligned}\phi(x) &= \sum_{i=0,2,4}\phi_{(i)}N_i(\xi) + \sum_{i=1,3,5}\phi'_{(i)}\overline{N}_i(\xi)\frac{dx}{d\xi},\\ \frac{d\phi(x)}{dx} &= \sum_{i=0,2,4}\phi_{(iel)}\frac{dN_i(\xi)}{d\xi}\frac{d\xi}{dx} + \sum_{i=1,3,5}\phi'_{(i)}\frac{d\overline{N}_i(\xi)}{d\xi}.\end{aligned} \tag{2.48}$$

In the first of Eqs. (2.48), observe that at the nodes, the function values are reproduced with $\overline{N}_i$ being zero there, while in the second equation, the function derivative values are reproduced correctly at the nodes with $dN_i/d\xi = 0$ there. The resulting form of the Lagrangian can be represented by

$$L = -\frac{1}{2}\phi_\alpha M_{\alpha\beta}\phi_\beta + \phi_\alpha R_\alpha.$$

Here α, β correspond to the degrees of freedom associated with $\{f_i, f_i'\}$ at each node, in that order. Now the principle of stationary action, expressed as $\delta_{\phi_\alpha}L = 0$, leads to

$$\begin{bmatrix} M_{00} & M_{01} & M_{02} & M_{03} & M_{04} & M_{05}\\ M_{10} & M_{11} & M_{12} & * & * & *\\ M_{20} & M_{21} & M_{22} & * & * & *\\ M_{30} & * & * & * & * & *\\ M_{40} & * & * & * & * & *\\ M_{50} & * & * & * & * & M_{55}\end{bmatrix}\begin{pmatrix}\phi_0\\ \phi_0'\\ \phi_1\\ \phi_1'\\ \phi_2\\ \phi_2'\end{pmatrix} = \begin{pmatrix}R_0\\ R_1\\ \vdots\\ \vdots\\ \vdots\\ R_5\end{pmatrix}. \tag{2.49}$$

The matrix elements are labeled by the node-number index *nnode* and the degree-of-freedom index *ndof*. The matrix is full because we have chosen to work with just one element in order to illustrate the details. The asterisk (∗) entries in the matrix are simply a shortcut to say that the corresponding matrix element is occupied. With just one element at hand, there is no issue of overlap of element matrices from adjacent elements. Implementing such overlap would require that we add all matrix elements associated with the first node in the current element to the corresponding matrix elements from the previous element. This will ensure that we have continuity of the solution *and* its derivative at the common nodes. The matrix elements here are from the integration of the Hermite interpolation polynomials in the action integral. Thus, for example, we have

$$\begin{aligned} M_{00} &= \int_{x_a}^{x_b} dx \left[\left(\frac{dN_1(\xi)}{d\xi} \frac{d\xi}{dx} \right) \varepsilon_0 \left(\frac{dN_1(\xi)}{d\xi} \frac{d\xi}{dx} \right) \right], \\ M_{01} &= \int_{x_a}^{x_b} dx \left[\left(\frac{dN_1(\xi)}{d\xi} \frac{d\xi}{dx} \right) \varepsilon_0 \left(\frac{d\overline{N}_1(\xi)}{d\xi} \right) \right], \end{aligned} \tag{2.50}$$

and so on.

Here ϕ_0 and ϕ_2 are known in the example we have been considering: $\phi_0 = V_a$; and $\phi_2 = V_b$. Applying benediction to rows/columns with indices 0 and 4, we obtain

$$\begin{bmatrix} 1 & 0 & 0 & 0 & 0 & 0 \\ 0 & M_{11} & M_{12} & M_{13} & 0 & M_{15} \\ 0 & M_{21} & M_{22} & M_{23} & 0 & M_{25} \\ 0 & M_{31} & M_{32} & M_{33} & 0 & M_{35} \\ 0 & 0 & 0 & 0 & 1 & 0 \\ 0 & M_{51} & M_{52} & M_{53} & 0 & M_{55} \end{bmatrix} \begin{pmatrix} \phi_0 \\ \phi_0' \\ \phi_1 \\ \phi_1' \\ \phi_2 \\ \phi_2' \end{pmatrix} = \begin{pmatrix} V_a \\ R_1 - M_{10}V_a - M_{50}V_b \\ \vdots \\ \vdots \\ V_b \\ R_5 - M_{04}V_a - M_{54}V_b \end{pmatrix}. \tag{2.51}$$

We can now proceed to solve for the quantities $\{\phi_0, \phi_0', \phi_1, \phi_1', \phi_2, \phi_2'\}$.

2.6.2 Hermite interpolation and action: Neumann BC

The action integral A has the added "surface terms" in the case of Neumann BC, as we had discussed earlier. The Lagrangian is given by

$$L = \int dx \left(-\frac{1}{2}\phi'(x)\varepsilon_0\phi'(x) + \phi(x)\rho(x) \right) + \mathcal{F}_b\phi(x_b) - \mathcal{F}_a\phi(x_a). \tag{2.52}$$

The substitution of the interpolation polynomials into the above equation generates the matrix relation

$$L = -\frac{1}{2}\phi_\alpha M_{\alpha\beta}\phi_\beta + \phi_\alpha R_\alpha + \mathcal{F}_b\phi_\alpha\delta_{\alpha 4} - \mathcal{F}_a\phi_\alpha\delta_{\alpha 0}. \tag{2.53}$$

Here the indices α, β range over [0, 5], with the even indices referring to ϕ_i, and the odd indices corresponding to ϕ_i'. Now we implement the principle of stationary action, setting $\delta_\phi A = 0$ by varying the nodal variables in the Lagrangian. We obtain the matrix equation, with M being the same as in Eq. (2.49),

$$\begin{bmatrix} \\ \\ & M & \\ \\ \\ \end{bmatrix} \begin{pmatrix} \phi_0 \\ \phi_0' \\ \phi_1 \\ \phi_1' \\ \phi_2 \\ \phi_2' \end{pmatrix} = \begin{pmatrix} R_0 \\ R_1 \\ R_2 \\ R_3 \\ R_4 \\ R_5 \end{pmatrix} + \begin{bmatrix} 0 & -1 & 0 & 0 & 0 & 0 \\ & 0 & & & & \\ & & 0 & & & \\ & & & 0 & & \\ & & & & 0 & +1 \\ & & & & & 0 \end{bmatrix} \begin{bmatrix} \phi_0 \\ \mathcal{F}_a \\ \phi_1 \\ \phi_1' \\ \phi_2 \\ \mathcal{F}_b \end{bmatrix}. \tag{2.54}$$

We note that

- The nodal variables $\phi'_0 = \mathcal{F}_a$ and $\phi'_2 = \mathcal{F}_b$ are known quantities. We have therefore moved terms containing them to the right side of the set of simultaneous equations.
- We have to allow one of the nodal values of ϕ (at node 0, 2, or 4) to take on a given fixed potential in order to have a reference potential. We do not show this in the above relation.
- The surface terms lead to $\delta_\phi\left(\phi'(x_b)\phi(x_b) - \phi'(x_a)\phi(x_a)\right)$. These terms have to be recast in terms of shape functions. Recall that for Hermite functions, we have ϕ and $d\phi/dx$ given by Eqs. (2.48). We obtain

$$
\begin{aligned}
\phi'(x_b)\phi(x_b) &= \left(\overline{N}'_3(\xi=1)\phi'_3 + 0\cdots\right)\cdot\left(N_3(\xi=1)\phi_3\right) \\
&= \delta_{\alpha 4}\phi_3 \cdot \delta_{\beta 5}\phi'_3, \\
\phi'(x_a)\phi(x_a) &= \delta_{\alpha 0}\phi_0 \cdot \delta_{\beta 1}\phi_3,
\end{aligned}
\tag{2.55}
$$

  Only the above terms survive because the Hermite interpolation polynomials satisfy the following conditions (see Appendix A)

$$
\begin{aligned}
N_i(\xi_j) = \delta_{ij}; \quad N'_i(\xi_j) = 0; \\
\overline{N}_i(\xi_j) = 0; \quad \overline{N}'_i(\xi_j) = \delta_{ij}.
\end{aligned}
$$

  The above terms can be cast in the form of a matrix as shown on the right side of Eq. (2.54).
- During the overlay of the element matrices into the global matrix, we have to overlap the last ϕ_i and ϕ'_i nodal values at the edge of an element with the corresponding ones from the first node of the next adjacent element.
- On applying benediction to the matrix M for ϕ'_0 and ϕ'_2, we find

$$
\begin{bmatrix}
M_{00} & 0 & M_{02} & M_{03} & M_{04} & 0 \\
0 & 1 & 0 & 0 & 0 & 0 \\
M_{20} & 0 & M_{22} & M_{23} & M_{24} & 0 \\
M_{30} & 0 & M_{32} & M_{33} & M_{34} & 0 \\
M_{40} & 0 & M_{42} & M_{43} & M_{44} & 0 \\
0 & 0 & 0 & 0 & 0 & 0
\end{bmatrix}
\begin{pmatrix}
\phi_0 \\ \phi'_0 \\ \phi_1 \\ \phi'_1 \\ \phi_2 \\ \phi'_2
\end{pmatrix}
=
\begin{bmatrix}
R_0 - \mathcal{F}_a - M_{01}\mathcal{F}_a - M_{05}\mathcal{F}_b \\
\mathcal{F}_a \\
R_2 - M_{21}\mathcal{F}_a - M_{25}\mathcal{F}_b \\
R_3 - M_{31}\mathcal{F}_a - M_{35}\mathcal{F}_b \\
R_4 - \mathcal{F}_b - M_{41}\mathcal{F}_a - M_{45}\mathcal{F}_b \\
\mathcal{F}_b
\end{bmatrix}.
\tag{2.56}
$$

  Thus, the case of Hermite interpolation requires additional care when implementing the Neumann BCs.

2.6.3 Galerkin's method and Hermite interpolation

Let us again consider just a single quintic Hermite element with 3 nodes with 2 degrees of freedom at each node. We have a 6×6 element matrix which provides a nontrivial example matrix to illustrate the implementation of the BCs.

Let us label the nodal values of the function as in Fig. 2.4. We can simplify the notation by writing the set of polynomials in a compact form

$$\begin{aligned} \phi(x) &= \phi_\alpha N_\alpha(\xi(x)) = \phi_\alpha N_\alpha(\xi(x)); \quad \alpha = 0, 1, \ldots, 5; \\ \frac{d\phi}{dx} &= \phi_\alpha \frac{d}{dx}\overline{N}_\alpha(\xi(x)), \end{aligned} \tag{2.57}$$

and include the factors of the Jacobian of transformation from the global coordinate to the local element coordinate with the $\overline{N}(\xi)$ polynomials (see Eqs. (2.48)). This makes the notation more compact, and *we keep track of the scale factors implicitly in the above*. The shape functions are implicitly functions of the global coordinate x and are defined in each element; here we have just one element.

Hermite elements with Dirichlet BC in Galerkin's method

Writing the residuals associated with projection of the differential equation onto shape functions, we have

$$\int_{x_a}^{x_b} dx N_\alpha(\xi) \left[\frac{d}{dx} \left(\varepsilon_0 N'_\beta(\xi) \right) \phi_\beta + \rho(x) \right] = 0,$$

and supplying an integration by parts, we obtain

$$\int_{x_a}^{x_b} dx \left[-N'_\alpha \varepsilon_0 N'_\beta \right] \phi_\beta + R_a + \left[N_\alpha \varepsilon_0 N'_\beta \phi_\beta \right] \Big|_{x_a}^{x_b} = 0. \tag{2.58}$$

Here $R_\alpha = \int dx N_\alpha(x)\rho(x)$. With the Dirichlet BCs, $\phi_a = V_a$ and $\phi_b = V_b$, we can render Eq. (2.58) to the form

$$-M_{\alpha\beta}\phi_\beta + R_\alpha + \delta_{\alpha,4}\delta_{\beta,5}\phi'_b - \delta_{\alpha,0}\delta_{\beta 1}\phi'_a = 0. \tag{2.59}$$

Here, $N_4(x_b) = 1$ and $N'_5(x_b) = 1$, and so on. On rearranging terms, we obtain

$$\begin{bmatrix} M_{00} & M_{01}+1 & M_{02} & M_{03} & M_{04} & M_{05} \\ M_{10} & \ddots & & & & \\ \vdots & & & & & \vdots \\ & & & & & M_{45}-1 \\ & & & & & M_{55} \end{bmatrix} \begin{pmatrix} \phi_0 \\ \phi'_0 \\ \phi_1 \\ \phi'_1 \\ \phi_2 \\ \phi'_2 \end{pmatrix} = \begin{pmatrix} R_0 \\ \vdots \\ \vdots \\ \vdots \\ R_5 \end{pmatrix}. \tag{2.60}$$

Since $\phi_a = V_a$ and $\phi_b = V_b$, we perform benediction as usual and obtain

$$\begin{bmatrix} 1 & 0 & 0 & 0 & 0 & 0 \\ 0 & M_{11} & M_{12} & M_{13} & 0 & M_{15} \\ 0 & M_{21} & M_{22} & M_{23} & 0 & M_{25} \\ 0 & M_{31} & M_{32} & M_{33} & 0 & M_{35} \\ 0 & 0 & 0 & 0 & 1 & 0 \\ 0 & M_{51} & M_{52} & M_{53} & 0 & M_{55} \end{bmatrix} \begin{pmatrix} \phi_0 \\ \phi'_0 \\ \phi_1 \\ \phi'_1 \\ \phi_2 \\ \phi'_2 \end{pmatrix} = \begin{pmatrix} V_a \\ R_1 - M_{10}V_a - M_{14}V_b \\ R_2 - M_{20}V_a - M_{24}V_b \\ R_3 - M_{30}V_a - M_{34}V_b \\ V_b \\ R_5 - M_{50}V_a - M_{54}V_b \end{pmatrix}. \tag{2.61}$$

This is the same as Eqs. (2.51) and (2.54) obtained from the action integral and variational principles.

Hermite elements with Neumann BC in Galerkin's method

Now the Galerkin procedure leads to Eq. (2.59) again, but with $\phi'(x_a) = \mathcal{F}_a$, and $\phi'(x_b) = \mathcal{F}_b$, and a rearrangement of terms leads to

$$\begin{bmatrix} M_{00} & 0 & M_{02} & M_{03} & M_{04} & 0 \\ 0 & 1 & 0 & 0 & 0 & 0 \\ M_{20} & 0 & M_{22} & M_{23} & M_{24} & 0 \\ M_{30} & 0 & M_{32} & M_{33} & M_{34} & 0 \\ M_{40} & 0 & M_{42} & M_{43} & M_{44} & 0 \\ 0 & 0 & 0 & 0 & 0 & 0 \end{bmatrix} \begin{pmatrix} \phi_0 \\ \phi_0' \\ \phi_1 \\ \phi_1' \\ \phi_2 \\ \phi_2' \end{pmatrix} = \begin{bmatrix} R_0 - \mathcal{F}_a - M_{01}\mathcal{F}_a - M_{05}\mathcal{F}_b \\ \mathcal{F}_a \\ R_2 - M_{21}\mathcal{F}_a - M_{25}\mathcal{F}_b \\ R_3 - M_{31}\mathcal{F}_a - M_{35}\mathcal{F}_b \\ R_4 - \mathcal{F}_b - M_{41}\mathcal{F}_a - M_{45}\mathcal{F}_b \\ \mathcal{F}_b \end{bmatrix}. \tag{2.62}$$

This equation is the same as Eq. (2.56).

It may be noted that the Neumann BC and the Dirichlet BC are "equivalent" in the Hermite interpolation scheme: in the case of Neumann BC, we work with the derivative nodal variables at the boundary, and in the Dirichlet BC, we work with the function nodal variables at the boundary. The Neumann BCs have to be incorporated into the action, as discussed earlier in detail.

2.7 Schrödinger's equation and 1D tunneling

Let us return to Schrödinger's equation of Chap. 1, Eq. (1.12), for the wavefunction of a particle, in which the solution $\Psi(x,t)$ allows us to describe the wave nature of the particle. Quantum mechanical particles can penetrate into potential energy barriers, while a classical particle cannot penetrate such a barrier region. If the potential barriers are thin enough, a quantum mechanical particle can tunnel through the barrier and escape into the region across the barrier without loss of energy; this is a manifestation of the wave nature of the particle. In this section, we consider a simple 1D scattering problem with a flux of particles of fixed energy $E = \hbar\omega$ incident on a potential barrier from the region $x < x_a$. The waves representing the particles are reflected and also penetrate into the region on the other side of the barrier, as in Fig. 2.5. The particles are then said to tunnel through the barrier. The potential in this simple example is

$$V(x) = \begin{cases} 0, & x < x_a; \\ V_0, & x_a \le x \le x_b; \\ 0, & x > x_b. \end{cases} \tag{2.63}$$

In the steady state, Schrödinger's equation is separable in x and t. Letting $\Psi(x,t) = \psi(x)e^{-i\omega t}$, we obtain

$$-\frac{\hbar^2}{2m}\frac{d^2\psi(x)}{dx^2} + [V(x) - E]\psi(x) = 0. \tag{2.64}$$

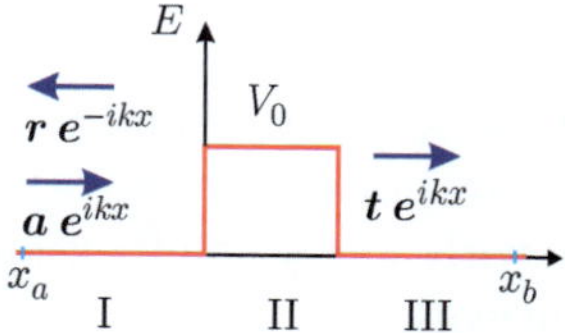

Figure 2.5 The simple rectangular barrier of height V_0 is shown here with the incident, reflected, and transmitted wavefunctions in regions I and III.

Solving this equation in the two regions labeled I and III in Fig. 2.5, where $V(x) = 0$, we have

$$\psi_I(x) = a \exp(ikx) + r \exp(-ikx), \tag{2.65}$$

$$\psi_{III}(x) = t \exp(ikx). \tag{2.66}$$

Here $k = \sqrt{(2mE)/\hbar^2}$ is the wavevector of the wave in regions I and III. The amplitude r of the reflected wave in region I and the amplitude t in region III are related to the incident amplitude a from the left in region I. Only an outgoing solution is acceptable in region III on physical grounds since we have assumed that there is no source of particles on the right side.

With a steady flux of particles, the probability current density for the incident wave (see Chap. 1) is given by Eq. (1.6) to be $j_{inc} = \hbar k |a|^2/m$. Here we wish to set up a finite element approach to calculating the transmission amplitude, t and the reflection amplitude r. While the solution in region II is known to be a linear combination of right and left exponential functions ($\exp(\pm\kappa x)$, $E < V_0$) or propagating ($E > V_0$) plane waves, the FEM will automatically generate the solution there based on the BCs in the outer regions.

The Schrödinger action in 1D is given by the time integral over the Lagrangian, L. In the case of a steady current density j_{inc}, incident from the left on the barrier, we consider the time-independent Schrödinger's equation and write

$$\begin{aligned} A_0 &= \int_0^T dt\, L \\ &= T \int_{x_a}^{x_b} dx \left[\left(\frac{d\psi^*(x)}{dx} \right) \frac{\hbar^2}{2m} \left(\frac{d\psi(x)}{dx} \right) + \psi^*(x) \left[V(x) - E \right] \psi(x) \right]. \end{aligned} \tag{2.67}$$

Note the absence of a factor of 1/2 in the above terms appearing in the Lagrangian. As mentioned in Chap. 1, this is because the wavefunction $\psi(x)$ is a complex function so that variations with respect to ψ^* correspond to independent variation of the real and of the imaginary part; instead of varying them separately, we can perform the variation with the complex function $\psi^*(x)$.

2.7.1 Tunneling BCs

It is not usually appreciated that scattering can be cast in a variational form so that we can apply the principle of stationary action approach to such problems. Most books on quantum mechanics, with a very small exception, do not consider scattering problems in a variational framework.

The BCs are obtained as follows. First we note that on the left, in region I, we have

$$ik\psi(x_a) = ik\left[ae^{ikx_a} + re^{-ikx_a}\right], \tag{2.68}$$

$$\psi'(x_a) = ik\left[ae^{ikx_a} - re^{-ikx_a}\right]. \tag{2.69}$$

We are then led to the relation

$$\psi'(x_a) + ik\psi(x_a) = 2i\, ak\, e^{ikx_a}, \tag{2.70}$$

at $x = x_a$. On the right side, in region III, the relation between ψ and its derivative is given by

$$\psi'(x_b) - ik\psi(x_b) = 0, \tag{2.71}$$

at $x = x_b$.

Such BCs relating the function and its derivative at the boundary are said to be Cauchy BCs. The same type of BCs are needed in 2D and 3D scattering also.

Note that the wavevectors in the regions I and III depend on the energy E and the potential energy $V(x)$. In principle, k_I and k_{III} could differ, depending on the form of $V(x)$. Here we are assuming that V is the same (zero) on both sides. If k were different on the two sides, we can go through similar steps and keep track of the differences in them with an index on k.

2.7.2 The Schrödinger action with surface terms

As with all derivative BC problems, we modify the standard action by additional BC terms. These are needed to cancel what we call surface terms that are generated when the $\psi^{*\prime}\psi'$ term in the action is integrated by parts in the variational procedure to derive the equation of motion. With the added surface terms, we have

$$A = \int dt\, L = T\left[\int_{x_a}^{x_b} dx\left(\psi^{*\prime}\frac{\hbar^2}{2m}\psi' + \psi^*(V-E)\psi\right) - \left(\psi^*\frac{\hbar^2}{2m}\Psi'\right)\Bigg|_{x_a}^{x_b}\right]. \tag{2.72}$$

The added surface term has the boundary value $\Psi' = \psi'(x_b)$, which is the boundary value of $\psi'(x)$ at $x = x_b$. Similarly, at x_a, we have $\Psi' = \psi'(x_a)$, as the boundary value of $\psi'(x)$ at that location.

First, let us make the action dimensionless by making the following substitutions. Let us scale the coordinate by ℓ_0 with a convenient choice of ℓ_0 to be 1 nm. We multiply and divide $\hbar$ by c, the velocity of light, recognizing that [11] $\hbar c \simeq 197$ eV nm and $mc^2 \simeq 0.511 \times 10^6$ eV for the electron. We find

$$x = \tilde{x}\ell_0; \quad d/dx = (1/\ell_0)(d/d\tilde{x})$$

$$\mathcal{C} \equiv \frac{\hbar^2}{2m\ell_0^2} = \frac{\hbar^2 c^2}{2mc^2\ell_0^2} \simeq 3.81 \times 10^{-2}\,\text{eV}. \tag{2.73}$$

Here, the energy unit is labeled by $\mathcal{C}$ and is 3.81×10^{-2} eV. The scale of length that is chosen determines the numerical value [12] of the constant $\mathcal{C}$. We also remove the units of the wavevector to write $k = \tilde{k}/\ell_0$. *In the following, we drop the tilde on x and on k for convenience.*

On dividing the action by $T\mathcal{C}$, we obtain a dimensionless quantity for the scaled action integral in which the derivatives are with respect to $\tilde{x}$, and the wavefunctions $\psi(\tilde{x})$ are in units of $1/\sqrt{\ell_0}$,

$$\tilde{A} = \frac{A}{(\mathcal{C}T)} = \int d\tilde{x}\ell_0 \left[\psi^{*\prime}\psi' + \psi^* \left(\frac{V(x) - E}{\mathcal{C}}\right)\psi\right] - \left[\psi^* \ell_0 \tilde{\Psi}'\right]\Bigg|_{x_a}^{x_b}. \tag{2.74}$$

We vary this action with respect to ψ^* and set $\delta_{\psi^*}\tilde{A} = 0$. Once more, we invoke the arguments presented earlier to derive the equation of motion using the variational principle with Neumann BCs for the present case to show that

$$-\psi''(x) + \frac{V - E}{\mathcal{C}}\psi(x) = 0, \tag{2.75}$$

$$\psi'(x)\Bigg|_{x=x_a} = \Psi'(x_a) = -ik\,\psi(x_a) + 2i\,ak\,e^{ikx_a}, \tag{2.76}$$

$$\psi'(x)\Bigg|_{x=x_b} = \Psi'(x_b) = ik\psi(x_b), \tag{2.77}$$

with the above equations obtained by (a) suitably choosing $\delta_{\psi^*}\psi^*(x)$ arbitrary within $x_a \le x \le x_b$ and zero at the two ends, (b) choosing $\delta_{\psi^*}\psi^*(x)$ to be zero everywhere except at $x = x_a$, and finally, (c) choosing $\delta_{\psi^*}\psi^*(x)$ to be zero everywhere except at $x = x_b$, where it is otherwise arbitrary. Now that we are assured of our choices for the additional surface terms in the action, we can proceed with the finite element implementation for the tunneling problem. We treat both the Lagrange and Hermite interpolation cases in the following.

2.7.3 FEM and tunneling: linear interpolation

We begin with

$$\frac{L}{\ell_0\mathcal{C}} = \int_{x_a}^{x_b} d\tilde{x}\left(\psi^{*\prime}\psi' + \psi^*\frac{V - E}{\mathcal{C}}\psi\right) - \left[\psi^*\Psi'\right]\Bigg|_{x_a}^{x_b}. \tag{2.78}$$

and substitute the wavefunction in terms of nodal values and linear shape functions, Eq. (2.11), in element iel

$$\psi = \psi_{(iel)}N_0(\xi) + \psi_{(iel+1)}N_1(\xi).$$

We can then integrate out the spatial dependence to obtain a 2×2 matrix representation of the local elemental action; these are overlaid to get a global matrix representation for L, as discussed earlier in this chapter. In the following, we note that in the scattering process, we have incoming particles with a specific energy E. We can represent the above in the form

$$\frac{L}{\ell_0 \mathcal{C}} = \{\psi_0^*, \ldots, \psi_N^*\} \begin{bmatrix} & & \\ & M & \\ & & \end{bmatrix} \begin{pmatrix} \psi_0 \\ \vdots \\ \psi_N \end{pmatrix} - \psi^*(x_b)\Psi'(x_b) + \psi^*(x_a)\Psi'(x_a). \quad (2.79)$$

Here, as in earlier calculations, M is given by integrals over the shape functions of each element. The difference is that now M has the integrals over products of the derivatives of the shape function, coming from the first term in Eq. (2.78), and also integrals with arguments that are products of the shape functions multiplied by $(V - E)/\mathcal{C}$. Note that the shape functions are real. Only the nodal coefficients ψ_i^* and ψ_i are complex. The first term in the above can be written compactly as $\psi_\alpha^* M_{\alpha\beta}\psi_\beta$, where α, β go over $[0, N]$. Note that the wavevectors and coordinates are dimensionless; we then have

$$\begin{aligned} \psi^*(x_b) &= \psi_N^*; \quad \ell_0\Psi'(x_b) = ik\psi_N, \\ \psi^*(x_a) &= \psi_0^*; \quad \ell_0\Psi'(x_a) = -ik\psi_0 + 2aike^{ikx_a}. \end{aligned} \quad (2.80)$$

A variation of L with respect to ψ_α^* leads to

$$\begin{bmatrix} M_{00} - ik & M_{01} & 0 & & & \\ M_{10} & M_{11} & M_{12} & & & \\ 0 & M_{21} & M_{22} & & & \\ & & \ddots & \ddots & & \\ & & & \ddots & M_{N-1,N} & \\ & & & M_{N,N-1} & M_{N,N} - ik \end{bmatrix} \begin{pmatrix} \psi_0 \\ \psi_1 \\ \vdots \\ \vdots \\ \vdots \\ \psi_N \end{pmatrix} + \begin{pmatrix} 2aike^{ikx_a} \\ 0 \\ 0 \\ \vdots \\ \vdots \\ 0 \end{pmatrix} = 0. \quad (2.81)$$

Notice that $ik\psi^*(x_b)\psi(x_b)$ can be expressed as $\psi_\alpha^* \left(\delta_{(\alpha N)}\, ik\delta_{(\beta N)}\right) \psi_\beta$. Here introducing factors of Kronecker δ-functions helps place the surface terms at their appropriate locations in the matrix. The surface term at x_a can be written as

$$\psi_\alpha^* \left[\delta_{(\alpha 0)}(-ik)\delta_{(\beta 0)}\right] \psi_\beta + \psi_\alpha^* \left[2i\, ak\, e^{ikx_a}\right]\delta_{(\alpha 0)}. \quad (2.82)$$

The quantities ψ_N and ψ_0 are expressible in terms of transmission and reflection coefficients.

$$\psi_0 = (ae^{ikx_a} + re^{-ikx_a}); \quad \psi_N = te^{ikx_b}. \quad (2.83)$$

Substitution of these values into the matrix equation leads to

$$\begin{bmatrix} M_{00}-ik & M_{01} & & & \\ M_{10} & M_{11} & & & \\ & & \ddots & & \\ & & & \ddots & \\ & & & & M_{N,N}-ik \end{bmatrix} \begin{pmatrix} re^{-ikx_a} \\ \psi_1 \\ \vdots \\ \psi_{N-1} \\ te^{ikx_b} \end{pmatrix} + \begin{bmatrix} ae^{ikx_a}(M_{00}-ik) \\ ae^{ikx_a}(M_{10}) \\ 0 \\ \vdots \\ 0 \end{bmatrix} + \begin{bmatrix} ae^{ikx_a}(2ik) \\ 0 \\ 0 \\ \vdots \\ 0 \end{bmatrix} = 0. \quad (2.84)$$

We have thus isolated all the "driving terms" associated with the incident wave amplitude a, and we can proceed with solving the simultaneous equations. We have

$$\begin{bmatrix} M_{00}-ik & M_{01} & & & \\ M_{10} & M_{11} & M_{12} & & \\ & M_{21} & M_{22} & & \\ & & & \ddots & M_{N-1,N} \\ & & & M_{N,N-1} & M_{N,N}-ik \end{bmatrix} \cdot \begin{bmatrix} re^{-ikx_a} \\ \psi_1 \\ \vdots \\ \psi_{N-1} \\ te^{ikx_b} \end{bmatrix} = \begin{bmatrix} -(M_{00}+ik)ae^{ikx_a} \\ -M_{10}ae^{ikx_a} \\ 0 \\ \vdots \\ 0 \end{bmatrix}. \quad (2.85)$$

Once the solution is obtained, we isolate the values of r and t, the two quantities of interest here. We should also reconstruct the nodal "vector" by adding back (ae^{ikx_a}) to the r-term in the solution vector in order to reconstruct the wavefunction.

The transmission coefficient T is defined by

$$\begin{aligned} T &= \frac{\text{Transmitted flux}}{\text{Incident flux}}, \\ &= \frac{\hat{x}\cdot\mathbf{j}_{trans}}{\hat{x}\cdot\mathbf{j}_{inc}} = \frac{\hbar k_{III}|t|^2}{\hbar k_I|a|^2}, \end{aligned} \quad (2.86)$$

in which the presence of the wave vector factors k_I and k_{III} is important. They arise from the definition of the current density. In our treatment, we have let the k values at the two ends be the same, so that $k_I = k_{III}$ in the above equations. The flux is the current through a standard area, and the area as a vector is directed in this case toward the positive x-direction for both the currents.

The reflection coefficient R is defined by

$$R = \frac{-\hat{x} \cdot \mathbf{j}_{refl}}{\hat{x} \cdot \mathbf{j}_{inc}} = \frac{\hbar k_I |r|^2}{\hbar k_I |a|^2}, \tag{2.87}$$

where the wavevectors are the same. Note that the current j_{refl} is directed toward the negative x-direction. But the flux is the current density dotted into the outward-directed normal, and is positive in the numerator of the above equation.

2.7.4 Galerkin's method and tunneling

In the Galerkin approach to tunneling, we set the projection of the differential equation onto the basis functions, the shape functions in our case, to zero. In this weak formulation, we have

$$\int_{x_a}^{x_b} dx N_\alpha(\xi) \left[-N_\beta''(\xi)\psi_\beta + \left(\frac{V-E}{\mathcal{C}}\right) N_\beta \psi_\beta \right] = 0.$$

Here we have directly expressed the local node indices in terms of the global node numbers. An integration by parts and overlay of equations to obtain N + 1 equations for the N + 1 nodal variables $\{\psi_0, \ldots, \psi_N\}$ leads to the matrix equation

$$+M_{\alpha\beta}\psi_\beta - \left[N_\alpha N_\beta' \psi_\beta\right]\Big|_{x_a}^{x_b} = 0.$$

Only the shape function N_0 survives in the first element at x_a, and N_1 survives at x_b, in the last element in the surface terms. We obtain

$$M_{a\beta}\psi_\beta - \Psi'(x_b)\delta_{\alpha N} + \Psi'(x_a)\delta_{\alpha 0} = 0, \tag{2.88}$$

and with

$$\begin{aligned} \Psi'(x_a) &= -ik\psi_0 + 2i\,ka\,e^{ik}, \\ \Psi'(x_b) &= ik\psi_N, \end{aligned} \tag{2.89}$$

we have

$$\begin{bmatrix} M_{00} - ik & M_{01} & & & \\ M_{10} & M_{11} & M_{12} & & \\ & M_{21} & M_{22} & \ddots & \\ & & \ddots & \ddots & M_{N-1,N} \\ & & & M_{N,N-1} & M_{N,N} \end{bmatrix} \begin{bmatrix} \psi_0 \\ \psi_1 \\ \vdots \\ \psi_{N-1} \\ \psi_N \end{bmatrix} + \begin{bmatrix} 2ik\,ae^{ikx_a} \\ 0 \\ \vdots \\ 0 \\ 0 \end{bmatrix} = 0. \tag{2.90}$$

This is the same as Eq. (2.81), and it can be reduced to Eq. (2.85) as before.

Note: If we employ quadratic Lagrange interpolation functions in the elements, we apply BCs as follows. For a wave incident from the left, with amplitude a, we have

$$\begin{aligned} \psi(x_a) &= ae^{ikx_a} + re^{-ikx_a}, \\ \psi'(x_a) &= ik(ae^{ikx_a} - re^{-ikx_a}), \\ \psi''(x_a) &= -k^2(ae^{ikx_a} + re^{-ikx_a}), \end{aligned} \tag{2.91}$$

where the reflected wave has amplitude r. We now expand the left-hand side in terms of the local shape functions in order to transform the nodal values and nodal derivatives into the incident and reflected amplitudes, which are the physically interesting quantities at the extremities. In the case of two nodes per element and one degree of freedom, we have

$$\begin{pmatrix} N_1(x_a) & N_2(x_a) \\ N_1'(x_a) & N_2'(x_a) \end{pmatrix} \begin{pmatrix} \psi_{x0} \\ \psi_{x1} \end{pmatrix} = \begin{pmatrix} 1 & 1 \\ ik & -ik \end{pmatrix} \begin{pmatrix} ae^{ikx_a} \\ re^{-ikx_a} \end{pmatrix}. \tag{2.92}$$

Similarly, we have for three nodes per element and one degree of freedom at the nodes

$$\begin{pmatrix} N_1(x_a) & N_2(x_a) & N_3(x_a) \\ N_1'(x_a) & N_2'(x_a) & N_3'(x_a) \\ N_1''(x_a) & N_2''(x_a) & N_3''(x_a) \end{pmatrix} \begin{pmatrix} \psi_{x1} \\ \psi_{x2} \\ \psi_{x3} \end{pmatrix} = \begin{pmatrix} 1 & 1 & 0 \\ ik & -ik & 0 \\ -k^2 & -k^2 & 1 \end{pmatrix} \begin{pmatrix} ae^{ikx_a} \\ re^{-ikx_a} \\ s \end{pmatrix}, \tag{2.93}$$

where s is an unknown that serves as a temporary variable which is used again when we wish to reconstruct $\psi(x)$ near the two extremities.

2.7.5 Hermite interpolation and tunneling

The use of Hermite interpolation polynomials in tunneling within the action integral and the Galerkin approaches has been discussed earlier, [4, 13] and these topics are left as an exercise for the reader. Once again, we remark that the use of these polynomials improves accuracy substantially.

2.8 Reconstruction of the solution from the nodal values

When the solution nodal vector ϕ_α is obtained after the required matrix computations, we reconstruct the vector to ensure that all the nodal values are present, including boundary values. To obtain the interpolated solution at any global coordinate x, we determine which element contains the point at x and use the nodal values for that element and the original shape functions employed in evaluating the element matrix once more to reconstruct $\phi(x)$. We can generate values for the function for a graphical representation of the solution at any required number of points.

We emphasize once more that it is important for the reader to fully appreciate the details of the book-keeping in the connectivity between elements and nodes, and between element matrices and global matrices in the FEM at the level of 1D problems before venturing into 2D and 3D applications. This chapter has focused on symbolically setting up the element matrix, inserting its element entries into a global matrix, and on implementing the BCs.

2.9 Final comments

The implementation of BCs within FEM is not typically treated at the level of detail presented here. Usually, the focus is on the construction of the element matrices and their overlay into a global matrix. The above discussion does not give explicit forms of the element matrix for all the different types of interpolation schemes; only the linear element evaluation is given as a concrete example. On the other hand, we focus on the occupancy pattern in the global matrix, corresponding to the distribution pattern of the non-zero matrix elements, and the connectivity of the elements. This is sufficient to show how to implement the Dirichlet, Neumann, and Cauchy BCs. The discretization of the action integral provides a simple way to implement all these three BCs in a very straightforward manner. This part of the finite element programming is usually considered challenging, and it is hoped that the somewhat lengthy presentation here makes the application of BCs easier to understand.

2.10 Problems

1. The action integral for the Helmholtz equation is

$$A = \int_{x_a}^{x_b} \left[\frac{1}{2} f'(x)\, f'(x) - \frac{k^2}{2}\, f(x)\, f(x) \right].$$

   Suppose the integral is discretized into 6 finite elements.
   (i) Determine the global matrix occupancy pattern, for this problem, when linear or quadratic Lagrange interpolation polynomials are employed.
   (ii) Apply the variational principle to the discretized action to obtain the discretized Helmholtz equation.
   (iii) Apply the Dirichlet boundary conditions for the problem to obtain the matrix equation for the nodal wavefunction.
2. (a) Construct the action for the quantum harmonic oscillator, Problem 3, Ch. 1, and consider the range of integration (physical region) to be cut off at $\tilde{x} = \pm 5$. Divide the region into 10 finite elements and evaluate the element matrix *iel*, where $0 \leq iel \leq 9$.
   (b) Use linear shape functions to represent the wavefunctions: $\psi(\tilde{x}) = \Psi_0 N_0(\xi) + \Psi_1 N_1(\xi)$, where ξ is the local coordinate variable in the element and $-1 \leq \xi \leq +1$. The notation is explained in Eqs. (2.10, 2.11).
   (c) For boundary conditions, set the wavefunctions to zero at the coordinates $\tilde{x} = \pm 5$.
   (d) Display the final set of simultaneous equations to be solved in matrix form.
3. Consider a physical region $x_\ell \leq x \leq x_r$ that is divided uniformly into *nelem* elements. Let us number the elements from 0 to $nelem - 1$, and let the size of

each element be $h = (x_r - x_\ell)/nelem$. Suppose that each element has *nodelem* equally spaced nodes including at the beginning and the end of the element. The elements use Lagrange interpolation polynomials of degree $nodelem - 1$, which are said to be C_0-continuous. Thus there is only one degree of freedom at each node to represent a function in the physical domain.

(a) How many nodes are there in the entire region globally?

(b) Determine the global node numbers *iglobal* of the nodes in the element *iel*.

(c) Obtain an analytical formula for these node numbers. Suppose (*i*) that the elements are linear elements with two nodes per element. (*ii*) Derive the analytic formula for quadratic elements, and for (*iii*) cubic elements. The relation between the local and global node numbers is called the connectivity for the system.

(d) Obtain the coordinates of the *nodlem* nodes in the element *iel*.

4. The physical region, $x_\ell \leq x \leq x_r$, is divided into *nelem* elements, and each element contains *nodlem* nodes. Suppose that the nodes have *ndof* degrees of freedom associated with them, and a given function is represented in terms of Hermite interpolation polynomials obeying $C_{(ndof-1)}$-continuity. We now have to provide the values of $f(x), f', f^{('')}, \ldots, f^{(ndof-1)}$, at each node in order to represent the function over the element.

(a) Suppose that the elements are numbered from 0 to $nelem - 1$, and the nodes in each element are numbered from 0 to *nodlem*. Determine the number of nodes in the entire region.

(b) Determine the size of the element matrix for element *iel* taking note of the *ndof* degrees of freedom at each node.

(c) Obtain the size of the global matrix.

(d) Obtain an algebraic formula for the location in the global matrix, i.e. (*ig*, *jg*) for the row and column labels, for each of the elements (*ie*, *je*) of the element matrix.

5. A steady stream of electrons is incident from the left onto the potential energy barrier shown in Fig. 2.5.

(i) Consider the action, Eq. (2.72), for this scattering problem and employ Hermite interpolation polynomials with $C_{(\infty)}$-continuity in the finite element discretization of the physical region. Use quintic polynomials corresponding to three nodes per element for the construction of the global matrix. Show how the boundary conditions of Eqs. (2.70) and (2.71) can be imposed on the global matrix.

(ii) Compare the above form of the final global matrix with that obtained for the scattering problem using Galerkin's approach.

6. Consider a rectangular region $a \times b$ with *nelemx* elements along the x-direction and *nelemy* elements along the y-direction. The total number of elements is thus $nelem = nelemx \times nelemy$.

(a) Determine the coordinates of the corners of a general element *iel*.

(b) Suppose that each element has $nodelmx \times nodelmy$ nodes and the interpolation polynomials are Lagrange polynomials along x and

y. To represent the solution, we use a product form of the shape functions in each element $N_\alpha = N_i(\xi) \times N_j(\eta)$, with the index $\alpha = (0, 1, \ldots, nelemx \times nelemy - 1)$. Determine the total number of nodes in the system.

(c) What is the size of the global matrix in this case.

(d) Obtain the connectivity, the relation between local (element) node numbers and the global nodes.

References

[1] K.-J. Bathe and E. Wilson, *Numerical Methods in Finite Element Analysis* (Prentice Hall, Englewood Cliffs, NJ, 1976); K.-J. Bathe, *Finite Element Procedures in Engineering Analysis* (Prentice Hall, Englewood Cliffs, NJ, 1982); K.-J. Bathe, *Finite Element Procedures* (Prentice Hall, Englewood Cliffs, NJ, 1996).

[2] O. C. Zienkiewicz and Y. K. Cheung, *Finite Element Methods in Structural and Continuum Mechanics* (McGraw-Hill, New York, 1967); O. C. Zienkiewicz, *The Finite Element Method* (McGraw-Hill, New York, 1977).

[3] T. J. R. Hughes, *The Finite Element Method* (Prentice Hall, Englewood Cliffs, NJ, 1987).

[4] L. R. Ram-Mohan, *Finite Element and Boundary Element Applications to Quantum Mechanics* (Oxford University Press, Oxford, UK, 2002).

[5] It is convenient to label variables with commonly used names in finite element programming. Thus, we use *nelem* for the total number of elements in the domain, *iel* is a running index for elements, and so on. It is usual in finite element literature to label variables with names that are usual in computer programming; hence the use of *nelem* and the other examples of such usage that follow.

[6] Interpolation polynomials typically used in FEM are presented in Appendices A and B, and in other chapters for ease of reference.

[7] Gauss quadrature is discussed in Ref. [4] in Appendix A. Also see: J. Stoer and R. Bulirsch, *Introduction to Numerical Analysis*, translated by R. Bartels, W. Gautschi, and C. Witzgall (Springer-Verlag, New York, 2002), and F. B. Hildebrand, *Introduction to Numerical Analysis* 2nd edition (Dover, New York, 1987). An extensive collection of Gauss points and weights is given in A. H. Stroud and D. Secrest, *Gaussian Quadrature Formulas* (Prentice Hall, Englewood Cliffs, NJ, 1966).

[8] E. Anderson, Z. Bai, C. Bischof, S. Blackford, J. Demmel, J. Dongarra, J. Du Croz, A. Greenbaum, S. Hammarling, A. McKenney, and Sorensen, D., *LAPACK Users' Guide LAPACK Users' Guide* 3rd edition. The software is available at the website: http://www.netlib.org/lapack/.

[9] The nomenclature "benediction" provides a very visual and succinct representation of the sequential removal of the column and row in a matrix and inserting unity for the diagonal entry. It was introduced in Ref. [4], and is now used by other practitioners of FEM.

[10] R. Courant and D. Hilbert, *Methods of Mathematical Physics*, Vol. I (J. Wiley, NY, 1991).

[11] For more significant figures for the determination of these constants, we note that $\hbar c = 197.326963\,\text{eV}\ nm$ and $m_e c^2 = 0.510998910 \times 10^6\,\text{eV}$. Other physical constants are available at the website supported by the US National Institute of Standards and Technology: http://physics.nist.gov/cuu/Constants/

[12] Note that with the length scale chosen to be $\ell_0 = 1\,\text{Å}$ we have $\mathcal{C} = 3.81\,hboxeV$. For the length scale given by the Bohr radius of the hydrogen atom, a_B, we have $\ell_0 = a_B = 0.0528\,\text{nm}$, and the energy scale coincides with the familiar value of the Rydberg, 13.6 eV. For $\ell_0 = 1\,\text{nm} = 10\,\text{Å}$, the scale of energy is $\mathcal{C} = 38.1\,\text{meV}$.

[13] R. Goloskie, J. W. Kramer, and L. R. Ram-Mohan, Comp. Phys. **8**, 679–686 (1994); "Quantum mechanical tunneling and finite elements."

3

Element geometries for 2D and 3D

In this chapter:

- We derive the interpolation polynomials for typical finite element shapes. We consider quadrilateral and triangular elements in 2D and bricks and tetrahedral elements in 3D for the breakup of the physical region into finite elements.
- The transformation of the shapes from the local "standard" element shapes to the corresponding element in the global geometry is obtained. This mapping involves the determination of the Jacobian of transformation and the evaluation of the element matrices using Gauss quadrature. We consider the case of a general quadrilateral element being mapped to a square specifically as an example of a transformation in which the Jacobian is a function of the coordinates; the other example considered is with the use of circular coordinates to transform circular regions into rectangular ones where again we have a Jacobian that is dependent on r.
- When derivative degrees of freedom (DoFs) are considered at the nodes, we have to pay careful attention to the transformation of the derivatives in the local coordinates to the derivative DoFs in the global coordinates.

3.1 Quadrilateral elements

Consider a quadrilateral physical region as in Fig. 3.1(a). The quadrilateral region has been discretized into smaller quadrilateral finite elements by dividing opposite sides of the quadrilateral into the same number of elements with equidistant nodes. These nodes are then connected to the corresponding nodes on the opposite side as seen in Fig. 3.1(a). In order to apply the finite element method (FEM), we first map the quadrilateral region linearly into a rectangular region with the same number of elements. As a result, every element in the quadrilateral region has a corresponding element in the rectangular region.

Finite Elements in Action. L. Ramdas Ram-Mohan, Oxford University Press. © L. Ramdas Ram-Mohan (2026).
DOI: 10.1093/oso/9780199563487.003.0003

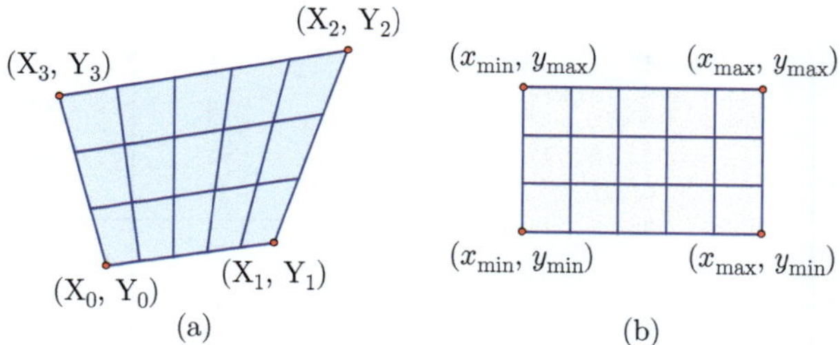

Figure 3.1 The physical quadrilateral region is shown (a), where the global geometry in (X,Y) coordinates is discretized into smaller quadrilateral elements. A linear mapping is performed of the physical region into a rectangular region in (x, y) coordinates, shown in (b), so that the elements in the rectangular region are easily constructed in the same manner.

Let us suppose that we have to solve the Poisson's equation

$$\nabla(\epsilon \nabla \phi(x, y)) = -\rho(x, y), \tag{3.1}$$

on the quadrilateral region with some suitable Dirichlet boundary conditions along the edges. The function $\rho(x, y)$ represents the distribution of the charge density in the physical domain, and ϵ is the dielectric constant of the medium. Within the framework of the FEM, we can readily accommodate the possibility of ϵ being a function of the coordinates. In applying the FEM to solve this problem, we consider the minimization of the electrostatic energy† and write

$$U = \iint dx\, dy\, \mathcal{L} = \iint dx\, dy \left(\frac{1}{2} \nabla \phi(x, y) \cdot \epsilon \cdot \nabla \phi(x, y) - \rho(x, y)\, \phi(x, y) \right). \tag{3.2}$$

As we have emphasized in earlier chapters, the FEM is simply the discretization of the above integral in which we express $\phi(x, y)$ in terms of nodal values for $\phi(x, y)$ multiplied by piecewise, local polynomials over each element. It is natural to evaluate the integral over each element using the Gauss-Legendre multiplicative quadrature by first mapping the quadrilateral element of Fig. 3.2(a), given in global coordinates (x, y) having vertices (x_i, y_i), $i = 0, \ldots, 3$, into a standard square element represented on the local coordinates ξ, η, $-1 \leq \xi, \eta \leq +1$, as in Fig. 3.2(b). It is natural and convenient to consider linear mapping between the two geometries for the given element. This is readily accomplished by using shape functions N_i that depend linearly on ξ, η.

Now, since any function $g(x)$ can be approximated by the linear polynomials over each element, we can also represent the coordinates x, y in terms of the same polynomials multiplied by the nodal coordinates for the global element. We write

$$\begin{aligned} N_0(\xi, \eta) &= \frac{(1-\xi)(1-\eta)}{4}, \quad N_1(\xi, \eta) = \frac{(1+\xi)(1-\eta)}{4}, \\ N_2(\xi, \eta) &= \frac{(1+\xi)(1+\eta)}{4}, \quad N_3(\xi, \eta) = \frac{(1-\xi)(1+\eta)}{4}. \end{aligned} \tag{3.3}$$

†The action integral is the time integral of this quantity.

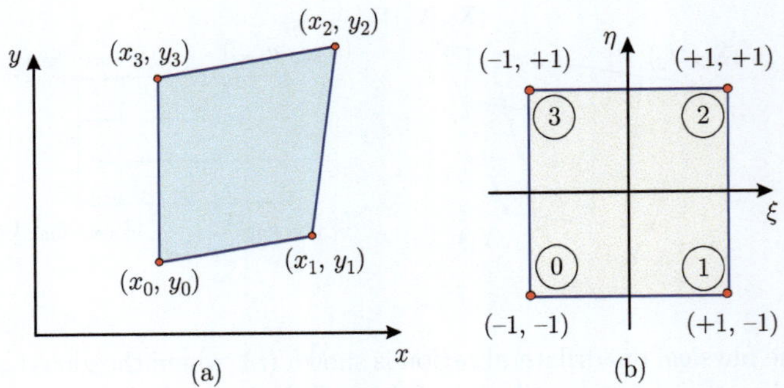

Figure 3.2 The integral over the quadrilateral finite element shown in (a) with vertices at the global coordinates (x_i, y_i), $i = 0, \ldots, 3$, is evaluated by transforming the quadrilateral into a standard square shown in (b) with the local coordinates (ξ, η), $-1 \le \xi, \eta \le +1$. The local node numbers are indicated in circles. Gauss quadrature is used for integration.

Here the index i corresponds to the node numbers indicated on the standard square in Fig. 3.2(b), and the shape functions satisfy the relation

$$N_i(\xi_j, \eta_j) = \delta_{ij}.$$

We have

$$x = \sum_i x_i N_i(\xi, \eta); \quad y = \sum_i y_i N_i(\xi, \eta). \tag{3.4}$$

With this linear mapping, it is straightforward to see that every coordinate (x, y) is represented by a unique point (ξ, η), with $-1 \le \xi, \eta \le +1$. The converse mapping from (x, y) to (ξ, η) requires a little algebra to show explicitly, and is left as an exercise.

The integral of any function $g(x, y)$ over a given quadrilateral element is evaluated by Gauss quadrature using the relations

$$\begin{aligned} \iint_{\text{quad}} dxdy\, g(x, y) &= \int_{-1}^{+1} d\xi \int_{-1}^{+1} d\eta \,|\det[\mathbf{J}(x, y, \xi, \eta)]|\, g(x(\xi, \eta), y(\xi, \eta)) \\ &\simeq \sum_{ig}^{Ng} W_{ig}^{(\xi)} \sum_{jg}^{Ng} W_{jg}^{(\eta)} |\det[\mathbf{J}(\xi_{ig}, \eta_{jg})]|\; g(x(\xi_{ig}, \eta_{jg}), y(\xi_{ig}, \eta_{jg})). \end{aligned} \tag{3.5}$$

Here $W_{ig}^{(\xi)}, W_{jg}^{(\eta)}$, with indices ig, jg, are the weight functions for each of the two one-dimensional Gauss integrations in the variables ξ and η, respectively. The Gauss points on the standard square element are located at the local coordinates (ξ_{ig}, η_{jg}). The Jacobian, or the determinant of the Jacobian matrix is included in the factor $|\det[\mathbf{J}]|$.

The calculation of the Jacobian matrix **J** is of interest here. From Eqs. (3.4), we have

$$\frac{\partial x}{\partial \xi} = \sum_i x_i \frac{\partial N_i}{\partial \xi} = \frac{1}{4}(-x_0 + x_1 - x_2 + x_3) + \frac{\eta}{4}(x_0 - x_1 - x_2 + x_3);$$
$$\frac{\partial x}{\partial \eta} = \sum_i x_i \frac{\partial N_i}{\partial \eta} = \frac{1}{4}(-x_0 - x_1 + x_2 + x_3) + \frac{\xi}{4}(x_0 - x_1 - x_2 + x_3), \tag{3.6}$$

and similar expressions hold for $\frac{\partial y}{\partial \xi}$ and $\frac{\partial y}{\partial \eta}$. We obtain

$$\begin{pmatrix} \partial_\xi \\ \partial_\eta \end{pmatrix} = \underbrace{\begin{pmatrix} \frac{\partial x}{\partial \xi} & \frac{\partial y}{\partial \xi} \\ \frac{\partial x}{\partial \eta} & \frac{\partial y}{\partial \eta} \end{pmatrix}}_{\text{Jacobian } \mathbf{J}} \begin{pmatrix} \partial_x \\ \partial_y \end{pmatrix} = \mathbf{J} \cdot \begin{pmatrix} \partial_x \\ \partial_y \end{pmatrix}, \tag{3.7}$$

The Jacobian is

$$\mathbf{J} = \frac{1}{4} \begin{bmatrix} (-x_0 + x_1 - x_2 + x_3) & (-y_0 + y_1 - y_2 + y_3) \\ +\eta(x_0 - x_1 - x_2 + x_3), & +\eta(y_0 - y_1 - y_2 + y_3) \\ & \\ (-x_0 - x_1 + x_2 + x_3) & (-y_0 - y_1 + y_2 + y_3) \\ +\xi(x_0 - x_1 - x_2 + x_3), & +\xi(y_0 - y_1 - y_2 + y_3) \end{bmatrix}. \tag{3.8}$$

The determinant of the Jacobian is of the form

$$\det[\mathbf{J}] = \frac{1}{16}(\alpha + \xi\,\beta + \gamma\,\eta), \tag{3.9}$$

and the coefficients α, β, γ are given by

$$\alpha = \frac{1}{8}\left[(x_0 - x_3)(y_1 - y_2) - (x_1 - x_2)(y_0 - y_3)\right],$$
$$\beta = \frac{1}{8}\left[(x_0 - x_1)(y_2 - y_3) - (x_2 - x_3)(y_1 - y_0)\right], \tag{3.10}$$
$$\gamma = \frac{1}{8}\left[(x_3 - x_1)(y_2 - y_0) - (x_2 - x_0)(y_3 - y_1)\right].$$

For a rectangle, as opposed to an arbitrary quadrilateral, we have $x_0 = x_2$, and $y_0 = y_2$, and also $x_1 = x_3$, $y_1 = y_3$, so that the Jacobian matrix reduces to

$$\mathbf{J}_{\text{rect}} = \begin{bmatrix} \frac{1}{2}(x_1 - x_0) & 0 \\ 0 & \frac{1}{2}(y_3 - y_1) \end{bmatrix}. \tag{3.11}$$

The determinant of $\mathbf{J}_{\text{rect}}$ equals a quarter of the area of the rectangle. For the standard square, the sides are of length 2 and the area is 4, and the Jacobian has a value 1.

In order to evaluate the energy functional for the Poisson problem, we require the derivatives of the potential function. Using the interpolation functions in a given element, we write

$$\frac{\partial \phi}{\partial x} = \sum_i \phi_i \left(\frac{\partial N_i}{\partial \xi}\frac{\partial \xi}{\partial x} + \frac{\partial N_i}{\partial \eta}\frac{\partial \eta}{\partial x} \right), \tag{3.12}$$

and a similar expression for the derivative with respect to y.

Now, the differentials in the global and local coordinates are related by

$$\begin{pmatrix} dx \\ dy \end{pmatrix} = \underbrace{\begin{bmatrix} \dfrac{\partial x}{\partial \xi} & \dfrac{\partial x}{\partial \eta} \\ \dfrac{\partial y}{\partial \xi} & \dfrac{\partial y}{\partial \eta} \end{bmatrix}}_{\text{Jacobian } \mathbf{J}} \begin{pmatrix} d\xi \\ d\eta \end{pmatrix} = \mathbf{J} \cdot \begin{pmatrix} d\xi \\ d\eta \end{pmatrix}, \tag{3.13}$$

and hence

$$\begin{pmatrix} d\xi \\ d\eta \end{pmatrix} = \mathbf{J}^{-1} \cdot \begin{pmatrix} dx \\ dy \end{pmatrix} = \frac{1}{|\det[\mathbf{J}]|} \begin{bmatrix} \dfrac{\partial y}{\partial \eta} & -\dfrac{\partial x}{\partial \eta} \\ -\dfrac{\partial y}{\partial \xi} & \dfrac{\partial x}{\partial \xi} \end{bmatrix} \begin{pmatrix} dx \\ dy \end{pmatrix}. \tag{3.14}$$

On the other hand, by the definition of the differentials, we have

$$\begin{pmatrix} d\xi \\ d\eta \end{pmatrix} = \begin{bmatrix} \dfrac{\partial \xi}{\partial x} & \dfrac{\partial \xi}{\partial y} \\ \dfrac{\partial \eta}{\partial x} & \dfrac{\partial \eta}{\partial y} \end{bmatrix} \begin{pmatrix} dx \\ dy \end{pmatrix}. \tag{3.15}$$

By comparing Eqs. (3.14) and (3.15), we find that

$$\begin{aligned} \frac{\partial \xi}{\partial x} &= \frac{1}{|\det[\mathbf{J}]} \frac{\partial y}{\partial \eta}; \quad \frac{\partial \xi}{\partial y} = -\frac{1}{|\det[\mathbf{J}]|} \frac{\partial x}{\partial \eta}; \\ \frac{\partial \eta}{\partial x} &= -\frac{1}{|\det[\mathbf{J}]|} \frac{\partial y}{\partial \xi}; \quad \frac{\partial \eta}{\partial y} = \frac{1}{|\det[\mathbf{J}]|} \frac{\partial x}{\partial \xi}. \end{aligned} \tag{3.16}$$

Equation (3.12) can now be expressed as

$$\begin{aligned} \frac{\partial \phi}{\partial x} &= \sum_i \phi_i \left(\frac{\partial N_i}{\partial \xi}\frac{\partial \xi}{\partial x} + \frac{\partial N_i}{\partial \eta}\frac{\partial \eta}{\partial x} \right) \\ &= \frac{1}{\det[\mathbf{J}]} \sum_i \sum_j \phi_i y_j \left(\frac{\partial N_i}{\partial \xi}\frac{\partial N_j}{\partial \eta} - \frac{\partial N_i}{\partial \eta}\frac{\partial N_j}{\partial \xi} \right). \end{aligned} \tag{3.17}$$

Here the sum over j is over the four nodes of the element in which y is represented by the y-coordinates of the nodes multiplied by the interpolation polynomials used in the linear transformation from the global to the local coordinates. On the other hand, the sum over i extends over the number of nodes in the standard element (4- or 9-nodal element) used to represent the solution $\phi(x, y)$. The two sums are independent of each other; one refers to the linear mapping of (x, y) to (ξ, η), while the other corresponds to the representation of the fields ϕ for which we can use any interpolation scheme we choose. We limit ourselves usually to linear elements for the mapping and employ linear, quadratic, or higher elements for the solution.

In a similar manner, we have

$$\begin{aligned}\frac{\partial \phi}{\partial y} &= \sum_i \phi_i \left(\frac{\partial N_i}{\partial \xi} \frac{\partial \xi}{\partial y} + \frac{\partial N_i}{\partial \eta} \frac{\partial \eta}{\partial y} \right) \\ &= \frac{1}{\det[\mathbf{J}]} \sum_i \sum_j \phi_i x_j \left(-\frac{\partial N_i}{\partial \xi} \frac{\partial N_j}{\partial \eta} + \frac{\partial N_i}{\partial \eta} \frac{\partial N_j}{\partial \xi} \right). \end{aligned} \tag{3.18}$$

With the above manipulations, we have reduced the derivatives of the potential function to derivatives of the shape functions. We are now in a position to evaluate the contribution to the total energy U from a given element *iel* by evaluating the integrals in Eq. (3.2) over the element *iel*. The rest of the steps are straightforward and standard to the FEM calculations since we have the spatial derivatives of $\phi(x, y)$ in terms of global variables (x, y) that appear in the form

$$\begin{aligned}U_{iel} &= \iint_{iel} dx dy \left(\frac{1}{2} \nabla \phi(x, y) \cdot \epsilon \nabla \phi(x, y) - f(x, y) \phi(x, y) \right) \\ &= \iint_{iel} d\xi d\eta \,|\det[\mathbf{J}]|\, \frac{1}{2} \left(\epsilon \frac{\partial \phi}{\partial x} \frac{\partial \phi}{\partial x} + \epsilon \frac{\partial \phi}{\partial y} \frac{\partial \phi}{\partial y} \right) \\ &\quad - \iint_{iel} d\xi d\eta \,|\det[\mathbf{J}]|\, f(x(\xi), y(\xi)) \sum_i \phi_i N_i(\xi, \eta), \end{aligned} \tag{3.19}$$

and we have expressions for $\partial\phi/\partial x$ and $\partial\phi/\partial y$ in terms of the derivatives of the shape functions N_i in local coordinates, as given in Eqs. (3.12) and (3.17).

We now turn to the connectivity relation for the global geometry. The method employed in mapping the global coordinates from a given element *iel* to the standard square region can once again be exploited in *mapping the entire global quadrilateral region onto a standard square region.* The advantage of doing so is that on the square region, the individual elements are readily identified by simple bookkeeping, the global node numbers for a given element can be obtained by an analytical formula, and the connectivity of the elements can be readily implemented while forming the global matrix from element matrices. We now linearly map the entire quadrilateral global region into a square region, with both regions divided up into the same number of elements. With a linear mapping, every coordinate in the rectangular region can be assigned a coordinate in the quadrilateral region, and *vice versa.*

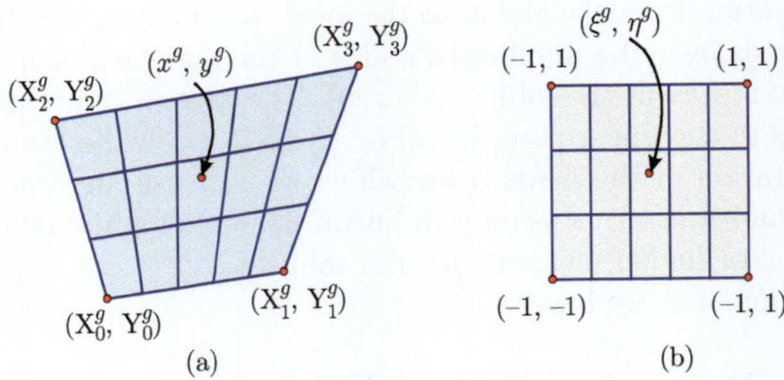

Figure 3.3 The physical quadrilateral region is shown (a), where the (global) quadrilateral is discretized into smaller quadrilateral elements. A linear mapping is performed of the physical region into a rectangular region, shown in (b), so that the elements in the rectangular region are easily constructed in the same manner to obtain rectangular finite elements.

The global coordinates $\left(x^{(g)}, y^{(g)}\right)$ in the quadrilateral region are given by

$$
\begin{aligned}
x^{(g)} &= \sum_{i=1}^{4} X_i N_i \left[\xi^{(g)}, \eta^{(g)}\right], \\
y^{(g)} &= \sum_{i=1}^{4} Y_i N_i \left[\xi^{(g)}, \eta^{(g)}\right].
\end{aligned}
\tag{3.20}
$$

Here the quantities $N_i \left[\xi^{(g)}, \eta^{(g)}\right]$ are bilinear interpolation polynomials given by Eq. (3.3). The square region is broken up into elements such that the nodes along the sides are distributed in the same ratio as the nodes in the quadrilateral. Now, given the vertex or corner coordinates $\xi_i^{(g)}, \eta_i^{(g)}$ of a particular element in the square region, as determined by connectivity relations for example, the corresponding coordinates in the quadrilateral region can be found rather easily from Eq. (3.20). Again, the integrals appearing in the energy functional can be performed by mapping the element coordinates in the quadrilateral region into those of the standard square element.

The main advantage of the present mapping of the global region into a square region is in the bookkeeping that is intrinsic to the FEM. Through this mapping, we have effectively changed our focus from the quadrilateral system to the rectangular region which has essentially simple analytic expressions for the node numbering and mapping from local node numbers to the global node numbers. This simplification helps with the issues we typically face in ensuring connectivity of elements.

We emphasize again that the evaluation of the element matrices for the calculation of the energy functional is still done by mapping the given quadrilateral element directly into a standard square element, as discussed previously.

3.2 Polar coordinates in 2D

Consider a circular region of radius R on which we wish to solve the Laplace equation, $\nabla^2\phi(r,\theta) = 0$ with the boundary conditions on the periphery given by

$$\phi(R,\theta) = V_0 \sin(\theta). \tag{3.21}$$

In the FEM, we start with the functional U given by

$$U = \iint_{r<R} r\, dr\, d\theta \left[\frac{1}{2}\nabla\phi(r,\theta)\nabla\phi(r,\theta)\right]. \tag{3.22}$$

The circular physical region is discretized in (r,θ) coordinates as a rectangle. The function $\phi(r,\theta)$ is interpolated in each element with linear shape functions or with Hermite interpolation polynomials defined as a direct product of 1D Hermite polynomials. The integral is evaluated over each rectangular (r,θ) element using the mapping

$$\begin{aligned} r &= \frac{(r_{iel+1} + r_{iel})}{2} + \frac{(r_{iel+1} - r_{iel})}{2}\xi, \\ \theta &= \frac{(\theta_{iel+1} + \theta_{iel})}{2} + \frac{(\theta_{iel+1} - \theta_{iel})}{2}\eta, \end{aligned} \tag{3.23}$$

with $-1 \le \xi \le 1,\ -1 \le \eta \le 1$. The integral is performed using Gauss-Legendre quadrature. Each element matrix is overlaid into a global matrix as usual, keeping in mind the connectivity of the elements. So far, with such an overlay of the element matrices, we have generated the global matrix representing the rectangular region of Fig. 3.4(a). We can "circle the square" by overlaying the nodes along side 3 on top of the corresponding nodes along side 1. The boundary conditions can be implemented at nodes along side 2. We now have the global matrix corresponding to the annular region displayed in Fig. 3.4(b). Since at $r = 0$, the value of θ is irrelevant, we overlay the rows and columns associated with the nodes along side 4 on to the first node at $(r = 0, \theta = 0)$. This then closes the "hole" at the center of the annulus, and we have thereby created the global matrix for the physical circular domain, as in Fig. 3.4(c).

We now discuss the absence of singularity at $r = 0$ in the energy functional U. For the circular region at hand, let us consider a circular region of radius R_0 with the boundary at R_0 having the BC $\phi(R_0,\theta) = V_0 \cos\theta$. We start with the functional

$$A = \frac{1}{2}\iint r dr d\theta \left[\vec{\nabla}\phi \cdot \vec{\nabla}\phi\right] \tag{3.24}$$

and use the variational principle to obtain

$$\delta_\phi A = 0 = -\iint_\Omega r dr d\theta \delta\phi \cdot \nabla^2\phi + \oint_\Gamma dl\ \delta\phi \left(\vec{\nabla}\phi \cdot \hat{n}\right), \tag{3.25}$$

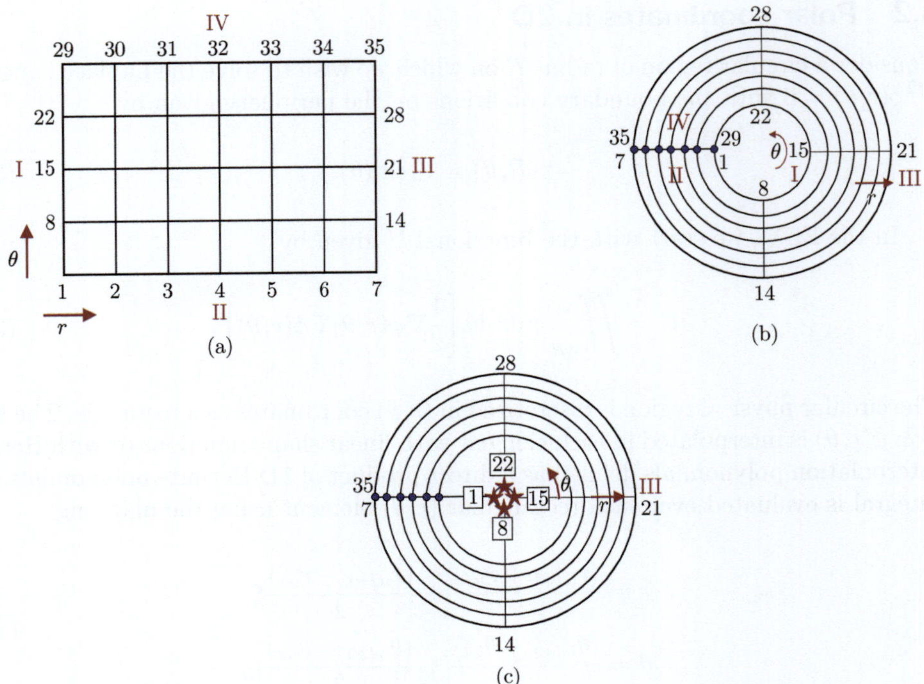

Figure 3.4 The global matrix calculated for the rectangular area, shown in (a), in (r, θ) coordinates, is mapped into an annular region shown in (b) by superposing nodal values associated with nodes located on side IV on to those on side II in the global matrix. A superposition of the nodal values associated with the inner circle onto the node at $(r = 0, \theta = 0)$ closes the annulus to represent a circular region.

where $\hat{n}$ is the outward drawn normal on the outer periphery and $\Gamma = \Gamma_0 + \Gamma_1 + \Gamma_2 + \Gamma_3$, as in Fig. 3.5.[‡]

In polar coordinates, $\Gamma = \Gamma_0 + \Gamma_1 + \Gamma_2 + \Gamma_3$ can be considered initially to be as in Fig. 3.5(a), and we treat the circle as a rectangle in the (r, θ) coordinates. So the circular geometry is reproduced by overlaying contour Γ_2 on Γ_0 (this first gives an annular geometry), and then shrinking Γ_3 to zero.

Note that the radial terms in $\nabla^2\phi$ have no singularity, as shown through evaluations in the following,

$$\iint \not{r}\, dr d\theta \left\{ \frac{1}{\not{r}} \frac{\partial}{\partial r} \left(r \frac{\partial \phi}{\partial r} \right) \right\}, \tag{3.26}$$

[‡]The same trick is used in the finite element calculation by setting up a rectangular mesh and overlaying the nodes from Γ_2 on the nodes on Γ_0. The nodes on Γ_3 are overlaid on the node at the center, since at $r = 0$ the variation of the angle is not defined.

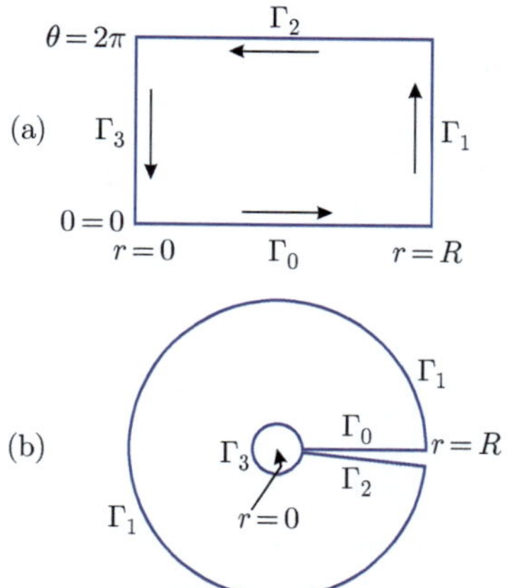

Figure 3.5 A rectangular region (a) in (r, θ) is transformed into the physical circular region (b) by overlapping the contour path Γ_0 and Γ_2, as shown.

is very well-behaved at $r = 0$. The annular term is of the form

$$\iint r dr d\theta \frac{1}{r^2} \frac{\partial^2 \phi}{\partial \theta^2}, \tag{3.27}$$

and we might think that it has a singularity at $r = 0$. However, we note that at $r = 0$, along the contour Γ_3 (which collapses to a point later on), we must have a unique $\phi(r = 0, \theta)$ for all θ. This means that $(\partial\phi/\partial\theta)|_{r=0} = 0$, which makes the angular term

$$\iint dr d\theta \not{r} \frac{1}{\not{r}\, r} \frac{\partial}{\partial \theta} \left(\frac{\partial \phi}{\partial \theta} \right). \tag{3.28}$$

This is of the form $(^{\text{zero}}/_{\text{zero}}) \xrightarrow{?} \text{zero}$.

Note 1: If we work with the action integral, we have

$$\begin{aligned} A &= \frac{1}{2} \iint r dr d\theta \left[\vec{\nabla}\phi \cdot \vec{\nabla}\phi \right] \\ &= \frac{1}{2} \iint r dr d\theta \left[\frac{\partial \phi}{\partial r} \cdot \frac{\partial \phi}{\partial r} + \left(\frac{1}{r} \frac{\partial \phi}{\partial \theta} \right) \left(\frac{1}{r} \frac{\partial \phi}{\partial \theta} \right) \right], \end{aligned} \tag{3.29}$$

which again shows the same issues. Again, the $\left(\frac{\partial\phi}{\partial r}\right)^2$ term poses no problem and the $\frac{r}{r^2}\left(\frac{\partial\phi}{\partial\theta}\right)^2$ term goes to zero as $\left(\frac{1}{r}\right)\left(\frac{\partial\phi}{\partial\theta}\right)$ vanishes (or could even be constant).

Note 2: *If we had a physical region in the form of an annulus, we do not have the same issue to address. The inner radius of the annulus will not be zero.*

Returning to $\delta_\phi A = 0$, we examine the contour integrals and their behavior. On Γ_0, $\hat{n} = -\hat{\theta}$. Hence,

$$\int_{r=0}^{R} dr\ \delta\phi \left(\frac{1}{r}\frac{\partial\phi}{\partial\theta}\right)(-1) = -\int_{r=0}^{R_0} dr\ \delta\phi \cdot \frac{\partial\phi}{r\partial\theta}. \tag{3.30}$$

On Γ_2, we have $\hat{n} = +\hat{\theta}$. However, $dl = -dr$. Now we have

$$\int_{r=R_0}^{R} (-dr)\ \delta\phi \left(\frac{1}{r}\frac{\partial\phi}{\partial\theta}\right)(+1) = +\int_{r=0}^{R_0} dr\ \delta\phi \left(\frac{1}{r}\frac{\partial\phi}{\partial\theta}\right). \tag{3.31}$$

Therefore, when we overlay Γ_0 on Γ_2, the two line integrals cancel each other. The $\frac{1}{r}$ term poses no problem at $r = 0$, again because it is multiplied by $\frac{\partial\phi}{\partial\theta}(r = 0, \theta) = 0$. In other words, on overlay, these contours became interior paths and cancel. Now along Γ_3, we have $dl = -rd\theta$ and $\hat{n} = -\hat{r}$. Hence,

$$\Gamma_3 = \lim_{r\to 0}\int_{\theta=2\pi}^{\theta=0} (-rd\theta)\left(\frac{\partial\phi}{\partial r}\right)(-1) = \lim_{r\to 0}\left[-\int_{\theta=0}^{2\pi} rd\theta\frac{\partial\phi}{\partial r}\right]. \tag{3.32}$$

If $(\partial\phi/\partial r)$ is "well-behaved" at $r = 0$, then this term also vanishes.

We are then left with the external line integral at $r = R_0$. This has $dl = +R_0 d\theta$ and $\hat{n} = \hat{r}$. Thus, on this contour

$$\Gamma_1 = \int_0^{2\pi} R_0 d\theta\ \delta\phi\left(\frac{\partial\phi}{\partial r}\right). \tag{3.33}$$

We then obtain

$$\delta_\phi A = 0 = -\iint r dr d\theta \delta\phi\left(\nabla^2\phi\right) + \int_{\substack{0\\ \Gamma_1}}^{2\pi} R_0 d\theta\ \delta\phi\frac{\partial\phi}{\partial r}. \tag{3.34}$$

Now if ϕ is specified on Γ_1, we have $\delta\phi = 0$ on Γ_1, and the line integral vanishes. We then have $\nabla^2\phi = 0$ from the first term since $\delta\phi$ is arbitrary in the interior. The above discussion is included here to show how the terms in the Laplacian behave.

Let us now consider the implementation of BCs. In the FEM, we evaluate element matrices using

$$A^{\mathrm{iel}} = \int_{\theta_1}^{\theta_2} d\theta \int_{r_1}^{r_2} rdr\left[\frac{1}{2}\Phi_a\left(\frac{\partial N_a(r,\theta)}{\partial r}\frac{\partial N_\beta(r,\theta)}{\partial r} + \frac{1}{r^2}\frac{\partial N_a(r,\theta)}{\partial\theta}\frac{\partial N_\beta(r,\theta)}{\partial\theta}\right)\Phi_\beta\right]. \tag{3.35}$$

We overlay the element matrices as usual.

1. The element matrices with one side at $r = 0$ are calculated without any problem at $r = 0$ since we use Gauss-Legendre quadrature. Even for a 60-point Gauss quadrature scheme, the root value closest to $r = 0$ is at $r = 10^{-4}$ only. We are then able to evaluate the element matrix even with the $r\left(\frac{1}{r}\frac{\partial}{\partial\theta}\phi\right)^2$ integrand in

the integral since r does not go to zero, in the calculation, in the denominator of the second term. Moreover, $\partial_\theta \phi = 0$ at $r = 0$, and this too helps.

2. To construct a circular region from the above diagram, we overlay the nodes from side 2 onto side 0. In effect, this ensures a seam-less annular region. Next, we look at nodes along r_3 to make a circular region. The overlay of element matrices in this fashion is done during the assembly of the global matrix.
3. Next, we overlay nodes at $(r = 0, \theta \neq 0)$ onto $(r = 0, \theta = 0)$ node. This is done during the assembly of the global matrix.
4. We now apply "pseudo-benediction" to the global matrix in order to keep the matrix size the same even after the above overlays for the operation.[§] Once the overlays associated with the boundaries Γ_0, Γ_2, and Γ_3 are complete, we perform the following operation:

   Suppose node n_1 is overlaid onto node n_2. This means that
   (a) all elements in row n_1 of the global matrix are added to row n_2, and
   (b) all elements in column n_1 are added to column n_2. Now the row and column n_1 are zero-ed out *after* this overlay.
   (c) Next, to preserve the matrix size, we insert a value 1 on the (n_1, n_1) diagonal and put a (-1) on the column n_2 in the same row.

   In other words, put $M_{n_1,n_1} = +1$ and set $M_{n_1,n_2} = -1$ (all other elements in the n_1 row are zero). This then leads to an equation

$$\phi_{n_1} - \phi_{n_2} = 0 \tag{3.36}$$

   on solving the system of simultaneous equations!

   Thus, once the global matrix equation is solved, we obtain ϕ_{n_2} as a component of the solution. Now ϕ_{n_1} is made equal to ϕ_{n_2}. We are thereby constructing the original mesh of Figure 3.4 and entering every nodal value in the solution vector automatically! This preserves the size of the global matrix even through the overlay procedure!

As an example, consider the matrix shown in Fig. 3.6(a), in which the row and column n_1 is added to row and column n_2. Next zero out row n_1 and column n_1. Put (1) at the (n_1, n_1) element and (-1) at the (n_1, n_2) element. We then have transformed the matrix shown in Fig. 3.6(a) into the matrix Fig. 3.6(b).

Lagrange interpolation versus Hermite interpolation:
With Lagrange interpolation, the terms in the action (Eq. (3.37)),

$$A = \frac{1}{2} \iint dxdy \left[\partial_a \phi (x, y)\right] \sigma_{a\beta} \; \left[\partial_\beta \phi (x, y)\right] - \int_{\Delta_1} dl \delta \phi j_{\text{in}} + \int_{\Delta_2} dl \delta \phi j_{\text{out}} \tag{3.37}$$

are directly evaluated in each element and inserted into a global matrix. The terms involving the current are readily included as integrals over a single factor of the interpolation polynomial. With the variation of the nodal values, the current terms at the

[§]**Note**: The overlays of operation (ii) ensure the finite element connectivity.

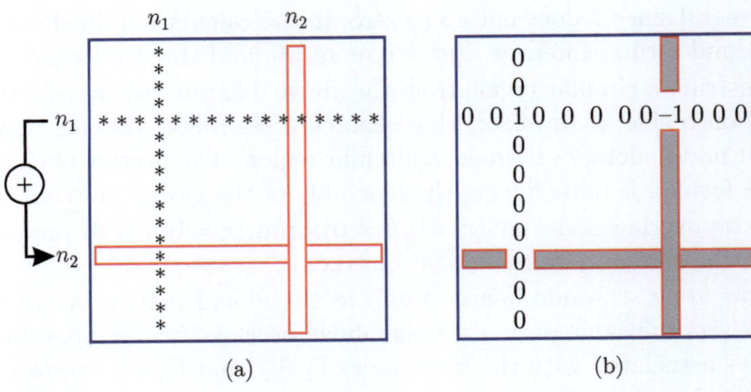

Figure 3.6 In (a), the variable ϕ_{n_1} is set *equal* to the variable ϕ_{n_2}. This is implemented by the step 4 discussed in the text. This yields the transformed matrix of figure (b) in which row and column n_1 is set to zero except for the diagonal element which is set to +1, and the element n_2 in the row n_1 which is set to -1. This leads to an equation $\phi_{n_1} - \phi_{n_2} = 0$.

ports are put into the right side of the matrix equation as the "driving terms" and the equations solved for the nodal values.

In the case of Hermite interpolation, we have explicit nodal variables corresponding to $\phi_i, (\partial\phi_i/\partial r), (\partial\phi_i/\partial\theta), (\partial^2\phi_i/\partial\theta\partial r)$ at each node i. At $r = 0$, we demand that $\phi(0,\theta) = \phi(0,0)$, which corresponds to nodal overlay, as in the case of Lagrange interpolation. We also require that $(\partial\phi_i/\partial\theta) = 0$, ensuring that we do not have singularities at the origin, as discussed earlier, and put $(\partial\phi_i/\partial r) = 0$ and $(\partial^2\phi_i/\partial\theta\partial r) = 0$ along the side $r = 0$ as a boundary condition on these derivatives. In principle, we should allow $(\partial\phi_i/\partial r), (\partial^2\phi_i/\partial\theta\partial r)$ to "float" and not apply the boundary conditions. After all, we need only one BC along each node. The nodal overlay for nodes along $r = 0$ is to ensure that we have just one unique node at the origin. The boundary condition, $(\partial\phi_i/\partial\theta) = 0$, is ensuring that the Laplacian is well-behaved at the origin.

We advocate the use of polar coordinates and higher order Lagrange interpolation or cubic/quintic Hermite interpolation with the above schemes to convert the physical region to a circular geometry through nodal overlays. This is far superior to the use of triangular elements to break up the physical region into unstructured triangles in order to conform to the circular geometry at hand. We need fewer elements and directly

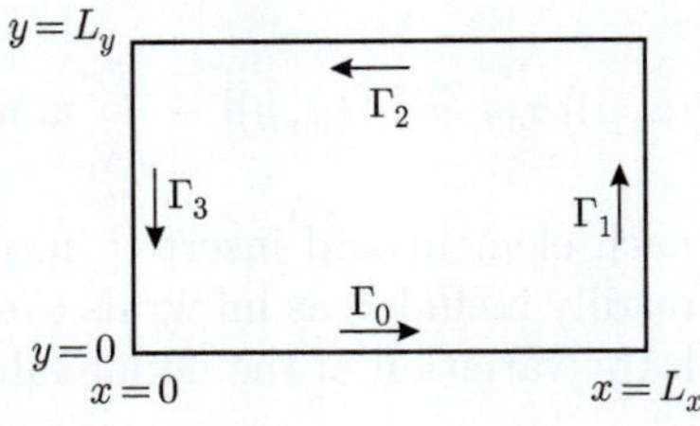

Figure 3.7 The determination of the unit vectors for segments of the contour and the line integral along a rectangle.

obtain the circular geometry. Improving accuracy through higher order polynomials is effective, and when combined with finer discretization, we can achieve excellent results. This can be tested using simple problems such as solving Laplace's equation for a potential problem on a circular geometry. The FEM results can be directly compared with analytical results which might require a summation of series.

3.2.1 The line integral and the signs of dl and $\hat{n}$

For a simple rectangular region of dimension $L_x \times l_y$, consider the line integral $\Gamma_{0+1+2+3} = \oint dl \stackrel{?}{=} 2L_x + 2L_y =$ perimeter of the rectangle. Now

$$\begin{aligned} \Gamma_0 &= \int_{L_x}^{0} dl = \int_0^{L_x} dx = L_x \\ \Gamma_1 &= \int_{L_y}^{0} dl = \int_0^{L_y} dy = L_y \end{aligned} \tag{3.38}$$

and

$$\begin{aligned} \Gamma_2 &= \int_{L_x}^{0} dl = -\int_0^{L_x} dl = -\int_0^{L_x} (-dx) = L_x \\ \Gamma_3 &= \int_{L_y}^{0} dl = -\int_0^{L_y} dl = -\int_0^{L_y} (-dy) = L_y. \end{aligned} \tag{3.39}$$

Note how the dl's are reinterpreted in terms of dx and dy. The result is $\oint dl = 2L_x + 2L_y$, as expected. The signs of the outward drawn normal $\tilde{n}$ on each segment is fairly straightforward:

$$\begin{aligned} \text{On } \Gamma_0 &: \hat{n} = -\hat{y} \\ \Gamma_1 &: \hat{n} = \hat{x} \\ \Gamma_2 &: \hat{n} = \hat{y} \\ \Gamma_3 &: \hat{n} = -\hat{x} \end{aligned} \tag{3.40}$$

We use Stokes' theorem to obtain Gauss' Law in 2D. Stokes' theorem states

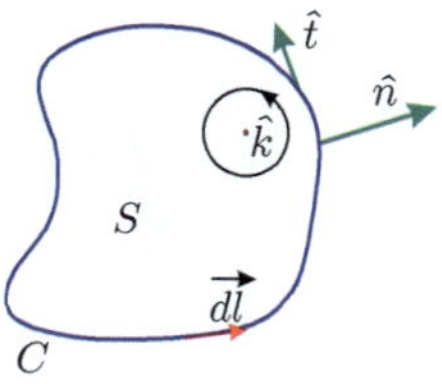

Figure 3.8 The area **S** enclosed by the contour **C** is shown. The outward drawn unit vector normal to the contour is **n** and the tangent is **t**.

$$\oint \vec{A} \cdot d\vec{l} = \iint \left(\vec{\nabla} \times \vec{A}\right) \cdot d\vec{s}. \tag{3.41}$$

This is of course valid in 3D also.

Consider the case of a plane on which we have a closed path defining the contour c. Let us define

$$\vec{a}\,(x, y) = M\,(x, y)\,\hat{x} + N\,(x, y)\,\hat{y}. \tag{3.42}$$

Then

$$\left(\vec{\nabla} \times \vec{A}\right)_z \hat{k} = \begin{vmatrix} 0 & 0 & \hat{k} \\ \partial_x & \partial_y & 0 \\ M & N & 0 \end{vmatrix} = \hat{k}\left(\frac{\partial}{\partial x} N\,(x, y) - \frac{\partial}{\partial y} M\,(x, y)\right). \tag{3.43}$$

We have Eq. (3.41) in the form

$$\oint_{\mathrm{C}} (Mdx + Ndy) = \iint_{\mathrm{S}} dxdy\,(\partial_x N - \partial_y M)\,. \tag{3.44}$$

This is Green's theorem.

Now let

$$Mdx + Ndy = \vec{A} \cdot d\vec{r} = \vec{A} \cdot \frac{d\vec{r}}{dl} \cdot dl. \tag{3.45}$$

Here $(d\vec{r}/dl) = \hat{t}$, is a unit vector along the tangent. Therefore,

$$\begin{aligned} \left(\vec{A} \cdot \hat{t}\right) dl &= \vec{A} \cdot \left(\hat{k} \times \hat{n}\right) dl \\ &= \hat{n} \cdot \left(\vec{A} \times \hat{k}\right) dl. \end{aligned} \tag{3.46}$$

Let

$$\vec{A} \times \hat{k} \equiv \vec{B} = (M\hat{x} + N\hat{y}) \times \hat{k} = (N\hat{x} - M\hat{y})\,. \tag{3.47}$$

Then $\vec{A} \cdot \hat{t}dl = \vec{B} \cdot \hat{n}dl$, and we have

$$\oint_{\mathrm{C}} \vec{B} \cdot \hat{n}dl = \iint_{\mathrm{S}} ds\,(\partial_x N - \partial_y M) = \iint_{\mathrm{S}} ds \left(\vec{\nabla} \cdot \vec{B}\right). \tag{3.48}$$

Thus, in 2D, we have

$$\iint_{\mathrm{S}} ds \vec{\nabla} \cdot \vec{B} = \oint_{\mathrm{C}} \left(\vec{B} \cdot \hat{n}\right) dl. \tag{3.49}$$

This is Gauss' Theorem in 2D.

Note that in 3D, we have $\iiint_{\mathrm{V}} \left(\vec{\nabla} \cdot \vec{E}\right) dV = \oiint_{\mathrm{S}} \vec{E} \cdot \hat{n}ds$ as Gauss' Theorem.

Comments:

1. Physically $\oint \vec{A} \cdot d\vec{l}$ may be thought of as the "work done" around a closed path, and in 2D it is equal to $\iint \vec{\nabla} \times \vec{A} \cdot d\vec{s}$. Hence, if for all c we have $\oint_c \vec{A} \cdot d\vec{l} = 0$, then $\vec{\nabla} \times \vec{A} = 0$ and $\vec{A} \equiv -\vec{\nabla}\phi$ as is the case for conservative forces.
2. In the plane $\partial_y M - \partial_x N$ is the same as $\vec{\nabla} \times \vec{A} = 0$.

3.2.2 Integration by parts in 2D and 3D

We use the vector identity for a scalar function f and a vector $\vec{u}\,(x, y)$

$$\vec{\nabla} \cdot (f\,(x, y)\,\vec{u}\,(x, y)) = f\,(x, y)\,\vec{\nabla} \cdot \vec{u} + \vec{\nabla} f \cdot \vec{u}. \tag{3.50}$$

Now choose $\vec{u} = \vec{\nabla} g\,(x, y)$. Then

$$\vec{\nabla} \cdot \left(f\vec{\nabla} g\right) = f\nabla^2 g + \vec{\nabla} f \cdot \vec{\nabla} g. \tag{3.51}$$

On integration we have

$$\begin{aligned} \iint dxdy \left(\vec{\nabla} f \cdot \vec{\nabla} u\right) &= -\iint dxdy \left(f\nabla^2 g\right) + \iint dxdy \vec{\nabla} \cdot \left(f\vec{\nabla} g\right) \\ &= -\iint dxdy \left(f\nabla^2 g\right) + \oint \left(f\vec{\nabla} g\right) \cdot \hat{n} dl. \end{aligned} \tag{3.52}$$

This is used in the integration by parts in the action integral formulation while deriving the differential equation.

3.3 Triangular elements

Linear triangular elements were first considered by Courant [1] in the context of vibration analysis and computational mechanics.

Let us suppose that the triangular element has vertices located at coordinates (x_i, y_i), $i = 1, 2, 3$, in the global coordinate system. This triangle is linearly mapped into a standard right-angled triangle on the ξ, η plane. The usual labeling of the nodes is shown in Fig. 3.9. In linear triangular elements, the nodal shape functions are given by

$$\begin{aligned} N_1(\xi, \eta) &= 1 - \xi - \eta, \\ N_2(\xi, \eta) &= \xi, \\ N_3(\xi, \eta) &= \eta. \end{aligned} \tag{3.53}$$

These shape functions are sometimes expressed in area coordinates for a more symmetric representation. [2]

A general function $f(x, y)$, with given values f_i at the three vertices of the triangle, can be linearly interpolated inside the triangle using

$$f(x, y) \simeq \sum_{i}^{3} f_i N_i(\xi, \eta). \tag{3.54}$$

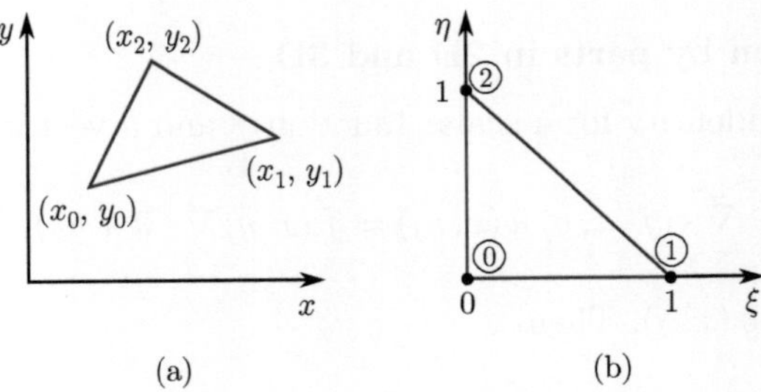

Figure 3.9 The arbitrary triangle (a) and its mapping into the standard triangle (b). The circled numbers are the node numbers.

We can also represent the global coordinates x and y themselves in terms of shape functions:

$$\begin{aligned} x &= x_1 N_1(\xi, \eta) + x_2 N_2(\xi, \eta) + x_3 N_3(\xi, \eta), \\ y &= y_1 N_1(\xi, \eta) + y_2 N_2(\xi, \eta) + y_3 N_3(\xi, \eta). \end{aligned} \tag{3.55}$$

This is an example of isoparametric mapping, in which the coordinate transformation to the standard triangle is the same for the coordinates as for the function being interpolated. Equation (3.55) can be written in a 2×2 matrix form

$$\begin{pmatrix} x - x_1 \\ y - y_1 \end{pmatrix} = \begin{pmatrix} x_2 - x_1 & x_3 - x_1 \\ y_2 - y_1 & y_3 - y_1 \end{pmatrix} \begin{pmatrix} \xi \\ \eta \end{pmatrix},$$
$$(\mathbf{r} - \mathbf{r}_1) \equiv \mathbf{J} \cdot \boldsymbol{\rho}. \tag{3.56}$$

This is a linear mapping between the variables (ξ, η) and (x, y). The determinant of the matrix $\mathbf{J}$ defined in equation (3.56) is the Jacobian of the transformation of the area element from $dx\, dy$ to $d\xi\, d\eta$. The Jacobian is twice the area of the global triangle

$$\mathbf{S} = \frac{1}{2}(\mathbf{r}_2 - \mathbf{r}_1) \times (\mathbf{r}_3 - \mathbf{r}_1 \equiv J/2).$$

The integrand in the quantum mechanical action will in general contain kinetic energy terms. We therefore determine the gradient of the wavefunction expressed in terms of the shape functions. Consider the partial derivative $\partial_x \psi(x, y)$. We have

$$\frac{\partial}{\partial x}\psi(x,y) = \frac{\partial}{\partial x}\left(\sum_i \psi_i N_i(\xi,\eta)\right)$$
$$= \sum_i \psi_i \left(\frac{\partial N_i}{\partial \xi}\frac{\partial \xi}{\partial x} + \frac{\partial N_i}{\partial \eta}\frac{\partial \eta}{\partial x}\right). \tag{3.57}$$

The inverse of the relation

$$\begin{pmatrix} dx \\ dy \end{pmatrix} = \begin{pmatrix} \dfrac{\partial x}{\partial \xi} & \dfrac{\partial x}{\partial \eta} \\ \dfrac{\partial y}{\partial \xi} & \dfrac{\partial y}{\partial \eta} \end{pmatrix} \begin{pmatrix} d\xi \\ d\eta \end{pmatrix}$$
$$= \mathbf{J}\begin{pmatrix} d\xi \\ d\eta \end{pmatrix} \tag{3.58}$$

is given by

$$\begin{pmatrix} d\xi \\ d\eta \end{pmatrix} = \frac{1}{2S}\begin{pmatrix} \dfrac{\partial y}{\partial \eta} & -\dfrac{\partial x}{\partial \eta} \\ -\dfrac{\partial y}{\partial \xi} & \dfrac{\partial x}{\partial \xi} \end{pmatrix} \begin{pmatrix} dx \\ dy \end{pmatrix}. \tag{3.59}$$

We can therefore identify, for example, $\partial\xi/\partial x = (1/2S)\partial y/\partial\eta$. Proceeding in this manner, we find the derivatives of the wavefunction:

$$\begin{aligned} \frac{\partial}{\partial x}\psi(x,y) &= \frac{1}{2S}\sum_i \psi_i \left(\frac{\partial N_i}{\partial \xi}\cdot\frac{\partial y}{\partial \eta} - \frac{\partial N_i}{\partial \eta}\cdot\frac{\partial y}{\partial \xi}\right) \\ \frac{\partial}{\partial y}\psi(x,y) &= \frac{1}{2S}\sum_i \psi_i \left(-\frac{\partial N_i}{\partial \xi}\cdot\frac{\partial x}{\partial \eta} + \frac{\partial N_i}{\partial \eta}\cdot\frac{\partial x}{\partial \xi}\right). \end{aligned} \tag{3.60}$$

Since the coordinates (x, y) themselves are also expressed in terms of shape functions, we have a systematic way of obtaining the gradient of the wavefunction in terms of products of the derivatives of the shape functions,

$$\begin{aligned} \frac{\partial}{\partial x}\psi(x,y) &= \frac{1}{2S}\sum_{ij} \psi_i y_j \left(\frac{\partial N_i}{\partial \xi}\cdot\frac{\partial N_j}{\partial \eta} - \frac{\partial N_i}{\partial \eta}\cdot\frac{\partial N_j}{\partial \xi}\right) \\ \frac{\partial}{\partial y}\psi(x,y) &= \frac{1}{2S}\sum_{ij} \psi_i x_j \left(-\frac{\partial N_i}{\partial \xi}\cdot\frac{\partial N_j}{\partial \eta} + \frac{\partial N_i}{\partial \eta}\cdot\frac{\partial N_j}{\partial \xi}\right). \end{aligned} \tag{3.61}$$

Thus the global derivatives of $\psi(x, y)$ can be cast in terms of the local derivatives of the shape functions. They can be calculated *a priori* and stored once and for all, to be used with all global elements.

Table 3.1 The shape functions for the Hermite interpolation polynomials for a triangle with 18 degrees of freedom are displayed. At the three nodal vertices, the ordering of the shape functions corresponds to the degrees of freedom $\{f(\xi,\eta),\ \partial_\xi f,\ \partial_\eta f,\ \partial^2_{\xi\xi} f,\ \partial^2_{\xi\eta} f,\ \partial^2_{\eta\eta} f\}$. Here, $\sigma = 1-\xi-\eta$.

Node	Shape function
Node 1	$N_1 = \sigma^2(10\sigma - 15\sigma^2 + 6\sigma^3 + 30\xi\eta(\xi+\eta))$
	$N_2 = \xi\sigma^2(3 - 2\sigma - 3\xi^2 + 6\xi\eta)$
	$N_3 = \eta\sigma^2(3 - 2\sigma - 3\eta^2 + 6\xi\eta)$
	$N_4 = \xi^2\sigma^2(1 - \xi + 2\eta)/2$
	$N_5 = \xi\eta\sigma^2$
	$N_6 = \eta^2\sigma^2(1 + 2\xi - \eta)/2$
Node 2	$N_7 = \xi^2(10\xi - 15\xi^2 + 6\xi^3 + 15\eta^2\sigma)$
	$N_8 = \xi^2(-8\xi + 14\xi^2 - 6\xi^3 - 15\eta^2\sigma)/2$
	$N_9 = \xi^2\eta(6 - 4\xi - 3\eta - 3\eta^2 + 3\xi\eta)/2$
	$N_{10} = \xi^2(2\xi(1-\xi)^2 + 5\eta^2\sigma)/4$
	$N_{11} = \xi^2\eta(-2 + 2\xi + \eta + \eta^2 - \xi\eta)/2$
	$N_{12} = \frac{1}{4}\xi^2\eta^2\sigma + \frac{1}{2}\xi^3\eta^2$
Node 3	$N_{13} = \eta^2(10\eta - 15\eta^2 + 6\eta^3 + 15\xi^2\sigma)$
	$N_{14} = \eta^2\xi(6 - 4\eta - 3\xi - 3\xi^2 + 3\eta\xi)/2$
	$N_{15} = \eta^2(-8\eta + 14\eta^2 - 6\eta^3 - 15\xi^2\sigma)/2$
	$N_{16} = \frac{1}{4}\eta^2\xi^2\sigma + \frac{1}{2}\eta^3\xi^2$
	$N_{17} = \eta^2\xi(-2 + 2\eta + \xi + \xi^2 - \eta\xi)/2$
	$N_{18} = \eta^2(2\eta(1-\eta)^2 + 5\xi^2\sigma)/4$

Quadratic triangular elements contain three additional nodes located at the midpoints of the three sides. The standard element is shown in Fig. 3.9, and the corresponding shape functions are

$$
\begin{aligned}
N_1(\xi,\eta) &= (1-\xi-\eta)(1-2\xi-2\eta),\\
N_2(\xi,\eta) &= \xi(2\,\xi-1),\\
N_3(\xi,\eta) &= \eta(2\,\eta-1),\\
N_4(\xi,\eta) &= 4\,\xi(1-\xi-\eta),\\
N_5(\xi,\eta) &= 4\,\xi\eta,\\
N_6(\xi,\eta) &= 4\,\eta(1-\xi-\eta).
\end{aligned}
\tag{3.62}
$$

These shape functions can be used to define triangles with curved sides through an isoparametric mapping. In this case, the triangular edges correspond to quadratic functions. In general, the elements must be such that the transformation from the standard triangle to the actual element should not distort the edges, generating multiple crossings of curves representing the sides of the elements.

Triangular Hermite elements have been defined in the literature. These elements have additional derivative DoFs defined at the nodes. Since the mapping from the global triangle with vertices at (x_i, y_i) to the standard right triangle will entail the transformation of the derivatives as well, we have to provide for all cross-derivatives to a given order. For example, if we allow the degree of freedom $\partial^2_{xy} f(x, y)$ at a node, we must also provide for the DoFs corresponding to the values of $\partial^2_{xx} f(x, y)$ and $\partial^2_{yy} f(x, y)$ at the nodes. It is possible to define 21 complete quintic polynomials for shape functions associated with the six DoFs at each vertex for

$$f(x, y),\ \ \partial_x f,\ \ \partial_y f,\ \ \partial^2_{xx} f,\ \ \partial^2_{xy} f,\ \ \partial^2_{yy} f,$$

together with the normal derivatives $\partial_n f$ at nodes located at the midpoints of the sides. Such an interpolation scheme is C_1-continuous across the triangular network. Now the shape functions $N_i(\xi, \eta)$ are given by

$$N_i(\xi, \eta) = \sum_{p+q \leq 5} a_{i,pq}\, \xi^p\, \eta^q, \quad i = 1, \ldots, 21. \tag{3.63}$$

The DoFs at the mid-side nodes, the normal derivatives at those points, can be eliminated without losing the derivative $\mathcal{C}_{(\infty)}$-continuity by imposing constraints on the shape functions. These additional conditions correspond to equating the coefficients of $\xi^4 \eta$ and $\xi \eta^4$ and requiring that the coefficients of $\xi^5, \xi^3\eta^2, \xi^2, \eta^3, \eta^5$ add up to zero:

$$\begin{aligned} a_{i,41} &= a_{i,14}, \\ 5a_{i,50} + a_{i,32} \ + \ a_{i,23} + 5a_{i,05} &= 0. \end{aligned} \tag{3.64}$$

The normal derivative now has a cubic variation along the sides. We thus have six DoFs located only at the three vertices. The $\mathcal{C}_{(\infty)}$-continuous polynomials [3, 4] are given in Table 3.1.

For a unique derivation of the Hermite shape functions on triangles, [5] see Appendix A. Earlier attempts at deriving these polynomials are given by Bell [6] and by Argyris. [7] Use is made of transformation properties of the shape functions under the symmetry group C_{3V} of the triangle.

A final note about triangular elements concerns the use of hierarchical shape functions for triangular elements. This has been investigated by Webb and Abouchacra, [8] who employ Jacobi polynomials for the development of the hierarchical shape functions. Gauss quadrature techniques can be used with triangular elements, and these are elaborated on in Stroud and Secrest. [9, 10] Furthermore, Dunavant [11] has provided quadrature rules with just a single summation over Gauss points in triangles and squares.

Besides the usual lower order shape functions associated with nodes, we can have additional shape functions which vanish at the nodes, but provide additional DoFs for the element in order to represent a function over the triangle. In two dimensions, we have (i) edge functions in which the shape functions are zero at the three vertex nodes and have polynomial variation along a given edge, and also (ii) bubble functions which are zero at the vertex nodes and are zero also along the edges. One way to think of these additional DoFs is to view them as "Fourier-like" terms in the representation of a function in terms of a sum of these shape functions as well as the node-based shape functions. As we know, Fourier coefficients are not tied to any nodal DoFs.

3.4 Finite elements in 3D

3.4.1 Brick elements in 3D

It is straightforward to generate Lagrange interpolation polynomials for brick elements by multiplying polynomials of a given degree in x, y, z. For example, for an 8-noded element shown in Fig. 3.10, the polynomials are direct products of the linear interpolation polynomials in x, y, z, respectively. To illustrate the procedure, let us suppose the 1D linear shape functions belonging to the element with two nodes are $N_0(x) = (1-x)/2$ and $N_1(x) = (1+x)/2$. We can write this as $N_i(x),\ i = 0, 1$. The eight shape functions associated with the nodes at the vertices of the cube are given by

$$N_\alpha(x, y, z) = N_i(x) \times N_j(y) \times N_k(z); \quad i, j, k = 0, 1. \tag{3.65}$$

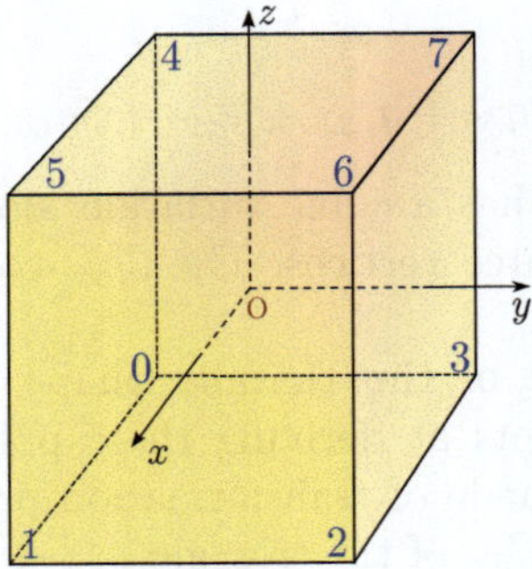

Figure 3.10 The linear 3D brick element is shown. The node numbers conform to the usual usage.

Care should be taken to verify that the value of α corresponds to one of the 8 nodes of the element where the choice of i, j, k gives a unit value for the functions N_α. The list of Lagrange and Hermite shape functions for brick, or hexahedral elements, is given in Appendix B.

3.4.2 Tetrahedral elements in 3D

Again, for linear tetrahedral elements we consider the right-tetrahedron defined by the axes ξ, η, ζ. We identify the four shape functions as

$$\begin{aligned} N_0(\xi, \eta, \zeta) &= \xi; \qquad & N_1(\xi, \eta, \zeta) &= \eta; \\ N_2(\xi, \eta, \zeta) &= \zeta; \qquad & N_4(\xi, \eta, \zeta) &= 1 - \xi - \eta - \zeta. \end{aligned} \tag{3.66}$$

The quadratic element is defined by setting six additional nodes at the midpoints of the six edges. The shape functions are listed in Appendix B.

We discuss $\mathcal{C}_{(1)}$- and $\mathcal{C}_{(2)}$-continuous symmetric Hermite tetrahedral finite elements in Appendix B. These are derived in Appendix A.

3.5 Problems

1. For the trapezoidal element, map the global coordinates (x, y) to (ξ, η) to show that it is a linear mapping.
2. (a) Determine the shape functions for a square Hermite finite element, which has $\mathcal{C}_{(1)}$-continuous interpolation polynomials and 4 degrees of freedom at each node. These polynomials can be obtained as a direct product of the 1D Hermite cubic shape functions. (This leads to f, f_x, f_y, and f_{xy} as the 4 DoFs at each node.) Note that the derivative DoFs will have Jacobian scaling factors, such as $d\xi/dx$ in 1D Hermite polynomials, to account for the size of the global elements as opposed to the standard element ranging over $-1 \leq \xi, \eta \leq 1$.
3. Consider the Hermite tetrahedral element with the DoFs corresponding to f, f'_x, f'_y, f'_z at the four vertices together with just one DoF, f_j at the four face centers. This leads to a total of 20 DoFs for this element. Determine the interpolation polynomials for (a) a right-angle tetrahedron, and for (b) a symmetric tetrahedron.

References

[1] R. Courant, Bull. Amer. Math. Soc. **49**, 1–23 (1943). "Variational methods for the solution of problems of equilibrium and vibrations."

[2] O. C. Zienkiewicz, *The Finite Element Method* (McGraw-Hill, New York, 1977).

[3] R. Wait, and A. R. Mitchell, *Finite Element Analysis and Applications* (John Wiley and Sons, Chichester, UK, 1985).

[4] G. Dhatt and G. Touzot, *The Finite Element Method Displayed* (Wiley, New York, 1984).

[5] P. G. Kassebaum, C. R. Boucher, and L. R. Ram-Mohan, J. Comp. Physics. **231**, 5747–5760 (2012). "Application of group representation theory to derive Hermite interpolation on a triangle."

[6] K. Bell, Int. J. Numer. Methods Eng. 1, 101–122 (1969); "A refined triangular plate bending finite element."

[7] J. H. Argyris, I. Fried, and D. W. Scharpf, Aeronaut. J. Royal Aeronaut. Soc. **72**, 701–709 (1968); "The TUBA family of plate elements for the matrix displacement method."

[8] J. P. Webb, and R. Abouchacra, *Int. J. Numer. Methods Eng.* **38**, 245–257 (1995); "Hierarchal triangular elements using orthogonal polynomials."

[9] A. H. Stroud and D. Secrest, *Gaussian Quadrature Formulas* (Prentice Hall, Englewood Cliffs, NJ, 1966).

[10] A. H. Stroud, *Approximate Calculation of Multiple Integrals* (Prentice Hall, Englewood Cliffs, NJ, 1971).

[11] D. A. Dunavant, *Int. J. Numer. Methods Eng.* **21**, 1129–1148 (1985); "High degree efficient symmetrical Gaussian quadrature rules for the triangle."

4

Boundary conditions at material interfaces

In this chapter:

- We analyze the boundary conditions (BCs) at material interfaces between two materials with different properties. For second-order partial differential equations we have the two conditions: the continuity of the solution function at the interface, and the continuity of the probability current for quantum systems. (This is equivalent to the familiar continuity of the normal displacement field in electrodynamics, where it manifests itself as a "jump condition" on the electric field, which is the negative gradient of the potential function.)
- For Lagrange interpolation, the jump condition is very simply implemented at the nodes common to the two materials. *Lagrange elements do not have derivative degrees of freedom.* So it is usually assumed that the jump condition is implicitly satisfied. We elaborate on this further below. This allows us to overlay element matrices without considering derivative degrees of freedom (DoF).
- With $\mathcal{C}_{(1)}$-continuous interpolation functions, in what we refer to as Hermite finite elements, first-order derivative DoF are included at the nodes. Now we have to develop an interface BC matrix in order to implement the jump condition. Overlaying the derivative DoF at a node common to two elements at the boundary then requires some care.
- When second-order derivative DoFs are included at nodes in elements, there are two choices that we elaborate on in the discussion below. One is to employ the differential equation itself to determine the corresponding jump condition. Alternately, we consider a new approach of locally smoothing out the material properties so that the material transitions from one to the other at the interface in a continuous fashion. This removes the complication of jump conditions. We argue as to why this is a very physical approach to the interface BCs.

Finite Elements in Action. L. Ramdas Ram-Mohan, Oxford University Press. © L. Ramdas Ram-Mohan (2026).
DOI: 10.1093/oso/9780199563487.003.0004

4.1 Interface boundary conditions in 1D

In Lagrange finite elements, the nodes have only the DoF associated with the value of the function. The overlay of element matrices, after each one of them is calculated, in the assembly of the global matrix is done as in Chap. 2, and illustrated for 1D in Fig. 2.3. For example, in 1D, we design the elements so that the material interface is at one of the nodes, say at *node3* between elements 2 and 3 in Fig. 4.2. For Lagrange elements, we implicitly assume the continuity of the derivative modified by the weight factor. For electrostatics, the continuity of the Maxwell displacement vector requires that $-\epsilon_1 \phi'(x_{n3}-\delta) = -\epsilon_2 \phi'(x_{n3}+\delta)$. If we inspect the evaluation of the derivative term in the energy functional, in 1D we have the integral

$$\int_{x=x_0}^{x=x_3-0^-} dx \frac{1}{2}\left(\frac{d\phi(x)}{dx}\epsilon_1(x)\frac{d\phi(x)}{dx}\right) + \int_{x=x_3+0^+}^{x=x_5} dx \frac{1}{2}\left(\frac{d\phi(x)}{dx}\epsilon_2(x)\frac{d\phi(x)}{dx}\right)$$

$$= -\int dx \phi(x)\left[\frac{d}{dx}\left(\epsilon_1(x)\frac{d\phi(x)}{dx}\right)\right] - \int dx \phi(x)\left[\frac{d}{dx}\left(\epsilon_2(x)\frac{d\phi(x)}{dx}\right)\right]$$

$$+ \left[\phi(x)\epsilon_1\phi'(x)\right]\Bigg|_{x=x_0}^{x=x_3-0^-} + \left[\phi(x)\epsilon_2\phi'(x)\right]\Bigg|_{x=_3+0^+}^{x=5} . \tag{4.1}$$

Hence, at the interface, if the continuity of the normal component of the Maxwell displacement vector **D** is assumed, then the upper limit of the "surface term" from the left cancels the lower limit of the surface term from the right. This condition is implicitly assumed to hold for Lagrange elements so that the element overlay can be continued across the interface as in Chap. 2, Fig. 2.3.

For Hermite finite elements, we have the assigned derivative DoF at each node, and each element matrix for Hermite shape functions on the element with two nodes now has a dimension of 4. The element matrix calculations are done such that we have the elements corresponding to the array $\{\phi_1, \phi_1'\}, \{\phi_2, \phi_2'\}, \ldots$, etc. The indices 1, 2 correspond to the node number. Then given the same geometry as above, we first

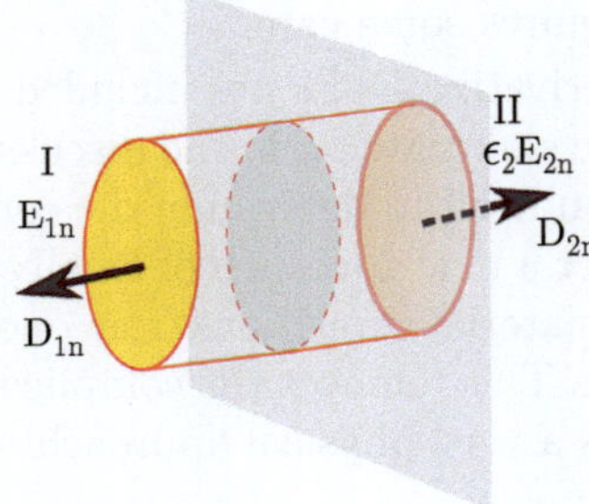

Figure 4.1 Boundary conditions in electrostatics given by Gauss's pill-box construction correspond to the continuity of the potential function and the continuity of the normal component of the displacement vector.

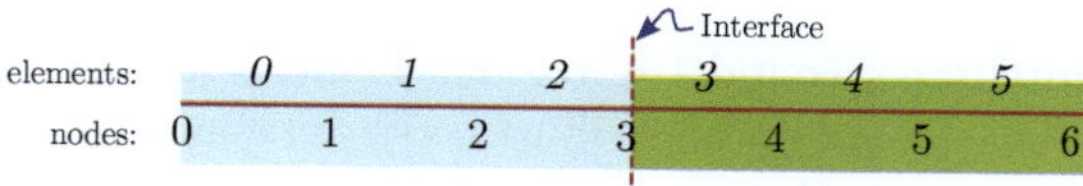

Figure 4.2 The interface boundary in 1D showing the node and elements on either side of the interface.

evaluate the element matrix for element 2 and put it into the global matrix. Now the right side element, element 3, is calculated and its derivative DoF for the third node is modified by a rescaling so that $\phi'_{3R} = (\epsilon_2/\epsilon_1)\,\phi'_{3L}$. Only then can it be inserted into the global matrix, since the DoF already in the global matrix corresponds to ϕ'_{3L}, and the quantity ϕ'_R is expressed in terms of the quantity from the left of the interface. Here, the subscripts L, R refer to the value of ϕ' approached from the element to the left (right) of the interface. This scaling can be implemented easily using a matrix multiplying the global matrix from the left and also from the right to modify the derivative DoFs loaded later into the global matrix.

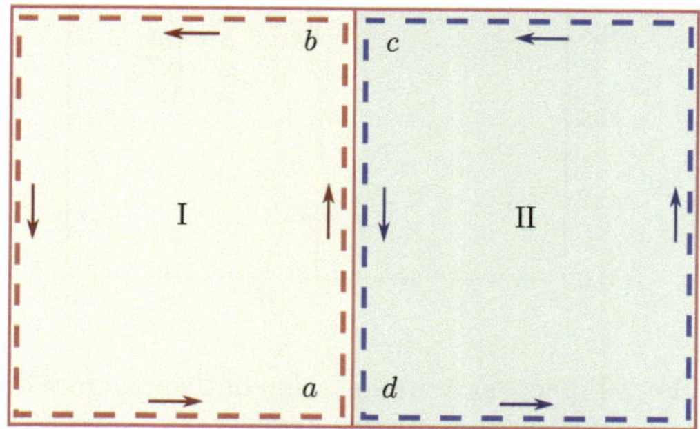

Figure 4.3 Boundary contour paths and interface continuity conditions in two dimensions.

4.2 Interface boundary conditions in 2D

In two dimensions, the derivative term in the energy functional over the domain $\Omega = \Omega_1 + \Omega_2$ (see Fig. 4.1) takes the form

$$
\begin{aligned}
&\iint_{\Omega_1+\Omega_2} dx\,dy \frac{1}{2}\Big[\nabla\phi(x,y)\cdot\epsilon(x,y)\cdot\nabla\phi(x,y)\Big] \\
&= -\iint_{\Omega_1+\Omega_2} dx\,dy\,\phi(x,y)\Big(\nabla\cdot\epsilon(x,y)\nabla\phi(x,y)\Big) \\
&\quad + \oint_{C_1} d\ell_1\,\phi(x,y)\hat{\mathbf{n}}\cdot\Big(\epsilon(x,y)\nabla\phi(x,y)\Big) \\
&\quad + \oint_{C_2} d\ell_2\,\phi(x,y)\hat{\mathbf{n}}\cdot\Big(\epsilon(x,y)\nabla\phi(x,y)\Big).
\end{aligned}
\tag{4.2}
$$

Here the contour integrals around the regions I and II in Fig. 4.3 are summed over increments of $d\ell_1$ and $d\ell_2$. The contributions from paths $a \to b$ and $c \to d$ in the two regions with ϵ_1, ϵ_2 cancel each other. For Lagrange finite elements, this cancellation is assumed to hold implicitly, and a straightforward overlay of element matrices is performed in the global matrix. The Lagrange elements do not have any derivative DoF.

For Hermite finite elements, the elements with a 1D line edge along the boundary have DoF that include derivatives. If elements from region I are first loaded into the global matrix, the elements from region II are modified to account for the rescaling of the derivative DoF due to the jump condition at the interface. The variation of the action integral and the global matrix equation, which represents the discretized differential equation.

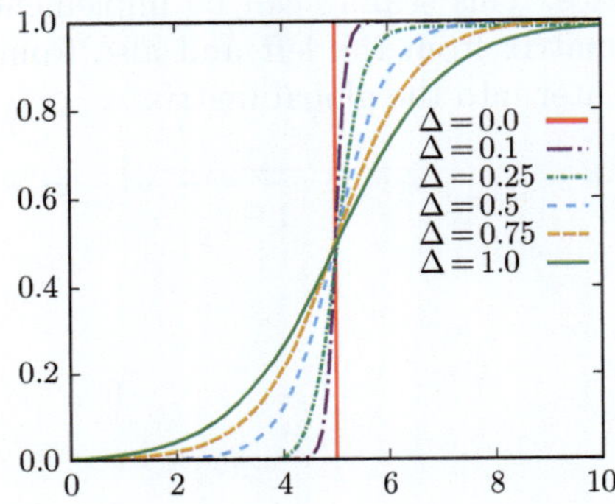

Figure 4.4 The use of Fermi function representation of the interface discontinuity is illustrated. By using a large number of finite elements, we are able to capture the effects of the interface with just simple overlay of elements in the global matrix, without any jump conditions for the derivative of the wavefunction. The potential function and its gradient are linked to the same dielectric function for a given coordinate x across the interface.

For the second derivative DoF, we could use the differential equation derived from the principle of stationary action to determine the jump condition in terms of continuity conditions for lower derivatives.

In the following, we introduce another approach that has fulfilled our tests in model calculations – the method of Fermi function smoothing of interfaces.

4.3 Interface smoothing using Fermi functions

It is now clear that the derivative interface BCs are difficult to implement. The situation is more cumbersome to resolve in terms of bookkeeping, if we had to ensure the jump conditions for second derivatives. For Hermite interpolation with $\mathcal{C}_{(n)}$-continuity and n being a reasonable integer, we consider a new alternative. We smooth out the abrupt changes in material properties using either a cubic Hermite polynomial to represent the interface, or a Fermi-Dirac function. In statistical mechanics, the distribution in energy of particles in equilibrium with a heat bath at temperature T is given by

$$F(E) = \frac{1}{\left[1 + \exp[(E - E_F)/K_B T]\right]}, \tag{4.3}$$

where E_F is the Fermi energy, K_B is the Boltzmann constant, and T is the temperature. As T tends to zero, the function F(E) represents a step function with a jump at E_F. For $E > E_F$, the denominator goes to infinity and $F(E)$ becomes zero. On the other hand, for $E < E_F$, the exponential function has an exponent that is negative infinity and $F(E)$ is unity. For finite T, the function smoothly decreases from 1 to 0, as E increases. The range in E over which this occurs is on the order of $\sim 2K_B T$. We now take over this picture in coordinate space to implement a smoothed out interface. We use the function†

$$F(x) = \frac{1}{\left[1 + \exp[(x - x_0)/\delta]\right]}. \tag{4.4}$$

Now we can use more finite elements which have derivative DoF around the interface region. The material property, such as the dielectric function in the case of electrostatics, or the effective mass $m^*(x)$ or quantum well potential function $V(x)$ in quantum mechanics, is correspondingly smoothed out. Now at any coordinate value, the left and right sides of the smoothed interface have the same material property. We can therefore overlay element matrices in assembling the global matrix without having to use any jump condition! The price we pay is in the computation of more finite element matrices needed to capture the hyperbolic-tangent-type behavior of the interface function.

We could also use a cubic Hermite function instead of $F(x)$ of Eq. (4.4) in similar situations. An example of this usage is in scattering, where we invoke stealth regions for absorbing outgoing waves. (See Chap. 6, and following chapters).

We have been conditioned to think of interfaces as being abrupt, with a jump in the physical property of the materials on the two sides of the interface. Actually, all material interfaces extend over one or two atomic bond-lengths. We can in fact consider the material properties to vary smoothly from their values from one side to those on the other over a distance of a few angstroms. This has a major impact on the way BCs are implemented. The smooth variation of the material properties allows us to (i) employ finite elements having higher derivative continuity and (ii) substantially simplify the procedure of overlaying element matrices in the global matrix.

†We could also use a single cubic Hermite function with zero slopes at either end and a cubic rise from the initial value to the final one.

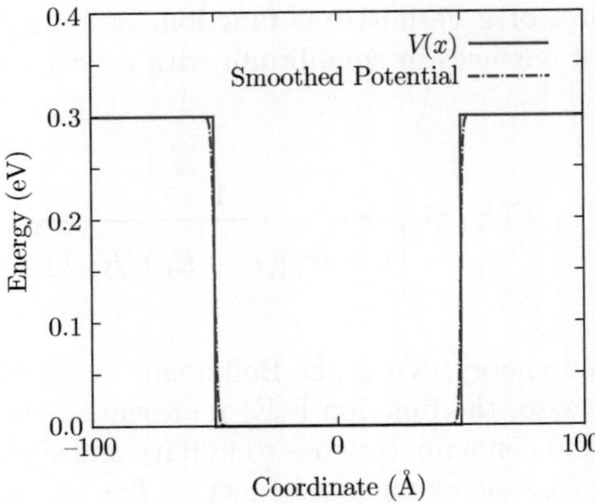

Figure 4.5 The quantum well-barrier interface is smoothed using a Fermi function representation for the potential well structure. The results using higher-order Hermite interpolation polynomials and simple overlay are compatible with the usual use of "mass derivative" continuity for the jump conditions.

4.4 Problems

1. Consider a global rectangular element of dimension $a \times b$ with the nodes located at $(x_0, y_0), (x_1, y_1), (x_2, y_2), (x_3, y_3)$, that is mapped on to a standard square with corners at

$$(\xi, \eta) = (-1, -1), (1, -1), (1, 1), \quad \text{and} \quad (-1, 1).$$

   (a) Map (x, y) to (ξ, η) in a linear manner.
   (b) Obtain expressions to relate (ξ, η) to (x, y) in a reverse mapping.
2. Derive the Jacobian of transformation from an arbitrary quadrilateral to a standard square element and explicitly show that:
   (a) the determinant has terms that depend on ξ and η in mapping a quadrilateral region into a square.
   (b) the Jacobian of transformation from a hexahedron to a standard cube contains terms that depend on ξ, η, ζ, and their bilinear products.
3. Obtain complete expressions for $\partial\phi/\partial x$ and $\partial\phi/\partial y$ purely in terms of derivatives of shape functions with respect to ξ, η as in Eqs. (3.12, 3.17).
4. Consider a circular annulus of inner radius $r_1 = 10$ cm and outer radius $r_2 = 100\,\text{cm}$.
   (a) Starting from a rectangle in (r, θ) variables, solve the Laplace equation $\nabla^2\phi = 0$ with boundary conditions $\phi(r_1, \theta) = 10 \sin 3\theta$, and $\phi(r_2, \theta) = 25 \sin 5\theta$.
   (b) Construct the global matrix for the annulus using linear elements in r and θ, as discussed in Chap. 3, and apply the given boundary conditions.
   (c) Solve and plot the solution for the potential $\phi(r, \theta)$.

(d) Use Hermite interpolation with $\{\phi, \phi_r, \phi_\theta, \phi_{r,\theta}\}$ as the degrees of freedom at nodes. Generate the global matrix, apply boundary conditions on the nodal degrees of freedom and obtain the solution. (Generate the Hermite interpolation on each square element as the direct product of 1D Hermite interpolation polynomials.)

5. Investigate the efficacy of a cubic Hermite interface function in contrast with the Fermi function in solving a static electric field with given potentials, $\phi_L = -1\text{V}$ and $\phi_R = +1\text{V}$, at two ends of a rectangular rod of length 2ℓ with a material interface at $x = 0$. The two sides of the interface have dielectrics ϵ_L and ϵ_R.
   (a) Compare the finite element solution using Hermite interpolation with the analytic solution.
   (b) Repeat part (a) above now using Laplace interpolation requiring the inclusion of the jump condition $\epsilon_L \phi'_L(x) = \epsilon_R \phi'_R$.

5

Accidental degeneracy in cubic semiconductor quantum dots

In this chapter†:

- We review level degeneracy and hidden symmetries which are known to be of importance in analyzing the electronic structure of atoms. The connection between dynamics and symmetry plays a central role in the construction of Mendeleev's Periodic Table of Elements, which initially had been designed by him on an empirical basis. We find a similar interplay between dynamics, symmetry, and degeneracy for energy levels of electrons in a cubic semiconductor quantum dot.
- We present a quantitative analysis of the energy levels and wavefunctions of carriers in a cubic quantum dot of GaAs embedded in $Ga_{(1-x)}Al_xAs$, as an example, with finite confining potential barriers at their interfaces.
- The energy spectrum in the semiconductor quantum dot has substantially reduced degeneracies compared to the analytically determined energy levels of the infinite barrier quantum box of the same dimensions. The level degeneracy of states is explained by group representations of the point group O_h, which allows only singlets, doublets, and triplets.

  On the other hand, the infinite barrier potential box allows a huge number of level degeneracies. This larger degeneracy is determined by the Pythagorean relation $n^2 = n_x^2 + n_y^2 + n_z^2$ among the quantum numbers used to describe the states in the box. We explain this discrepancy by showing that there are additional operators which permit many level degeneracies of O_h to collapse into one level.
- Projection operators for the irreducible representations provide a unique way of obtaining the linear combinations of the degenerate wavefunctions which form a basis set for the representations.

†This chapter is an elaboration of our published article: S. Bharadwaj, S. Pandey, and L. R. Ram-Mohan, Phys. Rev. B **96**, 195305 (2017). The author thanks AIP for permission to reuse our figures and tables here.

Finite Elements in Action. L. Ramdas Ram-Mohan, Oxford University Press. © L. Ramdas Ram-Mohan (2026).
DOI: 10.1093/oso/9780199563487.003.0005

- We conclude with an interesting comparison of the usual hydrogenic atoms, their level degeneracy, and the filling of levels with electrons to make the usual Periodic Table of Elements of Mendeleev, with the corresponding properties of cubic quantum dots.

5.1 Introduction

5.1.1 The Mendeleev Periodic Table of Elements and accidental degeneracy

In quantum mechanical systems, degeneracy in the energy spectrum arises from symmetries of the system that are either based on geometry or are internal symmetries. A classic example of such degeneracy is given by the hydrogen atom with its spherically symmetric Coulomb potential $-e^2/(4\pi\epsilon_0\, r)$ that binds the electron to the proton. Here $-e$ is the electron charge and ϵ_0 is the permittivity of free space. The orbital angular momentum is conserved for central potentials, and the spherical rotational symmetry $(O(3))$ of the Coulomb potential leads to a degeneracy of order $(2\ell+1)$ for the states, where ℓ is the orbital angular momentum quantum number. The surprise, from spectroscopic observations, was that for a given principal quantum number n, the different allowed orbital angular momentum states for $\ell = 0, 1, \ldots, n-1$, all have the same energy. [1] This additional degeneracy, referred to as "accidental degeneracy,"

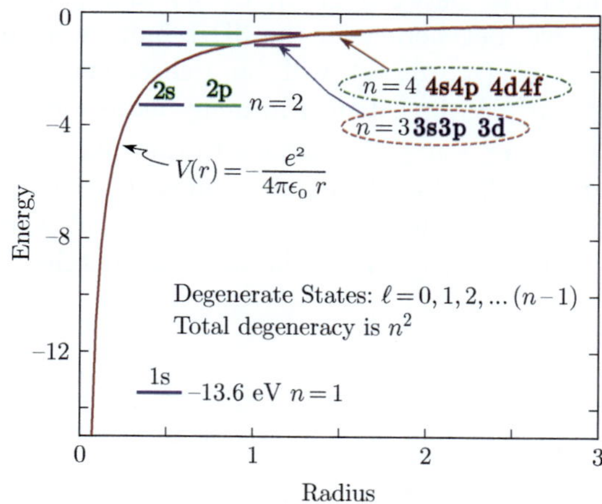

Figure 5.1 The energy levels of the bound electron in the hydrogen atom is shown. The state for $n = 1$ is the s-state with angular momentum zero. The states for $n = 2$ are the degenerate s- and p-states. The apparent symmetry, *i.e.* the rotational symmetry of the Coulomb potential does not demand that these be degenerate. It is the existence of the conserved Runge-Lenz-Laplace vector that makes the different orbital states degenerate in energy.

was explained by Pauli [2] by identifying a new conserved vector, the Runge-Lenz-Laplace vector. Fock [3] showed that the hydrogen atom has a symmetry higher than the 3D rotational symmetry, namely $O(4)$, in four dimensions. This beautiful application of group theory to the degeneracies of a physical system allows us to appreciate the fundamental role of symmetries and hidden symmetries in the quantum mechanics of a physical system. Further developments by Barut and Kleinert [4] showed that the dynamical symmetry group $O(4,2)$ for the hydrogen atom permits the algebraic evaluation of transition matrix elements and provides a complete realization of the dynamics of this problem directly in algebraic terms.

5.1.2 The cubic quantum dot

As early as in 1990, Shertzer and Ram-Mohan [5] observed that semiconductor quantum wires with a square cross-section exhibit the reduction of degeneracy normally associated with the infinite square well potential. An electron in a GaAs quantum wire embedded in $Ga_{(1-x)}Al_xAs$ is confined in the transverse direction by a "kitchen-sink" potential due to the finite band offsets [6] in the semiconductor heterostructure. Only some of the doubly degenerate states labeled by the quantum numbers (n_x, n_y) and their permuted partner (n_y, n_x) associated with the infinite well are split in energy when the well height is made finite. The 2D potential well has C_{4v} symmetry. Seven years after our work was reported, the infinite potential well was shown to have an additional symmetry corresponding to the semi-direct product of C_{4v} and a 1D continuous group of transformations generated by the dynamical operator $(\partial_x^2 - \partial_y^2)$, which commutes with the Hamiltonian. [7]

This unique example of accidental degeneracy and its removal provides a textbook case of the interplay between group theory and particle dynamics. We can anticipate that the level degeneracy in the 3D infinite well, of dimension L, given by the permutations of (n_x, n_y, n_z) will be richer, since the Pythagorean constraint $E \propto n_x^2 + n_y^2 + n_z^2$ can be satisfied in many more ways. The electron's energy $E = (\hbar^2\pi^2/2m^*L^2)(n_x^2 + n_y^2 + n_z^2)$ corresponds to the spherical surface of constant E over the positive octant defined in the number index space, and the degeneracy corresponds to the number of states that fall on such a surface. [8] The larger the energy, the larger is the level degeneracy. [8] Within the range of energies that we have explored,

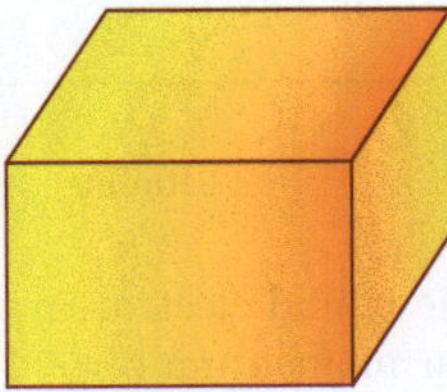

Figure 5.2 The metallic cubic cavity is shown in which the energy levels are determined for a single particle in the box.

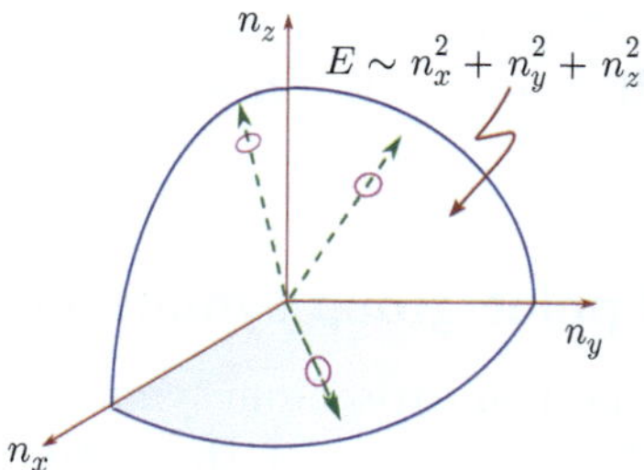

Figure 5.3 A constant-energy (frequency) surface is shown in the positive octant in the $\{n_x, n_y, n_z, n_i \geq 0\}$ space. The level degeneracy is given by the number of points occurring on this surface of constant energy [9].

we have observed degeneracies of ~ 3000. The reduction in this degeneracy for the cubic finite 3D potential well was already anticipated. [5]

In this chapter, the phenomenon of accidental degeneracy and its reduction for a physical system with point group symmetry is presented for a GaAs cubic QD embedded in AlGaAs. We note that the spatial confinement of charge carriers in semiconductor quantum dots (QDs) has led to several interesting electronic, optical, and transport properties in such systems. [10–12]

Here, we obtain numerically the energy levels of electrons and holes in cubic QDs of GaAs embedded in $Ga_{(1-x)}Al_xAs$. This is a prototypical system and our methods apply to any of the II-VI or III-V semiconductor material pairs with Type-I interfaces.

Almost all previous studies reported in the literature assume that the confining potential is infinite. Shertzer and Ram-Mohan have shown that for (2D) heterostructures of quantum wires, the infinite barrier approximation is not valid, and more importantly that the confining finite potential *is not separable.* Here we see that these findings from 2D carry over to 3D: for cubic QDs, the energy levels for the finite potential are significantly lower than those obtained with the infinite barriers. More conspicuous is the lifting of the degeneracy of certain energy levels in the energy spectrum of cubic QDs when a finite-well potential is used.

In Sec. 5.3, we use finite element analysis (FEA) to solve Schrödinger's equation in the effective-mass approximation. Here, the emphasis is on getting high accuracy in the eigenvalues and eigenfunctions. We show that the use of Hermite interpolation polynomials delivers this; the interface boundary conditions with the additional derivative degrees of freedom (DoF) in the Hermite interpolation can lead to serious bookkeeping issues while implementing jump conditions for the derivatives of wavefunctions. This is readily overcome by making a novel and unique use of Fermi-function smoothing described later in this section (also see Chap. 4). In Sec. 5.2, we use group representation theory to predict which of the accidental degeneracies present in the infinite cubic well are lifted when a finite barrier is used. In Sec. 5.4, we discuss results obtained through FEA and investigate splitting of energy levels in the presence of an external electric and magnetic field. With the level of accuracy delivered for actual physical structures discussed here, we mention further directions for research in the concluding remarks of Sec. 5.5. As a final note, FEA transcends geometrical constraints rather cleanly, so that eventually QDs of any shape may be considered for further applications.

5.2 Degeneracy and point group symmetry

Degeneracies in the energy spectrum arise from symmetries associated with the geometry. They are equal to the dimensions of the irreducible representations of the corresponding symmetry group. [13, 14] Any other additional degeneracy which cannot be explained by the obvious geometrical symmetry of the system is labeled as accidental degeneracy. Let $\tilde{G}$ be an infinitesimal transformation for the coordinate system. Given a Hamiltonian H, if $\left[\tilde{G}, H\right] = 0$ and $\tilde{G}$ does not explicitly depend on time, then we say that $\tilde{G}$ is a constant of motion. Such constants of motion generate symmetries since they transform one eigenstate to another of the same energy. We expect to find additional constants of motion whenever we observe accidental degeneracies as explained below. If the Hamiltonian is separable in a coordinate system, then the separation constants may be considered as constants of motion. [15] These are just the generators of the additional symmetry operations. Typically, accidental degeneracies are then rendered normal by identifying the hidden covering group.

In the H-atom problem, the conservation of the 3 components of the angular momentum provide us 3 constants of motion associated with $O(3)$. Equivalently, we consider $\{L_+, L_-, L_z\}$ as the set of 3 operators which commute with the Hamiltonian of the H-atom. We know that an eigenstate $|E, l, m\rangle$ of the H-atom transforms under the operation of ladder operators as

$$L_{\pm} |E, l, m\rangle = \sqrt{(l \mp m)(l \pm m + 1)}\, |E, l, m \pm 1\rangle \,. \tag{5.1}$$

Hence, angular momentum operators transform degenerate eigenstates of the same ℓ but of different azimuthal quantum numbers m into one another. Since the eigenstates of different allowed ℓ are degenerate for hydrogen, we expect to find additional operators (constants of motion) which commute with the Hamiltonian and connect eigenstates of different ℓ quantum numbers. Burkhardt and Leventhal [16] have shown that we can define the operators A_z, A_+, A_- corresponding to the Runge-Lenz vector $\boldsymbol{A}$ which transform degenerate eigenstates of different ℓ quantum numbers, analogous to Eq. (5.1). The components of the angular momentum $\boldsymbol{L}$ and the Runge-Lenz vector $\boldsymbol{A}$ generate the $O(4)$ symmetry group. It can be shown that the H-atom Hamiltonian is separable in parabolic coordinates. [17] We note that even though the components of $\boldsymbol{L}$ and $\boldsymbol{A}$ commute with the Hamiltonian, they will not mutually commute with each other. These components are subject to kinematic constraints of the Casimir operators for the group $O(4)$. Hence, the eigenstates of the Hamiltonian are represented by the complete set of commuting operators $\left\{H, L^2, L_z\right\}$.

We know that for a 3D infinite barrier cubic QD of length L, the energy eigenvalues are given by

$$E(n_1, n_2, n_3) = \left(\frac{\hbar^2 \pi^2}{2m^* L^2}\right) \left(n_1^2 + n_2^2 + n_3^2\right), \tag{5.2}$$

where $n_1, n_2, n_3 \in \mathbb{N}$. Energy eigenvalues reside on the first octant of the number sphere. From Fig. 5.5(a), we see that the degree of degeneracy increases with increase in energy; within the exploration done by us, it reaches fairly large values.

In the continuum limit, the density of states is equal to the area of positive octant in the number sphere which goes as $\sqrt{E}$. This semi-classical evaluation overestimates the true density (see Fig. 5.5(b)) for low energies. However, the theoretical continuum approximation is in excellent agreement already for intermediate energies. The averaging procedure to obtain the density of degenerate states is done as follows. For this problem, the states lie in the positive quadrant of the number space. The degeneracy is sorted by distributing the n values into bins of thickness $d(n^2) = 2n\,dn$, where $dn = 1$ or any suitable small number. The degeneracies of all the states in this shell are added, and divided by the size of the bin, which equals $d(n^2)$. The density of states versus energy are directly obtained by noting that $n^2 \propto E$. The analytical curve plotted in Fig. 5.5 is $\pi n/4$.

Note that there are two kinds of degeneracies present. The first kind is due to the exchange of quantum numbers that are listed in Table 5.1. The second kind is less transparent, occurring when the following relation is satisfied:

$$n_1^2 + n_2^2 + n_3^2 = m_1^2 + m_2^2 + m_3^2, \tag{5.3}$$

with $n_i \neq m_j$, for $i, j = 1, 2, 3$. Cubic QDs have geometrical octahedral symmetry (O_h). The character table for the point group O_h is given by Dresselhaus. [14] We see that O_h has only 1-, 2-, and 3-dimensional representations. The energy spectrum with the infinite barrier approximation on the other hand displays degeneracies of higher order (see Fig. 5.5). The accidental degeneracies due to the permutation of quantum numbers can be accounted for by finding the covering group which is a semi-direct product of the geometrical symmetry group O_h and a 2D compact continuous group. [18] The dynamical operators $(\partial_x^2 - \partial_y^2)$ and $(2\partial_z^2 - \partial_x^2 - \partial_y^2)$ connect the accidental degeneracies generated by the permutation of the quantum numbers. [19]

For finite barriers, however, the potential is *non-separable* and the accidental degeneracy present in the infinite barrier case is lifted. However, it suffices to use the infinite barrier quantum numbers for labeling the eigenstates. For the finite barrier, the states can be labeled with quantum numbers (n_1, n_2, n_3). We require degenerate symmetry adapted linear combinations of wavefunctions to form the basis for their irreducible representations. We shall thus classify the wavefunctions into their corresponding irreducible representations under O_h. Our approach is similar to that of Fernandez [19].

Let G be a group of order g and $\Gamma^{(i)}$ be an l_i-dimensional representation of G. For a group element R in G, its representation is given by an $l_i \times l_i$ square matrix $\Gamma^{(i)}(R)$. Then the projection operator [13] corresponding to $\Gamma^{(i)}$ is given by

$$\mathcal{P}^{(i)} = \frac{l_i}{g} \sum_R \chi^{(i)}(R) \cdot P_R, \tag{5.4}$$

where $\chi^{(i)}(R)$ is the character and P_R is the operator corresponding to the element R. The projection operator $\mathcal{P}^{(i)}$ projects out a function F onto a part $f^{(i)}$ belonging to the representation $\Gamma^{(i)}$. Let $\{\psi_i\}_{i=1}^n$ be a basis set of Hilbert space $\mathcal{H}$ which is in a representation Γ. In general, Γ is a reducible representation of the symmetry group of $\mathcal{H}$. Let, F be a general function in $\mathcal{H}$. Consider

$$\mathcal{P}^{(i)}F = f^{(i)} = \sum_{j=1}^{n} c_j^{(i)}\psi_j, \tag{5.5}$$

where the coefficients are given by

$$c_j^{(i)} = \int_V d^3r \; \psi_j^\dagger \left(\mathcal{P}^{(i)}F\right). \tag{5.6}$$

Similarly, for a basis function ψ_k, we write

$$c_{jk}^{(i)} = \int_V d^3r \; \psi_j^\dagger \left(\mathcal{P}^{(i)}\psi_k\right). \tag{5.7}$$

If the coefficient is nonzero, the wavefunction ψ_k has a component in the ith-representation and ψ_j is a partner. Once we determine all coefficients, from Eq. (5.5), we obtain a new basis function $f^{(i)}$ which is exclusively in the ith-representation. Using this procedure, we list out all basis functions and irreducible representations for different combinations of quantum numbers in Tables 5.2 and 5.3. These new functions that are listed in the 3rd column of Tables 5.2 and 5.3 are a more natural choice for the basis set, as they are compatible with the symmetry of irreducible representations. For instance, a basis function for A_{1g}, the symmetric irreducible representation should have total symmetry of the group under consideration.

5.3 FEM applied to cubic QDs

In the envelope function approximation (EFA), [6] the charge carrier's envelope function $\psi(x, y, z)$ satisfies Schrödinger's equation with a non-separable potential $V(x, y, z)$ corresponding to a finite barrier height,

$$\left[-\nabla\left(\frac{\hbar^2}{2m^*(x,y,z)}\nabla\right) + V(x,y,z)\right]\psi(x,y,z) = E\,\psi(x,y,z), \tag{5.8}$$

where m^* is the carrier effective mass m_w^* or m_b^* in the QD or the surrounding bulk barrier medium, respectively. The potential is

$$V(x,y,z) = \begin{cases} 0, & \text{for } |x| \le a/2; \quad |y| \le b/2; \quad |z| \le c/2, \\ V_0, & \text{outside.} \end{cases} \tag{5.9}$$

There is a discontinuity in the potential and the effective mass of the charged carriers at the interface between materials. The continuity of the probability current requires that the wavefunction $\psi(x)$ and the "mass-derivative" of the wavefunction $(1/m^*)\psi'(x)$ be continuous‡ as was shown by Ben-Daniels and Duke. [20] The input parameters for the effective masses and band offsets of conduction electrons, heavy holes, and light holes in GaAs and in $Ga_{(1-x)}Al_xAs$ are obtained from our review article. [21]. The FEA employing the action integral formulation [22] is used to evaluate

‡This result is directly obtained when we employ the Lagrangian formalism for the Schrödinger equation.

Table 5.1 Possible degree of degeneracies due to the permutation of quantum numbers in cubic QDs with infinite barrier approximation. The level degeneracy can be substantial for large energies, corresponding to the number of ways the Pythagorean sum $(n_x^2 + n_y^2 + n_z^2)$ corresponds to the given energy. The level degeneracy has no limit and increases with energy. We see in the table that degree of degeneracies are already not in accord with the geometrical symmetry group O_h in the low energy states.

Degree of degeneracy	Example
1	(1,1,1)
3	(1,2,2), (2,1,2), (2,2,1)
6	(1,2,3), (3,1,2), (2,3,1), (2,1,3), (3,2,1), (1,3,2)

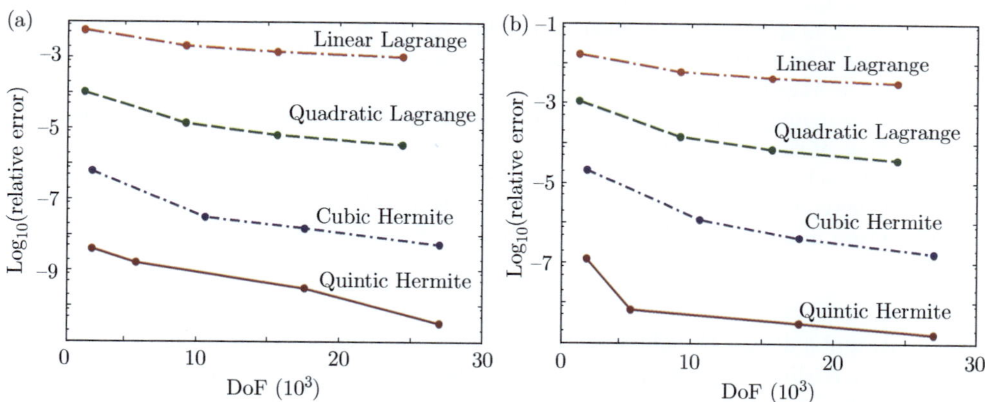

Figure 5.4 The convergence of the relative errors in eigenvalues of (a) the ground and (b) the first excited state in the QD for linear Lagrange, quadratic Lagrange, cubic Hermite, and quintic Hermite interpolation polynomials are shown for the case of an infinite quantum well of dimensions 200 Å × 200 Å × 200 Å. Using quintic Hermite interpolation polynomials, we can reduce the error to 10^{-8} with just 27 elements and 2744 degrees of freedom (DoF), with further reduction in error possible with mesh refinement. The total DoF corresponds to the dimension of the global matrix. [9]

the energies and eigenfunctions. Within each element, we express the envelope function as a linear combination of interpolation polynomials multiplied by as-yet undetermined coefficients that correspond to the value of the wavefunction at the vertices (nodes) of the elements that are usually tetrahedra or cubes in 3D. We use Hermite interpolation polynomials, as opposed to Lagrange polynomials, for which the expansion coefficients are function values and their derivatives at the nodes. [22] The additional derivative continuities required in this case substantially increase the accuracy of the eigenvalues. The global envelope function is constructed by summing contributions from all elements and ensuring the function value and its derivatives

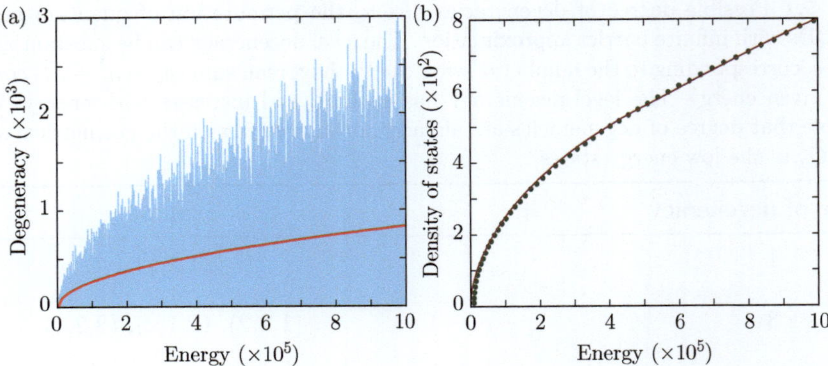

Figure 5.5 (a) A histogram for the degree of degeneracy as a function of energy for an infinite barrier cubic quantum dot is shown. Note that the degeneracy increases without limit, with the largest value shown being ~ 3000 in the plot. As the radius of the constant energy surface increases by 1 unit, the level degeneracy fluctuates considerably. The red curve shows the degeneracy averaged over small bins. (b) The calculated density of states (dots) is plotted along with the continuum approximation (continuous curve) of the density of states obtained as the local *average* of the density of states. We see that this evaluation overestimates the density for lowest energies but is found to be an excellent approximation already for intermediate energies. Here the energy is in units of $\hbar^2\pi^2/2m^*L^2 = 0.01413653$eV. This corresponds to a conduction band electron in an L = 200 Å cubic quantum dot. (From Ref. [9].)

are continuous across the element. Further details from a physical point of view may be seen in earlier chapters here, and elsewhere, [22] and we refrain from reporting more details of the method here. Since the finite element approach may be viewed as the discretization of the action integral, we define the action from which Eq. (5.8) is derivable to be

$$\mathcal{A}/T = \int_V d^3r\ \psi^* \left[\overleftarrow{\partial} \cdot \frac{\hbar^2}{2m^*}\ \overrightarrow{\partial} + V(x,y,z) - E\right] \psi. \tag{5.10}$$

We are here solving the time-independent problem so that the time integral over the range $[0, T]$ in the action is simply T. Dirichlet boundary conditions are implemented at the periphery of the physical region. In Figs. 5.4(a) and (b), we show the convergence of the error in the eigenvalues of the ground and the first excited state for different interpolation polynomials. Clearly, the Hermite interpolations yield more accurate results than the typical Lagrange interpolation polynomials. In the case of finite barriers, there is a discontinuity in the potential and effective mass of the charge carriers at the interface between materials. Traditionally, while using Hermite finite elements, the continuity of the effective mass derivatives is ensured by patching the corresponding row vectors. [5] But this is a computationally expensive and slow process especially in a parallel computing environment in 3D. We tackle this problem by representing the step function by a Fermi function. Let us define a function

$$\mathcal{F}(\xi, \alpha) = \left(1 + \exp\left(\frac{(\xi + \frac{\alpha}{2})(\xi - \frac{\alpha}{2})}{\delta^2}\right)\right)^{-1}, \quad (5.11)$$

where $\xi = \{x, y, \text{ or } z\}$ and $\alpha = \{a, b, \text{ or } c\}$. The parameters a, b, c are the dimensions of the brick. Thus, the effective mass is expressed as

$$m^*(x, y, z) = m_b^* - (m_b^* - m_w^*)\left(\mathcal{F}(x, a) \times \mathcal{F}(y, b) \times \mathcal{F}(z, c)\right). \quad (5.12)$$

Similarly, the confining potential is defined by

$$V(x, y, z) = V_0 \Big[1 - \mathcal{F}(x, a) \times \mathcal{F}(y, b) \times \mathcal{F}(z, c)\Big] \quad (5.13)$$

Here δ is the smoothing parameter, or the "temperature" in the Fermi function, which controls the smoothing of the interfaces. By systematically decreasing the parameter δ, we approach the finite barrier potential with a discontinuity. The most important advantage of this smoothing is that the properties of the function are the same on either side of the interface at any energy, so that there are no jump conditions to implement in the calculations. We have verified this and the accuracy obtained in several test calculations. We note that the Fermi-function smoothing may be argued as a more physical material interface, though here we employ it for its simplicity in implementation for computation with the Hermite interpolation polynomials.

Variation of the discretized action in Eq. (5.10) leads to a generalized eigenvalue matrix equation of the form $A\psi = EB\psi$, where, A and B are sparse matrices of dimension equal to the total DoF and ψ is the eigenvector corresponding to the eigenvalue E. We solve this equation within a parallel computing environment. [23–26] We use the Krylov-Schur algorithm as implemented in SLEPc. [27] Here, we employ a spectral transformation by a shift-invert operator. The solver works with an equation of the form $(A - \lambda B)^{-1} B\psi = \theta\psi$, where λ is the shift parameter. Energy eigenvalues are computed internally from the relation $E = \lambda + 1/\theta$. This procedure augments the convergence of eigenvalues near λ since the eigenvalues θ of the shift-invert operator are largest in magnitude for those energies in the vicinity of λ. [24] In the case of linear Lagrange interpolation polynomials, with 50 processors, matrices of dimensions 27000×27000 are assembled in 0.6 minutes and diagonalized in 2.9 minutes. On the other hand, using the quintic Hermite interpolation polynomials, with the same number of processors, matrices of the same dimensions are assembled in 70.0 minutes and diagonalized in 10.2 minutes. We pay the price for having to go through more loops while using the Hermite interpolation polynomials. This increase in time can be reduced by optimizing the number of processors used for the calculation.

The matrix bandwidth is defined as the sum of sub- and supra-diagonal arrays together with the main diagonal. For a total of 27000 DoF, the linear Lagrange polynomials utilize a bandwidth of 53987, while the quintic Hermite interpolation polynomials occupy a comparable bandwidth of 53903. The occupancy of a matrix is defined as the percentage of nonzero entries in the matrix. While going from the

Table 5.2 Different possible even and odd combinations of quantum numbers and their corresponding irreducible representations for the symmetry group O_h.

Quantum number	Irrep.	Basis function
$(2n-1, 2n-1, 2n-1)$	$1A_{1g}$	$\psi_{(2n-1,2n-1,2n-1)}$
$(2n, 2n, 2n)$	$1A_{2u}$	$\psi_{(2n,2n,2n)}$
$(2n-1, 2n-1, 2m)$	$1T_{1u}$	$\psi_{(2n-1,2n-1,2m)}$, $\psi_{(2n-1,2m,2n-1)}$, $\psi_{(2m,2n-1,2n-1)}$
$(2n, 2n, 2m-1)$	$1T_{2g}$	$\psi_{(2n,2n,2m-1)}$, $\psi_{(2n,2m-1,2n)}$, $\psi_{(2m-1,2n,2n)}$

linear Lagrange to quintic Hermite polynomials, there is only a nominal increase in the matrix occupancy from 0.214% to 0.929% for the B matrix and 0.06% to 0.928% for the A matrix. The increase in computational time with the Hermite interpolation calculations is compensated considerably by the higher accuracy delivered, as seen in Fig. 5.4.

5.4 Results and discussions

We have used the infinite barrier QD to determine the degeneracy of levels and how well the continuum approximation reproduces the discrete degeneracies. More importantly, we have determined the level of accuracy achieved in our numerical calculations with the infinite barrier QD for which the eigenvalues are known analytically. We expect that the eigenvalues obtained with a finite well will have almost the same level of accuracy given that the finite well QD calculations are not altering the input numerical values in any significant way.

In Tables 5.4–5.6, we list the calculated eigenvalues for conduction electrons, light holes, and heavy holes in GaAs/$Ga_{0.3}Al_{0.37}As$ cubic QDs of dimensions 200 Å × 200 Å × 200 Å.

We represent each eigenstate by a symmetry adapted linear combination of quantum numbers (n_1, n_2, n_3) associated with the infinite well. For comparison we include energy values obtained using the infinite barrier approximation. Note that all of the energy levels are lowered from the values obtained with infinite barriers. Additional accidental degeneracies observed in the case of an infinite barrier are lifted. Clearly, the infinite barrier approximation is invalid in this case. States with symmetric combinations of quantum numbers are more bound than others. For example, states corresponding to A_{1g} and A_{2u} have lower energies than that of E_g and E_u, respectively. As expected, we see only 1-, 2-, and 3-level degeneracies. States that were 6-level degenerate in the case of infinite barrier approximation are split into two triplet states for (2 even and 1 odd), or (2 odd and 1 even) combinations of quantum numbers. For all three different odd or even combinations, the states will split into two singlets and two doublets as expected (see Table 5.3). In Fig. 5.6, we have shown the wavefunction for the $(2,2,2)$ state projected along three perpendicular planes. We see that the wavefunction is odd with respect to the inversion center, since it is in the representation A_{2u}. The wavefunctions of $(1,1,3)$, $(1,3,1)$, and $(3,1,1)$ states split into a singlet and a degenerate doublet. In Fig 5.7(a), we see that the singlet which is in the A_{1g} representation has the complete O_h symmetry. The doublets belonging to the E_g representation are related by a proper C_4 rotation around x-axis (see Fig. 5.7(b) and (c)).

We can induce level splitting by having an additional perturbation. Thus the symmetry is reduced to one of the subgroups of O_h. The original Hamiltonian H_0 belongs to the irreducible representation A_1 of O_h. The additional perturbation H' may not have the complete O_h symmetry. We determine the new symmetry by finding the subgroup G of maximum dimension in which all the terms of H' form a basis for the irreducible representation A_1 in G. For example, consider an external constant electric field $H' = -eE_0 z$ applied to a cubic QD. We note that C_{4v} is the subgroup of maximum dimension in which z is a basis of the representation A_1. Hence, O_h symmetry is reduced to C_{4v} symmetry. We decompose all irreducible representations of O_h through direct products of irreducible representations of C_{4v}. [28] In Table 5.7, we list out numerically obtained eigenvalues for the cubic QD in an electric field. Level splitting is in accordance with group theoretical predictions. For an applied magnetic field $\boldsymbol{B}$ and corresponding vector potential $\boldsymbol{A}$, the perturbing Hamiltonian is given by

$$H' = \frac{e}{m}\boldsymbol{A}\cdot\boldsymbol{p} + \frac{e^2}{2m}|\boldsymbol{A}|^2 - \boldsymbol{\mu}\cdot\boldsymbol{B}, \tag{5.14}$$

where $\boldsymbol{\mu} = -(e/m)\boldsymbol{S}$ and $\boldsymbol{S}$ is the spin angular momentum. For a constant field $\boldsymbol{B} = B_0\hat{z}$, within the Landau gauge, the vector potential is given by $\boldsymbol{A} = B_0\, x\,\hat{y}$. Hence, H' has terms that transform as xy and x^2. The spin-orbit coupling term contributes only up to a constant value and it has the complete O_h symmetry. By inspection, we see that only the subgroup C_2 has all these terms belonging to its trivial representation A. We verify from Table 5.8 that all energy levels are split into non-degenerate states, as the group C_2 has only one dimensional representations. In Fig. 5.8, we see that in an applied magnetic field, the probability distribution of electrons retain only C_2 symmetry.

Finally, we study the dependence of level splitting on barrier height for heavy holes in GaAs/$Ga_{(1-x)}Al_xAs$ cubic QDs of dimensions 200 Å $\times$ 200 Å $\times$ 200 Å. As expected, level splitting between energy levels decreases (see Fig. 5.9(a)) and approaches that of an infinite barrier in the limit $V \to \infty$. In some cases, there is a level crossing between the states of different quantum numbers. In Fig. 5.9(b), we see that energy of the state $(5,5,3)+(5,3,5)+(3,5,5)$ reduces below the energy of state $2(3,3,7)-(3,7,3)-(7,3,3)$ for well depth $\geq 173\,\text{meV}$. Hence, even the ordering of energy levels is a function of well depth.

5.5 Conclusions

After Planck investigated the density of states available for the quantized electromagnetic field in a cube, researchers have not investigated the source of the large degeneracies in these energy levels. Given that the allowed degeneracies in a cube are singlets, doublets, and triplets associated with the group O_h for the cube, it is surprising that we have the extremely high level degeneracies for large energies (frequencies). This is due to accidental degeneracy, and we have identified the two continuous operators that commute with the Hamiltonian, and interlink different levels of the same energy belonging to different representations of the group O_h. This is analogous to the Laplace-Runge-Lenz vector for the hydrogen atom which links states of different angular momentum for the same energy. The level degeneracy in the hydrogen atom was identified, after the formulation of quantum mechanics, by Pauli, Fock, and others.

We have developed a formalism based on FEA which provides accurate energy spectra and symmetry adapted wavefunctions for a cubic QD. Such accuracy is pivotal to study linear and non-linear optical, electrical, and magnetic properties of a quantum dot. It is clear that the infinite barrier approximation leads to serious errors and additional (accidental) degeneracies. The group representation theory provides a robust method to predict degeneracies in the energy spectrum and symmetry adapted wavefunctions by just knowing the symmetries of the associated Hamiltonian. We have derived a coefficient formula which determines the basis for all irreducible representations of the symmetry group. Our method is general and can be easily extended to find the energy spectrum for a QD of any shape and size.

In layered semiconductor structures, FEA allows us to treat the issue of self-consistently solving the Schrödinger-Poisson equations to derive the changes in the band edge profile for arbitrary doping with impurities in arbitrary layers. [29] The ionized impurities are held in place, while the carriers that are freed are distributed in a self-consistent manner in the heterostructure. In modulation doped cubic QDs, the impurity concentration, typically in the barrier, leads to a self-consistent reconfiguration of the band edges. Lee et al., [30] proposed a numerical method to obtain self-consistent solutions to the Schrödinger-Poisson equations in cubic QDs with finite barriers. However, their expansion of wavefunctions in terms of separable global orthogonal periodic functions are expected to give inaccurate results since the finite potential itself is not separable. Extension of our method should provide a highly reliable solution, as we can systematically increase the accuracy through mesh size refinement (h-refinement), or by the use of higher order interpolation polynomials (p-refinement) for convergence within the FEA. Such freedom is lost using global basis functions.

In II-VI semiconductors, the inclusion of a dilute distribution of magnetic ions, such as Mn, leads to a dramatic change in the magnetic properties of the material. In CdMnTe, for example, the acceptor bound magnetic polarons are formed, and in the presence of the magnetic ions within the Bohr radius of the impurity state, leads to a large change in the binding energy of the acceptor. The hydrogenic acceptor state has a magnetic exchange contribution that leads to a highly non-linear Schrödinger equation. This is readily solved to obtain the new binding energy of the bound magnetic polaron. [31] A similar doping with Mn of a GaAs quantum well has the Mn entering the crystal as a p-type impurity, and releasing a very large number of holes in the well even for small stoichiometric concentrations. This has been calculated by us using FEA. [32] We note that doping diluted with Mn ions in the GaAs QDs again leads to a release of a large number of holes in the QD region. Such a structure also requires a self-consistent solution.

The optical second-order susceptibility in asymmetric quantum wells has been well-studied, [33] and the third-order optical susceptibility in checker-board superlattices of quantum wires has been investigated by us earlier. [34] In the literature, typically for QDs, the second- and third-order non-linear optical susceptibilities are obtained using the infinite barrier approximation. [35, 36] It will be of interest to examine the influence of finite barriers in the presence of external electric and magnetic fields, as delineated in this chapter in view of the more accurate solutions we obtain within our framework.

Our method can be adapted to study both spatial and spin entanglement of two or more electrons in double QDs. We require special techniques to evaluate computationally demanding Coulomb integrals. These calculations have now been done by the author who has an automated way of representing the Coulomb integrals using Gaussian quadrature techniques that allow fairly high accuracy.

We note that our method is well suited to study correlation and quantum confinement of excitons in QDs. FEA can be extended to solve Schrödinger's equation

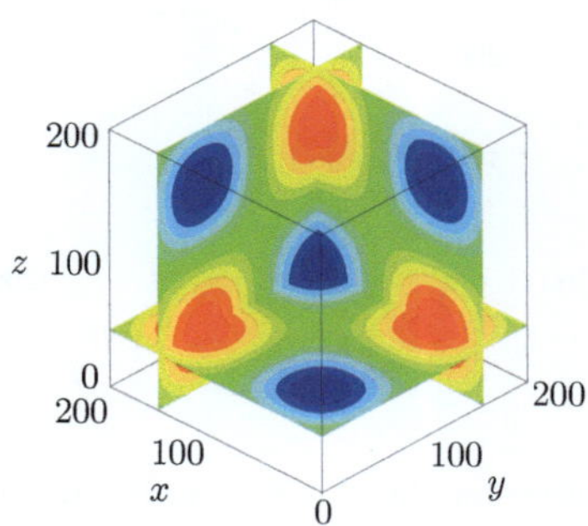

Figure 5.6 Wavefunction for the $(2, 2, 2)$ state of the cubic QD, which is in the representation A_{2u}. Notice that the wavefunction is odd with respect to inversion about the center. Darker regions correspond to wavefunction valleys (blue) and peaks (red), and lighter regions (green) correspond to values close to zero. (From Ref. [9].)

in 6 dimensions, though the bookkeeping aspects in 6D and computation time will increase.

Finally, it is very illuminating to consider the "periodic table" developed using the Pauli Exclusion Principle applied to the "cubic atom," governed by the filling of the energy levels labeled by (n_x, n_y, n_z). For students in undergraduate quantum mechanics courses, this can provide yet another simple illustration of the methods applied to the development of the usual periodic table based on the hydrogenic levels. They now face the challenge of giving names to each of their elements obtained from the cubic atoms and developing new rules for chemical reactions among them.

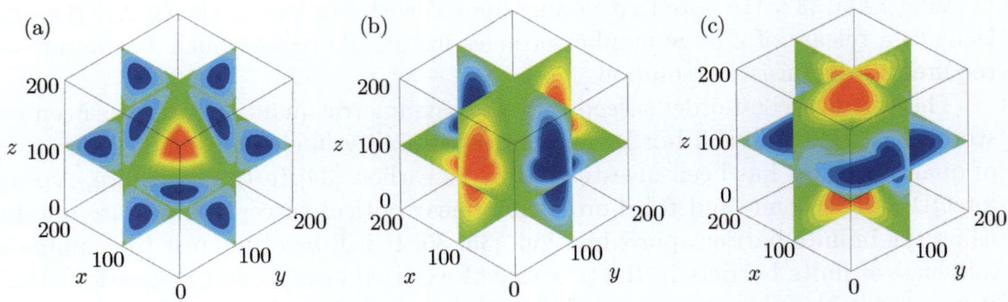

Figure 5.7 We show wavefunctions for (a) the state $(1,1,3)+(1,3,1)+(3,1,1)$ belonging to the representation A_{1g}, which has full O_h symmetry. We show in parts (b) the state $-(1,1,3)-(1,3,1)+2(3,1,1)$, and (c) the state $-(1,1,3)+2(1,3,1)+(3,1,1)$, which are in the representation E_g of the group O_h. These two states are related by a C_4 rotation about the x-axis. Darker regions correspond to wavefunction valleys (blue) and peaks (red), and lighter regions (green) correspond to values close to zero. (From Ref. [9].)

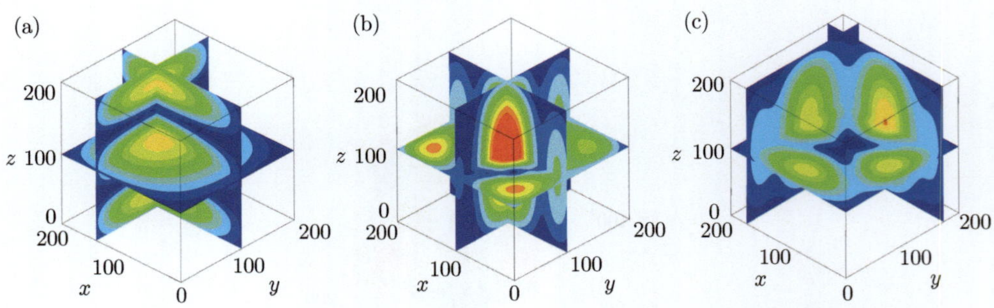

Figure 5.8 Probability densities for the states $(1,1,3)$, $(1,3,1)$, and $(3,1,1)$ state of the cubic QD in a $5\,T$ magnetic field along the $\hat{z}$ axis are shown. The magnetic field splits all states into singlets whose symmetry is reduced to the group C_2. Darker regions correspond to wavefunction nodes (blue) and antinodes (red), and lighter regions correspond to values in between. (From Ref. [9].)

Table 5.3 (continued from Table 5.2) Different possible even and odd combinations of quantum numbers and their corresponding irreducible representations for the symmetry group O_h.

Quantum number	Irrep.	Basis function
$(2n, 2m, 2k-1)$	$1T_{1g}$	$\frac{1}{\sqrt{2}}\Big(\psi_{(2k-1,2m,2n)} + \psi_{(2k-1,2n,2m)}\Big)$,
		$\frac{1}{\sqrt{2}}\Big(\psi_{(2m,2k-1,2n)} + \psi_{(2n,2k-1,2m)}\Big)$,
		$\frac{1}{\sqrt{2}}\Big(\psi_{(2m,2n,2k-1)} + \psi_{(2n,2m,2k-1)}\Big)$
	$1T_{2g}$	$\frac{1}{\sqrt{2}}\Big(\psi_{(2k-1,2m,2n)} - \psi_{(2k-1,2n,2m)}\Big)$,
		$\frac{1}{\sqrt{2}}\Big(\psi_{(2m,2k-1,2n)} - \psi_{(2n,2k-1,2m)}\Big)$,
		$\frac{1}{\sqrt{2}}\Big(\psi_{(2m,2n,2k-1)} - \psi_{(2n,2m,2k-1)}\Big)$
$(2m, 2n, 2k)$	$1A_{1u}$	$\frac{1}{\sqrt{6}}\Big(\psi_{(2m,2n,2k)} - \psi_{(2m,2k,2n)} + \psi_{(2n,2k,2m)}$
		$-\psi_{(2n,2m,2k)} + \psi_{(2k,2m,2n)} - \psi_{(2k,2n,2m)}\Big)$
	$1A_{2u}$	$\frac{1}{\sqrt{6}}\Big(\psi_{(2m,2n,2k)} + \psi_{(2m,2k,2n)} + \psi_{(2n,2k,2m)}$
		$+\psi_{(2n,2m,2k)} + \psi_{(2k,2m,2n)} + \psi_{(2k,2n,2m)}\Big)$
	$2E_u$	$\frac{1}{\sqrt{6}}\Big(2\psi_{(2m,2n,2k)} - \psi_{(2n,2k,2m)} - \psi_{(2k,2m,2n)}\Big)$
		$\frac{1}{\sqrt{6}}\Big(2\psi_{(2n,2k,2m)} - \psi_{(2k,2m,2n)} - \psi_{(2m,2n,2k)}\Big)$
		$\frac{1}{\sqrt{6}}\Big(2\psi_{(2m,2k,2n)} - \psi_{(2k,2n,2m)} - \psi_{(2n,2m,2k)},\Big)$
		$\frac{1}{\sqrt{6}}\Big(2\psi_{(2n,2m,2k)} - \psi_{(2m,2k,2n)} - \psi_{(2k,2n,2m)}\Big)$
$(2m-1, 2n-1, 2k-1)$	$1A_{1g}$	$\frac{1}{\sqrt{6}}\Big(\psi_{(2m-1,2n-1,2k-1)} - \psi_{(2m-1,2k-1,2n-1)}$
		$+\psi_{(2n-1,2k-1,2m-1)} - \psi_{(2n-1,2m-1,2k-1)}$
		$+\psi_{(2k-1,2m-1,2n-1)} - \psi_{(2k-1,2n-1,2m-1)}\Big)$
	$1A_{2g}$	$\frac{1}{\sqrt{6}}\Big(\psi_{(2m-1,2n-1,2k-1)} + \psi_{(2m-1,2k-1,2n-1)}$
		$+\psi_{(2n-1,2k-1,2m-1)} + \psi_{(2n-1,2m-1,2k-1)}$
		$+\psi_{(2k-1,2m-1,2n-1)} + \psi_{(2k-1,2n-1,2m-1)}\Big)$

Table 5.4 Conduction-electron energy levels in GaAs/$Ga_{0.63}Al_{0.37}As$ QDs, with $m_w^* = 0.0665m_0$ and $m_b^* = 0.0858m_0$. The typical simplification of using an infinite well to evaluate the energy levels is seen to be in significant error since the finite potential is not separable. (From Ref. [9].)

Quantum numbers	Irrep.	Energy (meV)	
		$V = 276$	$V = \infty$
$(1,1,1)$	$1A_{1g}$	27.6398458	42.4094382
$(1,1,2)$		55.1963195	84.8188763
$(1,2,1)$	$1T_{1u}$	55.1963195	84.8188764
$(2,1,1)$		55.1963195	84.8188766
$(1,2,2)$		82.5703718	127.2283147
$(2,1,2)$	$1T_{2g}$	82.5703718	127.2283147
$(2,2,1)$		82.5703718	127.2283147
$(1,1,3)+(1,3,1)+(3,1,1)$	$1A_{1g}$	100.9570475	155.5012738
$-(1,1,3)+2(1,3,1)-(3,1,1)$	$1E_g$	101.0083349	155.5012739
$2(1,1,3)-(1,3,1)-(3,1,1)$		101.0083349	155.5012739
$(2,2,2)$	$1A_{2u}$	109.7512732	169.6377529
$(2,3,1)-(2,1,3)$		128.0064988	197.9107120
$(3,2,1)-(1,2,3)$	$1T_{2u}$	128.0064988	197.9107121
$(1,3,2)-(3,1,2)$		128.0064988	197.9107121
$(2,3,1)+(2,1,3)$		128.0462602	197.9107121
$(3,2,1)+(1,2,3)$	$1T_{1u}$	128.0462602	197.9107121
$(1,3,2)+(3,1,2)$		128.0462602	197.9107121
$(1,2,4)-(1,4,2)$		190.8111029	296.8660777
$(2,1,4)-(4,1,2)$	$1T_{2g}$	190.8111029	296.8660777
$(2,4,1)-(4,2,1)$		190.8111029	296.8660777
$(1,2,4)+(1,4,2)$		191.3269094	296.8660777
$(2,1,4)+(4,1,2)$	$1T_{1g}$	191.3269094	296.8660777
$(2,4,1)+(4,2,1)$		191.3269094	296.8660778
$(2,2,4)+(2,4,2)+(4,2,2)$	$1A_{2u}$	216.6050003	339.2755159
$-(2,2,4)+2(2,4,2)-(4,2,2)$	$1E_u$	217.5244549	339.2755160
$2(2,2,4)-(2,4,2)-(4,2,2)$		217.5244549	339.2755160

Table 5.5 Light-hole energy levels in GaAs/$Ga_{0.63}Al_{0.37}As$ QDs, with $m_w^* = 0.0905m_0$ and $m_b^* = 0.1107m_0$. The typical simplification of using an infinite well to evaluate the energy levels is seen to be in significant error since the finite potential is not separable. (From Ref. [9].)

Quantum numbers	Irrep.	Energy (meV)	
		$V = 184$	$V = \infty$
$(1,1,1)$	$1A_{1g}$	20.2151693	31.1627363
$(1,1,2)$		40.3461623	62.3254726
$(1,2,1)$	$1T_{1u}$	40.3461623	62.3254727
$(2,1,1)$		40.3461623	62.3254728
$(1,2,2)$		60.3422579	93.4882091
$(2,1,2)$	$1T_{2g}$	60.3422579	93.4882091
$(2,2,1)$		60.3422579	93.4882091
$(1,1,3)+(1,3,1)+(3,1,1)$	$1A_{1g}$	73.6604738	114.2633670
$-(1,1,3)+2(1,3,1)-(3,1,1)$	$1E_g$	73.7321914	114.2633670
$2(1,1,3)-(1,3,1)-(3,1,1)$		73.7321914	114.2633670
$(2,2,2)$	$1A_{2u}$	80.1946351	124.6509455
$(2,3,1)-(2,1,3)$		93.4188421	145.4261033
$(3,2,1)-(1,2,3)$	$1T_{2u}$	93.4188421	145.4261033
$(1,3,2)-(3,1,2)$		93.4188421	145.4261033
$(2,3,1)+(2,1,3)$		93.4736301	145.4261034
$(3,2,1)+(1,2,3)$	$1T_{1u}$	93.4736301	145.4261034
$(1,3,2)+(3,1,2)$		93.4736301	145.4261034
$(1,2,4)-(1,4,2)$		138.6920158	218.1391620
$(2,1,4)-(4,1,2)$	$1T_{2g}$	138.6920158	218.1391620
$(2,4,1)-(4,2,1)$		138.6920158	218.1391620
$(1,2,4)+(1,4,2)$		139.3460826	218.1391621
$(2,1,4)+(4,1,2)$	$1T_{1g}$	139.3460826	218.1391621
$(2,4,1)+(4,2,1)$		139.3460826	218.1391621
$(2,2,4)+(2,4,2)+(4,2,2)$	$1A_{2u}$	157.2405642	249.3018984
$-(2,2,4)+2(2,4,2)-(4,2,2)$	$1E_u$	158.4166740	249.3018985
$2(2,2,4)-(2,4,2)-(4,2,2)$		158.4166740	249.3018985

Table 5.6 Heavy-hole energy levels in GaAs/$Ga_{0.63}Al_{0.37}$As QDs, with $m_w^* = 0.3774m_0$ and $m_b^* = 0.3865m_0$. The typical simplification of using an infinite well to evaluate the energy levels is seen to be in significant error since the finite potential is not separable. (From Ref. [9].)

Quantum numbers	Irrep.	Energy (meV)	
		$V = 184$	$V = \infty$
$(1,1,1)$	$1A_{1g}$	5.6339074	7.4727812
$(1,1,2)$		11.2717888	14.9455624
$(1,2,1)$	$1T_{1u}$	11.2717888	14.9455624
$(2,1,1)$		11.2717888	14.9455625
$(1,2,2)$		16.9084279	22.4183437
$(2,1,2)$	$1T_{2g}$	16.9084279	22.4183437
$(2,2,1)$		16.9084279	22.4183437
$(1,1,3)+(1,3,1)+(3,1,1)$	$1A_{1g}$	20.6978832	27.4001979
$-(1,1,3)+2(1,3,1)-(3,1,1)$	$1E_g$	20.7001304	27.4001979
$2(1,1,3)-(1,3,1)-(3,1,1)$		20.7001304	27.4001979
$(2,2,2)$	$1A_{2u}$	22.5438146	29.8911249
$(2,3,1)-(2,1,3)$		26.3330823	34.8729792
$(3,2,1)-(1,2,3)$	$1T_{2u}$	26.3330823	34.8729792
$(1,3,2)-(3,1,2)$		26.3330823	34.8729792
$(2,3,1)+(2,1,3)$		26.3345962	34.8729792
$(3,2,1)+(1,2,3)$	$1T_{1u}$	26.3345962	34.8729792
$(1,3,2)+(3,1,2)$		26.3345962	34.8729792
$(1,2,4)-(1,4,2)$		39.6765282	52.3094705
$(2,1,4)-(4,1,2)$	$1T_{2g}$	39.6765282	52.3094705
$(2,4,1)-(4,2,1)$		39.6765282	52.3094705
$(1,2,4)+(1,4,2)$		39.6883222	52.3094705
$(2,1,4)+(4,1,2)$	$1T_{1g}$	39.6883222	52.3094705
$(2,4,1)+(4,2,1)$		39.6883222	52.3094705
$(2,2,4)+(2,4,2)+(4,2,2)$	$1A_{2u}$	45.3002293	59.7822517
$-(2,2,4)+2(2,4,2)-(4,2,2)$	$1E_u$	45.3181042	59.7822517
$2(2,2,4)-(2,4,2)-(4,2,2)$		45.3181042	59.7822517
$(1,3,5)-(1,5,3)+(3,5,1)-$ $(3,1,5)+(5,1,3)-(5,3,1)$	$1A_{1g}$	67.0254522	87.1824696
$(1,3,5)+(1,5,3)+(3,5,1)+$ $(3,1,5)+(5,1,3)+(5,3,1)$	$1A_{2g}$	67.0857917	87.1824696

Table 5.7 Lowering of symmetry and splitting of degeneracy in the presence of 1kV/cm and 20kV/cm electric fields, in the conduction and valence bands. (From Ref. [9].)

O_h	C_2	Energy (meV)			
		Conduction-electron		Heavy-hole	
		$\mathcal{E}=1$	$\mathcal{E}=20$	$\mathcal{E}=1$	$\mathcal{E}=20$
A_{1g}	A_1	30.6325	84.8862	8.6034	33.1401
T_{1u}	E	58.1889	112.4161	14.2413	38.7771
		58.1889	112.4161	14.2413	38.7771
	A_1	58.1984	115.9042	14.2809	41.6810
T_{2g}	B_2	85.5629	139.7615	19.87799	44.4130
	E	85.5725	143.2800	19.9176	47.3171
		85.5725	143.2800	19.9176	47.3175
A_{1g}	A_1	76.6536	129.9254	23.6681	48.2027
E_g	A_1	104.0008	158.1774	23.66963	48.2041
	B_1	104.0069	161.4718	23.7051	51.2423
A_{2u}	B_2	112.7534	170.4624	25.5530	52.9528
T_{2u}	E	131.0028	185.1617	29.3033	53.8370
		131.0028	185.1617	29.3033	53.8370
	B_1	131.0086	188.7178	29.3422	56.7420
T_{1u}	E	131.0435	188.5047	29.3396	56.8769
		131.0435	188.5047	29.3396	56.8769
	A_1	131.0484	188.7579	29.3438	56.7435
T_{2g}	B_2	193.8032	247.8271	42.6460	67.1775
	E	193.8123	251.1427	42.6825	70.0745
		193.8123	251.1427	42.6825	70.0747
T_{1g}	A_2	194.3190	248.3613	42.6573	67.1893
	E	194.3283	251.8995	42.6954	70.0939
		194.3283	251.8995	42.6954	70.0932
A_{2u}	B_2	219.6064	276.9942	48.3073	75.7013
E_u	A_2	220.5255	277.9842	48.3248	75.7200
	B_2	220.5266	278.2286	48.3273	75.7267

Table 5.8 Lowering of symmetry and splitting of degeneracy in the presence of 1T and 5T magnetic fields, in the conduction and valence bands. (From Ref. [9].)

		Energy (meV)			
O_h	C_2	Conduction-electron		Heavy-hole	
		B = 1 T	B = 5 T	B = 1 T	B = 5 T
A_{1g}	A	26.4932	–4.8840	5.4796	0.2583
T_{1u}	A	53.2169	20.1339	10.9697	5.3889
	$2B$	54.0533	22.7498	11.1174	5.8963
		54.8778	28.5790	11.2641	6.8571
T_{2g}	A	80.5974	47.6065	16.6064	11.0255
	$2B$	81.2683	50.2308	16.7298	11.4770
		82.2530	56.0301	16.9008	12.4939
A_{1g}	A	99.8112	68.6822	20.5430	15.3240
E_g	$2A$	99.8679	69.8135	20.5456	15.5922
		100.0224	73.9183	20.5715	16.1463
A_{2u}	A	108.4532	77.5113	22.3652	17.1125
T_{2u}	A	125.2321	90.1747	25.8860	19.8888
	$2B$	126.0649	93.2403	26.0318	20.4512
		126.8611	96.9336	26.1780	21.2266
T_{1u}	A	127.0645	101.0549	26.2059	21.7809
	$2B$	127.7112	101.6253	26.3262	21.9191
		128.4123	104.9692	26.4558	22.6043
T_{2g}	A	189.0605	156.5853	39.3802	33.8002
	$2B$	189.6253	160.8458	39.5216	34.7193
		189.9392	161.5775	39.5322	34.8313
T_{1g}	A	190.0764	161.7920	39.5658	34.9417
	$2B$	190.5608	162.9376	39.6026	35.1853
		190.8418	164.3101	39.6751	35.2686
A_{2u}	A	215.3905	185.3730	45.1302	39.8812
E_u	$2A$	216.2796	187.1525	45.1545	40.3494
		216.7630	193.1348	45.2324	41.7345

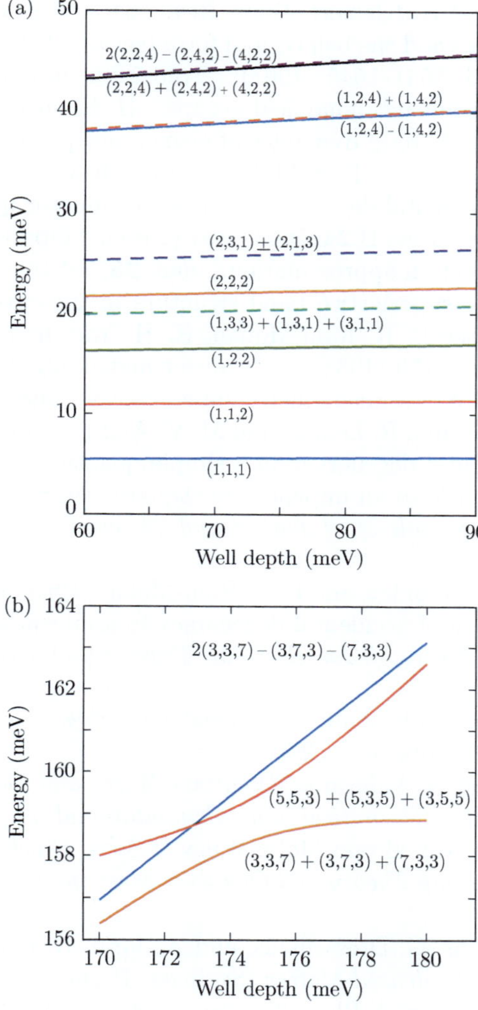

Figure 5.9 The dependence of heavy-hole energy levels on the barrier height in a 200 Å × 200 Å × 200 Å cubic QD for (a) the first 10 energy levels and (b) selected few higher states in which level crossing is observed. (Following Ref. [9]). This is very similar to the 2D case explored in Ref. [5] by us.

References

[1] J. J. Sakurai, *Modern Quantum Mechanics* (Addison-Wesley, Reading, MA, 1994).

[2] W. Pauli, Z. Phys. **36**, 336–363 (1926); "Über das Wasserstoffspektrum vom Standpunkt der neuen Quantenmechanik."

[3] V. Fock, Z. Phys. **98**, 145–154 (1935); "Zur Theorie des Wasserstoffatoms."

[4] A. O. Barut, and H. Kleinert, Phys. Rev. **161**, 1464–1466 (1967); "Dynamical group for baryons and the behavior of form factor;" A. O. Barut and H. Kleinert, Phys. Rev. **156**, 1541–1545 (1967); "Transition probabilities of the hydrogen atom from noncompact dynamical groups;" H. Kleinert, Fortschritte der Phys. **16**, 1–74 (1968); "Group dynamics of elementary particles."

[5] J. Shertzer and L. R. Ram-Mohan, Phys. Rev. B **41**, 9994–9999 (1990); "Removal of accidental degeneracies in semiconductor quantum wires."

[6] G. Bastard, Phys. Rev. B **24**, 5693–5697 (1981); "Superlattice band structure in the envelope-function approximation." *ibid.* **25**, 7584–1–25 (1982); "Theoretical investigations of superlattice band structure in the envelope-function approximation." Also see L. R. Ram-Mohan, K. H. Yoo, and R. L. Aggarwal, Phys. Rev. B **38**, 6151–6159 (1988); "A transfer matrix algorithm for the calculation of band structure of semiconductor superlattices," and references therein.

[7] F. Leyvarz, A. Frank, R. Lemus, and M. V. Andrez, Am. J. Phys. **65**, 1087–1094 (1997); "Accidental degeneracy in a simple quantum system: A new symmetry group for a particle in an impenetrable square-well potential."

[8] F. Reif, *Fundamentals of Statistical and Thermal Physics* (McGraw-Hill, New York, 1965).

[9] S. Bharadwaj, S. Pandey, and L. R. Ram-Mohan, Phys. Rev. B **96**, 195305–1–13 (2017); "Removal of accidental degeneracy in semiconductor quantum dots."

[10] S. L. Chuang, *Physics of Optoelectronic Devices* (John Wiley & Sons, New York, 1995).

[11] M. Balkanski and R. F. Wallis, *Semiconductor Physics and Applications* (Oxford University Press, New York, 2000).

[12] P. Harrison, and A. Valavanis, *Quantum Wires and Dots, in Quantum Wells, Wires and Dots: Theoretical and Computational Physics of Semiconductor Nanostructures*, 4th edition (John Wiley & Sons, Chichester, UK, 2016).

[13] M. Tinkham, *Group Theory and Quantum Mechanics* (Dover Publications, New York, 2003).

[14] M. S. Dresselhaus, G. Dresselhaus, and A. Jorio, *Group Theory: Application to the Physics of Condensed Matter* (Springer, Berlin, 2008).

[15] D. F. Greenberg, Am. J. Phys. **34**, 1101–1109 (1966); "Accidental degeneracy."

[16] C. E. Burkhardt and J. J. Leventhal, Am. J. Phys. **72**, 1013–1016 (2004); "Lenz vector operations on spherical hydrogen atom eigenfunctions."

[17] E. G. Kalnins, W. Miller, Jr., and P. Winternitz, SIAM J. Appl. Math. **30**, 630–664 (1976); "The Group $O(4)$, Separation of Variables and the Hydrogen Atom."

[18] A. O. Hernandez-Castillo and R. Lemus, J. Phys. A: Math. Theor. **46**, 465201–1–21 (2013). "Symmetry group of an impenetrable cubic well potential."

[19] F. M. Fernandez, arXiv:1310.5136 [quant-ph]. "On the symmetry of the quantum-mechanical particle in a cubic box."

[20] D. J. Ben-Daniels, and C. B. Duke, Phys. Rev. **152**, 683–692 (1966); "Space-charge effects on electron tunneling." In the multiband context, the probability current continuity condition can be obtained directly from the Lagrangian density, as in: L. R. Ram-Mohan, J. R. Meyer, I. Vurgaftman, Microelectron. J.

30, 1031–1042 (1999); "Wave function engineering of antimonide quantum-well lasers." L. R. Ram-Mohan, Proc. of the 31st Int. Symposium on Compound Semiconductors, Seoul, Korea, September 12–16, 2004, edited by Y.-S. Kwon, T. Yao, K.-H. Yoo, H. Hasegawa, J. C. Woo, Institute of Physics Conference Series Number **184**, pp. 1–8, (IoP, Bristol, UK, 2005); "A Lagrangian approach to wavefunction engineering of layered quantum semiconductor structures." L. R. Ram-Mohan and K. H. Yoo, J. Phys.: Condens. Matter **18**, R901 (2006); "Wavefunction engineering of layered semiconductors: Theoretical foundations."

[21] I. Vurgaftman, J. R. Meyer, and L. R. Ram-Mohan, J. Appl. Phys. **89**, 5815–5875 (2001); "Band parameters for III–V compound semiconductors and their alloys."

[22] L. R. Ram-Mohan, *Finite and Boundary Element Applications in Quantum Mechanics* (Oxford University Press, New York, 2002).

[23] Satish Balay et al., Argonne National Laboratory, ANL-95/11 - Revision 3.7, 2016. *PETSc Users Manual.*

[24] V. Hernandez, J. E. Roman, and V. Vidal, ACM Trans. Math. Software, **31**, 3, 351–362 (2005); "SLEPc: A scalable and flexible toolkit for the solution of eigenvalue problems."

[25] P. R. Amestoy, I. S. Duff, J.-Y. L'Excellent, and J. Koster, SIAM J. Matrix Anal. Appl. **23**, 1, 15–41 (2006); "A fully asynchronous multifrontal solver using distributed dynamic scheduling."

[26] Xiaoye S. Li, and James W. Demmel, ACM Trans. Math. Software, **29**, 2, 110–140 (2003); "SuperLU_DIST: A scalable distributed-memory sparse direct solver for unsymmetric linear systems."

[27] V. Hernandez, J. Roman, A. Tomas, and V. Vidal, Universidad Politécnica de Valencia, Tech. Rep. STR-7, (2007); "Krylov-Schur methods in SLEPc;" Also see, G. W. Stewart, SIAM J. Matrix Anal. Appl. **23**, 3, 601–614 (2001).

[28] H. Bethe, Ann. Phys., **3**, 133–206 (1929); "Termaufspaltung in Kristallen."

[29] L. R. Ram-Mohan, K. H. Yoo, and J. Moussa, J. Appl. Phys. **95**, 3081–3092 (2004); "The Schrödinger-Poisson self-consistency in layered quantum semiconductor structures."

[30] J. Lee, H. N. Spector, W. C. Chou, Phys. Status Solidi (b) **242**, 2846–2855 (2005); "Self-consistent calculation for energy band profiles and energy levels of cubic quantum dots."

[31] L. R. Ram-Mohan and P. A. Wolff, Phys. Rev. B **38**, 1330–1339 (1988); "The energetics of acceptor-bound magnetic polarons in diluted magnetic semiconductors."

[32] S. T. Jang, K. H. Yoo, and L. R. Ram-Mohan, J. Korean Phys. Soc. **50**, 834–838 (2007); "Spontaneous magnetization in diluted magnetic semiconductor quantum wells."

[33] I. Vurgaftman, J. R. Meyer, and L. R. Ram-Mohan, IEEE J. Quant. Electron. **32**, 1334–1346 (1996); "Optimized second-harmonic generation in asymmetric double quantum wells."

[34] L. R. Ram-Mohan and J. Shertzer, Appl. Phys. Lett. **57**, 282–284 (1990); "Electronic energy bands and optical nonlinearity of checker-board superlattices."

[35] S. Shao, K.-X. Guo, Z.-H. Zhang, N. Li, and C. Peng, Physica B **406**, 393–396 (2011); "Studies on the second-harmonic generations in cubical quantum dots with applied electric field."

[36] Z.-H. Zhang, K.-X. Guo, B. Chen, R.-Z. Wang, and M.-W. Kang, Superlattices Microstruct. **46**, 672–678 (2009); "Third-harmonic generation in cubical quantum dots."

Part II

Quantum Scattering

6

Quantum scattering in 1D revisited

In this chapter:

- A new approach to quantum scattering is developed by invoking sources and absorbers. In Chap. 2, we had shown that scattering phenomena can be represented as variational integrals with Cauchy boundary conditions (BCs). Almost all scattering problems require numerical work. Usually, global wavefunctions are assumed to provide answers. In a computational scheme, we have the freedom of considering any finite potential for the scattering center. Within the finite element method (FEM), we have seen how to implement the Cauchy BCs.
- In 1D, we have very simple modes to consider in regions away from the scattering center, namely, $\psi \sim \exp(\pm ikx)$. In tunneling problems with $V(x) > E$, it is necessary to include solutions of the type $\exp(\pm\kappa x)$, where κ is the imaginary wavevector in this case. In textbooks on quantum theory, this fact is exploited to obtain BCs at large distances, where $\kappa^2 = (2m/\hbar^2)(V - E)$. Matching the wavefunction and its derivative at interfaces and at $\pm\infty$ gives algebraic equations that are solved to obtain the solution everywhere in terms of the incoming amplitude.
- We employ absorbers at either end of the physical region so that the wavefunction needs to only satisfy simple BCs outside the absorbers, $\psi \equiv 0$, at these locations. In effect, we convert the Cauchy BC into Dirichlet BC. The absorbing finite elements are referred to as "stealth elements." We ensure that the stealth regions do not reflect waves that are incident on them. In this manner, we ensure that the wavefunctions in the scattering region are not affected by the absorbers.
- If the physical domain is closed off by such absorbers, we will need to introduce a source for the incoming particle current. We show how this is feasible, and demonstrate with examples that the resulting framework conforms to the usual scattering in all aspects, but with much simpler BCs to fulfill.

Finite Elements in Action. L. Ramdas Ram-Mohan, Oxford University Press. © L. Ramdas Ram-Mohan (2026).
DOI: 10.1093/oso/9780199563487.003.0006

6.1 Quantum scattering in 1D

Scattering phenomena are very well understood in terms of the specific observable effects they display. They are the main method of probing the properties of matter using photons, electrons, and other particle projectiles in many disciplines of research.

The first application of quantum mechanics to the theory of scattering was done by Gamow, [1] who was able to explain the variations in the lifetimes of decaying nuclei under alpha particle emission. Since then, the theory of quantum scattering has evolved very rapidly into scattering of particles from nuclei and elementary particle scattering, all due to sources of radioactive nuclei being isolated for experimental work and due to high-energy accelerators. Textbooks that are very representative of the treatments of scattering theory are by Bohm, [2] Sakurai, [3] Mott and Massey, [4] Goldberger and Watson, [5] Goldberger, M. L. Taylor, [6] Cohen-Tannoudji et al., [7] and R. G. Newton, [8] to name a few.

Here we focus on the scattering of electrons from potentials in 1D in order to develop a new approach to the theoretical modeling of quantum scattering. Note that this topic has a few analytic results, but otherwise requires numerical work to obtain quantitative results to compare with experiments.

6.2 Sources and absorbers: boundary conditions

Let us briefly discuss scattering by a standard 1D rectangular barrier and the corresponding BCs. We consider an incoming plane wave with a given energy E and an amplitude a from a source at $x = -\infty$ (see Fig. 6.1). The reflected and transmitted wavefunctions are given by

$$\begin{aligned} \psi_1 &= ae^{ikx} + r\, e^{-ikx}, \\ \psi_3 &= te^{ikx}, \end{aligned} \tag{6.1}$$

where, the wavevector $k = \sqrt{2m^* E/\hbar^2}$ and m^* is the effective mass of an electron.

We obtain the Cauchy (mixed) BCs at x_a and x_b given by

$$\begin{aligned} \left[\frac{d\psi_1(x)}{dx} + ik\psi_1(x)\right]_{x=x_a} &= 2i\, kae^{ikx}\Big|_{x=x_a}, \\ \left[\frac{d\psi_3(x)}{dx} - ik\psi_3(x)\right]_{x=x_b} &= 0. \end{aligned} \tag{6.2}$$

These conditions illustrate the complexity of BCs already at the 1D level. (These BCs are usually circumvented in 1D by there being only the modes $e^{\pm ikx}$.) In the analytical formulation of scattering, such BCs are built into the solution by starting with an expansion in terms of the incoming and outgoing basis wavefunctions. In 2D and 3D, a partial wave analysis is then performed to obtain amplitudes and angular dependencies in the asymptotic limit. This method was first developed by Rayleigh [9] in the context of sound waves, and later by Faxen and Holtsmark [10] for the scattering of electrons. For example, in 2D open domain scattering, the scattered

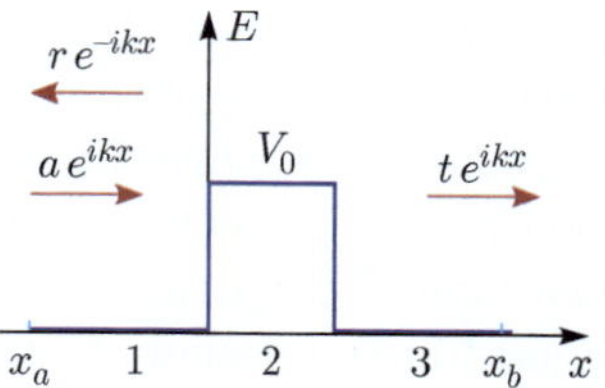

Figure 6.1 A schematic diagram of scattering by a 1D barrier of height V_0 is shown. We consider an incoming plane wave with an amplitude a and energy E. The corresponding wavevector is $k = \sqrt{2m^* E/\hbar^2}$. Here, r and t are the reflection and transmission amplitudes, respectively.

wavefunctions are expressed in terms of Hankel functions of the first kind $H_m^{(1)}(k\rho)$, all of which have the same *asymptotic* form [11] given by $e^{ik\rho}/\sqrt{\rho}$. The mixed BCs can be used in the variational treatment of the scattering [12–14] as the current continuity condition for the Hankel function of each order. However, it is complicated to use such BCs in calculation. This is because a derivative of the Hankel function of order n is linked to the Hankel functions of order $n \pm 1$ through a recursion relation. Hence, the asymptotic form of the BCs is invoked. The consequence of this choice is that evanescent modes are ignored; they do not survive in the asymptotic domain. The determination of the amplitudes of the Hankel functions are alone compared with experiment.

Discretization of the scattering boundary leads to inaccuracies and a non-zero reflection from the boundary. Moreover, such BCs have to be applied at each point on the boundary which is cumbersome in 2 and 3 dimensions. This is formally resolved in the usual formulation by going to the asymptotic regime, at the price of having a larger scattering region and also which do not survive as we go to the asymptotic region, sacrificing the evanescent modes in multi-band wavefunctions.

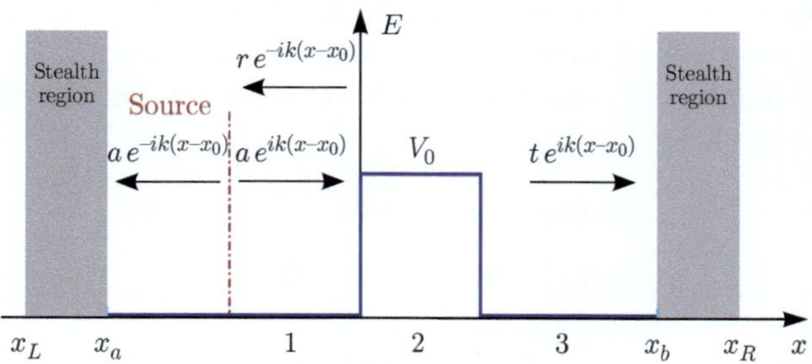

Figure 6.2 Scattering through a 1D barrier of height V_0 with stealth regions on either side of the barrier is shown. We consider a source at $x = x_0$ within the active region which injects an incoming wave with an amplitude a and the energy E in either direction. The corresponding wavevector is $k = \sqrt{2m^* E/\hbar^2}$. Here, r and t are the reflection and transmission coefficients, respectively.

Here we propose a method to reduce mixed BCs to simpler Dirichlet BCs by creating absorbers (stealth regions) on either side of the barrier as in Fig. 6.2. We optimize the parameters within these artificial absorbers in such a way that there is no reflection from any wave incident on them. The concept of stealth elements closely follows the idea of "perfectly matched layers" frequently used in electromagnetic scattering; [15] the same techniques are used in the design of reflection-less optical components.

We begin by considering a complex electron effective mass

$$\overline{m} = m^* \left(1 + i\ \alpha(x)\right), \tag{6.3}$$

where $\alpha(x)$ is a cubic Hermite interpolation polynomial varying smoothly from 0 to α_{max} within the stealth region and equal to 0 in the active (scattering) region. Let $\alpha(x_a) = \alpha(x_b) = 0$ and, $\alpha(x_L) = \alpha(x_R) = \alpha_{max}$ as shown in Fig. 6.3. The wave equation within stealth regions is given by

$$\frac{d}{dx}\left(\frac{1}{(1+i\alpha(x))}\frac{d}{dx}\psi(x)\right) + k^2(1+i\alpha(x))\psi(x) = 0. \tag{6.4}$$

Solutions to the above equation will have a highly damped behavior due to suitably chosen $\alpha(x)$. Therefore, we can ensure Dirichlet BCs on the outer boundaries of such stealth regions at $\psi(x_L) = \psi(x_R) = 0$.

6.2.1 No-reflection condition at the stealth interface

From optics, we know that any abrupt interfaces give rise to reflections. Fabry-Pérot interference effects are observed when we have multiple interfaces. [16] The stealth regions are designed to simplify the BCs, and are required not to alter the physics within the active region. Hence, we fix the parameters such that there is no reflection from any wave incident on the stealth interface.

In Fig. 6.4, we consider a 1D problem with a stealth interface at $x = 0$. In the active region $x < 0$, we have an incoming plane wave of amplitude a and energy E that emerged from a source at $x = -\infty$. A uniform stealth region is filled in the region $x > 0$. In general, at an interface, we will have both transmitted and reflected waves. Let the electron effective mass in region 2 be $\overline{m} = m^* (1 + i\,\alpha)$, where α is the absorption parameter considered to be a constant for the moment. The total wavefunction is given by

$$\begin{aligned} \psi_1(x) &= ae^{ik_1x} + re^{-k_1x}, \quad x < 0; \\ \psi_2(x) &= te^{ik_2x}, \quad x > 0, \end{aligned} \tag{6.5}$$

where the wavevector $k_1 = \sqrt{2m^*E/\hbar^2}$, and k_2 is as yet undetermined. The differential equations satisfied by the wavefunction in the two regions are given by

$$\begin{aligned} \frac{d^2}{dx^2}\psi_1(x) + k_1^2\psi_I(x) = 0, \quad x < 0; \\ \frac{d}{dx}\frac{m^*}{\overline{m}}\frac{d}{dx}\psi_2(x) + k_1^2\beta\psi_2(x) = 0, \quad x > 0, \end{aligned} \tag{6.6}$$

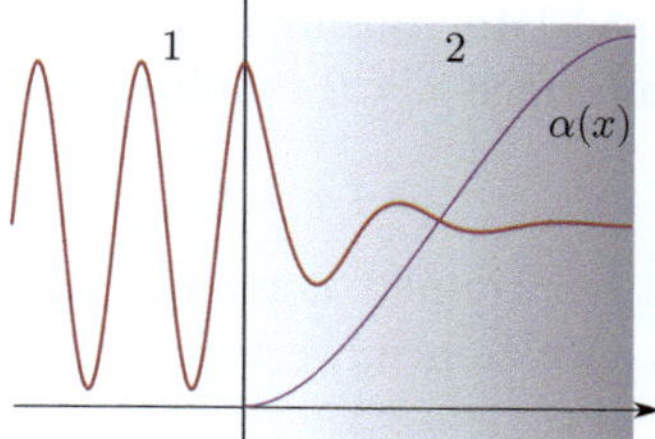

Figure 6.3 We have shown a plane wave incident from the active region decay rapidly to zero within the absorber (shaded region). The stealth function $\alpha(x)$ is a cubic Hermite polynomial that varies smoothly in the stealth region which is shown as a continuous curve.

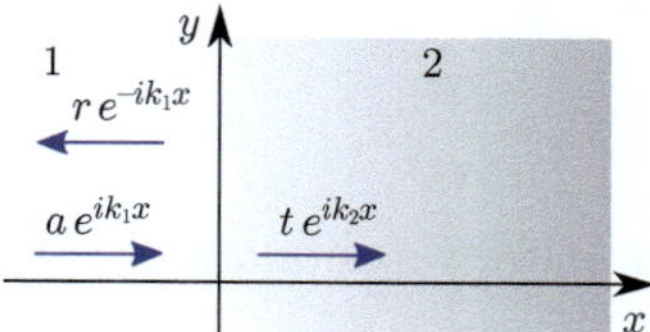

Figure 6.4 We show the transmission and reflection at the stealth region interface. Here, k_1 and k_2 are the corresponding wavevectors in the regions 1 and 2, respectively. Here, a, r, and t are the incident, reflected, and transmitted amplitudes, respectively.

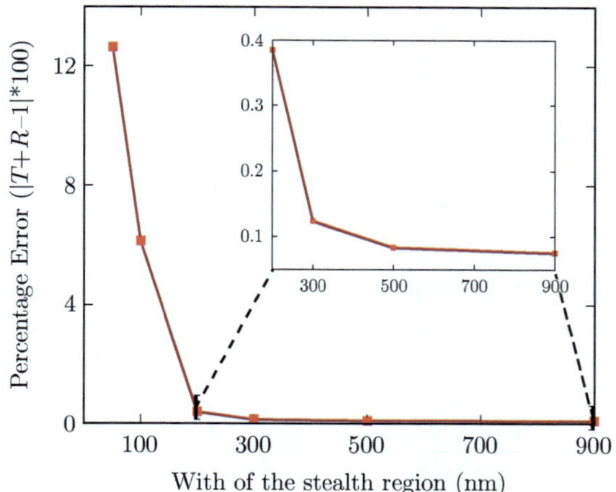

Figure 6.5 The percentage error $(T + R - 1) \times 100$ is plotted as a function of the width of the stealth region for scattering in a 2D waveguide. Here we choose the incoming energy to be 0.01 eV, $\alpha_0 = 16$, and the mesh size as ≈ 5000. We have employed $\mathcal{C}_{(1)}-$ continuous quintic Hermite interpolation polynomials on a triangle [17] for all calculations. The global matrix size is $\sim 30000 \times 30000$.

where the parameter β is fixed later through the no-reflection condition. Continuity of the wavefunction at the interface $x = 0$ requires that

$$a + r = t. \tag{6.7}$$

The probability current continuity demands that the mass-derivative of the wavefunction be continuous. [18] Hence, we have the condition

$$i\frac{k_1}{m^*}(a - r) = i\frac{k_2}{\overline{m}}t. \tag{6.8}$$

From Eqs. (6.7) and (6.8), the reflection coefficient is given by

$$r = a\left(\frac{k_1\overline{m} - k_2 m^*}{k_1\overline{m} + k_2 m^*}\right), \tag{6.9}$$

and *the no-reflection condition* is

$$k_2 = \frac{\overline{m}}{m^*}k_1 = (1 + i\alpha)\, k_1. \tag{6.10}$$

Substituting Eq. (6.5) in Eq. (6.6), we obtain a dispersion relation of the form

$$-\frac{m^*}{\overline{m}}k_2^2 + \beta k_1^2 = 0. \tag{6.11}$$

Hence, we obtain $\beta = (1 + i\,\alpha)$. In practice, we consider the absorption parameter α to increase smoothly over the stealth region so that there are no sharp interfaces or jump conditions to generate any reflections. Therefore, the wave equation in the stealth region is given by

$$-\frac{\hbar^2}{2m^*}\frac{d}{dx}\left(\frac{1}{(1 + i\alpha(x))}\frac{d\psi(x)}{dx}\right) - E\,(1 + i\alpha(x))\,\psi(x) = 0, \tag{6.12}$$

with solutions of the form

$$\psi(x) \sim \exp\left(\pm i k_1\, x - k_1 \int_0^{\pm x} dx'\alpha(x')\right). \tag{6.13}$$

We see that the solutions are highly damped, which allows us to apply Dirichlet BCs at the outer boundaries of the stealth regions. The mass-derivative continuity [18] in Eq. (6.6) for any interface at $x > 0$ requires $\alpha(x)$ to satisfy a condition of the form

$$\left.\frac{1}{1 + i\alpha(x)}\frac{d}{dx}\psi_2\right|_{x-0^+} = \left.\frac{1}{1 + i\alpha(x)}\frac{d}{dx}\psi_2\right|_{x+0^+}. \tag{6.14}$$

Hence, the absorption parameter $\alpha(x)$ has to be a smooth function. We choose $\alpha(x)$ to be a cubic Hermite interpolation polynomial which varies smoothly from 0 to α_{max} within the stealth region, as shown in Fig. 6.3. The width of the stealth region and α_{max} are optimized in such a way that the sum of reflection (R) and transmission (T) coefficients is closest to 1 up to a desired tolerance. In the absence of scatterers we predetermine these parameters in a convenient manner. The *no-reflection condition* similar to Eq. (6.10) is valid even for scattering in 2D waveguides. In Fig. 6.5 we see that the percentage error decreases monotonically as a function of the width of the stealth region. The error can be decreased further by increasing the thickness of the absorbers and increasing the mesh density. A choice of α_{max} can be made for all energies by having $\alpha_{max} \propto E\,\alpha_0$, where α_0 is a fixed constant. We note that creating absorbers around the scattering center provides a convenient way of specifying the region where we are interested in determining the transport properties.

6.2.2 Schrödinger's equation with a source term

Having enclosed the scattering center with totally absorbing stealth regions, we now require an electron "antenna" or source within the active region which can inject plane waves of given energy E and amplitude a. A "driving term" on the right-hand side is added to the standard Schrödinger's equation to create a δ-function source at a given location. The Schrödinger's equation with a source term in the active region is given by

$$\left[-\frac{d}{dx}\left(\frac{\hbar^2}{2m^*}\frac{d}{dx}\right) + V(x) - E\right]\psi(x) = -S\frac{\hbar^2}{2m^*}\delta(x - x_0), \tag{6.15}$$

where x_0 is the location of the source within the active region and S is the source parameter. In the absence of any potential, the equation for Green's function $G(x - x_0)$ in the active region is given by

$$\left[-\frac{d}{dx}\left(\frac{\hbar^2}{2m^*}\frac{d}{dx}\right) - E\right]G(x - x_0) = -S\frac{\hbar^2}{2m^*}\delta(x - x_0). \tag{6.16}$$

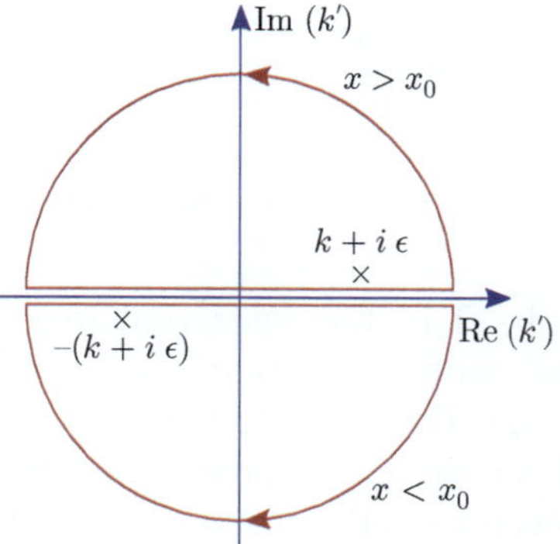

Figure 6.6 Poles and contours in the complex k'- space that are used to determine the integral in Eq. (6.17) are shown. For $x > x_0$ and $x < x_0$, we chose the contour and pole in the upper and lower half plane, respectively.

The Fourier transform of $G(x-x_0)$ is defined by

$$G(x-x_0) = \frac{1}{2\pi}\int_{-\infty}^{\infty} dk'\, g(k') \exp(ik'(x-x_0)). \tag{6.17}$$

Substituting Eq. (6.17) in Eq. (6.16) and representing the δ-function in the Fourier space we obtain

$$g(k') = -\frac{S}{2k}\left[\frac{1}{k'-k-i\epsilon} - \frac{1}{k'+k+i\epsilon}\right], \tag{6.18}$$

where we have used the "$i\epsilon$" prescription to specify poles in the 1st and 3rd quadrants. Now the integral in Eq. (6.17) is determined such that for $x > x_0$ ($x < x_0$), we evaluate the contour integral in the upper (lower) half plane as shown in Fig. 6.6 to obtain

$$G(x-x_0) = \frac{S}{2ik}\begin{cases} \exp(ik(x-x_0)), & x > x_0; \\ \exp(-ik(x-x_0)), & x < x_0. \end{cases} \tag{6.19}$$

We are then able to determine the source parameter to be $S = 2ika$, where a is the amplitude, and k is the wavevector for the incoming waves emerging on either side of the source located at $x = x_0$. Similar considerations in 2D and 3D lead to appropriate sources in either geometry.

6.3 Finite element analysis for scattering

The action integral for the 1D scattering problem with sources and absorbers is derived from Eq. (6.15) and is given by

$$\begin{aligned}\mathcal{A} = T \times \Bigg[\int_{x_L}^{x_R} & dx\, \psi^*(x) \Big[\overleftarrow{\partial}_x \frac{1}{(1+\alpha(x))} \overrightarrow{\partial}_x \\ & + \frac{2m^*}{\hbar^2}\left[V(x) - E\,(1+i\alpha(x))\right]\Big] \psi(x) \\ & + \int_{x_L}^{x_R} dx\, \psi^*(x)\, S\, \delta(x-x_0) \Bigg],\end{aligned} \tag{6.20}$$

where, $\alpha(x)$ is taken to be 0 in the scattering region. We are solving the time-independent (steady-state) problem so that the time integral over the range $[0, T]$ in the action is simply T, and Dirichlet BCs are implemented at $x = x_L$ and $x = x_R$.

FEA is a generalized variational approach in which we discretize the physical domain of interest into several small elements. [19] Within each element, we express the wavefunction as a linear combination of interpolation polynomials multiplied by as-yet undetermined coefficients that correspond to the value of the wavefunction at the vertices, called nodes of the elements that are line intervals in 1D and triangles or squares in 2D. We have used quintic Hermite interpolation polynomials for all calculations, for which the expansion coefficients are the wavefunction values and its derivatives at the nodes. [12] The additional derivative continuities allowed for in this case substantially increase the accuracy as compared with traditional Lagrange interpolation polynomials. [20] Discretization of the action integral in Eq. (6.20) within the finite element framework, and variation of the resulting action with respect to ψ^* leads to a system of linear equations of the form $M\,\Psi = b$, where Ψ is the column vector containing function values and derivatives of the wavefunction at the nodes, M is the corresponding coefficient matrix, and b is the column vector corresponding to the integral over the source term. We solve this matrix equation in a parallel computing environment [21–24] to obtain the total wavefunction throughout.

For scattering through a 1D single barrier, transmission coefficients determined through our formalism are found to be accurate within 10^{-5} when compared with analytical results. [25] We can systematically increase the accuracy through mesh size refinement (h-refinement), or by employing higher-order interpolation polynomials (p-refinement) for convergence within the FEA.

As an example, we consider an electron scattering through a 1D symmetric double barrier in GaAs. The effective mass of the electron equals $m* = 0.067\,m_e$, where m_e is the free electron mass. The barrier heights and widths are taken to be 0.3 eV and 10 nm, respectively. In Fig. 6.7, we plot the probability distribution for the first above-barrier resonant state corresponding to the energy 0.355 eV. Notice that the probability density reaches a maximum at each barrier due to Fabry-Pérot interference. Such resonances have been observed in several experiments involving superlattices and heterostructures. [26–29] Within the stealth regions (shaded) the wavefunction decays smoothly to zero as expected. In Fig. 6.8, we plot the well-known tunneling transmission coefficient [30, 31] as a function of incoming energy for

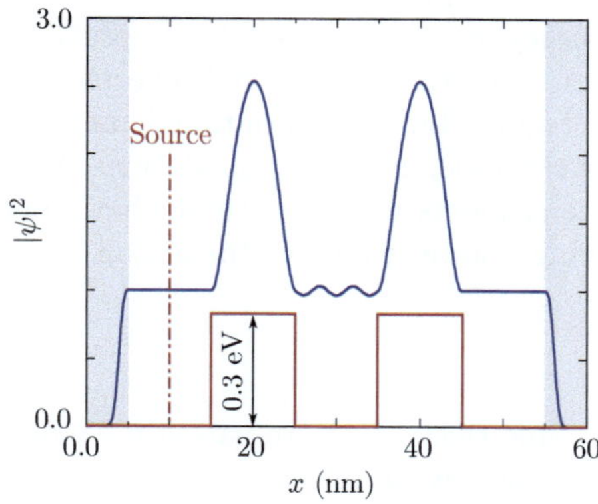

Figure 6.7 Probability distribution for the first above-barrier resonant state with incoming energy 0.355 eV is shown. The source is located at $x = 10$ nm. The stealth region is shaded in gray. The barrier heights and widths are taken to be 0.3 eV and 10 nm, respectively.

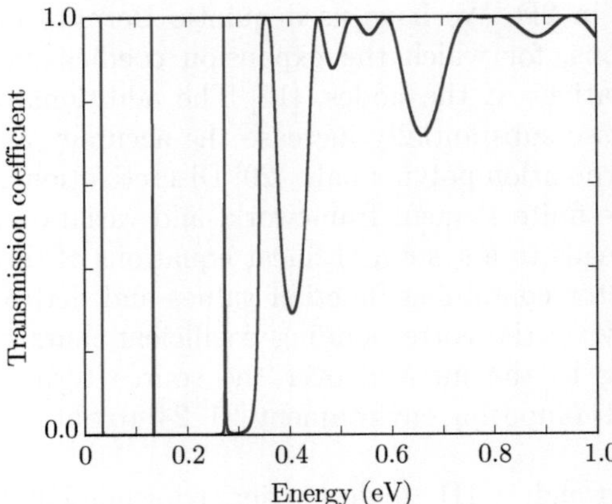

Figure 6.8 We have plotted the transmission coefficient as a function of energy for a double barrier potential. Each of the barriers is of height 0.3 eV with a width of 100 Å and are 100 Å apart.

the double barrier. Resonant peaks that are observed below the barrier (0.3 eV) are attributed to the electron trapped inside the well between the two barriers.

We typically employ ∼500 elements in 1D for these calculations, and the matrix size is ∼2000 × 2000 when we use quintic Hermite interpolation polynomials. At a given energy, the computational time is under 0.5 minutes and matrix condition number is $\sim 10^{-6}$.

6.4 Concluding remarks

Our introduction of absorbing elements, or stealth elements as we call them, for enclosing the active scattering region, provides a unique way of simplifying and applying BCs without decreasing accuracy in calculations. The absorption parameter is increased in a smooth fashion so that there are no unwanted reflected waves back from the stealth region into the active region. Being a variational method, the FEM allows us to improve the accuracy systematically in the calculations. We have remarkable freedom in putting in more finite elements where we think they could make the wavefunction more flexible for achieving a lower value for the action.

6.5 Problems

1. We have used the cubic Hermite interpolation polynomials in order to define the rise of absorption in the stealth region.
   (a) Consider the use of Fermi-function

$$F(x, x_0, \Delta) = C\left[\exp((x - x_0)/\Delta) + 1\right]^{-1}$$

which has the values

$$F(x, x_0, \Delta) = \begin{cases} C, & \text{for } x < x_0, \text{ with } \Delta \to 0; \\ 0, & \text{for } x > x_0, \text{ with } \Delta \to 0. \end{cases}$$

(b) Are other functions better for this purpose?

2. How should the source term be modified if we required a Gaussian wave packet to be incident on the scattering center? Elaborate on this theme.
3. Can other functional forms be developed for the source terms, such as a rectangular wave packet? Attempt a solution for this case.
4. One of the issues in using the absorber strategy in scattering is defining the depth of the stealth region. For incident waves with low energy, the depth has to be sufficient to reduce their amplitudes to about 1 part in 10^{-5}–10^{-7}. Make a table and also plot the values of absorber thicknesses versus the incoming wavevectors (or energy) in order to show that the thickness can be determined for low incident energy, and can be applied for higher incident energies.

References

[1] G. Gamow, Z. Phys. **51**, 204–212 (1928);"Zur Quantentheorie des Atomkernes;" Nature **122**, 805–806 (1928);"The quantum theory of nuclear disintegration;" Z. Phys. **52**, 510–515 (1929);"Zur Quantentheorie der Atomzertrümmerung;" *ibid.* **53**, 601–604 (1929);"Bemerkung zur Quantentheorie des radioaktiven Zerfalls."

[2] D. Bohm, *Quantum Theory* (Dover, NY, 1989).

[3] J. J. Sakurai, *Modern Quantum Mechanics* (Addison-Wesley, Reading, MA, 1994).

[4] N. F. Mott and H. S. W. Massey, *The Theory of Atomic Collisions* (Oxford University Press, London, 1949).

[5] M. L. Goldberger and K. M. Watson, *Collision Theory* (J. Wiley and Sons, NY, 1964).

[6] J. R. Taylor, *Scattering Theory: The Quantum Theory of Nonrelativistic Collisions* (Dover, NY, 2006).

[7] C. Cohen-Tannoudji, B. Diu, and F. Laloë, *Quantum Mechanics* Vol. 2 (Wiley Inter-Science, NY, 1977); translated from the French by S. R. Hemley, N. Ostrowsky and D. Ostrowsky.

[8] R. G. Newton, *Scattering Theory of Waves and Particles*, 2nd edition (Springer-Verlag, NY, 1982).

[9] J. W. Strutt, and B. Rayleigh, *The Theory of Sound*, Vol. **2** (Macmillan and Co., Ltd., New York, 1896); Ch. 17, pp. 236–284.

[10] H. Faxen and J. P. Holtsmark, Zeits. f. Physik **45**, 307–324 (1927); "Beitrag zur Theorie des Durchganges langsamer Elektronen durch Gase."

[11] A. Sommerfeld, *Partial Differential Equations in Physics* (Academic Press, New York, NY, 1949); Ch. 4, pp. 98–101.

[12] L. R. Ram-Mohan, *Finite Element and Boundary Element Applications in Quantum Mechanics* (Oxford University Press Inc., New York, 2002).

[13] Y. Li, and Z. L. Cendes, IEEE Trans. Magn. **29**, 1835–1838 (1993); "Modal expansion absorbing boundary conditions for two-dimensional electromagnetic scattering."

[14] R. Goloskie, J. W. Kramer, and L. R. Ram-Mohan, Comput. Phys. **8**, 679–686 (1994); "Quantum-mechanical tunneling and finite elements."

[15] J. Jin, *The Finite Element Method in Electromagnetics*, 2nd edition (Wiley, New York, 2002).

[16] M. Born, E. Wolf, A. B. Bhatia, P. C. Clemmow, D. Gabor, A. R. Stokes, A. M. Taylor, P. A. Wayman, and W. L. Wilcock, *Principles of Optics: Electromagnetic Theory of Propagation, Interference and Diffraction of Light*, 7th edition (Cambridge University Press, Cambridge, 1999).

[17] P. G. Kassebaum, C. R. Boucher, and L. R. Ram-Mohan, J. Comp. Physics. **231**, 5747–5760 (2012). "Application of group representation theory to derive Hermite interpolation on a triangle."

[18] D. BenDaniel, and C. Duke, Phys. Rev. **152**, 683–692 (1966);"Space-charge effects on electron tunneling."

[19] O. C. Zienkiewicz and Y. K. Cheung, *Finite Element Methods in Structural and Continuum Mechanics* (McGraw-Hill, New York, 1967); O. C. Zienkiewicz, *The Finite Element Method* (McGraw-Hill, New York, 1977).

[20] S. K. Adhikari, *Variational Principles and the Numerical Solution of Scattering Problems* (Wiley Inter-Science, NY, 1998).

[21] Satish Balay et al., Argonne National Laboratory, ANL-95/11 - Revision 3.7, 2016. *PETSc Users Manual.*

[22] P. R. Amestoy, I. S. Duff, J.-Y. L'Excellent, and J. Koster, SIAM J. Matrix Anal. Appl. **23**, 15–41 (2006); "A fully asynchronous multifrontal solver using distributed dynamic scheduling."

[23] P. R. Amestoy, A. Guermouche, J.-Y. L'Excellent, and S. Pralet, Parallel Comput. **32**, 136–156 (2006); "Hybrid scheduling for the parallel solution of linear systems."

[24] X. S. Li, and J. W. Demmel, ACM Trans. Math. Software, **29**, 2, 110–140 (2003); "SuperLU_DIST: A scalable distributed-memory sparse direct solver for unsymmetric linear systems."

[25] Sathwik Bharadwaj, Andrei Ilyashenko, Anthony Gianfrancesco, and L. R. Ram-Mohan, Trans. JASCOME, **17**, 22, 107–112 (2017); "Quantum Scattering and stealth finite element analysis."

[26] D. Dossa, Lok C. Lew Yan Voon, L. R. Ram-Mohan, C. Parks, R. G. Alonso, A. K. Ramdas, and M. R. Melloch, Appl. Phys. Lett. **59**, 2706–2708 (1991); "Observation of above-barrier quasi-bound states in asymmetric single quantum wells by piezomodulated reflectivity."

[27] F. C. Zhang, N. Dai, H. Luo, N. Samarth, M. Dobrowolska, J. K. Furdyna, and L. R. Ram-Mohan, Phys. Rev. Lett. **68**, 3220–3223 (1992); "Observation of localized above-barrier excitons in type-I superlattices."

[28] C. Parks, A. K. Ramdas, M. R. Melloch, G. Steblovsky, L. R. Ram-Mohan, and H. Luo, Solid State Commun. **92**, 563–567 (1994). See also, N. Dai, L. R. Ram-Mohan, H. Luo, G. L. Yang, F. C. Zhang, M. Dobrowolska, and J. K. Furdyna, Phys. Rev. B **50**, 18153–18166 (1994); "Observation of electronic states confined in surface quantum wells and above quantum barriers in modulated reflectivity."

[29] C. Parks, A. K. Ramdas, M. R. Melloch, G. Steblovsky, L. R. Ram-Mohan, and H. Luo, J. Vac. Sci. Technol. B **13**, 657–659 (1995); "Modulated reflectivity spectroscopy of electronic states confined in surface quantum wells and above quantum barriers."

[30] R. Tsu and L. Esaki, Appl. Phys. Lett. **22**, 562–564 (1973); "Tunneling in a finite superlattice."

[31] S. Datta, *Quantum Phenomena* (Addison-Wesley, Reading, 1989).

7
2D quantum waveguides

In this chapter:

- We treat the propagation of electrons in narrow channels or waveguides. Given a 2D distribution of carriers, the transport of electrons through the quantum mechanical system is studied.
- The principle of stationary action is employed to provide a variational solution to the 2D scattering. We restrict ourselves to waveguides and provide a finite element framework for the calculation of the transmission and reflection coefficients into the modes that are sustained in the waveguides.
- The implementation of current continuity boundary conditions for propagation through waveguides is described in detail.

7.1 Introduction

In Chap. 2, we investigated scattering from potential barriers in one dimension. The asymptotic form of the incoming and outgoing waves are readily obtained in 1D in terms of the two modes $e^{\pm ikx}$, where the wavevectors are defined by the kinetic energy and the mass of the particle. The "prepared" incoming wave is specified to be $a\,e^{ikx}$, where a is the amplitude of the incident wave. The outgoing waves are the reflected wave $r\,e^{-ikx}$ and the transmitted wave $t\,e^{ikx}$, with r and t being the corresponding amplitudes. Here, we extend these considerations to two-dimensional quantum mechanical systems. We are faced with a new set of issues to consider in the solution of Schrödinger's equation and a richer variety of physical effects than can be observed in 1D systems. The corresponding asymptotic forms of the wavefunctions in 2D are more complex. The application of scattering boundary conditions (BCs) in 2D is correspondingly more challenging.

It is straightforward to create a 2D distribution of electrons in layered semiconductor structures. When a crystal of the compound semiconductor GaAs is grown with Si, the impurity atoms of Si occupy the positions of Ga in the lattice. Here Ga is a

Finite Elements in Action. L. Ramdas Ram-Mohan, Oxford University Press. © L. Ramdas Ram-Mohan (2026).
DOI: 10.1093/oso/9780199563487.003.0007

group III element, whereas Si is a group IV element of the periodic table. The additional electron in Si is bound to it in a manner similar to the electron in the hydrogen atom, except that the effective mass of the electron in GaAs is $m^* = 0.063\,m$ at room temperature, and the dielectric constant of the host material is 12.9. The screening of the Coulomb potential and the smaller mass reduces the binding energy of the electron from that in hydrogen (13.6 eV) to 6 meV. The impurity atom is ionized at fairly low temperature and the electron is released into the conduction band of GaAs. The Si is said to be a donor in GaAs which is said to be n-doped, and we are able to alter the conductivity of GaAs through the doping. With layered semiconductor heterostructures, we have to take the energy band alignment between the layers into consideration. In the case of $\mathrm{Al}_x\mathrm{Ga}_{(1-x)}\mathrm{As}$, where x is the stoichiometric ratio of Al in GaAs, the conduction band of the AlGaAs is higher in energy relative to that of GaAs by about 300 meV for $x = 30\%$. When a layer of n-doped AlGaAs is grown on very lightly doped GaAs, as depicted in Fig 7.1(a), the bound electrons from the impurities in the AlGaAs are able to fall into the GaAs layer. They are still electrostatically coupled with the ionized impurities. This leads to a redistribution of charge in the layers, resulting in a bending of the conduction band edge, as shown in Fig. 7.1(b).

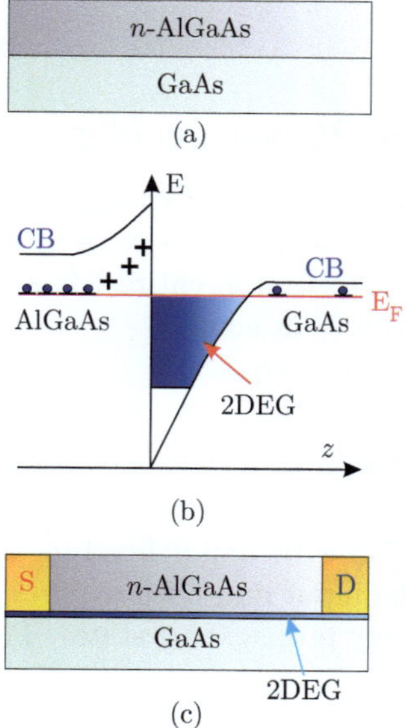

Figure 7.1 In (a), a layer of n-AlGaAs grown on a GaAs substrate is shown. The impurities in AlGaAs give up their electrons which fall into the GaAs layer. The charge redistribution leads to a bending of the conduction band edge, as shown in (b), with a common Fermi level E_F and the formation of a 2D electron gas (2DEG). (c) shows the inversion layer with the source, drain and gate contacts. On etching and putting contacts for a source S and a drain D, we have the 2DEG with electrons propagating in the thin potential region at the interface.

On attaching leads to the structure, under bias, the electrons can be made to propagate in the region of the GaAs/AlGaAs interface. We are interested in the transport properties of the electrons near the interface. If the doping of the AlGaAs layer is done with a set-back, away from the interface, the above-mentioned redistribution of the charges leaves the ionized impurities deeper in the AlGaAs layer. The electrons in the GaAs layer will then not experience ionized impurity scattering, leading to an extraordinarily high mobility, on the order of $10^6\,\mathrm{cm}^2/\mathrm{V\,s}$. Such structures are used in the fabrication of high-speed transistors and other devices. [1]

Here we are interested in investigating the effects of placing a third contact, the gate on the AlGaAs layer so that the gate potential can influence the propagation of electrons. In particular, when the gate contact is shaped so that the electrons in the 2D electron gas (2DEG) are forced to propagate through very narrow channels, quantum mechanical effects come into play, leading to the observation of quantized resistance. The geometry of the gates can be shaped so that it is possible to have a dynamic formation of quantum dots. By shaping the gate contact and applying gate potentials, it is possible to make single electron transistors and very far infrared photodetectors with such structures.

In the following, we will set up the formalism for the modeling of these quantum effects in carrier transport by considering a simple 2D electron waveguide created by an aligned gate in which the distance between the straight edges of the gates is on the order of 50–100 nm.

7.2 Action integral for 2D waveguides

We idealize the physical situation to be the straight, rectangular waveguide channeling of electrons between very high barriers, as shown in Fig. 7.2. The waveguide has one incident port, which is taken to be the entire left edge, and one transmission port, which is taken to be the entire right edge.

We wish to solve Schrödinger's equation,

$$-\frac{\hbar^2}{2m^*}\nabla^2\psi(x,y) + (V(x,y) - E_{\mathrm{inc}})\psi(x,y) = 0, \tag{7.1}$$

in the region shown in Fig. 7.3 with well-defined incoming and outgoing modes for a given incident energy. At the input and output ports, we set $V(x,y) = 0$ in order to permit our working with freely propagating states. The wavefunction at the incident

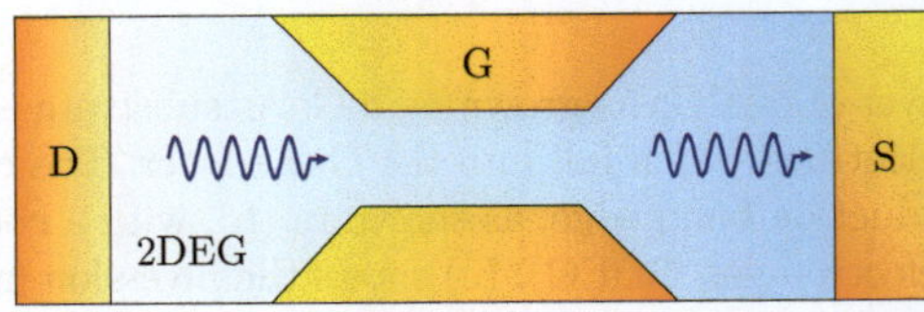

Figure 7.2 A schematic of the quantum waveguide formed by an aligned gate geometry is shown. Electrons propagate from the left to the right under applied positive bias.

port at x_{min} is a combination of the incoming and outgoing waves that can be expressed in terms of the allowed modes $\phi_\nu(y)$ in the form

$$\psi(x_{\text{min}}, y) = \sum_{\nu=1}^{N_{\text{mode}}} \left[a_\nu \exp(ik_\nu x_{\text{min}}) + r_\nu \exp(-ik_\nu x_{\text{min}})\right] \phi_\nu(y), \tag{7.2}$$

where a_ν and r_ν are the incident and reflected amplitudes of the νth transverse mode, respectively. The sums over a_ν and r_ν have been combined here; however, we could have different numbers of incident and reflection amplitudes. The quantity k_ν is the wavevector of the wave propagating in the x-direction of the νth transverse mode, which is defined in terms of the incident energy. Here, N_{mode} is the number of modes in the waveguide that we wish to work with in our calculations, and should be equal to or greater than the allowed modes for the given incident energy. The wavefunction at the transmission port located at x_{max} is

$$\psi(x_{\text{max}}, y) = \sum_{\mu=1}^{N_{\text{mode}}} t_\mu \exp(ik_\mu x_{\text{max}}) \phi_\mu(y), \tag{7.3}$$

where t_μ is the transmitted amplitude of the μth mode. The transverse modes are denoted by $\phi_\nu(y)$ and are the eigenfunctions of the infinite square well [2]

$$\phi_\nu(y) = \sqrt{\frac{2}{L_y}} \sin\left[\frac{\nu\pi(y - y_{\text{min}})}{L_y}\right], \quad L_y = y_{\text{max}} - y_{\text{min}}, \tag{7.4}$$

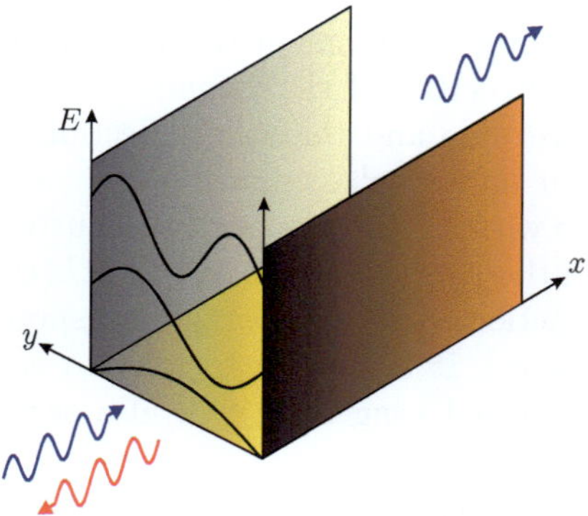

Figure 7.3 A rectangular waveguide showing the geometry and the incoming modes. The transverse wavefunctions are shown at the incident port.

which satisfy the BCs $\phi(0) = 0 = \phi(L_y)$. The integer indices μ, ν that label the modes are ≥ 1. In the y-direction, the infinite well provides a "transverse quantization" of the energy, and the energy spectrum is given by

$$E_\nu = \frac{\hbar^2}{2m^*} \frac{\nu^2 \pi^2}{L_y^2}, \tag{7.5}$$

where m^* is the effective mass of the electron in the semiconductor. While this effective mass is usually different in the various layers, we shall ignore its layer dependence since we are concerned with the transport along the active layer. The incident energy can be written as

$$E_{\text{inc}} = \frac{\hbar^2 k_0^2}{2m^*} = E_\nu + \frac{\hbar^2 k_\nu^2}{2m^*}. \tag{7.6}$$

The wavevector along the x-direction for the νth propagating mode can therefore be written† in the form

$$k_\nu = \sqrt{\left(\frac{2m^*}{\hbar^2}\right)(E_{\text{inc}} - E_\nu)} = \sqrt{k_0^2 - \frac{\nu^2 \pi^2}{L_y^2}}$$

$$= \sqrt{k_0^2 - k_{y,\nu}^2}. \tag{7.7}$$

In the cases where $E_{\text{inc}} > E_\nu$, the value of k_ν will be a real number and the mode is said to be a traveling wave. In the cases where $E_{\text{inc}} < E_\nu$, the value of k_ν will be purely imaginary, and we will have an evanescent mode. In Fig. 7.4, we show the energy versus wavevector relation for the various modes for an incident energy that is above and the energy of the third transverse mode, as an example. Then the waves associated with the first, second and the third modes can propagate down the waveguide. Their wave vectors are shown on the graph. Modes above the third will have energies $E_{\text{inc}} < E_\nu$, so that they will correspond to evanescent modes that will fall off rapidly in amplitude as these waves attempt to propagate down the waveguide.

The number of modes to be taken into consideration depends on the energy E_{inc} of the incident electron and the need to account for evanescent modes if we have scattering centers and absorbers present in the waveguide. In a straight waveguide, we expect that the mode structure of the incoming wave is maintained as the wave propagates down the waveguide and that the reflection coefficients in the reflected modes are zero, within our numerical accuracy.

We are concerned here with a steady state flow of current so that the action, which is an integral over time of the Lagrangian, is $A_0 = \int dt L(\psi, \psi^*) = T \times L(\psi, \psi^*)$. Most of the references to the action in the following correspond to A_0/T, i.e. the action is scaled by the time interval. It is simpler to refer to variational principles for the "action" since this is the usual usage. We will also scale the action by an energy

†It is useful to write k_ν with an additional label, $k_{\nu,x}$, to indicate that it is the wave propagating along the x-direction. However, we will not do so for ease of writing the equations. Similarly, it is useful to introduce the notation $\nu^2\pi^2/L_y^2 = k_{y,\nu}^2$, but for clarity we will write out the value.

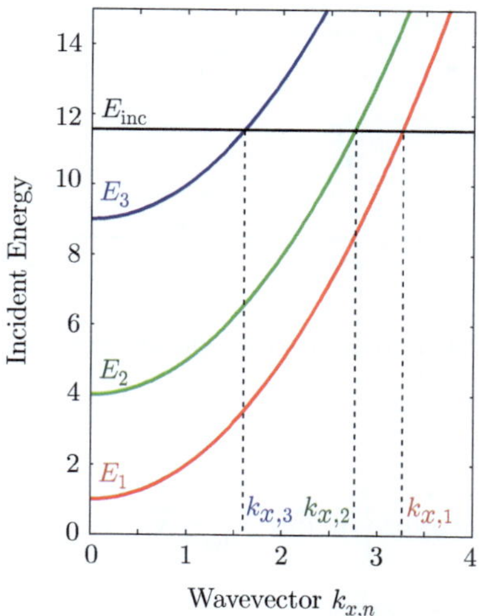

Figure 7.4 The energy versus wavevector plot for an energy above the third transverse modal energy E_3. The corresponding wavevectors for the propagating modes are shown.

factor in the following, so that it is then dimensionless. The action leading to the 2D Schrödinger's equation is given by

$$A_0 = \int_{x_{\min}}^{x_{\max}} \int_{y_{\min}}^{y_{\max}} dx\, dy \Big\{ \nabla\psi^*(x, y) \frac{\hbar^2}{2m^*} \nabla\psi(x, y) \\ + \psi^*(x, y) \left[V(x, y) - E_{\rm inc} \right] \psi(x, y) \Big\}. \tag{7.8}$$

The BCs on the wavefunction along the infinite barriers, see Fig. 7.7, require that they vanish along the sides Γ_0 from $(x_{\min}, y_{\min})$ to $(x_{\max}, y_{\min})$, and on the opposite side, Γ_2, from $(x_{\max}, y_{\max})$ to $(x_{\min}, y_{\max})$. We thus have

$$\psi(\Gamma_0) = \psi(\Gamma_2) = 0, \tag{7.9}$$

at the barriers on the sides of the waveguide, as can be correlated with the contour shown in Fig. 7.5. At the input and output ports, we have to treat the BCs with care since the wavefunctions there are given by Eqs. (7.2, 7.3). In other words the conditions on the wavefunctions are given in terms of the amplitudes of the waveguide modes that are coming in or going out in the waveguide.

Taking a variation of the above action with respect to ψ^* and invoking the principle of stationary action yields

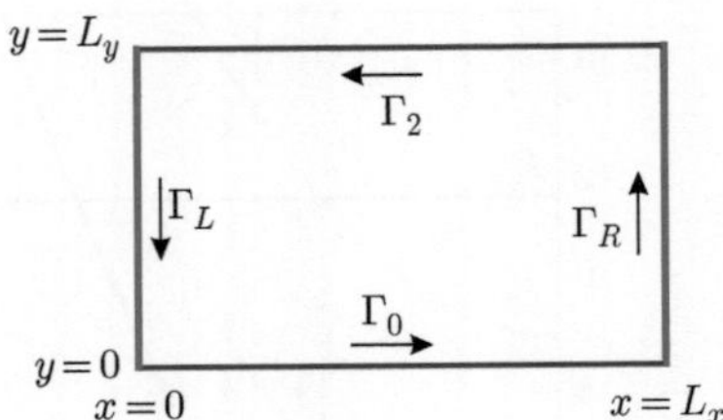

Figure 7.5 The contour path for the surface terms Γ_i of Eq. (7.11).

$$\delta_{\psi^*} A_0 = \int_{x_{\min}}^{x_{\max}} \int_{y_{\min}}^{y_{\max}} dx\, dy \left\{ \nabla(\delta\psi^*) \frac{\hbar^2}{2m^*} \nabla\psi + \delta\psi^* (V - E_{\text{inc}})\psi \right\} = 0, \qquad (7.10)$$

and an integration by parts leads to

$$\delta A_0 = \int_{x_{\min}}^{x_{\max}} \int_{y_{\min}}^{y_{\max}} dx\, dy\; \delta\psi^* \left[-\frac{\hbar^2}{2m^*} \nabla^2\psi + (V - E_{\text{inc}})\psi \right] + \left(\frac{\hbar^2}{2m^*} \right) \oint_{\Gamma_0+\Gamma_2+\Gamma_L+\Gamma_R} \delta\psi^* \nabla\psi \cdot \hat{n}\, d\ell = 0. \qquad (7.11)$$

The contour integral in Eq. (7.11) is zero along Γ_0 and Γ_2 since the wavefunction is specified there (to be zero), and hence its variation along these two sides is zero. However, the line integrals along Γ_L and Γ_R are non-zero at the ports since we are specifying the normal derivative of the wavefunction there while the wavefunction itself and hence $\delta\psi^*$ are arbitrary there. This implies that we cannot obtain the equation of motion, Eq. (7.1), as is usually derived were we have Dirichlet BCs. The currents at the ports are not zero, and this then requires that we modify the action itself by the addition of surface terms to the original action. The new term must cancel the outward currents represented by the surface term from integration by parts at the current ports. As shown below, we can then obtain the equation of motion in a clear manner. This is analogous to the modification to the action carried out in Chap. 1 for the Neumann and Cauchy BCs in 1D. Let us see how a modified action saves the situation, while at the same time builds in the derivative BCs at the ports.

Consider the modified action

$$A = A_0 - \left(\frac{\hbar^2}{2m} \right) \int_{\Gamma_L+\Gamma_R} \psi^* \frac{\partial \Psi_\Gamma}{\partial n} dl. \qquad (7.12)$$

The quantity $\nabla\Psi(\Gamma) \cdot \hat{n} \equiv \partial\Psi_\Gamma / \partial n$ is the derivative of the wavefunction in the direction normal to the contour. Here the normal derivative of the wavefunction at the periphery only along $\Gamma_L + \Gamma_R$ is employed in order to specify the BCs corresponding to the input and output currents in the system. Varying the action A with respect to ψ^* obtains

$$\delta_{\psi^*} A = \int_{x_{\min}}^{x_{\max}} \int_{y_{\min}}^{y_{\max}} dx\, dy\ (\delta\psi^*) \left[-\frac{\hbar^2}{2m} \nabla^2 \psi + (V - E_{\text{inc}})\psi \right]$$

$$+ \left(\frac{\hbar^2}{2m} \right) \int_{\Gamma_0 + \Gamma_2 + \Gamma_L + \Gamma_R} \delta\psi^* \frac{\partial \psi}{\partial n}\, dl$$

$$- \frac{\hbar^2}{2m} \int_{\Gamma_L + \Gamma_R} \delta\psi^* \frac{\partial \Psi_\Gamma}{\partial n}\, dl = 0. \tag{7.13}$$

We proceed as follows:

(1) As mentioned above, the line integrals along $\Gamma_0 + \Gamma_2$ vanish since the wavefunction is specified there, so that $\delta\psi^* = 0$ along these portions of the boundary.

(2) Since $\delta\psi^*$ is an arbitrary variation, we are free to choose it to be zero on the boundary and arbitrary inside the physical region, in which case the line integrals vanish and the quantity in square brackets in the following equation is zero

$$\delta A_{interior} = \int_{x_{\min}}^{x_{\max}} \int_{y_{\min}}^{y_{\max}} dx\, dy\ (\delta\psi^*) \left[-\frac{\hbar^2}{2m} \nabla^2 \psi + (V - E_{\text{inc}})\psi \right] = 0. \tag{7.14}$$

We have thus successfully derived the Schrödinger's equation from a variation of the action.

(3) Now, choosing in turn an arbitrary non-zero value for $\delta\psi^*$ on each of the ports, with the other port having $\delta\psi^* = 0$, and also setting $\delta\psi^* = 0$ in the interior, gives the relation

$$\delta A_{\Gamma_{L,R}} = \left(\frac{\hbar^2}{2m} \right) \int_{\Gamma_{L,R}} \delta\psi^* \left(\frac{\partial \psi}{\partial n} - \frac{\partial \Psi_\Gamma}{\partial n} \right) dl = 0, \tag{7.15}$$

which leads to the result that

$$(\partial\psi/\partial n)_{\Gamma_{L,R}} = \partial\Psi_{\Gamma_{L,R}}/\partial n \tag{7.16}$$

at the ports. This relation states that the BC on the (internally defined) currents given on the left side of the equation correspond to the ones we set at Γ_L and Γ_R. The quantities $\partial\Psi_\Gamma/\partial n$ represent the gradients used to define the currents associated with the incident, reflected, and transmitted modes at the boundary.

7.3 Discretization of the action in FEM

In order to apply the finite element method (FEM), we discretize the physical region into rectangular elements. Suppose that the ith element has the coordinates $x_a \le x \le x_b$ and $y_a \le y \le y_b$. We evaluate the action A by considering the contributions from each element i,

$$A = \sum_{i=0}^{nelem} A^{(i)}, \tag{7.17}$$

where $A^{(i)}$ is the action integral over element i. In addition, the wavefunction in element i is expressed as a sum of interpolation polynomials multiplied by coefficients $\psi_j^{(i)}$

$$\psi^{(i)}(x, y) = \sum_{j=1}^{nodelem} \psi_j^{(i)} N_j(\xi, \eta), \tag{7.18}$$

where $N_j(x, y)$ is the interpolation polynomial associated with the jth node of the element, and there are *nodelem* nodes in each element.

The elemental action $A^{(i)}$ is

$$A^{(i)} = \int_{x_a}^{x_b} \int_{y_a}^{y_b} dx\, dy \left\{ \frac{\hbar^2}{2m} \sum_{j=1}^{nodelem} \psi_j^{*(i)} \nabla N_j \cdot \sum_{k=1}^{nodelem} \psi_k^{(i)} \nabla N_k \right.$$
$$\left. + \sum_{j=1}^{nodelem} \psi_j^{*(i)} N_j \left[V(x, y) - E_{inc} \right] \sum_{k=1}^{nodelem} \psi_k^{(i)} N_k \right\}. \tag{7.19}$$

We suppose here that the incident particle has energy E_{inc}.

The integrals over each element are calculated using Gauss-Legendre quadrature along x and y directions, as a double summation over Gauss points labeled by $gx = 1, \ldots, ngx$ and $gy = 1, \ldots, ngy$. Typically, the same number of Gauss points, *ngaus*, will be employed in both directions. The Gauss-Legendre points and weights $\{\xi_{gx}, w_{gx}\}$ and $\{\eta_{gy}, w_{gy}\}$ are tabulated for an integral over the interval $[-1, 1]$. We therefore map the range over the element coordinates to the "standard element" by the transformations

$$x = \frac{(x_b + x_a)}{2} + \xi \frac{(x_b - x_a)}{2},$$
$$y = \frac{(y_b + y_a)}{2} + \eta \frac{(y_b - y_a)}{2}, \tag{7.20}$$

and write the integrals over x, y in terms of integrals over ξ, η multiplied by Jacobian factors

$$J_x^{(i)} = dx/d\xi = (x_b - x_a)/2,$$
$$J_y^{(i)} = dy/d\eta = (y_b - y_a)/2. \tag{7.21}$$

The summation over Gauss points can be brought inside the summations over nodes of the element so that

$$A^{(i)} = \sum_{(j=1,\,k=1)}^{nodelem} J_x^{(i)} J_y^{(i)} \psi_j^{*(i)} \Bigg[\sum_{(gx=1,\,gy=1)}^{ngaus} w_{gx} w_{gy}$$
$$\times \left\{ \frac{\hbar^2}{2m} \nabla N_j(\xi_{gx}, \eta_{gy}) \cdot \nabla N_k(\xi_{gx}, \eta_{gy}) \right.$$
$$\left. + N_j(\xi_{gx}, \eta_{gy}) \Big[V(\xi_l, \xi_m) - E_{inc} \Big] N_k(\xi_{gx}, \eta_{gy}) \right\} \Bigg] \psi_k^{(i)}. \tag{7.22}$$

The action integral for the element i can now be written in matrix notation, with implicit summation signs as

$$A^{(i)} = \psi_j^{*(i)} M_{jk}^{(i)} \psi_k^{(i)}. \tag{7.23}$$

The element matrices $M^{(i)}$ are then overlaid into a global matrix M such that the contributions to the integral from different elements which share the same nodal values are added to the same location in the global matrix. The total action can then be written as

$$A = \psi_\alpha^* M_{\alpha\beta} \psi_\beta. \tag{7.24}$$

The indices α, β range over the set of nodes in the entire physical domain. At this stage, the surface terms involving the incident, reflected, and transmitted currents have not been included in the action.

7.4 From nodal variables to modal amplitudes

The BCs at the incident port specify the incident amplitudes a_n. The action integral in the previous section is written in terms of the nodal values of the wavefunction. We specified the wavefunctions at the ports in terms of a superposition of well-defined modes at the two ports. So we can use a transformation to replace just those nodal values at the ports with the incident, reflected, and transmitted amplitudes. The transformation can be visually represented as

$$\begin{bmatrix} \psi_1 \\ \\ \vdots \\ \\ \psi_n \end{bmatrix} = \begin{bmatrix} \\ \\ T \\ \\ \\ \end{bmatrix} \begin{bmatrix} a_1 \\ \vdots \\ a_{N_{mode}} \\ \hdashline r_1 \\ \vdots \\ r_{N_{mode}} \\ \hdashline t_1 \\ \vdots \\ t_{N_{mode}} \\ \hdashline \tilde{\psi}_1 \\ \vdots \\ \tilde{\psi}_n \end{bmatrix}, \tag{7.25}$$

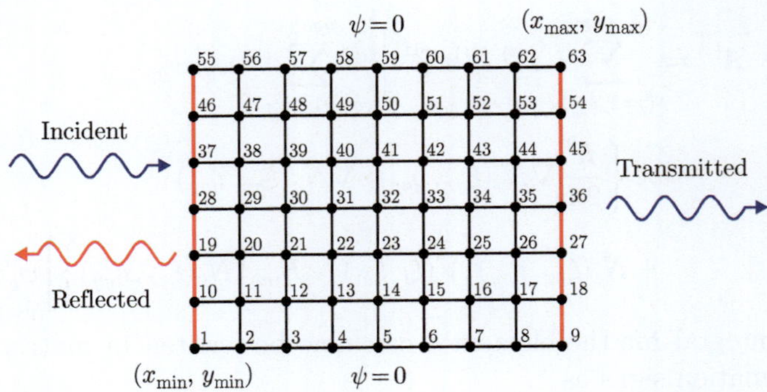

Figure 7.6 The physical region in the waveguide is divided into finite elements. The number labeling each node is shown in a 1-based scheme with the numbers initiating with 1.

where $\tilde{\psi}(1,\ldots,n)$ represents all of the nodal wavefunction values which are not on one of the ports. In the above notation, the elements of $\tilde{\psi}$ have been renumbered starting with one and skipping the nodes on the ports. Since the values $\tilde{\psi}$ are not part of the incoming or outgoing wavefunctions, they must be passed through the transformation without any modification. Consider the mesh shown in Fig. 7.1 containing nine nodes in the x-direction and seven nodes in the y-direction. This means that the vector $\tilde{\psi}$ will contain the nodal values

$$\left\{\tilde{\psi}_i\right\} = \{\psi_{2..8},\ \psi_{11..17},\ \psi_{29..35},\ \psi_{20..26},\ \psi_{38..44},\ \psi_{47..53},\ \psi_{56..62}\}\,. \tag{7.26}$$

This numbering defines which diagonal elements in the matrix T should contain unity. Now, consider the reduced transformation T_{ar} defined by

$$\begin{bmatrix} \\ \psi_{a,r} \\ \\ \end{bmatrix} \begin{bmatrix} \\ T_{a,r} \\ \\ \end{bmatrix} \begin{bmatrix} a_1 \\ \vdots \\ a_{N_{node}} \\ \hdashline r_1 \\ \vdots \\ r_{N_{node}} \end{bmatrix}. \tag{7.27}$$

The array vector $\psi_{a,r}$ is the set of nodal values which lie on the leftmost edge of the system. For example, the vector $\psi_{a,r}$ for the mesh shown in Fig. 7.1 is

$$\{\psi_{ar}\} = \{\psi_1\ \psi_{10}\ \psi_{19}\ \psi_{28}\ \psi_{37}\ \psi_{46}\ \psi_{55}\}\,. \tag{7.28}$$

Now, the matrix $T_{a,r}$ can be constructed using the expressions for the wavefunctions on the left side of the system

$$T_{ar} = \left[\begin{array}{ccc|ccc} \phi_1(y_1)e^{ik_1 x_a} & \cdots & \phi_{N_{mode}}(y_1)e^{ik_{N_{mode}} x_a} & 0 & \cdots & 0 \\ \vdots & \ddots & \vdots & \vdots & \ddots & \vdots \\ \phi_1(y_{55})e^{ik_1 x_a} & \cdots & \phi_{N_{mode}}(y_{55})e^{ik_{N_{mode}} x_a} & 0 & \cdots & 0 \end{array}\right] + \left[\begin{array}{ccc|ccc} 0 & \cdots & 0 & \phi_1(y_1)e^{-ik_1 x_a} & \cdots & \phi_{N_{mode}}(y_1)e^{-ik_{N_{mode}} x_a} \\ \vdots & \ddots & \vdots & \vdots & \ddots & \vdots \\ 0 & \cdots & 0 & \phi_1(y_{55})e^{-ik_1 x_a} & \cdots & \phi_{N_{mode}}(y_{55})e^{-ik_{N_{mode}} x_a} \end{array}\right]. \tag{7.29}$$

Here $x_a = x_{min}$ has been used for convenience. For the rightmost edge of the system, consider the reduced transformation T_t defined by

$$\left[\begin{array}{c} \psi_t \end{array}\right] = \left[\begin{array}{c} T_t \end{array}\right] \left[\begin{array}{c} t_1 \\ \vdots \\ t_{N_{mode}} \end{array}\right]. \tag{7.30}$$

The vector ψ_t is the set of nodal values which lie on the rightmost edge of the system.

The vector ψ_t for the mesh shown in Fig. 7.1 is

$$\{\psi_t\} = \{\psi_9,\ \psi_{18},\ \psi_{27},\ \psi_{36},\ \psi_{45},\ \psi_{54},\ \psi_{63}\}. \tag{7.31}$$

Now, the matrix T_t can be constructed using the expressions for the wavefunctions on the right side of the system

$$T_t = \left[\begin{array}{ccc} \phi_1(y_9)e^{ik_1 x_b} & \cdots & \phi_{N_{mode}}(y_9)e^{ik_{N_{mode}} x_b} \\ \cdots & \ddots & \cdots \\ \phi_1(y_{63})e^{ik_1 x_b} & \cdots & \phi_{N_{mode}}(y_{63})e^{ik_{N_{mode}} x_b} \end{array}\right]. \tag{7.32}$$

Here, $x_b = x_{max}$ has been used for compactness.

The elements of T_{ar} and T_t must be placed into the correct locations in the full transformation matrix T based on the renumbering of the nodal variables. After the transformation matrix is constructed, the action integral A_0 can be written in the form

$$A_0 = \left[a_i^*\ r_i^*\ t_i^*\ \tilde{\phi}^*\right] \left[\begin{array}{c} T^\dagger M T \end{array}\right] \left[\begin{array}{c} a_i \\ r_i \\ t_i \\ \tilde{\phi} \end{array}\right]. \tag{7.33}$$

The functional variation with respect to the ϕ^* nodal values is now a variation with respect to the $\tilde{\phi}^*$ nodal values and the amplitudes a_i^*, r_i^*, t_i^*. At this stage, the surface terms can be added to the global matrix, but first they must be written in terms of the amplitudes rather than the nodal variables.

7.5 Surface terms

The surface terms which need to be added to the global matrix are

$$\Gamma = -\left(\frac{\hbar^2}{2m}\right) \oint \phi^* \nabla \phi \cdot \hat{n} dl, \tag{7.34}$$

where the integration is to be performed over the entire boundary of the system. Since the wavefunction is zero on the top (Γ_2) and bottom (Γ_0) edges of the system, the surface integral is composed of two parts from the left and right sides of the system

$$\Gamma = -\left(\frac{\hbar^2}{2m}\right) \left[\underbrace{\int_{y_{min}}^{y_{max}} \phi_L^* \nabla \phi_L \cdot \hat{n} dl}_{\Gamma_L(x=0)} + \underbrace{\int_{y_{min}}^{y_{max}} \phi_R^* \nabla \phi_R \cdot \hat{n} dl}_{\Gamma_R(x=L)} \right]. \tag{7.35}$$

On the left side of the system $dl = -dy$ and $\hat{n} = -\hat{x}$, which leads to the following expression

$$\Gamma_L = \left[-\left(\frac{\hbar^2}{2m}\right) \int_{y_{min}}^{y_{max}} \phi_L^* \frac{\partial \phi_L}{\partial x} dy \right]_{x_{min}}. \tag{7.36}$$

The wavefunction on the left side is

$$\phi_L = \sum_{n=1}^{N_{mode}} \left[a_n \exp(i k_n x_{min}) + r_n \exp(-i k_n x_{min}) \right] \phi_n(y), \tag{7.37}$$

as defined previously. Substituting this into the expression for Γ_L yields the following relation

$$\Gamma_L = -\left(\frac{\hbar^2}{2m}\right) \int_{y_{min}}^{y_{max}} dy \sum_{n=1}^{N_{mode}} \sum_{m=1}^{N_{mode}} \Big[\left[a_n^* \exp(-i k_n^* x_{min}) + r_n^* \exp(i k_n^* x_{min}) \right]$$
$$\times (i k_m) \left[a_m \exp(i k_m x) - r_m \exp(-i k_m x) \right] \Big] \times \phi_n^*(y) \phi_m(y). \tag{7.38}$$

Bringing the integral into the summation and employing the orthogonality of the transverse modes $\phi_m(y)$, only terms for which $n = m$ survive in the summation

$$\Gamma_L = -\left(\frac{\hbar^2}{2m}\right) \sum_{n=1}^{N_{mode}} \sum_{m=1}^{N_{mode}} \sum_{n=1}^{N_{mode}} \Big\{ \left[a_n^* \exp(-i k_n^* x_{min}) + r_n^* \exp(i k_n^* x_{min}) \right]$$
$$\times (i k_n) \left[a_n \exp(i k_n x) - r_n \exp(-i k_n x) \right] \Big\}. \tag{7.39}$$

Expanding the product yields

$$\Gamma_L = -\left(\frac{\hbar^2}{2m}\right) \sum_{n=1}^{N_{mode}} ik_n \left[a_n^* a_n \exp\left[i(k_n - k_n^*)x_{min}\right] \right.$$
$$- a_n^* r_n \exp\left[-i(k_n + k_n^*)x_{min}\right]$$
$$+ r_n^* a_n \exp\left[i(k_n + k_n^*)x_{min}\right]$$
$$\left. - r_n^* r_n \exp\left[-i(k_n - k_n^*)x_{min}\right] \right]. \quad (7.40)$$

The left side surface terms can now be written in matrix notation as

$$\Gamma_L = -\left(\frac{\hbar^2}{2m}\right) [\vec{a}^* \ \vec{r}^*] \begin{bmatrix} L_{11} & L_{12} \\ L_{21} & L_{22} \end{bmatrix} \begin{bmatrix} \vec{a} \\ \vec{r} \end{bmatrix}, \quad (7.41)$$

where each of the entries in the 2×2 matrix L is a matrix of size $N_{mode} \times N_{mode}$. The vectors $\vec{a}$ and $\vec{r}$ contain the incident and reflected amplitudes. The four matrices L_{ij} are

$$L_{11} = \begin{bmatrix} ik_1 \exp\left[i(k_1 - k_1^*)x_{min}\right] & & 0 \\ & \ddots & \\ 0 & & ik_{N_{mode}} \exp\left[i(k_{N_{mode}} - k_{N_{mode}}^*)x_{min}\right] \end{bmatrix},$$

$$L_{22} = \begin{bmatrix} -ik_1 \exp\left[-i(k_1 - k_1^*)x_{min}\right] & & 0 \\ & \ddots & \\ 0 & & -ik_{N_{mode}} \exp\left[-i(k_{N_{mode}} - k_{N_{mode}}^*)x_{min}\right] \end{bmatrix},$$

and

$$L_{12} = \begin{bmatrix} -ik_1 \exp\left[i(k_1 + k_1^*)x_{min}\right] & & 0 \\ & \ddots & \\ 0 & & -ik_{N_{mode}} \exp\left[-i(k_{N_{mode}} + k_{N_{mode}}^*)x_{min}\right] \end{bmatrix}$$

$$L_{21} = \begin{bmatrix} ik_1 \exp\left[i(k_1 + k_1^*)x_{min}\right] & & 0 \\ & \ddots & \\ 0 & & ik_{N_{mode}} \exp\left[i(k_{N_{mode}} + k_{N_{mode}}^*)x_{min}\right] \end{bmatrix}.$$

The surface terms on the right-hand side are much simpler because there exists only a single type of amplitude (transmission amplitudes). On the right side of the system $dl = dy$ and $\hat{n} = \hat{x}$, which leads to the following expression

$$\Gamma_{\mathbb{R}} = \left[-\left(\frac{\hbar^2}{2m}\right) \int_{y_{min}}^{y_{max}} \phi_{\mathbb{R}}^* \frac{\partial \phi_{\mathbb{R}}}{\partial x} dy \right]_{x_{max}}. \quad (7.42)$$

The wavefunction on the right side is

$$\phi_{\mathbb{R}} = \sum_{n=1}^{N_{mode}} t_n \exp(ik_n x)\phi_n(y), \quad (7.43)$$

as defined previously. Substituting this into the expression for $\Gamma_{\mathbb{R}}$ yields the following expression

$$\Gamma_{\mathbb{R}} = -\left(\frac{\hbar^2}{2m}\right)\int_{y_{min}}^{y_{max}} \sum_{n=1}^{N_{mode}}\sum_{m=1}^{N_{mode}} t_n^* t_m ik_m \exp\left[i(k_m - k_n^*)x_{max}\right] \times \phi_n^*(y)\phi_m(y)dy. \tag{7.44}$$

Bringing the integral into the summation yields the following result

$$\Gamma_{\mathbb{R}} = -\left(\frac{\hbar^2}{2m}\right)\sum_{n=1}^{N_{mode}}\sum_{m=1}^{N_{mode}} t_n^* t_m ik_m \exp\left[i(k_m - k_n^*)x_{max}\right] \times \int_{y_{min}}^{y_{max}} \phi_n^*(y)\phi_m(y)dy. \tag{7.45}$$

Due to the orthogonality of the transverse modes, only terms for which $n = m$ remain in the summation

$$\Gamma_{\mathbb{R}} = -\left(\frac{\hbar^2}{2m}\right)\sum_{n=1}^{N_{mode}} t_n^* t_n ik_n \exp\left[i(k_n - k_n^*)x_{max}\right]. \tag{7.46}$$

The right side surface terms can now be written in matrix notation as

$$\Gamma_L = -\left(\frac{\hbar^2}{2m}\right)\left[\vec{t}^{\,*}\mathcal{R}\vec{t}\,\right], \tag{7.47}$$

where $\vec{t}$ is a vector with N_{mode} components, and the matrix $\mathcal{R}$ is a matrix of size $N_{mode} \times N_{mode}$ given by

$$\mathcal{R} = \begin{bmatrix} ik_1 \exp\left[i(k_n - k_n^*)x_{max}\right] & & 0 \\ & \ddots & \\ 0 & & ik_{N_{mode}} \exp\left[i(k_{N_{mode}} - k_{N_{mode}}^*)\right] \end{bmatrix}. \tag{7.48}$$

At this stage, both surface terms can be placed in a global matrix as follows

$$A_{surface} = \begin{bmatrix} a_i^* & r_i^* & t_i^* & \tilde{\phi}^* \end{bmatrix} \begin{bmatrix} L_{11} & L_{12} & 0 & 0 \\ L_{21} & L_{22} & 0 & 0 \\ 0 & 0 & \mathcal{R} & 0 \\ 0 & 0 & 0 & 0 \end{bmatrix} \begin{bmatrix} a_i \\ r_i \\ t_i \\ \tilde{\phi} \end{bmatrix}. \tag{7.49}$$

Denoting the matrix in the above expression $M_{surface}$, the total action can now be written as

$$\begin{aligned} A_0 &= A_0 + \mathcal{A}_{surface} \\ &= \begin{bmatrix} a_i^* & r_i^* & t_i^* & \tilde{\phi}^* \end{bmatrix} \begin{bmatrix} & T^\dagger MT + M_{surface} & \end{bmatrix} \begin{bmatrix} a_i \\ r_i \\ t_i \\ \tilde{\phi} \end{bmatrix}. \end{aligned} \tag{7.50}$$

Denoting the matrix in the above equation by K and taking a variation of the action with respect to a_i^*, r_i^*, t_i^*, and ϕ^* leads to the following system of equations

$$\begin{bmatrix} & & \\ & K & \\ & & \end{bmatrix} \begin{bmatrix} a_i \\ p_i \\ t_i \\ \tilde{\phi} \end{bmatrix} = 0. \tag{7.51}$$

7.6 Applying BCs on the nodal/modal variables

In order to solve the system of equations in the last section, we must supply the incident amplitudes a_i. The variables a_i are given and hence can be eliminated from the left side of the system of equations by transferring the information associated with them to the right-hand side. The system of linear equations is then

$$\left[\begin{array}{ccc|c} 1 & & & \\ & \ddots & & \mathbf{1} \\ & & 1 & \\ \hline & \mathbf{0} & & \tilde{\mathbf{K}} \end{array}\right], \tag{7.52}$$

where the variables A_j represent the specified value of the amplitude, and the matrix $\tilde{K}$ is the sub-matrix of K corresponding to the variables r_i, t_i, and $\tilde{\phi}$. The rows and columns corresponding to the variables a_i have been zeroed with ones placed in the diagonal entries. The same process can be used to set the wavefunction to zero on the top and bottom of the system. Since we are setting the wavefunction at those nodes to zero there is no addition to the right-hand side. The rows and columns must still be zeroed and ones placed in the corresponding diagonal entries.

7.7 Probability currents from nodal solutions

After applying BCs and solving the system of equations, the solution will contain the modal amplitudes a_i, r_i, and t_i, as well as the wavefunction values for those nodes which do not lie on the ports. In order to get the wavefunction values for all nodes in the system, the solution vector must be transformed using the same transformation as was used to transform the global matrix

$$\begin{bmatrix} \phi \end{bmatrix} \begin{bmatrix} & & \\ & T & \\ & & \end{bmatrix} \begin{bmatrix} a_i \\ r_i \\ t_i \\ \tilde{\phi} \end{bmatrix}. \tag{7.53}$$

Now that the nodal values of the wavefunction are known, the solution can be reconstructed using the same interpolation polynomials in each element. In addition, the probability current can be calculated using the equation

$$\vec{J}(x,y) = \frac{\hbar}{2mi}(\phi^* \nabla \phi - \phi \nabla \phi^*). \tag{7.54}$$

The incident current is given by

$$\begin{aligned} J_{\text{inc}} =& \frac{\hbar}{2mi} \sum_n |a_n|^2 \left[\int_{y_{min}}^{y_{max}} dy\, \phi^*(y)\phi(y) \right] \\ & \times \{ik_n \exp[i(k_n - k_n^*)x_{min}] + ik_n^* \exp[i(k_n - k_n^*)x_{min}]\}. \end{aligned} \tag{7.55}$$

Since the evanescent modes do not contribute to the current $k_n - k_n^* = 0$, and the integral over the transverse orthonormal eigenfunctions (the y-integral) gives unity, the above expression reduces to

$$J_{\text{inc}} = \frac{\hbar}{m} \sum_n k_n |a_n|^2, \tag{7.56}$$

where the above summation excludes the evanescent modes. In a similar manner, the reflected and transmitted currents are

$$J_{\text{refl}} = \frac{\hbar}{m} \sum_n k_n |r_n|^2, \tag{7.57}$$

and

$$J_{\text{trans}} = \frac{\hbar}{m} \sum_n k_n |t_n|^2. \tag{7.58}$$

The transmission coefficient is defined to be

$$T = \frac{J_{\text{trans}}}{J_{\text{inc}}} = \frac{\sum_n k_n |t_n|^2}{\sum_n k_n |a_n|^2}, \tag{7.59}$$

and the reflection coefficient is defined to be

$$R = \frac{J_{\text{ref}}}{J_{\text{inc}}} = \frac{\sum_n k_n |r_n|^2}{\sum_n k_n |a_n|^2}, \tag{7.60}$$

where again, the evanescent modes are not included in the summations.

7.8 Landauer conductance

It is useful to provide a brief derivation of the Landauer formula for quantized conductance in ballistic electron transport through very narrow structures. Electrons in the devices of dimensions ≈ 50 nm or less, so-called "mesoscopic" devices, start behaving more like waves than particles. With the mean free path becoming comparable to the dimensions of the electronic device, an electron is transmitted through the device ballistically with no scattering. The electron waves will display phase coherence, however, and undergo quantum interference effects as defined by the *geometry* of the mesoscopic device. These features of electron transport change the usual macroscopic nature of resistance, leading to a quantization of resistance. We give a qualitative derivation of the Landauer relation defining the quantized conductance in terms of the transmission coefficient of electron waves through the device. The electron waves are guided through the device in a manner similar to optical waveguides.

We can visualize the structure used in the original experiments. A 2DEG created at the interface between a thin layer of GaAs and n-doped AlGaAs was used (see Fig. 7.1). By placing contacts for gates on the surface of the GaAs layer, as in Fig. 7.2, it is possible to measure the electrical resistance, R, of the 2DEG in its flow through the split-gate channel. When the channel is constricted by the application of a gate voltage, the resistance of the device is essentially the resistance of the channel. In the absence of any scattering, one might anticipate that the carriers will go through ballistically with no resistance at all; however, this is not so, and the resistance of the device remains finite.

One of the results of Landauer's theory of conductance in mesoscopic devices is that the device may be thought of as two contact reservoirs of carriers maintained at constant electro-chemical potentials μ_1 and μ_2 with the "quantum device" in between, as shown in Fig. 7.2. The current I through the device is proportional to $\mu_1 - \mu_2$, and its conductance $G = 1/R$ is given by

$$G = \frac{2e^2}{\hbar} \sum_{\mu v} T_{\mu v}. \tag{7.61}$$

The device is treated as a waveguide with input and output modes labeled by μ and v with the transmission coefficients $T_{\mu v}$ at the Fermi energy being labeled accordingly.

Experiments performed independently by Wharam et al., [3] and by van Wees et al., [4] confirmed this remarkable quantization of resistance. Here we follow the arguments given by Wharam and by Sharvin [5] to derive this result. The waveguide of Fig. 7.2, which is an idealization of an actual experimental structure, has transverse modes, labeled by and index μ, as discussed earlier, having energy

$$E_y^{(\mu)} = \frac{\hbar^2 k_\mu^2}{2m^*} = \frac{\hbar^2}{2m^*} \frac{\pi^2 \mu^2}{L^2}. \tag{7.62}$$

Here L is the transverse width of the waveguide and m^* is the electron's effective mass in the medium. The current is as given in subband μ associated with the energy of motion along the lateral direction of the waveguide is given by

$$I_\mu = n_\mu e \delta v_\mu \tag{7.63}$$

where n_μ is *half* of the number of carriers per unit length in the subband, since only half the electrons have wavevectors in the $+x$ direction. In the above equation, e is the electron charge and δv_μ is the increase in the velocity. The number of carriers n_μ

contributing to the current, including a spin degeneracy factor of 2, is

$$n_\mu = \frac{2}{2\pi}\int_0^{k_{max}} dk = \frac{2m^* v_F}{2\pi\hbar}. \tag{7.64}$$

The change in kinetic energy is given in terms of the applied voltage as

$$eV = \mu_1 - \mu_2 = \frac{1}{2}m^*(v_F + \delta v_{\mu F})^2 - \frac{1}{2}m^* v_F^* \tag{7.65}$$

so that for $\delta v_{\mu F} \ll v_F$ we have

$$\delta v_{\mu F} = \frac{eV}{m^* v_{\mu F}}. \tag{7.66}$$

Now the current in the channel is given by

$$I = \frac{2e^2}{2\pi\hbar}V \tag{7.67}$$

and the conductance corresponding to carriers in one subband is

$$G_0 = \frac{2e^2}{h}. \tag{7.68}$$

This conductance is dependent only on fundamental physical constants and is the same for any transmission mode. The resistance standard defined by this relation has the value: $R = 1/G_0 = 25,812.8056 \pm 0.0012$ ohm. For transmission in more than one transverse mode of the waveguide, we can estimate the total conductance assuming that the transport of carriers in the various modes is independent of each other. In this case, the double sum over incoming and outgoing modes reduces to just one and we have

$$G = \frac{2e^2}{h}N \tag{7.69}$$

for N channels.

The above derivation leads to a resistance quantization essentially because there is a cancellation of the velocity dependence of the number density and the change in the velocity, in the expression for the current in one dimension. The resistance of the system may be thought of as arising from the contacts between the waveguide and the reservoirs at the two ends.

7.9 Further considerations

With the scattering from the straight waveguide being accounted for within the FEM, we can now extend the considerations to resonant 2D cavities with more than two ports. This is the advantage of the FEM in that the geometrical complexities are easily overcome through the discretization of the physical region. If we assume that the cavities are connected to straight waveguides where we know the modal structure of the incoming and outgoing waves, we now have a framework for calculating the electron propagation through such structures. An example of such a structure is shown in Fig. 7.7.

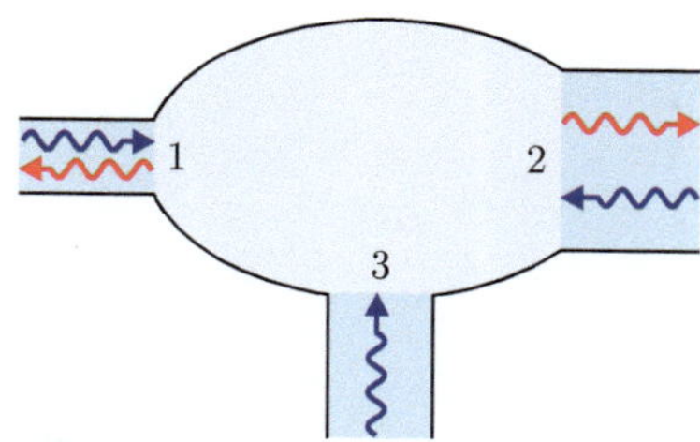

Figure 7.7 A resonant cavity with three ports. The electron waves enter via ports 1, 2, and 3, and are reflected back and transmitted out of ports 1 and 2.

References

[1] T. Ando, A. B. Fowler, and F. Stern, Rev. Mod. Phys. **54**, 437–672 (1982); "Electronic properties of two-dimensional systems."

[2] L. R. Ram-Mohan, *Finite Element and Boundary Element Applications to Quantum Mechanics* (Oxford UP, Oxford, UK, 2002).

[3] D. A. Wharam, T. J. Thronton, R. Newbury, M. Pepper, H. Ahmed, J. E. F. Frost, D. G. Hasko, D. C. Peacock, D. A. Ritchie, and G. A. C. Jones, *J. Phys. C:Solid State Phys.*, **21**, L209–L214 (1988); "One-dimensional transport and the quantisation of the ballistic resistance."

[4] B. J. van Wees, H. van Houten, C. W. J. Beenakker, J. G. Williamson, L. P. Kouwenhoven, D. van der Marel, and C. T. Foxon, *Phys. Rev. Lett*, **60**, 848–850 (1988); "Quantized conductance of point contacts in a two-dimensional electron gas;" B. J. van Wees, et al., *Phys. Rev. B*, **43**, 12431–12454 (1991); "Quantum ballistic and adiabatic electron transport studied with quantum point contacts."

[5] Yu V. Sharvin, *Zh. Eksp. Teor. Fiz.* **48**, 984 (1965) [Sov. Phys. *JETP* **21**, 655–656 (1965)]; "A possible method for studying Fermi surfaces." 655–656.

8
Quantum scattering in 2D waveguides

In this chapter:

- We explore the physical phenomena that can occur in 2D waveguides.
- We continue to develop the methodology presented in Chap. 6. The quantum waveguide, and the features associated with modes propagating in it, discussed in Chap. 7, are considered here using the method of sources and absorbers.
- The transverse confinement of the waves gives rise to energy quantization for that direction; the dispersion associated with subband edges defined by this quantization, act as thresholds for propagating waves down the waveguide. For a given incoming wave of energy E the energy subband edges *below* that energy give propagating modes; those subbands with energy *higher* than E give rise to evanescent modes.
- We show that in quantum waveguides the Fano resonance profile associated with propagating modes interacting with bound states in quantum wells or impurities in the waveguide have their analogs with evanescent modes as well. This is found to be an unusual and a universal effect for any attractive potential.
- Further, we show that quantum dots or attractive impurity potentials embedded in the interior of a quantum waveguide yields significantly large Seebeck coefficient (thermopower) and power factor. Hence, they are good candidates for enhancing the thermoelectric energy conversion efficiency.
- We study the effect of waveguide tapering on transport properties and on the transmission coefficients. We propose a nanoscale current rectification device in two-dimensions using tapered quantum waveguides.
- We show the quantitative effects of the curvature on the 2D waveguide on transmission coefficients.

Finite Elements in Action. L. Ramdas Ram-Mohan, Oxford University Press. © L. Ramdas Ram-Mohan (2026).
DOI: 10.1093/oso/9780199563487.003.0008

8.1 Introduction

This chapter should be considered as a direct continuation of Chap. 6 in terms of the method of sources and absorbers. There we had developed a non-asymptotic, variational description for quantum scattering theory with sources and absorbers, to study transport properties in one dimension. Here we consider scattering in quantum waveguides. The introduction of absorbing regions, "stealth regions" as we call them, for enclosing the active scattering region provides a unique way of simplifying and applying Dirichlet boundary conditions (BCs). This is in contrast with the need for applying Cauchy BCs for scattering phenomena. The absorption parameter in the stealth region is smoothly increased so that there is no unwanted back-scattering from the stealth regions. This scheme is analogous to the setting up of no-reflection coatings in optics and to the use of the so-called "perfectly matched layers" in electromagnetic field propagation.

Note that a straight waveguide is effectively a 1D system with modifications corresponding to the transverse confinement and a quantization of this confinement energy. While the outcomes will be richer than in the 1D scattering, the physical issues are similar. A more challenging consideration is to model the open domain using a circular (spherical) source and an annular (spherical) absorber in 2D and 3D, respectively. We comment on the open domain scattering and the issues associated with it in Sec. 8.8. The advantages of the use of sources and absorbers in the fuller theory of scattering in the open domain are very substantial.

In this method, the presence of near-field evanescent solutions are not neglected from consideration. Furthermore, the calculations can be focused on the physical region without going to the asymptotic limit. We show that it is straightforward to obtain accurate solutions throughout the region under consideration.

In the following, we show that for quantum waveguides with attractive impurity potentials, the phenomenon of Fano resonance in the transmission profiles is observed for both propagating and evanescent modes. These resonances arise from the interaction between the bound states and these modes. We are able to demonstrate that the evanescent modes also interact with the bound states to generate the Fano profiles.

We present results on transport in tapered waveguides with different tapering angles, and study their dependency on the directionality of incoming waves. The transmission coefficients behave asymmetrically with respect to the injection ports. This configuration is readily achievable using split-gate geometry on semiconductor inversion layers. Hence our predictions can be verified experimentally.

We calculate the conductance, Seebeck coefficient, and power factors of waveguides with impurities, providing experimentally measurable quantities. We have shown that quantum dots or attractive impurity potentials embedded in the interior of a waveguide are good candidates for enhancing the thermoelectric energy conversion efficiency since they yield large power factors around the subband minimum.

In the following, in Sec. 8.2, we calculate the Fano interference profile arising due to the interaction between bound states in attractive potentials with the propagating and *evanescent* waves. The dependence of transmission and reflection coefficients on the variation of the scattering potential is discussed in Sec. 8.3. Section 8.4 is devoted to calculations of the conductance and Seebeck coefficient for scattering in quantum waveguides. A novel result is the calculation of transmission coefficients in tapered waveguides presented in Sec. 8.5. The effect of curvature on scattering in bent waveguides is studied in Sec. 8.6. We compare our method with the Feshbach coupled-channel theory in Sec. 8.7. Concluding remarks are provided in Sec. 8.8.

8.2 Attractive potentials and Fano resonances

Fascinating physics arises when we apply our method of sources and absorbers [1, 2] to study scattering in quantum waveguides with attractive potential scattering centers. Let us consider a straight two-dimensional (2D) mesoscopic waveguide of width w with wave propagation along the x-axis. A hard wall confinement in the transverse y-direction is considered to be present, so that the energy for a wave in the nth subband is given by

$$E_n = E_{x,n} + \frac{n^2\pi^2\hbar^2}{2m^*w^2}, \tag{8.1}$$

where n is the integer mode number for the incoming wave, as discussed in Chap. 7. We refer to $E_{y,n} = n^2\pi^2\hbar^2/(2m^*w^2)$ as the subband minimum corresponding to the band index n. The dispersion relation for the wavevector k is given by

$$k^2 = k_{xn}^2 + \frac{n^2\pi^2}{w^2}, \tag{8.2}$$

where $k_{xn} = \sqrt{2m^*E_{x,n}/\hbar^2}$. The transmission and reflection coefficients between a propagating mode m and the incoming mode n are defined as [1, 2]

$$T_{nm} = \frac{k_{xm}\left|t_{nm}\right|^2}{k_{xn}a_n^2}; \quad R_{nm} = \frac{k_{xm}\left|r_{nm}\right|^2}{k_{xn}a_n^2}, \tag{8.3}$$

where t_{nm}, r_{nm}, and a_n are the transmitted, reflected, and incoming amplitudes, respectively. In the presence of scattering centers, the total scattered wavefunction will have contributions from both propagating and evanescent modes. To study the contributions from evanescent modes, we define [1, 2] the reflection and transmission coefficients in an analogous manner as

$$\begin{aligned}\tilde{T}_{nm} &= \frac{t_{nm}^2}{a_n^2} \sim \exp\left(-2K_{xm}\left|x\right|\right);\\ \tilde{R}_{nm} &= \frac{r_{nm}^2}{a_n^2} \sim \exp\left(-2K_{xm}\left|x\right|\right),\end{aligned} \tag{8.4}$$

where $K_{xn} = \sqrt{n^2\pi^2/w^2 - k^2}$ are the evanescent wavevectors. These coefficients represent the probability strength of each evanescent mode.

The waveguide empty of any scattering centers allows us to identify the range over which the absorption coefficient for the stealth regions are to be increased from zero to a maximum value. The criterion is that the amplitude of the wavefunction decreases to $\sim 10^{-8}$, corresponding to a probability current reducing to the level of machine precision in our double-precision calculations. The lowest energy mode with a larger wavelength requires the largest length for the wave to attenuate to "zero." This calibration is applicable for all higher energies.

Only the propagating modes contribute to the probability current, and the evanescent modes being purely real functions will have vanishing current contributions at large distances from the scattering centers. Therefore, through the conservation of probability current, we obtain the relation $\sum_m (T_{nm} + R_{nm}) = 1$. We note that the transmission and reflection coefficients for the evanescent modes (given in Eq. (8.4)) do not satisfy any summation rule of this type.

As an example of a multiple scattering problem, we consider a 2D waveguide with three circular impurities each of radius 30 Å and a constant potential, $V_0 = -50\,\text{meV}$. In Fig. 8.1, we observe an asymmetric Fano resonance [3] transmission profile for the propagating modes due to interference between scattering states in one subband and a bound state supported by a different subband, [4–6] an effect analogous to that of atomic autoionization. [7] Multiple Fano resonances observed in Fig. 8.1 are a special feature of multiple scattering.

The Fano resonance profile in waveguides has been qualitatively explained for the propagating states with an attractive δ-potential scatterer through two-band models. [4, 5] Such Fano resonances have been reported for the special cases of scattering from a rectangular impurity potential, [8, 9] from two antidots, [10] the Pöschl-Teller [11]

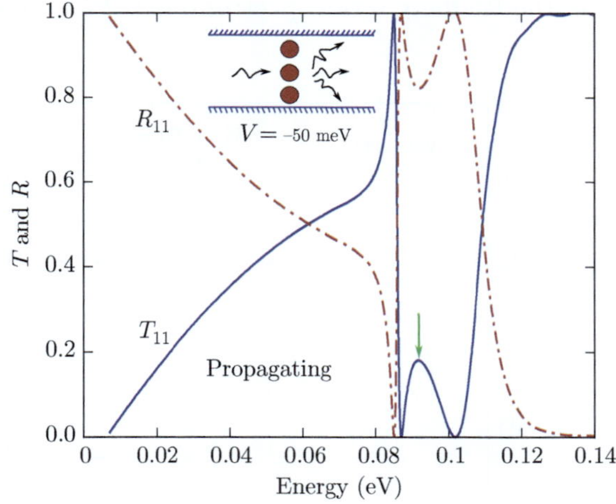

Figure 8.1 The transmission (T_{11}) and reflection (R_{11}) coefficients are shown as a function of energy for the propagating mode $n = 1$, with an incoming wave from the same mode, for scattering through three vertically aligned circular impurities of radius 3 nm whose centers are 10 nm apart, and potential depth of $-50\,\text{meV}$. The zero-transmission energy for T_{11} (E_R), the corresponding Fano q factor, and the line width Γ are given by $E_{R_1} = 0.087\,\text{eV}$; $q_{R_1} = -1.1607$; $\Gamma_{R_1} = 0.001\,\text{eV}$, and $E_{R_2} = 0.102\,\text{eV}$; $q_{R_2} = 0.3229$; $\Gamma_{R_2} = 0.009\,\text{eV}$, respectively. (As in Ref. [1].)

attractive impurity potential, [12, 13] and an impurity in the bottleneck conduction channel. [14] Experimentally, asymmetric Fano resonances have been observed in single-electron transistors with droplets of confined electrons, [15] crossed carbon nanotubes, [16] and in quantum wires coupled to a quantum dot. [17, 18] Through the Feshbach coupled-channel [8, 9, 19] theory and an S-matrix approach, it has been shown that one can derive a Breit-Wigner type formula with an additional asymmetry parameter for transmission coefficients of propagating modes. [20] However, we observe a characteristic resonance profile *even for the evanescent modes in either direction* which reach a maximum at the resonance as seen in Fig. 8.2. This is due to the interference between the quasi-bound metastable state and the evanescent modes. Notice that only the odd numbered modes contribute since the potential has C_{1h} symmetry, and the evanescent coefficients reach a maximum at the 5th subband minimum (155.94 meV), as expected. [1] Such a profile was not found in all earlier theoretical simulations since they used asymptotic BCs. [21] In Sec. 8.7, we briefly discuss the Feshbach coupled-channel approach to derive the transmission coefficients with the Fano form and compare the results with our method. We find that there are small errors for the Fano parameters obtained through the Feshbach approach even for the simplest geometries. Moreover, the Feshbach approach is not amenable to a closed form solution, other than for very few potentials generated by simple geometrical structures. [9]

In Fig. 8.1, we observe the Fano profile reversal at energy $E = 0.0917\,\text{eV}$ (marked by an arrow in the plot). This can be quantified by noting the change of sign in the q-factor from -1.1607 to 0.3229. This q-reversal phenomenon has been widely studied in atomic physics, where the weak mixing of the interloper levels are attributed to such q-reversal observed in the Rydberg series spectrum. [22, 23] Connerade and Lane [23]

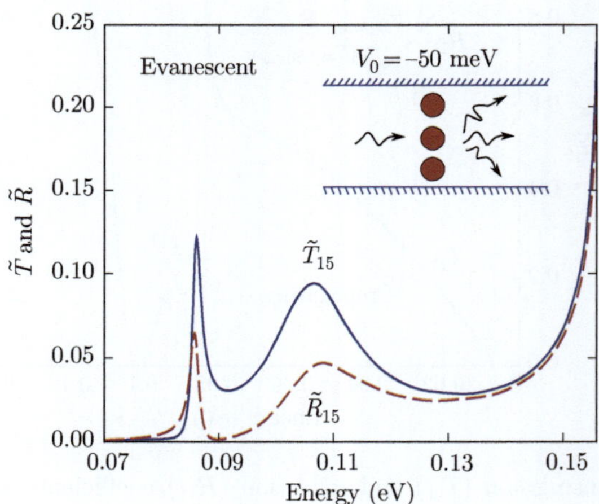

Figure 8.2 The transmission ($\tilde{T}_{15}$) and reflection ($\tilde{R}_{15}$) coefficients for evanescent modes are shown as a function of energy for the evanescent mode $n = 5$, with an incoming wave from the mode 1 at a distance $l = 10\,\text{nm}$, for scattering through vertically aligned three circular impurities of radius 3 nm whose centers are 10 nm apart, and whose potential depth is $V_0 = -50\,\text{meV}$. (From Ref. [1].)

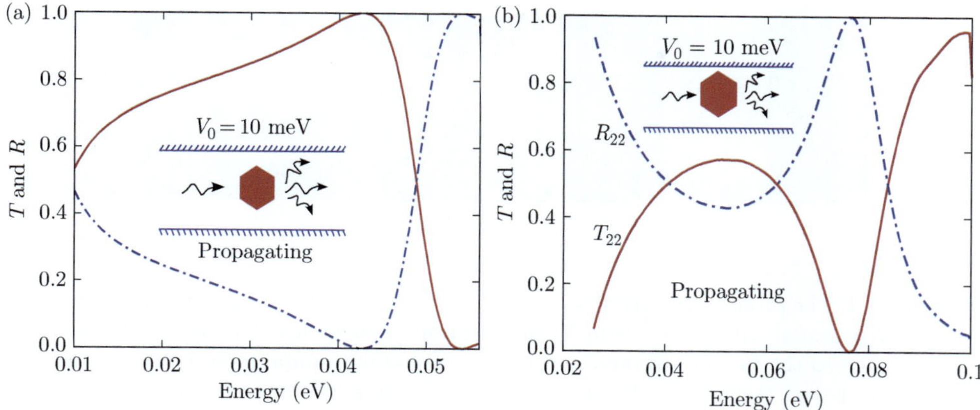

Figure 8.3 The transmission and reflection coefficients for propagating modes (a) T_{11} and R_{11}, and (b) T_{22} and R_{22} in transport through a hexagonal impurity with $V_0 = -10\,\text{meV}$ are plotted. Each side of the hexagon equals 5 nm. The zero-transmission energy (E_R), the Fano q-factor, and the line width Γ are given by (a) $E_R = 0.055\,\text{eV}$; $q = -1.1384$; $\Gamma = 0.006\,\text{eV}$, and (b) $E_R = 0.0764\,\text{eV}$; $q = 0.3161$; $\Gamma = 0.008\,\text{eV}$. (From Ref. [1].)

have derived several conditions for such q-reversal. This phenomenon has also been identified in the transmission spectrum of quantum dots connected with 1D-channels. [18, 24] Such a change in parity is not observed for the evanescent modes which reach a maximum around the energy corresponding to a zero transmission.

In Fig. 8.3, we notice that for a hexagonal impurity,† the Fano resonance profile changes parity while switching from the incoming mode 1 to the incoming mode 2. This is yet another type of q-reversal which occurs due to the change in symmetry of the incoming mode.

In Fig. 8.4, we plot the probability current at $E = 47\,\text{meV}$ for scattering through a hexagonal impurity with potential $V_0 = -10\,\text{meV}$ for an incoming mode 1. We observe that the current direction is unaltered since $T_{11} = 1$ at this energy due to the Fano resonance, as seen in Fig. 8.3(a). The magnitude of the probability current density has *diffraction spikes* due to the hexagonal shape of the impurity scattering potential, as seen in Fig. 8.4. Notice that when we have a repulsive hexagonal barrier, the probability current density will flow around the barrier. [1] On the other hand, with an attractive hexagonal potential, the current will reach a maximum at the center of the potential and form diffraction spikes emanating from its vertices.

In Fig. 8.5, we show the decay of evanescent modes $\tilde{T}_{24}$ and $\tilde{R}_{24}$ with distance for scattering through a hexagonal defect with $V_0 = -10\,\text{meV}$, at energy $E = 76\,\text{meV}$ at which $T_{22} = 0$ (see Fig. 8.3 (b)). Through exponential curve fitting, we notice that the evanescent modes have large amplitudes around the scattering center. Note that the

†All phenomena discussed in this chapter are valid essentially irrespective of the geometry of the potential. We have used a nontrivial hexagonal, circular, or rectangular geometry to illustrate the basic behavior through our numerical work.

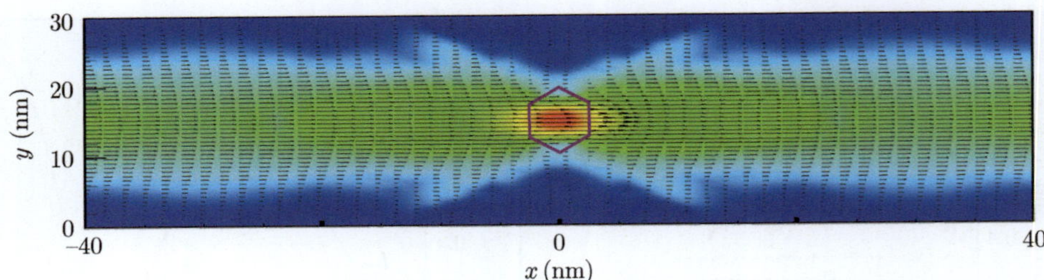

Figure 8.4 The probability current density at $E = 47\,\text{meV}$ for transport through a hexagonal impurity potential with $V_0 = -10\,\text{meV}$ for an incoming mode $n = 1$ is shown. At this energy, $T_{11} = 1$ due to the Fano resonance, and hence we obtain an unperturbed current flow in the forward direction throughout. In this plot, the magnitude of the probability current density ranges from 0 to 0.065, which is represented by a continuous contour color coding varying from blue to red. (As in Ref. [1].)

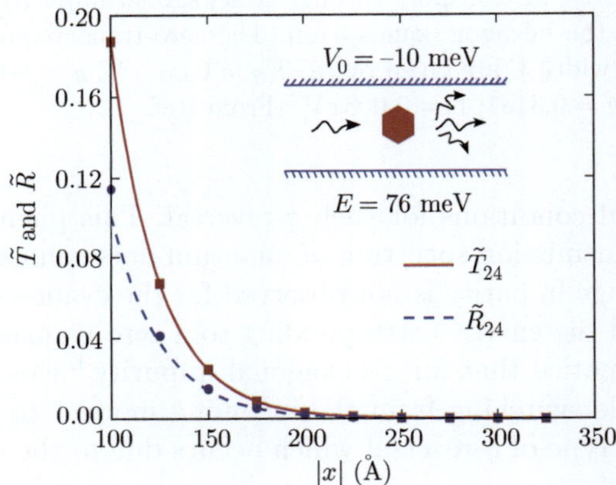

Figure 8.5 Decay of transmission ($\tilde{T}_{24}$) and reflection ($\tilde{R}_{24}$) coefficients for evanescent modes are shown as a function of x(Å) for $E = 76\,\text{meV}$ for scattering through a regular hexagonal impurity of side $= 50\,\text{Å}$. The incoming mode is $n = 2$ and the scattering potential is $V_0 = -10\,\text{meV}$. Here we carried out the modal analysis [1] at every 25 Å. Equations from the curve fitting are given by $\tilde{T}_{24} = 11.18650 \exp(-2 \times 0.02045\, x)$, and $\tilde{R}_{24} = 5.18618 \exp(-2 \times 0.01969\, x)$. Theoretically determined $K_{x5} = 0.02045/\text{Å}$, which is close to the curve fit values. (From Ref. [1].)

decaying behavior of the transmission and reflection coefficient observed in Fig. 8.5 is consistent with Eq. (8.4).

8.3 The effect of varying the potential on transmission

In this section, we discuss the effects of varying the potential on the transmission profile for both propagating and evanescent modes. We include the case of both attractive and repulsive scatterers. For a rectangular barrier, the coefficient T_{11} attains a resonance maximum which will shift toward the next allowed subband minimum ($E_{y,3} = 0.056\,\text{eV}$) with increase in the barrier height as shown in Fig. 8.6. We effectively have slowed down the waves in the barrier region by increasing the barrier height. Whereas for the evanescent waves, the maximum will be still at the next allowed subband minimum ($E_{y,3}$), and its maximum will increase with barrier height.

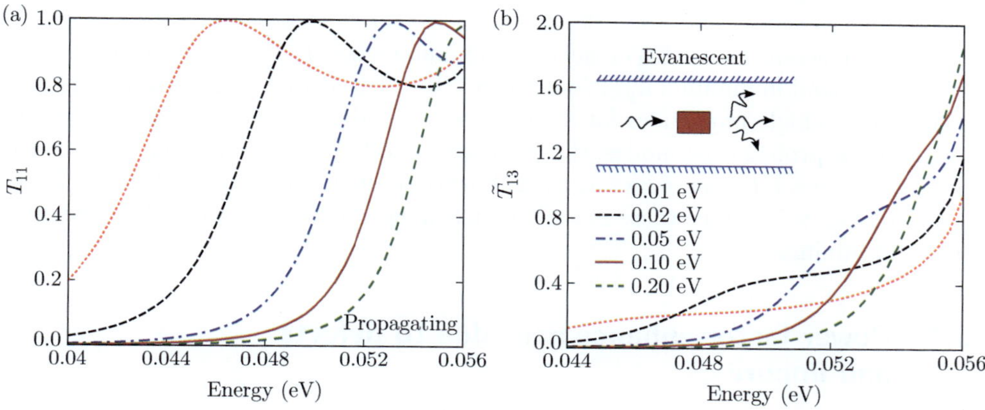

Figure 8.6 The transmission coefficients T_{11} and $\tilde{T}_{13}$ as a function of energy for a rectangular barrier with varying height are shown. A modal analysis [1] is done at $|x| = 10\,\text{nm}$ for the scattering coefficients. (From Ref. [2])

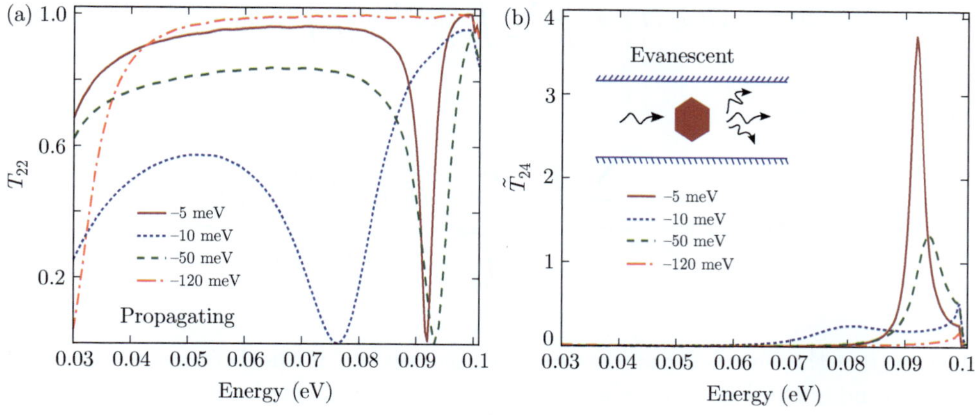

Figure 8.7 The transmission coefficients T_{22} and $\tilde{T}_{24}$ as a function of energy for a regular hexagonal impurity with a side length of 5 nm for different values of the potential are shown. A modal analysis is done at $|x| = 10\,\text{nm}$ for the scattering coefficients. (From Ref. [1].)

In Sec. 8.2, we discussed the emergence of the Fano profile for the transmission coefficients in the case of attractive scatterers. Here, we study the effect of varying the impurity potential for such a resonance. To illustrate this, we study the scattering through a hexagonal impurity. In Fig. 8.7, we plot coefficients T_{22} and $\tilde{T}_{24}$

as a function of energy for different values of the potential barrier. We see that for weaker attractive potentials ($V_0 = -5\,\text{meV}$), we have a strong resonance with $T_{22} = 1$ near $E_{y,4} = 99.80\,\text{meV}$ (see Fig. 8.7). The resonance value decreases for more negative potential values, since the bound states move further below the continuum. For $V_0 = -120\,\text{meV}$, the Fano resonance profile disappears. Additional bound states may appear, which can interact with scattering modes if we further deepen the well.

The evanescent mode ($\tilde{T}_{24}$) reaches a maximum when $T_{24} \approx 0$ which is much before the 4th subband minimum ($E_{y,4}$) (see Fig. 8.7(b)). There are two governing conditions which lead to high amplitudes for the evanescent modes; (i) at the transmission minimum for the propagating modes, the evanescent modes will have a peak, and (ii) they also tend to reach higher values at the subband minima. For $V_0 = -120\,\text{meV}$, $\tilde{T}_{24}$ has a much lower value since there is no Fano resonance, with the maximum value at the subband minimum.

8.4 Power factor enhancement due to defects and impurities

Consider transport through a multichannel lead connected to reservoirs at equilibrium attached to the waveguide on either end. The conductance G and the Seebeck coefficient (thermopower) S, for a given chemical potential μ_F and temperature T, calculated using the Landauer-Büttiker formalism [25–31] are given by

$$G(\mu_F, T) = \frac{2e^2}{h} \sum_i \int_0^\infty dE \left(-\frac{df}{dE}\right) T_i(E), \tag{8.5}$$

and

$$S(\mu_F, T) = \frac{k_B}{e} \frac{\sum_i \int_0^\infty dE \left(-\frac{df}{dE}\right) T_i(E) \left(\frac{E-\mu_F}{k_B T}\right)}{\sum_i \int_0^\infty dE \left(-\frac{df}{dE}\right) T_i(E)}, \tag{8.6}$$

where, f is the Fermi-Dirac distribution function, k_B is the Boltzmann constant, e is the fundamental electron charge, and T_i is the transmission probability from all channels to the channel i given by $T_i = \sum_j T_{ij}$. In the low temperature limit, these expressions reduce to a simpler form [32, 33] given by

$$\begin{aligned} G(\mu_F, T=0) &= \frac{2e^2}{h} \textstyle\sum_i T_i(\mu_F) = \frac{2e^2}{h} \sum_{i,j} T_{ij}(\mu_F), \\ S(\mu_F, T) &= \frac{(k_B \pi)^2}{3e} T \frac{d}{dE} (\ln G(E,0)) \bigg|_{E=\mu_F}. \end{aligned} \tag{8.7}$$

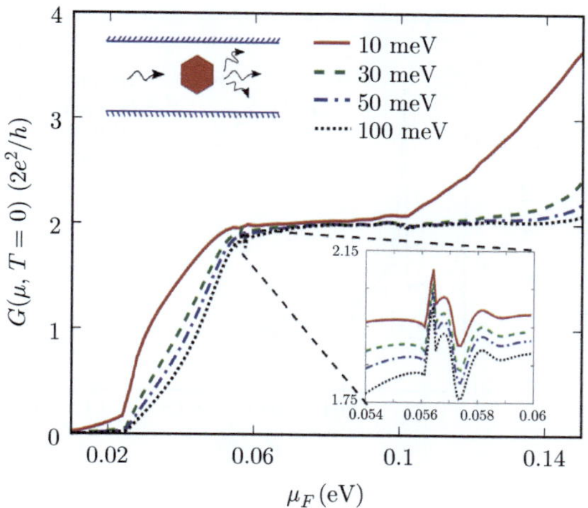

Figure 8.8 The conductance $G(\mu_F, T = 0)$ as a function of the chemical potential μ_F for scattering from a regular hexagonal barrier defect with a side of length 5 nm. We have shown the conductance for four different potential values. Within the inset, we have shown $G(\mu_F, T = 0)$ between the range $0.054 - 0.06$ eV. (From Ref. [1].)

We evaluate the conductance and the Seebeck coefficient as functions of the chemical potential in the case of attractive and repulsive scatterers for different values of the potential. It is essential to determine both G and S to characterize the energy conversion efficiency by the dimensionless figure of merit (ZT) given by

$$ZT = \frac{GS^2}{\kappa} T, \tag{8.8}$$

where κ is the thermal conductance of the material which includes phonon and electron contributions. In Fig. 8.8, we plot the conductance as a function of chemical potential for different heights of a hexagonal barrier. Conductance will no longer preserve the quantization, and it will be suppressed as we increase the barrier height. Near the subband minima, we observe the Ramsauer-Townsend type resonance due to the high probability of evanescent modes. [34, 35] In Fig. 8.9, we plot the thermopower as a function of the chemical potential for different heights of a hexagonal barrier. We see that the thermopower reaches high positive values at low energies around the subband minima. Increase in the barrier height can slightly increase the values of the Seebeck coefficient though its profile remains the same. The power factor (GS^2) will be close to 1 around the second subband minimum, and will not vary much with the barrier height (see Fig. 8.10).

In the case of attractive scatterers, resonant dips are observed for the conductance (see Fig. 8.11) at the subband minima, which are attributed to the formation of quasi-bound states whose energies are determined by finding poles of the Green's function for the system. [36, 37] We see that the conductance decreases and the resonant dips disappear as we deepen the negative potential. The quasi-bound states lead to both

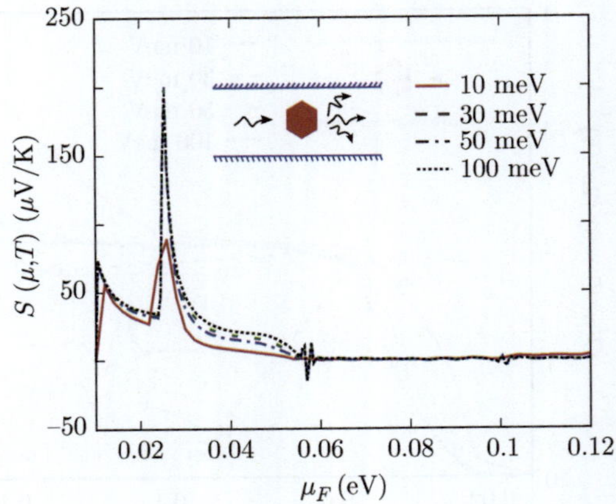

Figure 8.9 The Seebeck coefficient $S(\mu_F, T)$ (in units of μV/ K) as a function of the chemical potential μ_F for scattering from a regular hexagonal barrier defect with side length 5 nm evaluated at $T = 10$ K is shown. The thermopower for four different potential values are displayed. (From Ref. [1].)

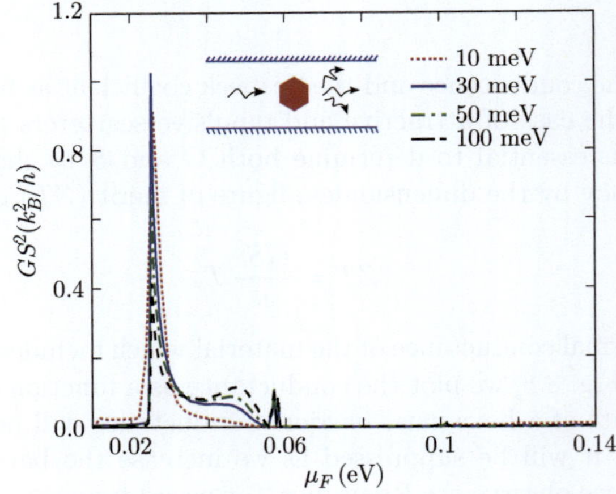

Figure 8.10 The power factor GS^2 (in units of k_B^2/h) as a function of the chemical potential μ_F for scattering from a regular hexagonal barrier defect with side length 5 nm evaluated at temperature $T = 10\,K$. We have shown the power factor for four different potential values. (From Ref. [1].)

positive and negative values for the Seebeck coefficient around the subband minima (see Fig. 8.12). Notice that the maximum value of the thermopower decreases as we deepen the negative potential.

From Fig. 8.13, we see that the power factor can reach large values around the subband minima in the case of impurity scatterers. We obtain multiple peaks at each subband minimum since the Seebeck coefficient has multiple resonances.

Transport in a waveguide, with attractive impurities or quantum dots being present, provides the means of obtaining large Seebeck coefficients and the power factor. This will be of interest in thermoelectric applications, [38, 39] and these can be extended to spatially confined nanostructures. Having an attractive impurity distribution surpasses the limits on the power factor asserted earlier [40] in the case of ballistic transport.

We note that the Landauer-Büttiker conductance formula in the present form will not take into account changes in the potential arising from a charge accumulation due to the evanescent modes. It would require a self-consistent solution [41] of the Schrödinger-Poisson equation to calculate the additional potential which depends on the probability density of carriers in the evanescent modes. This issue will be addressed separately in the near future.

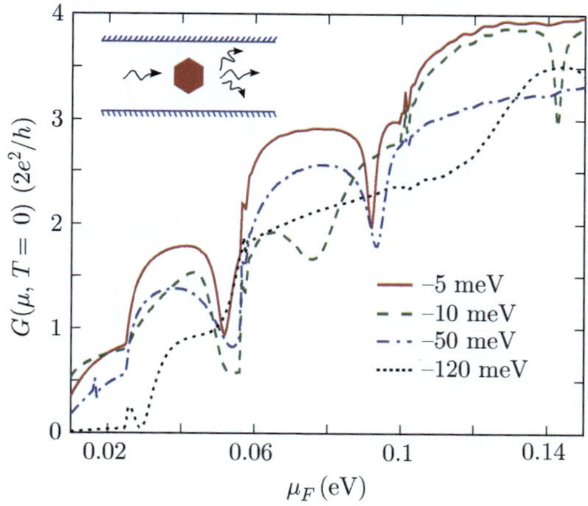

Figure 8.11 The conductance $G(\mu_F, T = 0)$ as a function of the chemical potential μ_F for scattering from a regular hexagonal impurity with a side length 5 nm is shown. The conductance for four different potential values is displayed. We observe the formation of resonant dips at the subband minimum which disappear for $V_0 = -120\,\text{meV}$. (From Ref. [1].)

8.5 Tapered waveguides

Next, we study the ballistic transport in a tapered waveguide. In this case, the energy levels for local subband minima at one end are lower or higher than the subbands at the other end. Hence, new propagating and evanescent modes will emerge in the scattered wavefunction. A tapered waveguide will still preserve the C_{1h} symmetry if the tapering angle is the same in either direction. Therefore, contributions from different modes still follow the selection rules. [1]

In Fig. 8.14, we plot the transmission and reflection coefficients for the tapered waveguide with a left end of width 50 nm, and the right end of width 10 nm. We choose our incoming energy below the first subband minimum (56.14 meV) of the right end. We perform the modal analysis [1] at either end of the waveguide to

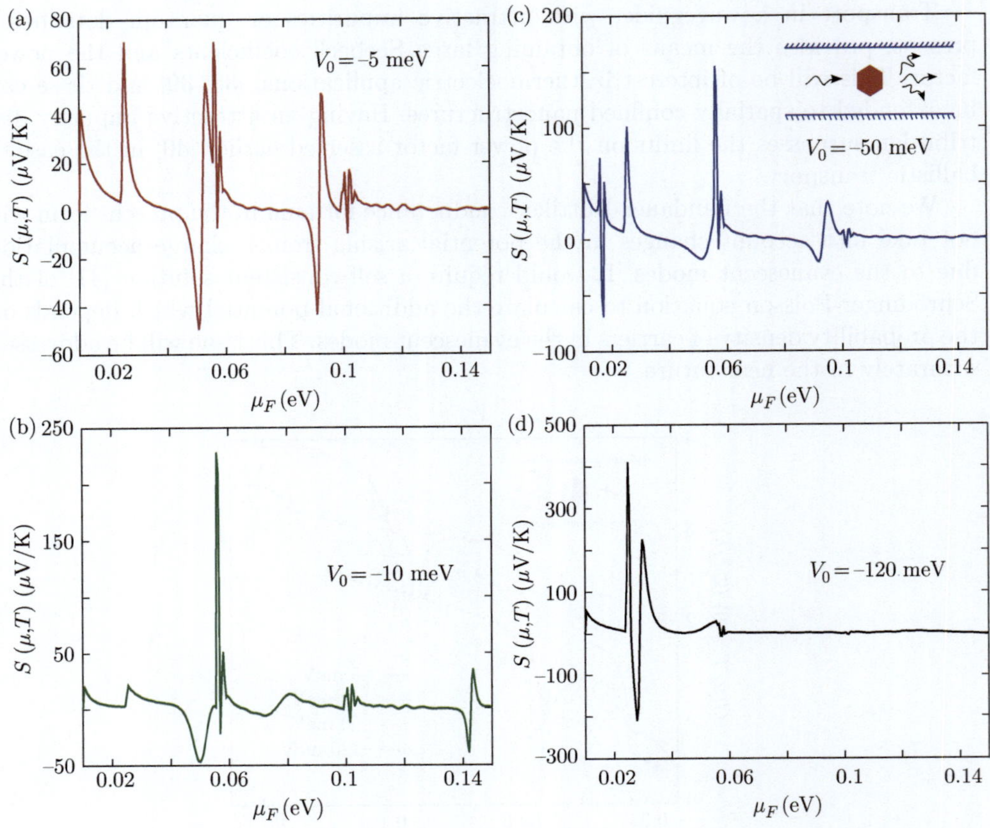

Figure 8.12 The Seebeck coefficient $S(\mu_F, T)$ (in units of μV/K) as a function of the chemical potential μ_F for scattering from a regular hexagonal impurity with side length 5 nm evaluated at $T = 10$ K is shown. $S(\mu_F, T)$ for four different potential values are displayed. We can obtain large Seebeck coefficient by deepening the value of negative potential. (From Ref. [1].)

obtain the scattering amplitudes. Since this set up does not allow any transmission of propagating waves below 56.14 meV, we observe only contributions for the current coming from the reflection coefficients R_{11} and R_{13}. However, we observe a fairly high amplitude for the evanescent transmission coefficient $\tilde{T}_{11}$ due to the tapering. Evanescent modes lead to electron localization in the waveguide, thereby altering the potential through a change in the local electron distribution. *Since we have evanescent modes only in the transmission direction, tapered waveguides provide the means of changing the conductance while keeping the current constant.*

Next, we study the transport from a smaller to a wider cross-sectional waveguide. In Fig. 8.15, we plot the transmission coefficients for three different tapering angles. For our choice of incoming energies, we observe only propagating modes since a wider cross-section has all subband minima below the incoming subband minimum for the smaller cross-section. We see that only the 1st, 3rd, and 5th modes contribute, again as per the selection rules. [1]

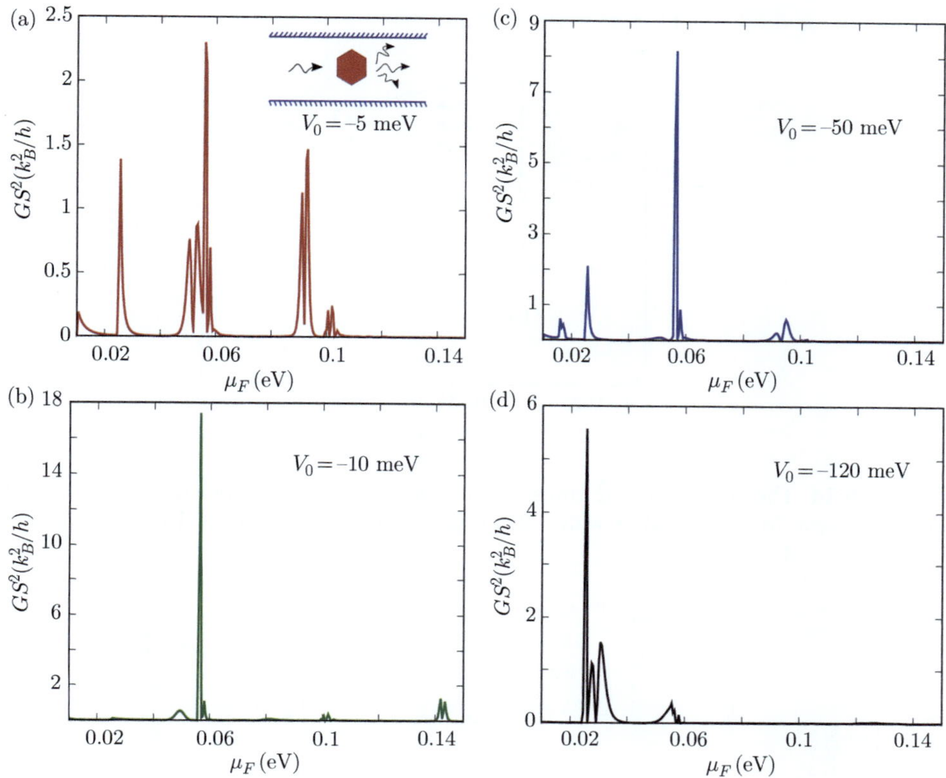

Figure 8.13 The power factor GS^2 (in units of (k_B^2/h)) as a function of the chemical potential μ_F for scattering from a regular hexagonal impurity with side length 5 nm evaluated at a temperature $T = 10\,K$ is shown. We have shown the power factor for four different potential values. We can obtain large power factors around the subband minima. (From Ref. [1].)

Though our incoming mode is $n = 1$, we observe other odd number modes because of the coupling generated to these modes from the tapering boundaries of the waveguide.

At $E_{y,3} = 56.14$ meV for the incoming mode, we observe a strong resonance in the transmission profiles. By increasing the tapering angle we can decrease the contribution from mode 1, while contributions from the 3rd and the 5th mode increase. *The tapered waveguide provides a new way of developing a nanoscale rectifier since the currents through it that are initiated at either end are inequivalent.* This is seen in Figs. 8.14 and 8.15.

8.6 Curved waveguides

While designing electronic circuits, invariably we will have channels with bends connecting different straight waveguides. Goldstone and Jaffe [42] have shown that the curved channels in 2D and twisted tubes in 3D can support bound states. Schult et al. [43] found the presence of bound states in a classically unbound system of

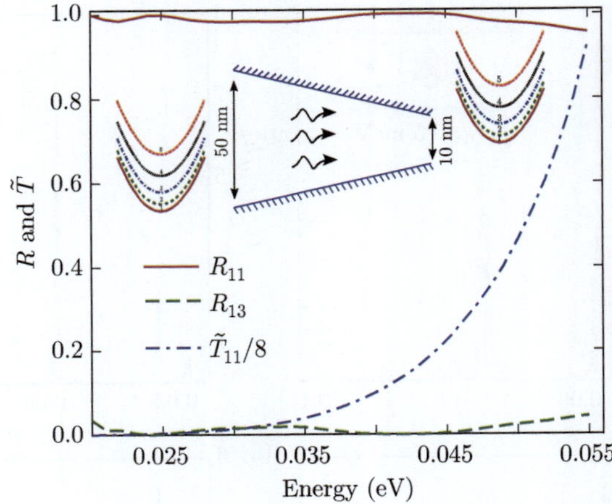

Figure 8.14 The reflection and transmission coefficients for a tapered waveguide with an incoming wave from the end of width 50 nm to the outgoing end of width 10 nm are shown. (From Ref. [1].)

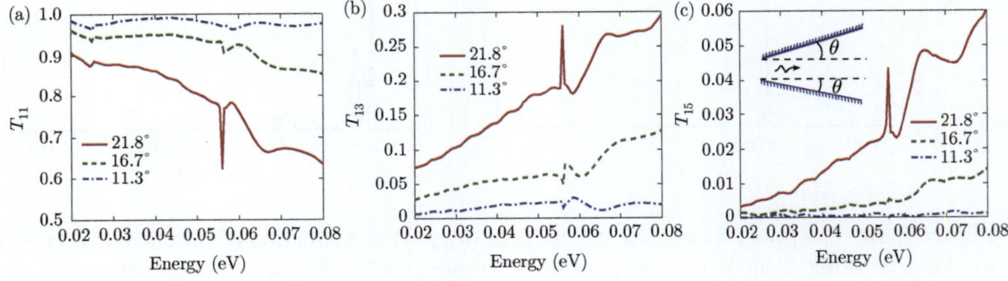

Figure 8.15 The transmission coefficients for the propagating modes (a) T_{11}, (b) T_{13}, and (c) T_{15} as a function of energy are plotted for a tapered waveguide going from the smaller (30 nm) to a wider cross-section for three different tapering angles. (From Ref. [1].)

crossed wires. We expect to have evanescent modes even in the ballistic transport through empty curved or crossed channels since the translational symmetry is broken. It has been shown that we can obtain a series of Fano lines in an empty curved waveguide due to curvature effects. [44] Olendski and Mikhailovska [46] have shown the presence of a Fano resonance profile when a quantum dot is embedded in a uniformly curved waveguide. Such calculations are done by solving the wave equation in the circular coordinate system, and the potential is taken to be circularly symmetric. In our method, we have no such constraints since we work with a discretized space. As an example we obtain the transmission coefficients for different bending angles in a curved waveguide with an embedded square quantum dot.

In Fig. 8.16, we plot the transmission coefficients T_{11}, $\tilde{T}_{12}$, $\tilde{T}_{13}$, and $\tilde{T}_{14}$ for scattering from a square quantum dot of dimensions 20×20 nm and a potential $V_0 = -10$ meV. The bent geometry of the waveguide does not have C_{1h} symmetry, hence all modes contribute to the wavefunction. Since the energy is taken to be below $E_{y,2} = 0.0249$ eV, we have only one propagating mode ($n = 1$). We see several transmission dips for T_{11} due to the Fano effect. As explained earlier, such a dip in the propagating mode leads to a high probability for evanescent modes. For a curvature angle of $\theta = 30^o$, $\tilde{T}_{12}$ will be as large as 85 at $E = 0.0126\,\text{eV}$. The probability of evanescent modes decreases with the curvature angle as seen in Figs. 8.16(b), (c), and (d). We note the formation of Fano line profiles even for the evanescent modes around the transmission minima.

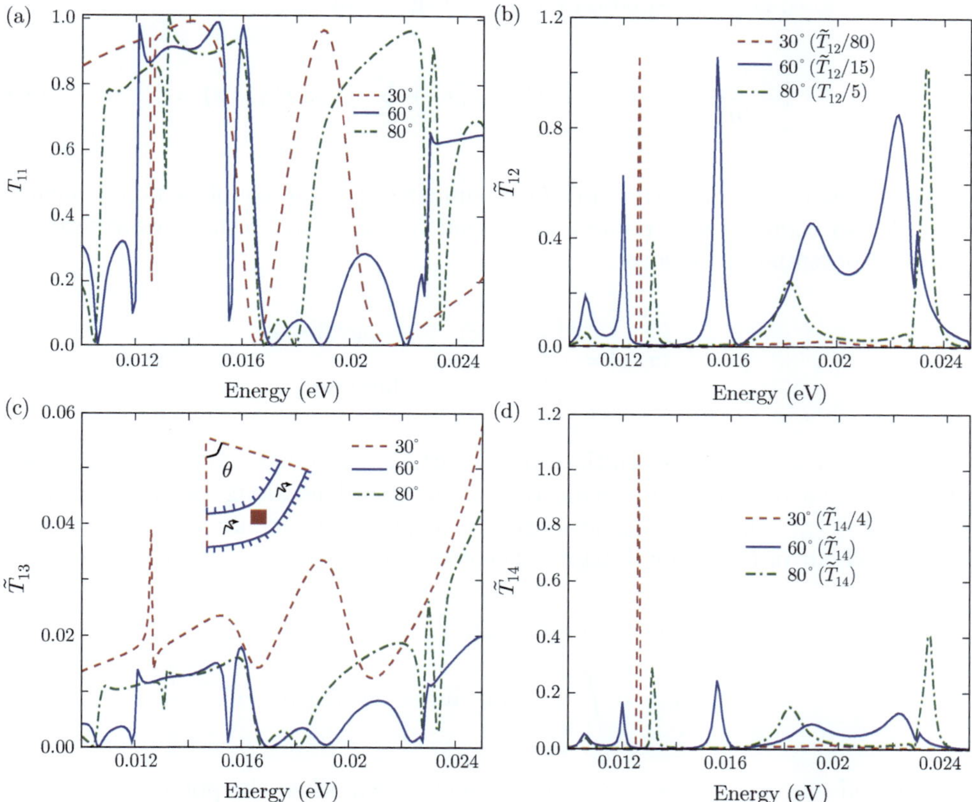

Figure 8.16 The transmission coefficients for the propagating mode (a) T_{11}, and for the evanescent modes (b) $\tilde{T}_{12}$, (c) $\tilde{T}_{13}$, and (d) $\tilde{T}_{14}$ are plotted as a function of energy for a curved waveguide of width 30 nm embedded with a square quantum dot of dimensions $20 \times 20\,\text{nm}$. The potential within the quantum dot $V_0 = -10\,\text{meV}$. (From Ref. [1].)

8.7 Comparison with Feshbach coupled-channel theory for Fano resonances

The Fano resonance for waveguide scattering with attractive scatterers are typically treated [8, 9, 20] within the Feshbach approach. [19] We compare the analytical expression for transmission coefficients obtained through this analysis and the numerically calculated ones with our approach. In a 2D straight waveguide, we expand the scattered wavefunction in the form

$$\psi(x, y) = \sum_{n=1}^{\infty} \chi_n(x) \sin\left(\frac{n\pi y}{w}\right), \tag{8.9}$$

and the coupled-channel equation for $\chi_n(x)$ is given by

$$\left[-\frac{\hbar^2}{2m}\frac{d^2}{dx^2} + (E_{y,n} - E)\right] \chi_n(x) + \sum_{m=1}^{\infty} U_{nm}(x)\chi_m(x) = 0, \tag{8.10}$$

where $U_{nm}(x) = \int_0^w dy \sin(n\pi/w)V_0(x, y)\sin(m\pi/w)$ is the coupling matrix element. In order to compare our method with other methods and results, we consider the *simplest* potential of the form

$$V_0(x, y) = \begin{cases} -\alpha\,\delta(x), & |y - w/2| \le y_s, \\ 0, & \text{otherwise}, \end{cases} \tag{8.11}$$

where w is the waveguide width and $y_s < w/2$ is the range. Numerically, we specify the above potential by having several nodal values along the line $x = 0$, and $|y - w/2| < y_s$. Through this procedure, the $\delta(x)$ dependence of the potential is integrated out in the action integral. The coupling matrix element is

$$\begin{aligned} U_{nm}(x) &= -\alpha\delta(x)\, u_{nm}; \\ u_{nm} &= \int_{-y_s+w/2}^{y_s+w/2} dy \sin(n\pi/w)\sin(m\pi/w). \end{aligned} \tag{8.12}$$

The off-diagonal (in m, n) matrix elements occur when the potential has a finite range $2y_s$ that is less than the width w of the waveguide. The potential has C_{1h} symmetry, and hence only the odd-modes are relevant to our analysis. For an incoming wave in mode 1, let us study the scattering within the energy interval $E_{y,1} < E < E_{y,3}$. We retain only the first two bands, the propagating ($n = 1$) and the evanescent ($n = 3$) in Eq. (8.10). If the coupling matrix element U_{13} is absent, we write

$$\left[-\frac{\hbar^2}{2m}\frac{d^2}{dx^2} + U_{11}(x)\right] \chi_1^0(x) = (E - E_{y,1})\chi_1^0(x). \tag{8.13}$$

Hence, the formal solution to $|\chi_1\rangle$ is given by

$$|\chi_1\rangle = |\chi_1^0\rangle + \hat{G}_1 U_{13} |\chi_3\rangle, \tag{8.14}$$

where $\langle x|\chi_1^0\rangle = \chi_1^0(x)$, is the solution to Eq. (8.13), and $\hat{G}_1$ is the Green's function operator given by

$$\hat{G}_1 = \left[\frac{\hbar^2}{2m}\frac{d^2}{dx^2} + E - E_{y,1} - U_{13} + i\epsilon^+\right]^{-1}. \tag{8.15}$$

We use the *ansatz* [20, 45] $\chi_3(x) = \tilde{t}_{13}\,\Phi_0(x)$, where $\tilde{t}_{13}$ is the evanescent amplitude, and $\Phi_0(x)$ is the bound state solution to the equation:

$$\left[-\frac{\hbar^2}{2m}\frac{d^2}{dx^2} + U_{33}(x)\right]\Phi_0(x) = (\mathcal{E} - E_{y,3})\Phi_0(x), \tag{8.16}$$

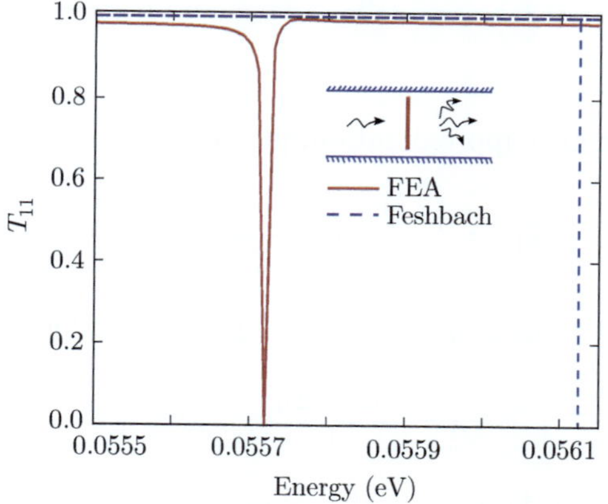

Figure 8.17 The transmission coefficient T_{11} is plotted as a function of energy obtained using our method (continuous curve) and using the Feshbach approach (dashed curve) for $\alpha = -0.01\,\text{eV m}$ and $y_s = 10\,\text{nm}$ for a straight waveguide of width 30 nm. The Fano parameters determined through (i) Feshbach calculations are $E_R = 0.05612\,\text{eV}$, $q = 0.0046$, and $\Gamma = 0\,\text{eV}$, (ii) our formalism are $E_R = 0.05572\,\text{eV}$, $q = 0.0047$, and $\Gamma = 3.4 \times 10^{-6}\,\text{eV}$. The difference between the parameters obtained through our method and the Feshbach approach is within 3%. (From Ref. [1].)

where $\mathcal{E}$ is the bound state energy. Solving Eqs. (8.13), (8.15) and (8.16) we obtain the transmission amplitudes in the form

$$\begin{aligned} t_{11} &= t\,\frac{(E-\mathcal{E})}{(E-(\mathcal{E}+\Delta)+i\Gamma)}, \\ \tilde{t}_{13} &= \frac{\mathcal{F}}{(E-\mathcal{E}+i\,\eta)}, \end{aligned} \tag{8.17}$$

where the corresponding parameters are listed in Table 8.1.

Table 8.1 The Fano parameters corresponding to the amplitudes defined in Eq. (8.17) are listed, where $k_1 = \sqrt{2m(E_R - E_1)/\hbar^2}$.

$\mathcal{E} = E_{y,2} - \dfrac{m\alpha^2 u_{33}^2}{2\hbar^2}$,	$t = \dfrac{ik_1}{\left(ik_1 + \dfrac{m\alpha\, u_{11}}{\hbar^2}\right)}$,
$\Delta = \left(\dfrac{\alpha^4 m^3}{\hbar^6}\right)\left(\dfrac{u_{33}u_{11}u_{13}^2}{k_1^2 + \dfrac{m^2\alpha^2 u_{11}^2}{\hbar^4}}\right)$,	$\mathcal{F} = -\alpha\, t\, u_{13}\sqrt{\dfrac{u_{33}m\alpha}{\hbar^2}}$,
$\Gamma = \left(\dfrac{\alpha^3 m^2}{\hbar^4}\right)\left(\dfrac{u_{33}u_{13}^2 k_1}{k_1^2 + \dfrac{m^2\alpha^2 u_{11}^2}{\hbar^4}}\right)$,	$\eta = \alpha^3 u_{13}^2 u_{33}\, t\, \dfrac{m^2}{k_1\hbar^4}$.

Within the resonance approximation, [9] we have the transmission coefficient

$$T_{11} = |t_{11}|^2 = |t|^2 \frac{(E - \mathcal{E})^2}{(E - \mathcal{E} - \Delta)^2 + \Gamma^2}. \tag{8.18}$$

Let the quasi-bound state energy $E_R = \mathcal{E} + \Delta$, and we define the reduced variables

$$\epsilon = \frac{(E - E_R)}{\Gamma}; \quad q = \frac{\Delta}{\Gamma}, \tag{8.19}$$

where the Fano q-factor determines the asymmetry of the line shape, and Γ represents the line width. With this reduction, we obtain

$$T_{11} = |t|^2 \frac{(\epsilon + q)^2}{\epsilon^2 + 1}, \tag{8.20}$$

which has the standard Fano form. [3, 7, 20] When $T_{11} = 0$ (at $E = \mathcal{E}$), we see that the evanescent amplitude $\tilde{t}_{13}$ reaches a maximum, as seen in Fig. 8.7.

In Fig. 8.17, we compare the transmission coefficient T_{11} obtained using our method and the Feshbach approach. We observe that the Feshbach approach has a small error in predicting the energy value of the transmission minimum, and the Fano line width Γ is under-estimated, as compared with our approach. This is evident already for the simplest δ-potential. We expect to have significant errors with such coupled-channel theory when we consider geometrically more complex potentials. This is because the Feshbach approach is essentially a two-band model, whereas our method incorporates contributions from all the subbands. Moreover, such analytical methods are not amenable to a solution, other than for a very few potentials generated by simple geometrical structures. We note that our method transcends any such geometrical complications and is expected to yield accurate results.

8.8 Conclusions

We have shown that in quantum waveguides with attractive scatterers, there are two dominant characteristics of the evanescent modes: (i) the evanescent modes reach a maximum at the subband minimum, and (ii) the evanescent modes have large amplitudes around the zero-transmission resonance energy. A universal way of calculating amplitudes of both propagating and evanescent modes is described. We have displayed the Fano resonance profile *for both propagating and evanescent modes* for the case of scattering from impurities. This effect can be explicitly evaluated for any attractive scattering potential. The resonance formula for the two cases are derived in Sec. 8.7.

Earlier theoretical developments were constrained by the need for analytical techniques, limiting considerations to δ-function potential. With our method, we can easily overcome this artificial limitation. Moreover, electron transport in a waveguide with multiple scattering centers are not an issue anymore. Since we are espousing a numerical calculation, there is no additional step required as in earlier considerations, such as repeatedly invoking the Born approximation. [47]

We have described the scattering mechanism in tapered and curved waveguides. These 2D waveguides can be fabricated on inversion layers of hetero-interfaces which hosts 2D electron gas, and by designing appropriate gate geometries. Thus, the results can be readily confronted with experimental outcomes. In tapered waveguides, the direction of the incoming current has a broken inversion symmetry. This provides a unique situation of possible current rectification throughout the device geometry.

We can include the effect of an external magnetic field using a local gauge which acts as a multiplicative term for the matrix elements. [48] This generates rapidly converging results even in the presence of a magnetic field. The usual gauge-invariant substitution for the momentum operator leads to a quadratic diamagnetic potential which is troublesome to evaluate at large distances from the orbit center. The Ueta method [48] circumvents this problem. Similarly, the Dresselhaus [49] and Rashba [50] spin-orbit effects can be incorporated as additive terms in the action integral. Absorption effects can be deftly included in our method by introducing a negative imaginary part to the potential. [51] Such a scheme will account for gate-contacts in 2D inversion layers, where the gate-contacts siphons off the probability current going across the device.

Calculations of conductance, Seebeck coefficient, and power factors can directly predict experimentally measurable quantities. Lastly, we have shown that quantum dots or attractive impurity potentials embedded in the interior of a waveguide are good candidates for fabricating thermoelectric devices since they yield large power factors around the subband minima.

We conclude with comments about the modeling of open-domain scattering. [38, 39] Any detailed analysis of open-domain scattering has invariably used the asymptotic BCs, leading to a loss of information about the near-field solutions. In a one-component Schrödinger's equation, the evanescent modes do not occur, so this is not an issue, and we may follow the theory detailed in the large number of books on scattering theory and on quantum mechanics generally. However, in the case of multi-component Schrödinger's equations or in the case of the relativistic Dirac equation, it is possible to generate evanescent modes in scattering from potential discontinuities. This would become relevant for semiconductor physics, for example. We have succeeded in setting up the circular (2D) and spherical (3D) sources and absorbers that can again convert the Cauchy BCs occurring in scattering theory into Dirichlet BCs. Once the solutions to the wavefunctions are determined within the confines of the absorbers, we can follow the usual partial wave decomposition to determine all information about the scattering amplitudes, evanescent mode amplitudes, scattering lengths, and cross-sections. This is the advantage of the method of sources and absorbers proposed here.

References

[1] Sathwik Bharadwaj and L. R. Ram-Mohan, J. Appl. Phys. **125**, 164306–1–11 (2019); "Electron scattering in quantum waveguides with sources and absorbers: I Theoretical Formalism."

[2] Sathwik Bharadwaj and L. R. Ram-Mohan, J. Appl. Phys. **125**, 164307–1–12 (2019); "Electron scattering in quantum waveguides with sources and absorbers: II Applications."

[3] U. Fano, Phys. Rev. **124**, 1866–1878 (1961); "Effects of configuration interaction on intensities and phase shifts;" U. Fano and J. W. Cooper, *ibid* **137**, A1364–A1379 (1965); "Line profiles in the far-UV absorption spectra of the rare gases."

[4] E. Tekman and P. F. Bagwell, Phys. Rev. B. **48**, 2553–2559 (1993); "Fano resonances in quasi-one-dimensional electron waveguides."

[5] Y. S. Joe, A. M. Satanin, and C. S. Kim, Phys. Scr. **74**, 259–266 (2006); "Classical analogy of Fano resonances."

[6] A. E. Miroshnichenko, S. Flach, and Y. S. Kivshar, Rev. Mod. Phys. **82**, 2257–2298 (2010); "Fano resonances in nanoscale structures."

[7] A. R. P. Rau, Phys. Scr. **69**, 1, C10–C13 (2004); "Perspectives on the Fano resonance formula."

[8] C. S. Kim, A. M. Satanin, Y. S. Joe, and R. M. Cosby, Phys. Rev. B **60**, 10962–10970 (1999); "Resonant tunneling in a quantum waveguide: Effect of a finite-size attractive impurity."

[9] C. S. Kim, O. N. Roznova, A. M. Satanin, and V. B. Stenberg, J. Exp. Theor. Phys. **94**, 992–1007 (2002); "Interference of quantum states in electronic waveguides with impurities."

[10] Y. S. Joe and A. M. Satanin, J. Supercond. Nov. Magn. **20**, 179–182 (2007); "Fano-interference of decaying states with continuum in nanostructures."

[11] S. Flügge, *Practical Quantum Mechanics* (Springer-Verlag, Berlin, 1971).

[12] V. Vargiamidis, P. Komninou, and H. M. Polatoglou, Phys. Stat. Sol. (c) **5**, 3813–3817 (2008); "Fano effect in quasi-one-dimensional wires with short- or finite-range impurities."

[13] V. Vargiamidis and H. M. Polatoglou, Phys. Rev. B **72**, 195333–1–12 (2005); "Resonances in electronic transport through a quantum wire with impurities and variable cross-sectional shape."

[14] S. A. Gurvitz, and Y. B. Levinson, Phys. Rev. B **47**, 10578–10587 (1993); "Resonant reflection and transmission in a conducting channel with a single impurity." See also, A. A. Clerk, X. Waintal, and P. W. Brouwer, Phys. Rev. Lett. **86**, 4636–4639 (2001); "Fano resonances as a probe of phase coherence in quantum dots."

[15] J. Göres, D. Goldhaber-Gordon, S. Heemeyer, and M. A. Kastner, Phys. Rev. B **62**, 2188–2194 (2000); "Fano resonances in electronic transport through a single-electron transistor."

[16] J. Kim, J.-R. Kim, J.-O. Lee, J. W. Park, H. M. So, N. Kim, K. Kang, K.-H. Yoo, and J.-J. Kim, Phys. Rev. Lett. **90**, 166403–1–4 (2003); "Fano resonance in crossed carbon nanotubes."

[17] Kensuke Kobayashi, Hisashi Aikawa, Akira Sano, Shingo Katsumoto, and Yasuhiro Iye, Phys. Rev. B **70**, 035319–1–6 (2004); "Fano resonance in a quantum wire with a side-coupled quantum dot."

[18] A. C. Johnson, C. M. Marcus, M. P. Hanson, and A. C. Gossard, Phys. Rev. Lett. **93**, 106803–1–4 (2004); "Coulomb-modified Fano resonance in a one-lead quantum dot."

[19] H. Feshbach, Ann. Phys. **5**, 357–390 (1958); "Unified theory of nuclear reactions;" Ann. Phys. **19**, 287–313 (1962); "A unified theory of nuclear reactions. II."

[20] J. U. Nöckel, and A. D. Stone, Phys. Rev. B **50**, 17415–17432 (1994); "Resonance line shapes in quasi-one-dimensional scattering."

[21] C. Berthod, F. Gagel, and K. Maschke, Phys. Rev. B **50**, 18299–18311 (1994); "DC transport in perturbed multichannel quantum wires."

[22] J.-P. Connerade and A. M. Lane, Rep. Prog. Phys. **51**, 1439–1478 (1988), and references within; "Interacting resonances in atomic spectroscopy."

[23] J.-P Connerade, and A M Lane, J. Phys. B: At. Mol. Phys. **20**, 1757–1769 (1987); "Q reversals in a Rydberg series caused by an interloper level."

[24] S. Klaiman, N. Moiseyev, and H. R. Sadeghpour, Phys. Rev. B **75**, 113305–1–4 (2007); "Interpretation of the Fano lineshape reversal in quantum waveguides."

[25] R. Landauer, IBM J. Res. Dev. **1**, 223–231 (1957); "Spatial variation of currents and fields due to localized scatterers in metallic conduction."

[26] R. Landauer, Philos. Mag. **21**, 863–867 (1970); "Electrical resistance of disordered one-dimensional lattices."

[27] M. Büttiker, Phys. Rev. Lett. **57**, 1761–1764 (1986); "Four-terminal phase-coherent conductance;" see also, IBM J. Res. Dev. **32**, 317–334 (1988); "Symmetry of electrical conduction."

[28] M. Büttiker, Y. Imry, R. Landauer, and S. Pinhas, Phys. Rev. B **31**, 6207–6215 (1985); "Generalized many-channel conductance formula with application to small rings."

[29] D. S. Fisher and P. A. Lee, Phys. Rev. B **23**, 6851–6854(R) (1981); "Relation between conductivity and transmission matrix."
[30] U. Sivan and Y. Imry, Phys. Rev. B **33**, 1, 551–558 (1986); "Multichannel Landauer formula for thermoelectric transport with application to thermopower near the mobility edge."
[31] S. Datta, *Electronic Transport in Mesoscopic Systems* (Cambridge University Press, Cambridge, 1995).
[32] C. R. Proetto, Phys. Rev. B **44**, 16, 9096–9099 (1991); "Thermopower oscillations of a quantum-point contact."
[33] N. F. Mott, and E. A. Davis, *Electronic Processes in Non-Crystalline Materials* (Oxford University Press, Oxford, 2012).
[34] P. F. Bagwell, Phys. Rev. B. **41**, 10354–10357 (1990); "Evanescent modes and scattering in quasi-one-dimensional wires."
[35] A. A. Gorbatsevich and V. V. Kapaev, Russ. Microelectron. **36**, 1, 1–13 (2005); "Waveguide nanoelectronics."
[36] D. Boese, M. Lischka, and L. E. Reichl, Phys. Rev. B **61**, 5632–5636 (2000); "Resonances in a two-dimensional electron waveguide with a single δ-function scatterer."
[37] C. S. Chu and R. S. Sorbello, Phys. Rev. B **40**, 5941–5949 (1989); "Effect of impurities on the quantized conductance of narrow channels."
[38] M. Zebarjadi, K. Esfarjani, A. Shakouri, J.-H. Bahk, Z. Bian, G. Zeng, J. Bowers, H. Lu, J. Zide, and A. Gossard, Appl. Phys. Lett. **94**, 202105–1–3 (2009); "Effect of nanoparticle scattering on thermoelectric power factor."
[39] T. Zhu, K. Swaminathan-Gopalan, K. Stephani, and E. Ertekin, Phys. Rev. B **97**, 174201–1–9 (2018); "Thermoelectric phonon-glass electron-crystal via ion beam patterning of silicon."
[40] B. Wang, J. Zhou, R. Yang, and B. Li, New J. Phys. **16**, 065018–1–16 (2014); "Ballistic thermoelectric transport in structured nanowires."
[41] L. R. Ram-Mohan, K. H. Yoo, J. Moussa, J. Appl. Phys. **95**, 3081–3092 (2004); "The Schrödinger–Poisson self-consistency in layered quantum semiconductor structures."
[42] J. Goldstone and R. L. Jaffe, Phys. Rev. B **45**, 14100–14107 (1992); "Bound states in twisting tubes."
[43] R. L. Schult, D. G. Ravenhall, and H. W. Wyld, Phys. Rev. B **39**, 5476–5479 (1989); "Quantum bound states in a classically unbound system of crossed wires."
[44] Z.-M. Bai, X. Chen, X.-T. An, Physica B **407**, 4293–4297 (2012); "Electron transport through curved two-dimensional quantum waveguide."
[45] H. Friedrich, *Theoretical Atomic Physics* (Springer-Verlag, Berlin, 1990).
[46] O. Olendski and L. Mikhailovska, Phys. Rev. B **67**, 035310–1–13 (2003); "Fano resonances of a curved waveguide with an embedded quantum dot."
[47] J. D. Walls, J. Huang, R. M. Westervelt, and E. J. Heller, Phys. Rev. B **73**, 035325–1–17 (2006); "Multiple-scattering theory for two-dimensional electron gases in the presence of spin-orbit coupling."

[48] T. Ueta and Y. Miyagawa, Phys. Rev. E **86**, 026707–1–8 (2012); "Local-gauge finite-element method for electron waves in magnetic fields."

[49] G. Dresselhaus, Phys. Rev. **100**, 580–586 (1955); "Spin-orbit coupling effects in zinc blende structures."

[50] E. I. Rashba, Sov. Phys. Solid. State, **1**, 368–380 (1959); "Symmetry of bands in wurzite-type crystals–1;" See also, G. Bihlmayer, O. Rader, and R. Winkler, New J. Phys. **17**, 050202–1–8 (2015); "Focus on the Rashba effect."

[51] N. F. Mott and H. S. W. Massey, *The Theory of Atomic Collisions* (Oxford University Press, London, 1949), Ch. 8, pp. 184–194.

9

Open domain quantum scattering with sources and absorbers

In this chapter:

- We note that results of first-principles calculations, such as using density functional theory (DFT), can be used as input to obtain accurately the carrier transport properties in nanoelectronic devices. First-principles calculations can effectively determine the atomistic potentials associated with defects and impurities. However, such first-principles calculations are not feasible at present for direct modeling of carrier transport properties at length scales relevant for device applications.
- We develop a scalable quantum transport theory to investigate the carrier transport properties of two-dimensional materials. First, we derive a new quantum scattering framework which does not entail going to the asymptotic limit. We obtain the transport properties at the proximity of the scattering centers also. We then bridge our scattering theory framework with the $\boldsymbol{k} \cdot \boldsymbol{P}$ perturbation theory, with inputs from *ab-initio* electronic structure calculations; we thereby provide a versatile multiscale formalism for scattering calculations. This makes use of the source and absorber theory for scattering within the framework of novel finite element methods employing Hermite shape functions for high accuracy outcomes. This approach preserves the crucial contributions of decaying evanescent solutions that arise in multi-band scattering across heterointerfaces and at localized potentials. We employ this method to study electron transport in lateral transition-metal dichalcogenide heterostructures (TMDCs).
- We discuss the emergence of Fano resonances due to material inclusions forming heterostructures, and show that such material encapsulations offer the enhancement of mobility by an order of magnitude larger than the phonon-limited mobility.
- We demonstrate that such analysis is of great interest to realize high-mobility devices.

Finite Elements in Action. L. Ramdas Ram-Mohan, Oxford University Press. © L. Ramdas Ram-Mohan (2026).
DOI: 10.1093/oso/9780199563487.003.0009

9.1 Introduction

We are interested in two-dimensional (2D) and three-dimensional (3D) scattering of electron waves from a potential center in the open domain. The open domain here means that there are no geometrical constraints, such as waveguides, that could channel the scattered waves. The picture is that sources of the incoming waves at large distances will emit spherical waves that can be considered locally near the scattering center to be plane waves. The scattered waves then proceed from the scattering center as circular waves (2D) or as spherical waves in the case of 3D scattering. The open domain description is used for the device modeling in typical calculations.

Nanoelectronic devices built on low-dimensional materials offer approaches to construct integrated circuits to go beyond Moore's law [1, 2]. Atomically thin monolayers such as graphene [3], silicene [4], phospherene [5], transition metal dichalcogenides (TMDCs), [6–8] have all been explored to construct high-performance field effect transistors, and they have shown great potential to succeed silicon-based devices in next generation computers. Accurate prediction of the transport properties in these materials is a crucial challenge to overcome before they can be used in practical devices [1].

Electron scattering due to imperfections, additive atoms, and phonons determine the carrier dynamics in 2D materials. Carrier transport properties can also be engineered by creating potential patterns and quantum confinements, such as through the application of local potentials via STM/AFM tips [9, 10], surface functionalization by organic molecules [11, 12], or through the geometrical confinement in lateral heterostructures [13]. For all such applications, we need to develop an *ab-initio* quantum transport formalism.

In the following, density functional theory (DFT) [14] provides a "parameter-free" method for the electronic structure calculations, and accurately determines the atomistic potentials associated with heterointerfaces, defects, and impurities. However, they are as yet ineffective for direct modeling of carrier transport properties at length scales relevant to device applications; this is due to the limitations of computational resources. The difference in the local potential energy of the electrons with and without the defects or impurities yield the *in situ* potential energy that represents the scattering potential. This then is used in our finite element scattering calculations.

There have been efforts to provide atomistic quantum transport framework based on the tight-binding methods [15], plane-wave representation of the empirical pseudopotentials [16], and DFT-based non-equilibrium Green's function techniques [17, 18]. However, such simulations are limited to a small number of atoms, ranging from a few hundred to thousands, and can be computationally expensive.

In this chapter alternatively, we develop a first-principles-informed continuum quantum transport theory. We first develop a novel non-asymptotic quantum scattering theory. We show that bridging our scattering theory framework with the $\boldsymbol{k} \cdot \boldsymbol{P}$ perturbation theory, with inputs from *ab-initio* electronic structure calculations, as mentioned above, provides a versatile multiscale formalism. This formalism can easily be scaled into device dimensions. Hence, it will be very useful to simulate and design electron optics, and nanoelectronics platforms. In Fig. 9.1, we display the flowchart

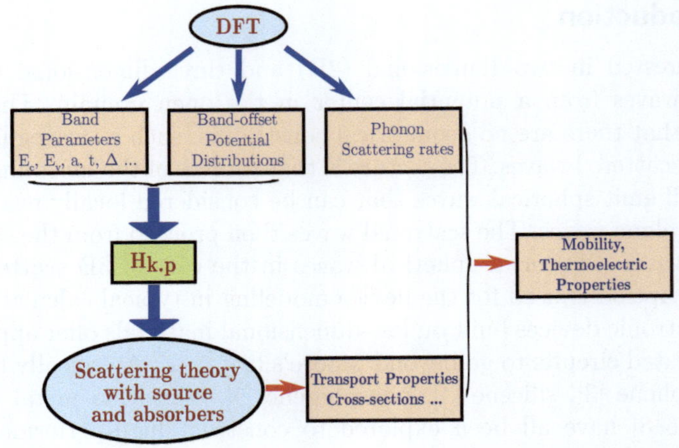

Figure 9.1 Flowchart of the quantum transport framework is discussed here.

of the quantum transport framework discussed in this chapter. The band parameters, potential distributions, and phonon scattering rates are obtained through the DFT calculations. These quantities will be an input to the $\boldsymbol{k} \cdot \boldsymbol{P}$ Hamiltonian, which is then solved using quantum scattering theory. By combining the carrier scattering rate obtained through our calculations with the phonon scattering rate obtained through first-principles calculations, we can accurately obtain the mobility and thermoelectric performance of the devices.

The traditional treatment of scattering considers an *ad-hoc* condition that the distance between scattering centers and the observer is infinite. At large distances, the solution wavefunction is expressed as a linear combination of the incoming and outgoing basis functions with undetermined amplitudes. A partial wave analysis is then performed in the asymptotic limit to obtain these amplitudes in terms of the phase-shift parameters. For example, in the 3D open domain, the wavefunction is expressed in terms of the spherical Hankel functions of the first kind $h_m^{(1)}(kr)$ [19]. We can then easily deduce the scattering amplitudes since all the Hankel functions have the same asymptotic form given by e^{ikr}/r, which represents the outgoing spherical wave [20]. Lord Rayleigh [21] developed this method for the first time in the context of acoustic scattering, and later it was extended to quantum mechanics by Faxen and Holtsmark [22].

In the asymptotic limit, only the radial component of the scattered current contributes to the cross-section. As a result, information about the angular current contributions [23], and the distance between target and observer is lost in this analysis. In reports by Dai et al., [24, 25] it has been shown that a rigorous scattering treatment without the asymptotic approximations will introduce a modification factor to the scattering amplitude, and an additional phase factor to the scattered wavefunction. Moreover, we can show that in multi-band scattering processes, the wavefunction will have contributions from not only the propagating waves but also from the exponentially decaying evanescent waves [26]. This will be particularly prevalent in transport across heterointerfaces. For example, in Fig. 9.2, we consider heterointerface formed between MoS_2 and WS_2 monolayers. From the electronic structure calculations, we know that MoS_2 ($E_c = -4.31\,\text{eV}$) has a lower conduction band minimum than WS_2 ($E_c = -3.97\,\text{eV}$) monolayer. Hence, when an electron with energy $-4.31\,\text{eV} \leq E < -3.97\,\text{eV}$ is injected from the MoS_2 to WS_2 monolayer, transport occurs only through the evanescent modes. These evanescent modes are placed within the bandgap and will have purely complex wavevectors. This results in exponentially decaying contributions to the scattered wavefunction. Traditional scattering calculations will not be able to account for these crucial contributions, since the probability of evanescent modes vanish at the asymptotic limit, where the boundary conditions (BCs) are applied to determine the scattering amplitudes. We then require a new quantum scattering framework that will accurately compute both the evanescent mode and angular current contributions in the near-field of the scattering center.

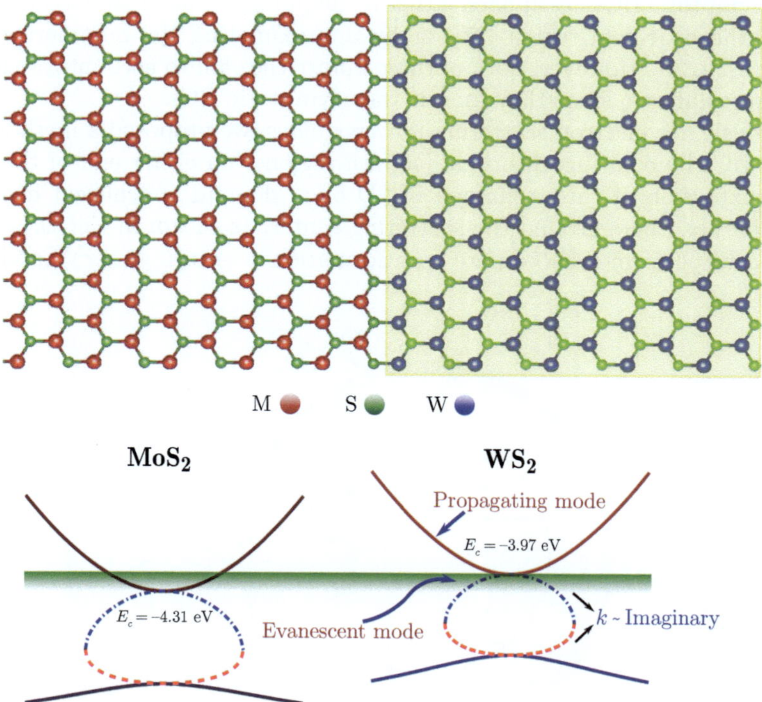

Figure 9.2 Schematic representation of the heterointerface between MoS_2 and WS_2 monolayers is shown. We also display their corresponding conduction and valence bands near the K-point. MoS_2 ($E_c = -4.31\,\text{eV}$) has a lower conduction band minimum than WS_2 ($E_c = -3.97\,\text{eV}$). Hence, below the energy $-3.97\,\text{eV}$, electron transport across the interface occurs only through evanescent modes.

This chapter is organized as follows. In Sec. 9.2, we develop a "near-field" quantum scattering theory based on sources and absorbers. With this setup, we obtain the total wavefunction that includes both propagating and evanescent band contributions. This formalism is extended to a multi-band framework, by integrating it with the $\boldsymbol{k} \cdot \boldsymbol{P}$ perturbation theory with inputs from DFT calculations. This will be explained further in Sec. 9.3. In that section we will also explain the method for including the thermal properties in our formalism through deformation potentials obtained through DFT calculations. In Sec. 9.4, we apply our formalism to study transport properties in lateral TMDC heterostructures. We observe the emergence of Fano resonance in 2D materials with material encapsulations. We will also study the mobility in lateral heterostructures for a family of TMDC monolayers. Concluding remarks are given in Sec. 9.6.

9.2 Construction of sources and absorbers

Our first aim is to formulate a framework for the scattering properties without applying asymptotic BCs. In the conventional variational formulation of the scattering theory, asymptotic BCs are applied either by mapping the far-field partial wave expansion into the near-field region [27], or by approximating the Sommerfeld radiation conditions [20, 28]. However, both of these approaches fail to account for the decaying evanescent solutions, since they inherit the asymptotic BCs.

Previously, for electron scattering in waveguides, we proposed a method to reduce Dirichlet BCs by creating absorbers (stealth regions) on either end of the waveguide [26]. An extension of this technique in 2D open domain is achieved here by creating circular absorbers around the scattering center, as shown in Fig. 9.3. Within the absorber, we perform a coordinate transformation $\rho \to \rho(1 + i\alpha(r))$, and an energy

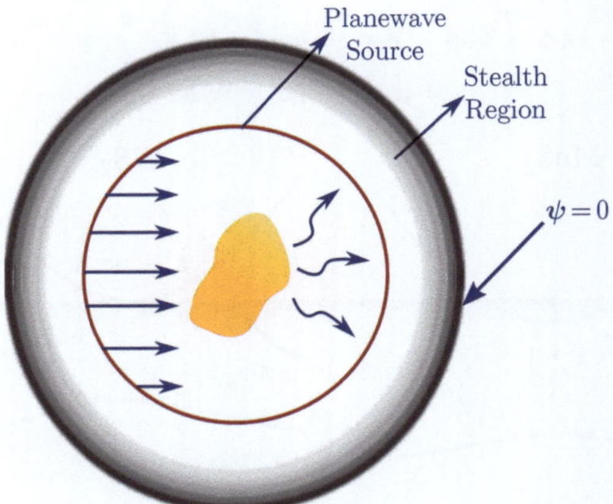

Figure 9.3 Schematic representation of the construction of a circular source (red line) and absorber (gray region, also known as the stealth region) around a scattering center (green shaded region) is shown. We can apply Dirichlet boundary conditions at the exterior boundary of the absorber. Amplitudes on the source circle are chosen as per Eq. (9.8), so that we inject plane waves into the region confining the scattering center.

transformation $E \to E\,(1 + i\alpha(r))$. Here, $\alpha(r)$ is the cubic Hermite interpolation polynomial varying smoothly from 0 to α_{max}. Through such transformation, it has been shown that the no-reflection condition is satisfied [26], and the absorber will not reflect any wave back into the active region confined within the absorber. Therefore, scattering properties in the active region remain unaffected with the presence of the absorber. As a result of such a transformation, the wavefunction will decay rapidly within the absorber. Hence, we can apply the Dirichlet BCs at the boundaries of the absorber. Through this technique, we can confine the computational domain, and preserve the near-field nature of the wavefunction in the region confined by the absorber (scattering region), that are otherwise lost due to the application of asymptotic BCs.

Once we enclose the active scattering region with the absorber, we require a source (an electron antenna, as it were) in the active region to initiate the incoming plane waves (see Fig. 9.3). This will be achieved by introducing a source term in the Schrödinger's equation. In order to derive this source term, we start with the Green's function equation for the Schrödinger operator in the 2D circular coordinate system given by

$$\left[\frac{1}{\rho}\frac{\partial}{\partial \rho}\left(\rho\frac{\partial}{\partial \rho}\right) + \frac{1}{\rho^2}\frac{\partial^2}{\partial \phi^2} + k^2\right] G(\boldsymbol{\rho}, \boldsymbol{\rho}') = S(\rho', \phi')\,\frac{\delta\,(\rho - \rho')}{\rho'}\,\delta\,(\phi - \phi')\,, \tag{9.1}$$

where $S(\rho', \phi')$ is the source term yet to be determined, the wavevector $k = \sqrt{2m^* E/\hbar^2}$, and E is the incoming energy, the scattering potential is not included in the above equation, and m^* is the effective mass. We expand the Green's function in the Fourier form given by

$$G(\boldsymbol{\rho}, \boldsymbol{\rho}') = \frac{1}{2\pi}\sum_{m=-\infty}^{\infty} e^{im(\phi-\phi')} g_m(\rho, \rho'). \tag{9.2}$$

Inserting this relation into Eq. (9.1), we obtain

$$\begin{aligned}\frac{1}{2\pi}\sum_{m=-\infty}^{\infty} e^{im(\phi-\phi')}&\left[\frac{1}{\rho}\frac{\partial}{\partial \rho}\left(\rho\frac{\partial g_m}{\partial \rho}\right) + \left(k^2 - \frac{m^2}{\rho^2}\right) g_m\right] \\ &= S(\rho', \phi')\,\frac{\delta\,(\rho - \rho')}{\rho'}\,\delta\,(\phi - \phi')\,.\end{aligned}$$

We multiply the above equation by $e^{-im\phi}$ and integrate over ϕ from 0 to 2π, to obtain

$$\left[\frac{1}{\rho}\frac{\partial}{\partial \rho}\left(\rho\frac{\partial g_m}{\partial \rho}\right) + \left(k^2 - \frac{m^2}{\rho^2}\right) g_m\right] = S(\rho', \phi')\,\frac{\delta\,(\rho - \rho')}{\rho'}. \tag{9.3}$$

The radial part $g_m(\rho, \rho')$ is the Green's function for the one-dimensional Sturm-Liouville operator [29], and will be of the form

$$g_m(\rho, \rho') = A_m \begin{cases} J_m(k\rho) H_m(k\rho'), & \rho \le \rho' \\ J_m(k\rho') H_m(k\rho), & \rho > \rho' \end{cases}, \tag{9.4}$$

where, $J_m(k\rho)$ and $H_m(k\rho)$ are the Bessel and Hankel function of first kind, respectively. We determine the coefficient A_m by applying the jump condition for the derivative of the Green's function at $\rho = \rho'$ given by

$$\left.\frac{\partial g_m}{\partial \rho}\right|_{\rho'+\epsilon} - \left.\frac{\partial g_m}{\partial \rho}\right|_{\rho'-\epsilon} = \frac{S(\rho',\phi')}{\rho'},$$

$$A_m \mathcal{W}\left[J_m(k\rho'), H_m(k\rho')\right] = \frac{S(\rho',\phi')}{\rho'}, \tag{9.5}$$

where, the Wronskian $\mathcal{W}\left[J_m(k\rho'), H_m(k\rho')\right] = 2i/\pi\rho'$. Thus, A_m is determined to be

$$A_m = -\frac{i\pi}{2}\, S(\rho',\phi'). \tag{9.6}$$

Substituting Eqs. (9.4) and (9.6) into Eq. (9.2), we obtain the total Green's function of the form

$$\begin{aligned} G(\boldsymbol{\rho},\boldsymbol{\rho}') &= -\frac{i}{4}\, S(\rho',\phi') \\ &\times \sum_{m=-\infty}^{\infty} e^{im(\phi-\phi')} \begin{cases} J_m(k\rho)H_m(k\rho'), & \rho \le \rho' \\ J_m(k\rho')H_m(k\rho), & \rho > \rho' \end{cases}, \\ &= -\frac{i}{4}\, S(\rho',\phi')\, H_0(k\,|\boldsymbol{\rho}-\boldsymbol{\rho}'|). \end{aligned}$$

We note that the $H_0(k\,|\boldsymbol{\rho}-\boldsymbol{\rho}'|)$ represents the wave originating from the point source at $\boldsymbol{\rho}'$. As shown in Fig. 9.3, to obtain a circular source we integrate $G(\boldsymbol{\rho},\boldsymbol{\rho}')$ over the angular coordinate ϕ' from 0 to 2π. Hence, the wavefunction emerging from the circular source is given by

$$\psi_{in}(\rho,\rho') = -\frac{i}{4}\int_0^{2\pi} d\phi' S(\rho',\phi')\, H_0(k\,|\boldsymbol{\rho}-\boldsymbol{\rho}'|). \tag{9.7}$$

Our aim is to generate plane waves propagating in the forward direction from the circular source within the region $\rho \le \rho'$. Hence, we choose the source term to be

$$S(\rho',\phi') = \frac{2i}{\pi}\sum_{n=-\infty}^{\infty} \frac{e^{in\phi'}\, i^n}{H_n(k\rho')}. \tag{9.8}$$

Now, substituting the above form into Eq. (9.7), we obtain

$$\psi_{in} = \sum_{m=-\infty}^{\infty} e^{im\phi} i^m \begin{cases} J_m(k\rho) & , \quad \rho \le \rho'; \\ \dfrac{J_m(k\rho')}{H_m(k\rho')} H_m(k\rho), & \rho > \rho', \end{cases} \tag{9.9}$$

where we have utilized the orthogonal properties of the function $e^{in\phi'}$.

We know the expansion of plane wave is

$$e^{ikx} = e^{ik\rho\cos\phi} = \sum_{m=-\infty}^{\infty} e^{im\phi} i^m J_m(k\rho). \tag{9.10}$$

Hence,

$$\psi_{in} = \begin{cases} e^{ikx} & , \quad \rho \leq \rho'; \\ \sum_{m=-\infty}^{\infty} e^{im\phi} i^m \dfrac{J_m(k\rho')}{H_m(k\rho')} H_m(k\rho), & \rho > \rho', \end{cases} \tag{9.11}$$

we obtain plane waves impinging on the scattering center from a circular source at $\rho = \rho'$, as shown in Fig. 9.3. The outgoing wavefunction in the $\rho > \rho'$ region will simply get absorbed by the stealth region.

Once we formulate the quantum scattering theory with the *source* and *absorber*, we can employ numerical methods to obtain the total wavefunction ψ. In the presence of a scattering potential, we obtain the total wavefunction $\psi = \psi_{in} + \psi_{sc}$. Here, ψ_{sc} will include both the radial and angular current contributions, and also the evanescent solutions in case of an absorbing scattering potential or in case of multi-band scattering processes, discussed in the next section.

Throughout this chapter, we have solved the Hamiltonian equation by casting it into an action integral. This action integral is solved using the finite element method (FEM) [30–33]. In FEA, we discretize the physical domain of interest into small elements. Within each element, we express the wavefunction as a linear combination of interpolation polynomials multiplied by undetermined coefficients. These coefficients correspond to the value of the wavefunction and their derivatives at the vertices (nodes) of the element. With this approach, one can systematically increase the accuracy through mesh size refinement (h-refinement) or by employing higher-order interpolation polynomials (p-refinement).

To validate our scattering formalism, we consider the case of 2D scattering from a hard circle. The hard circle potential is given by

$$V(\rho) = \begin{cases} 0, & \rho \leq a \\ \infty, & \rho > a \end{cases}, \tag{9.12}$$

where, a is the radius of the circle. Through partial wave analysis, we can obtain a closed form result for the scattering cross-section length σ measured at the asymptotic distance (see Sec 9.4.2). In Fig. 9.4(b), we compare the analytical result (see Eq. (9.36)), and the result obtained through our formalism. We see that these results match accurately. Behavior of σ at low and high energy limits observed in Fig. 9.4(b) are discussed in Sec. 9.4.2.

In Sec. 9.4.1, we outline the scattering theory without applying large-distance asymptotes [24, 25]. In Fig. 9.4(a), we plot the cross-section length obtained from an observer at a very large distance (asymptotic limit), and at a finite distance 10 nm.

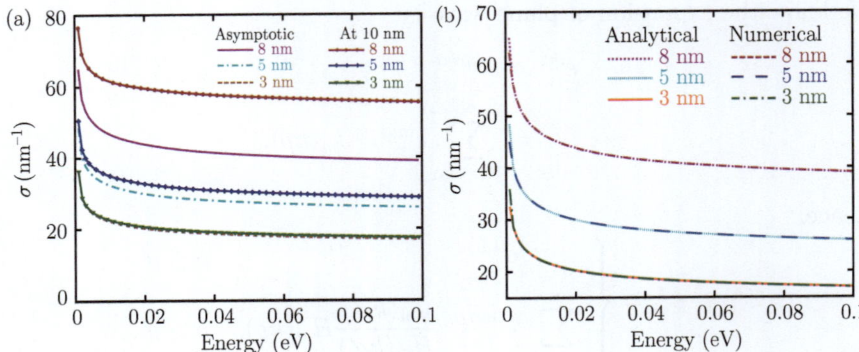

Figure 9.4 (a) Scattering cross-section length obtained through our scattering formalism is plotted as a function of incoming energy. Here, the dotted line represents the cross-section measured by an observer at infinity, whereas the continuous line represents the cross-section measured by an observer at a finite distance 10 nm. Radius of the hard circle potential considered here are equal to 8 nm, 5 nm, and 3 nm. (b) The analytical result through partial wave analysis, and the numerical result obtained through our calculation for the scattering cross-section in hard circle potential is plotted against energy.

We see a significant difference between their quantitative values. As we increase the radius from 3 nm to 8 nm, this difference will grow, as will the distance between the observer at 10 nm, and the hard circle potential will also reduce. Hence, validating that in nanoscale systems, it is especially important to employ the non-asymptotic scattering theory developed here to obtain accurate transport properties. In Sec. 9.4.3, we derive the source term for the non-asymptotic scattering theory in three-dimensions (3D). Such analysis in 3D can be employed to obtain transport properties in bulk materials.

9.3 Envelope function theory with sources and absorbers

In materials, the low energy dynamics of the charge carriers can be described by the $\boldsymbol{k} \cdot \boldsymbol{P}$ Hamiltonian [34–37]. This Hamiltonian provides an accurate characterization of the energy dispersion around the high-symmetry points of the Brillouin Zone, in terms of a small number of input band parameters. These band parameters are taken from the database [38] obtained from the DFT and many-body perturbation theory calculations. In this section, we describe the scattering theory for the envelope functions operated on the $\boldsymbol{k} \cdot \boldsymbol{P}$ Hamiltonian.

The $\boldsymbol{k} \cdot \boldsymbol{P}$ Hamiltonian effectively captures the coupled dynamics of the conduction and valence bands around the high-symmetry points of the Brillouin Zone. This is also the region in which we are most interested in determining the transport properties. Effects of remote bands can be easily included by using the Löwdin perturbation theory [39]. Effects due to strain, magnetic field, and spin-orbit coupling are included as additional input parameters. Being a long wavelength theory, it is well-suited to simulate transport properties in device-relevant scales.

Let us consider the Hamiltonian equation of the form

$$(H_0 + V_0 + V_d - E) \cdot \psi_{nk} = S(\rho', \phi') \, \eta_{\text{scale}} \, \mathbb{A}_{nk} \, \frac{\delta(\rho - \rho')}{\rho'}, \tag{9.13}$$

where $S(\rho', \phi')$ is the source term defined in Eq. (9.8), $\mathbb{A}_{nk}$ is the incoming amplitude, n is the band-index, k is the wavevector, and η_{scale} is the scale factor (derived in Sec. 9.5.1). Here, V_0 is the cell-periodic potential and V_d is the defect/impurity potential obtained from the DFT calculations. In the envelope function approximation (EFA), the general form of the wavefunction and the incoming amplitude will be considered as a linear combination of a finite number of bands [34, 40, 41], given by

$$\begin{aligned} \psi_{nk} &= \sum_{m=0}^{\text{nband}} F_m \, u_{mk}, \\ \mathbb{A}_{nk} &= \sum_{m=0}^{\text{nband}} a_m \, u_{mk}, \end{aligned} \tag{9.14}$$

where F_m is the slowly-varying envelope function, u_{mk} is the cell-periodic Bloch function, a_m is the incoming amplitude, and nband is the number of bands considered. Here, the expansion is at a fixed k value, and we will assume from now on that the expansion is around $k = 0$. Within the framework of EFA, we perform 'cell-averaging' by integrating over each unit cell in the crystal. Bloch functions satisfy Schrödinger's equation with band-edge energies, and result in:

$$\begin{aligned} \int_{\text{cell}} d^2\rho \, u_{n0}^{\dagger} \cdot u_{m0} &= \delta_{mn}, \\ \int_{\text{cell}} d^2\rho \, u_{n0}^{\dagger} \, (H_0 + V_0) \cdot u_{m0} &= E_m \delta_{mn}, \\ \int_{\text{cell}} d^2\rho \, u_{n0}^{\dagger} \, \boldsymbol{p} \cdot u_{m0} &= \boldsymbol{p}_{mn}, \end{aligned} \tag{9.15}$$

where, $\boldsymbol{p}_{mn}$ is the momentum matrix. This averaging procedure results in the equation of the form

$$[\mathbf{H}_{kp} + (V_d - E) \, \mathbf{1}] \cdot \mathbb{F} = S(\rho', \phi') \, \eta_{\text{scale}} \, \mathbb{A} \, \frac{\delta(\rho - \rho')}{\rho'}, \tag{9.16}$$

where, $\mathbf{H}_{kp}$ is the $\boldsymbol{k} \cdot \boldsymbol{P}$ Hamiltonian, and $\mathbf{1}$ is the identity matrix. The envelope function $\mathbb{F} = \begin{pmatrix} F_1 & F_2 \ldots \end{pmatrix}^T$, and the incoming amplitude $\mathbb{A} = \begin{pmatrix} a_1 & a_2 \ldots \end{pmatrix}^T$. We consider that in the absence of the defect potential V_d, the incoming amplitudes satisfy an equation of the form

$$[\mathbf{H}_{kp} - E \, \mathbf{1}] \cdot \mathbb{A} = 0. \tag{9.17}$$

Hence, for a given energy we can obtain $\mathbb{A}$ by solving the above matrix equation. Note that Eq. (9.17) represents the standard $\boldsymbol{k} \cdot \boldsymbol{P}$ Hamiltonian equation for the bulk material.

We cast Eq. (9.16) into an action integral equation, and this will allow us to solve the equation in the variational approach. The advantages of the action integral formulation are that there is no ambiguity about reordering of operators for symmetrization [42], and we can easily evaluate the conserved current using the gauge-variational approach introduced by Gell-Mann and Levy [43] and used extensively by us [42] (see Sec. (9.5)). This action integral is solved using the FEA.

Within the Boltzmann transport formalism, important transport properties such as the mobility, conductance, and Seebeck coefficient are expressed in terms of the total relaxation time $\tau(E)$. To determine $\tau(E)$, we need to consider both the intrinsic and extrinsic scattering rate, and the extrinsic scattering rate is arising from the potential scattering. According to the Matthiessen's law

$$\frac{1}{\tau(E)} = \frac{1}{\tau_e(E)} + \frac{1}{\tau_{ph}(E)}, \tag{9.18}$$

where, τ_e is the extrinsic carrier scattering time, and τ_{ph} is the total intrinsic scattering time arising from all the acoustic and optical phonon mode contributions.

9.3.1 Carrier scattering time

Through kinetic theory [44], for elastic scattering processes in 2D, τ_e is given by

$$\tau_e(E) = \frac{1}{n_d \sigma_m \langle v \rangle}, \tag{9.19}$$

where n_d is the disorder density, $\langle v \rangle$ is the average velocity, and σ_m is the momentum scattering cross-section defined as

$$\sigma_m = \int_0^{2\pi} \frac{d\sigma}{d\theta} \left(1 - \cos\theta\right) d\theta, \tag{9.20}$$

where, $d\sigma/d\theta$ is the differential cross-section length. The average incident velocity

$$\langle v \rangle = \frac{1}{\pi} \int_{-\pi/2}^{\pi/2} d\beta \, v \cos\beta, \tag{9.21}$$

where $v = \left|\nabla_k E_n(k)\right| / \hbar$ is the carrier velocity, and β is the angle between the velocity vector and the longitudinal axis. For example, for a uniform velocity distribution, $\langle v \rangle = 2v/\pi$.

9.3.2 Phonon scattering time

First-principles-calculations provide a convenient way of determining the phonon scattering times by computing the deformation potentials of the crystal lattice [45]. In 2D materials, the phonon scattering rate from the longitudinal and transverse acoustic phonons is given by [46–48]

$$\frac{1}{\tau_{ph,0}} = \frac{m^* D_1^2 \, k_B T}{\hbar^3 \, \rho_{2D} \, c_s^2}, \tag{9.22}$$

where m^* is the effective mass, D_1 is the first-order deformation potential, T is the absolute temperature, ρ_{2D} is the 2D ion mass density, and c_s is the sound velocity of the acoustic phonons. The inter-valley acoustic phonon and optical phonon rates are expressed in terms of the zero-order deformation potential D_0, and is given by

$$\frac{1}{\tau_{ph,1}} = \frac{m^* D_0^2}{2\hbar^2 \omega_\xi \rho_{2D}} \left(e^{\hbar\omega_\xi / k_B T} \Theta(E - \hbar\omega_\xi) + 1 \right) f_\xi^{BE}, \tag{9.23}$$

where Θ is the Heaviside step function, ω_ξ is the phonon frequency, and f_ξ^{BE} is the Bose-Einstein distribution function for the phonon mode ξ. Effect of the Fröhlich interaction can be implicitly added to the deformation potential [47]. We neglect the first-order optical deformation potential contributions, that are typically small while compared to the values from Eqs. (9.22) and (9.23). Hence, the total intrinsic phonon scattering rate is computed by summing over the all phonon modes, $\tau_{ph}^{-1} = \sum_\xi \tau_\xi^{-1}$.

In Fig. 9.5, we have plotted the total intrinsic phonon scattering rate at room temperature for TMDC monolayers. We have included deformation potentials for the acoustic and optical phonon modes corresponding to the transition

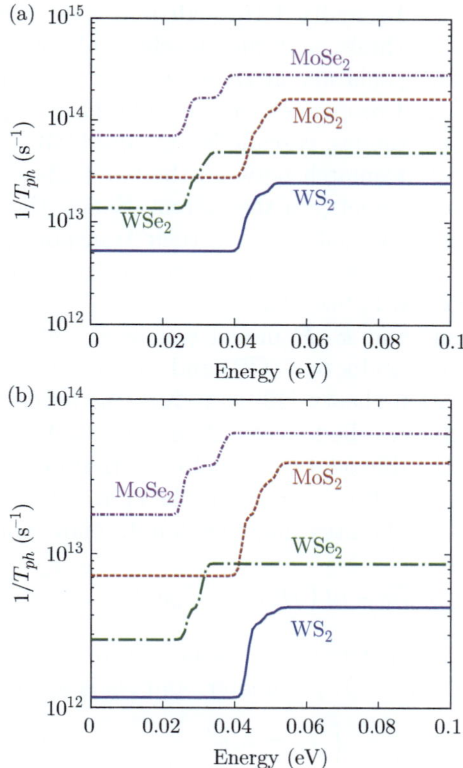

Figure 9.5 The total phonon scattering rates for the K-valley electrons are plotted as a function of energy for TMDCs at temperature $T = 300\,\text{K}$. Scattering rates are calculated using the deformation potentials listed in Ref. [46–48].

$K \rightarrow \{K, K', Q, Q'\}$. Emergence of optical phonon modes are observed as steps in the scattering rate. We observe that $MoSe_2$ (WS_2) has the strongest (weakest) interaction with the phonons in the family of TMDC monolayers. WX_2 has a greater electrical conductivity than MoX_2. These observations and scattering rate values are consistent with previous first-principles study in the literature [46, 47].

9.4 Transport properties in TMDC heterostructures

Transition metal dichalcogenides have attracted tremendous recent interest in fabricating high-mobility nanoelectric circuits [49–51]. In contrast to graphene, TMDCs host massive Dirac fermions due to their direct band gap, which negates the probability of Klein tunneling [52]. Strong coupling between spin and valley degrees of freedom are observed in TMDCs due to their lack of inversion symmetry [53]. This makes them interesting candidates to realize "valleytronic" devices as well [54].

Similar crystal structure and comparable lattice constants observed in MX_2 (M = Mo, W; X = S, Se) monolayers have motivated the study of TMDC-based quantum confinement and heterostructures [13, 55–60]. Experimentally, such lateral heterostructures can be realized through multistep chemical vapor deposition techniques, bottom up synthesis, and micromechanical techniques [61–70]. Ability to pattern the material encapsulation in these systems provides an enhanced level of control over their electrical properties. Hence, they have a great potential to realize in-plane transistors, photodiodes, cascade lasers, and CMOS circuits.

As shown in Fig. 9.2, mismatch between the band-offsets between the two layers breaks the translational symmetry of the system. Hence, the evanescent modes (with complex wavevector) heavily contribute to their transport properties. Therefore, in 2D lateral heterostructures, the carrier scattering calculations have to be carried out using the method described in Sec. 9.3.

In TMDCs, a seven-band $\boldsymbol{k} \cdot \boldsymbol{P}$ model can accurately capture most important dispersion features of the conduction (CB) and valence (VB) bands [71]. We can then apply Löwdin perturbation theory [39] to reduce the seven-band model into a two-band model corresponding to the lowest CB and highest VB. Here, we keep the terms up to second-order in the diagonal element in the Hamiltonian corresponding to the basis set $\{|\phi_c, s\rangle, |\phi_v, s, \eta\rangle\}$. Here, $s = \pm 1$ is the spin index, and $\eta = \pm 1$ denotes the valley $\pm K$. C_{3h} symmetry dictates that the Hamiltonian should have the form

$$H_{kp} = H_0 + at\,(\eta\, k_x \hat{\sigma}_x + i k_y \hat{\sigma}_y) - \lambda\,\eta \frac{(\hat{\sigma}_z - 1)}{2} s, \tag{9.24}$$

where $\hat{\sigma}$ denotes the Pauli matrices, a is the lattice constant, t is the effective hopping integral, λ is the spin-orbit (SO) parameter, and H_0 is given by

$$H_0 = \begin{bmatrix} E_c + \alpha_s k^2 & 0 \\ 0 & E_v + \beta_s k^2 \end{bmatrix}. \tag{9.25}$$

Here, α_s, β_s are the material parameters, and E_c and E_v are the CB minima and the VB maxima, respectively. The $\boldsymbol{k} \cdot \boldsymbol{P}$ theory is corrupted with the occurrence of spurious solutions. We have utilized the Foreman transformations [72] to eliminate the spurious solutions, and hence, we consider $\beta_s = 0$. In Table 9.1 we have listed all the $\boldsymbol{k} \cdot \boldsymbol{P}$ parameters used in our calculations for MoS_2, WS_2, $MoSe_2$, and WSe_2 monolayers. We have neglected strain matrix elements since MoS_2 ($MoSe_2$) and WS_2 (WSe_2) have comparable lattice constants. Material parameters between the two layers are smoothly interpolated through the self-consistent potential distribution obtained from the DFT calculations.

In Fig. 9.6, we have plotted the electron scattering rate for a 2D triangular superlattice between MoS_2 and WS_2 monolayers. There have been experimental efforts to realize such structures through lateral epitaxy [69]. MoS_2 has a lower CB minimum while compared to WS_2. Hence, below $E < -3.97\,\text{eV}$ (marked by a green arrow in Fig. 9.6), transport in WS_2 layers can occur only through the evanescent modes.

Table 9.1 The $\boldsymbol{k} \cdot \boldsymbol{P}$ parameters used in our calculations are listed here. These parameters were obtained from the previously reported first-principles study [38, 71].

	E_c (eV)	E_v (eV)	a (Å)	t (eV)	λ (eV)	$\alpha_+(\alpha_-)$ (eV·Å^2)
MoS_2	−4.31	−5.89	3.184	1.059	0.073	−5.97 (−6.43)
WS_2	−5.97	−5.50	3.186	1.075	0.211	−6.14 (−7.95)
$MoSe_2$	−3.91	−5.23	3.283	0.940	0.090	−5.34 (−5.71)
WSe_2	−3.61	−4.85	3.297	1.190	0.230	−5.25 (−6.93)

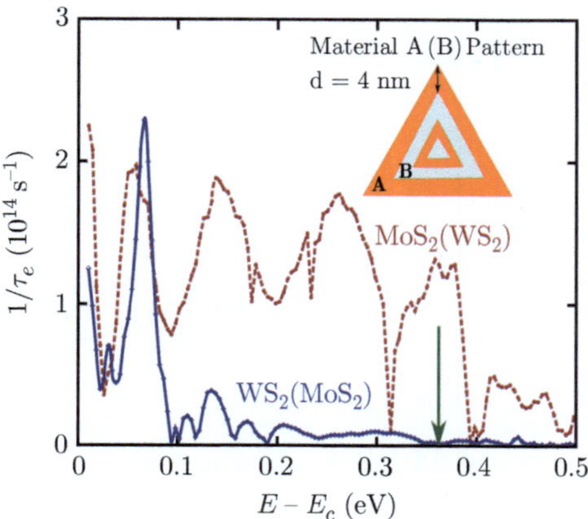

Figure 9.6 Electron scattering rate is plotted as a function of energy for triangular heterostructures of $MoS_2(WS_2)$ and $WS_2(MoS_2)$. A(B) material pattern is displayed in the inset. The material parameters employed here are listed in Table 9.1 and $n_d = 10^{12}\,\text{cm}^{-2}$. Width of each layer is taken to be 4 nm. Scattering rates are calculated using the envelope function scattering theory discussed in Sec. 9.3.

Evanescent wavevector are the purely imaginary k roots of the $\boldsymbol{k} \cdot \boldsymbol{P}$ Hamiltonian defined in Eq. (9.24). First, we observe that the WS_2 (MoS_2) structure has a lower scattering rate (higher lifetime) while compared to MoS_2 (WS_2) structure. This is because an electron injected from WS_2 layer can scatter off through the additional propagating conduction channel available from MoS_2 encapsulations. We observe that for transport in the MoS_2 (WS_2) structure, above the incoming energy $E > -3.97\,\text{eV}$ ($E - E_c = 0.34\,\text{eV}$, marked by an arrow in Fig. 9.6), corresponding to the CB minima of WS_2, the electron scattering rate decreases significantly, since the evanescent channel disappears above this energy. The resonances observed in the MoS_2 (WS_2) structure, below the energy $E < -3.97\,\text{eV}$ are due to the interaction with the trapped electron states within the WS_2 barrier.

For transport in WS_2 (MoS_2) structure, we observe a maxima and minima in the electron scattering rate around $E - E_c = 0.1\,\text{eV}$ (see Fig. 9.6). This marks the occurrence of Fano resonance in the transmission profile, an effect analogous to that of atomic autoionization [73], first observed in the context of inelastic electron scattering by a helium atom. Analogous Fano profiles are observed around the subband minima of a quantum waveguide with attractive potentials [74]. Here, the MoS_2 encapsulation acts as a quantum well and leads to the formation of quasi-bound metastable states. Interaction between the propagating modes and the quasi-bound states leads to the Fano resonance. Multiple Fano resonances are observed when we have a superlattice between WS_2 and MoS_2 layers as shown in Fig. 9.7. Formation of Fano resonances leads to an enhancement in the thermoelectric figure-of-merit far beyond pure 2D monolayers.

Lastly, we study the mobility in the TMDC heterostructures. Around the $K-$valley, CB is isotropic and parabolic in nature. Hence, the mobility can be defined as

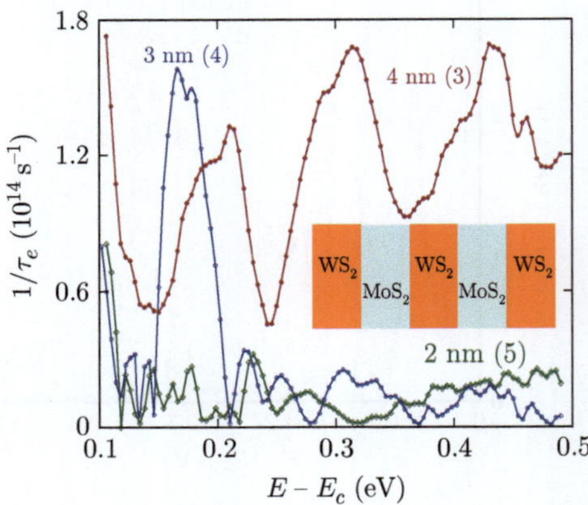

Figure 9.7 Electron scattering rate is plotted as a function of energy for WS_2-MoS_2 lateral heterostructures. Electron will be initiated from the WS_2 layer. Here, the notation 4 nm(3) represents that there are 3 periods of WS_2 and MoS_2 layers, and each layer has a width 4 nm. The material parameters employed here are listed in Table 9.1 and $n_d = 10^{12}\,\text{cm}^{-2}$. Scattering rates are calculated using the envelope function scattering theory discussed in Sec. 9.3.

$$\mu = \frac{e}{m^*}\frac{1}{n}\int dE\, g(E)\, E\, \tau(E)\left(-\frac{\partial f(E)}{\partial E}\right), \tag{9.26}$$

where, the 2D carrier density $n = \int dE\, g(E)\, f(E)$, $g(E)$ is the density of states, and the Fermi-Dirac distribution is given by $f(E) = (1 + \exp[(E - E_F)/k_B T])^{-1}$.

Due to the parabolic nature of the CB, we consider a constant $g(E) = g_s g_v m_K^*/2\pi\hbar^2$, where g_s and g_v ($g_v = 1, 2$ for VB, CB) are the spin and valley degeneracies, respectively.

In Fig. 9.8, we plot the total mobility (including both phonon and electron contributions) at room temperature for the n-type TMDC monolayers with triangular encapsulations. The density of triangular regions is considered to be $n_d = 10^{12}\,\mathrm{cm}^{-2}$. We observe an enhancement in the mobility at low carrier density. WS_2 monolayer with MoS_2 encapsulation will have the highest mobility among the considered combinations. This can be attributed to two main factors: (i) additional CB channels available from MoS_2 encapsulations enhance the electron scattering lifetime, (ii) WS_2 monolayer has the highest phonon scattering lifetime among the family of TMDC monolayers (see Fig. 9.5). MoS_2 monolayer with WS_2 encapsulation has the lowest mobility. This is because the evanescent modes, being real functions will not contribute to the probability current, and hence, suppress the overall mobility. We also note that the mobility observed here in WX_2 monolayers with triangular encapsulations is an order of magnitude larger than the phonon-limited mobility, and mobility in layers with charged vacancies. In charged vacancies, Coulomb contributions suppress the mobility [75]. Hence, short-range potentials such as the encapsulations considered here are promising candidates to realize high-mobility devices.

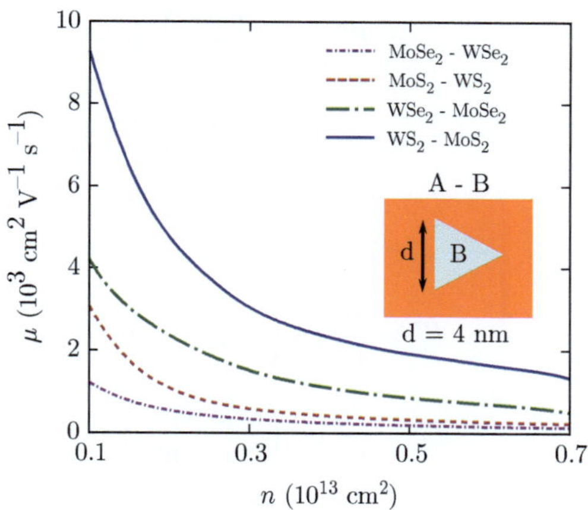

Figure 9.8 Electron mobility at room temperature is plotted as a function of 2D carrier density for the triangular encapsulations in TMDC heterostructures. The material parameters employed here are listed in Table 9.1 and $n_d = 10^{12}\,\mathrm{cm}^{-2}$.

9.4.1 Scattering cross-section length at a finite distance

The differential cross-section is determined by dividing the scattered flux by the incoming flux, given by

$$\frac{d\sigma}{d\theta} = \frac{\mathbf{J}^{\text{sc}} \cdot d\mathbf{S}}{J_{\text{in}}}, \tag{9.27}$$

where $\mathbf{J}^{\text{sc}}$ is the scattered current, and $J_{\text{in}} = \hbar k/m^*$ is the magnitude of the current from the incoming plane wave. In the traditional scattering theory, the observer is at infinity, and the outgoing wave is a circular wave of the form $e^{ik\rho}/\sqrt{\rho}$. When the observer is at a finite distance ρ, the outgoing wave-front incident on the observer will have both radial and angular components. The surface element $d\mathbf{S}$ is considered orthogonal to the outgoing current [24, 25]. In 2D scattering,

$$\begin{aligned} \mathbf{J}^{\text{sc}} \cdot d\mathbf{S} &= J^{\text{sc}}\mathbf{n} \cdot dS\mathbf{n}, \\ &= J^{\text{sc}}\sqrt{g}d\theta, \end{aligned} \tag{9.28}$$

where g is the determinant of the metric $g_{\mu\nu}$. In 2D circular coordinate system

$$\sqrt{g} = \rho\sqrt{1 + \frac{1}{\rho^2}\left(\frac{d\rho}{d\phi}\right)^2}. \tag{9.29}$$

Hence,

$$\begin{aligned} \mathbf{J}^{\text{sc}} \cdot d\mathbf{S} &= J^{\text{sc}}\sqrt{1 + \frac{1}{\rho^2}\left(\frac{d\rho}{d\phi}\right)^2}\,\rho\, d\theta, \\ &= J^{\text{sc}}\sqrt{1 + \left(\frac{J^{\text{sc}}_\theta}{J^{\text{sc}}_\rho}\right)^2}\,\rho\, d\theta, \\ &= J^{\text{sc}}_\rho\left[1 + \left(\frac{J^{\text{sc}}_\theta}{J^{\text{sc}}_\rho}\right)^2\right]\rho\, d\theta, \end{aligned} \tag{9.30}$$

where,

$$\begin{aligned} J^{\text{sc}}_\rho &= \frac{\hbar}{m}\,\text{Im}\left(\psi^\dagger_{sc}\frac{\partial\psi_{sc}}{\partial\rho}\right), \\ J^{\text{sc}}_\theta &= \frac{\hbar}{m}\,\text{Im}\left(\frac{\psi^\dagger_{sc}}{\rho}\frac{\partial\psi_{sc}}{\partial\theta}\right). \end{aligned}$$

The differential cross-section length is given by

$$\frac{d\sigma}{d\theta} = \frac{J^{\text{sc}}_\rho\left[1 + \left(\frac{J^{\text{sc}}_\theta}{J^{\text{sc}}_\rho}\right)^2\right]\rho}{J_{\text{in}}}. \tag{9.31}$$

Note that as the source and observer are pushed to infinity ($\rho \to \infty$),

$$
\begin{aligned}
J_{\theta}^{\text{sc}} &\to 0, \\
J^{\text{sc}} &\to \frac{\hbar}{m^*}\frac{|f|^2}{\rho}, \\
\frac{d\sigma}{d\theta} &\to \frac{|f|^2}{k},
\end{aligned}
\tag{9.32}
$$

where f is scattering amplitude. Hence, we obtain the standard definition of the differential cross-section length [19] in the limit $\rho \to \infty$.

9.4.2 Example: scattering from a hard circle potential

In this section, we discuss the analytical results for the scattering from a hard circle potential in 2D. Even though, this problem has been addressed before in the literature [76, 77], they have not discussed the limiting behaviors. Furthermore, some of their work contains minor errors.

The hard circle potential of radius a in a 2D space is given by

$$
V = \begin{cases} \infty, & \rho \leq a \\ 0, & \rho > a \end{cases}. \tag{9.33}
$$

The solution wavefunction for the above potential will be of the form

$$
\psi(\rho, \phi) = \begin{cases} 0, & \rho \leq a \\ e^{ik\rho\,\cos\phi} + \sum\limits_{n=-\infty}^{\infty} c_n\, e^{in\phi} H_n(k\rho), & \rho > a \end{cases}. \tag{9.34}
$$

To determine the coefficient c_n, we utilize the BC $\psi(a, \phi) = 0$. Hence,

$$
c_n = \frac{-i^n J_n(ka)}{H_n(ka)}, \tag{9.35}
$$

where we have used the expansion of the plane wave

$e^{ik\rho cos\phi} = \sum\limits_{n=-\infty}^{\infty} e^{in\phi} i^n J_n(k\rho)$.

Through partial wave analysis, we obtain the total cross-section length given by

$$
\sigma = \frac{4}{k} \sum_{n=-\infty}^{\infty} \frac{|J_n(ka)|^2}{|H_n(ka)|^2}. \tag{9.36}
$$

To compare the quantum mechanical calculations with classical mechanics predictions, we consider the cross-section length in the high energy limit $ka >> 1$. We know that,

$$J_n(ka) \xrightarrow[ka\to\infty]{} \sqrt{\tfrac{2}{\pi ka}} \cos\left(ka - \tfrac{n\pi}{2} - \tfrac{\pi}{4}\right),$$
$$H_n(ka) \xrightarrow[ka\to\infty]{} \sqrt{\tfrac{2}{\pi ka}}\, e^{i\left(ka - \frac{n\pi}{2} - \frac{\pi}{4}\right)}. \tag{9.37}$$

Therefore,

$$\sigma \simeq \frac{4}{k} \sum_{n=-\infty}^{\infty} \cos^2\left(ka - \frac{n\pi}{2} - \frac{\pi}{4}\right). \tag{9.38}$$

Classically, the particle will not be deflected by the potential if the impact parameter, $b > a$. The angular momentum will follow the limit $|L_\phi| = n\hbar \le \hbar ka$. Hence, we can restrict the sum in Eq. (9.38) between the limits $-ka \le n \le ka$. With this simplification, we obtain the cross-section length

$$\sigma \simeq 4a, \tag{9.39}$$

which is twice the classical mechanics prediction [76].

To obtain the low energy limit ($ka << 1$) of the cross-section length, we retain only the first term ($n = 0$) in the series given in Eq. (9.36). In this limit,

$$J_n(ka) \xrightarrow[ka\to 0]{} 1,$$
$$H_n(ka) \xrightarrow[ka\to 0]{} \frac{2}{\pi} \ln(ka). \tag{9.40}$$

Hence,

$$\sigma \simeq \frac{\pi^2}{k} \frac{1}{(\ln(ka))^2}. \tag{9.41}$$

This is not an analytic function. Therefore, as $k \to 0$, cross-section length will diverge even though the scattering current vanishes.

9.4.3 Scattering in 3D using sources and absorbers

In this section, we set up the framework for the quantum scattering theory in 3D, using our source and absorber scheme. In 3D, we consider a spherical source, and the absorber will be a spherical shell. Incoming plane wave will be of the form

$$e^{i\mathbf{k}\cdot\mathbf{r}} = 4\pi \sum_{l=0}^{\infty} \sum_{m=-l}^{l} i^l\, Y^*_{ml}(k_\theta, k_\phi) Y_{ml}(\theta, \phi) j_l(kr), \tag{9.42}$$

where, Y_{ml} is the Laplace spherical harmonics, $j_l(kr)$ is the spherical Bessel function, and (k_θ, k_ϕ) define the direction of the wavevector $\mathbf{k}$. Following a similar scheme as in Sec. 9.2, we can derive the corresponding source term

$$S(r',\theta',\phi') = \frac{4\pi i}{k^2}\sum_{l=0}^{\infty}\sum_{m=-l}^{l}\frac{i^l\,Y^*_{ml}(k_\theta,k_\phi)Y_{ml}(\theta',\phi')}{h_l(kr')}, \tag{9.43}$$

where, $h_l(kr')$ is the spherical Hankel function. With this set up, incoming wavefunction in the absence of external potential is given by

$$\psi_{in} = \begin{cases} e^{i\mathbf{k}\cdot\mathbf{r}} & ,r \le r'; \\ 4\pi\displaystyle\sum_{l=0}^{\infty}\sum_{m=-l}^{l} i^l\,Y^*_{ml}(k_\theta,k_\phi)Y_{ml}(\theta',\phi')\frac{j_l(kr')h_l(kr)}{h_l(kr')} & \\ & ,r > r'. \end{cases}$$

Hence, we obtain plane waves injecting on the scattering center from a spherical source at $r = r'$.

9.5 Action integral and open domain scattering

In this section, we explain the action integral formulation employed to solve Eq. (9.16). Let us consider a general form of the $\boldsymbol{k}\cdot\boldsymbol{P}$ Hamiltonian in a 2D material given by

$$\mathbf{H}_{kp} = \mathbf{A}_{xx}\partial_x^2 + \mathbf{A}_{yy}\partial_y^2 + \mathbf{A}_{xy}\partial_{xy}^2 + \mathbf{B}_x\partial_x + \mathbf{B}_y\partial_y + \mathbf{C}, \tag{9.44}$$

where $\mathbf{A}_{xx}, \mathbf{A}_{yy}, \mathbf{A}_{xy}, \mathbf{B}_x, \mathbf{B}_y, \mathbf{C}$ are the coefficient matrices. For simplicity, we have considered only the terms up to the second-order derivatives. With this Hamiltonian, the action integral corresponding to Eq. (9.16) is given by

$$\begin{aligned}\mathcal{A} &= \int d^2\rho\,\mathbb{F}^\dagger\cdot\mathbf{L}_{kp}\cdot\mathbb{F} \\ &\quad + \int d^2\rho\,S(\rho',\phi')\,\eta_{\text{scale}}\,\mathbb{F}^\dagger\cdot\mathbb{A}\,\frac{\delta(\rho-\rho')}{\rho'},\end{aligned} \tag{9.45}$$

where $\mathbf{L}_{kp}$ is the Lagrangian operator given by

$$\begin{aligned}\mathbf{L}_{kp} = \Big[&\overleftarrow{\partial}_x\mathbf{A}_{xx}\vec{\partial}_x + \overleftarrow{\partial}_y\mathbf{A}_{yy}\vec{\partial}_y + \frac{1}{2}\left(\overleftarrow{\partial}_x\mathbf{A}_{xy}\vec{\partial}_y + \overleftarrow{\partial}_y\mathbf{A}_{xy}\vec{\partial}_x\right) \\ &+\frac{i}{2}\left(\overleftarrow{\partial}_x\mathbf{B}_x - \mathbf{B}_x\vec{\partial}_x\right) + \frac{i}{2}\left(\overleftarrow{\partial}_y\mathbf{B}_y - \mathbf{B}_y\vec{\partial}_y\right) + \mathbf{C}\Big],\end{aligned}$$

where $\overleftarrow{\partial}$ and $\vec{\partial}$ represent the derivative operator acting on the functions appearing to the left and right sides, respectively. This will ensure the correct operator ordering at the material interface [42].

Across a material interface, the envelope wavefunction, and the probability current has to be continuous. The continuity of the envelope function is ensured by using Hermite interpolation polynomials in our calculations, which will have both first and second derivative continuity [78]. The probability current is evaluated using a gauge-variational approach that is commonly used in high energy/particle physics. Following Gell-Mann and Levy [43], we perform a transformation of the envelope function with respect to an arbitrary gauge function of the form, $\mathbb{F} \to \mathbb{F}\, e^{i\Lambda(\rho)}$. With this transformation, we obtain the conserved current

$$\begin{aligned} J_x &= \mathbb{F}^\dagger \cdot \frac{\delta\, \mathbf{L}_{kp}}{\delta\, \partial_x \Lambda} \cdot \mathbb{F}, \\ J_y &= \mathbb{F}^\dagger \cdot \frac{\delta\, \mathbf{L}_{kp}}{\delta\, \partial_y \Lambda} \cdot \mathbb{F}. \end{aligned} \tag{9.46}$$

The interfacial potential is obtained from the DFT calculations, and will be continuous across the interface. Hence, this conserved current will also be continuous.

9.5.1 The scale factor

In this section, we obtain the expression for the scale factor η_{scale} employed in Eq. (9.16). For simplicity, let us consider the $\boldsymbol{k} \cdot \boldsymbol{P}$ Hamiltonian equation in one-dimension (1D). The expression we obtain for 1D will translate into higher dimensions as well. Let us consider the equation

$$[\mathbf{H}_{kp}(x) - E\mathbf{1}] \cdot \mathbb{F} = S\, \eta_{\text{scale}} \mathbb{A}\, \delta(x), \tag{9.47}$$

where, the source term in 1D is given by $S = 2ik_1$ [26, 30], where k_1 is the incoming wavevector. The solution to the above equation in terms of the Fourier components is given by

$$\mathbb{F}(x) = \frac{1}{2\pi} \int dk\, e^{ikx} 2ik_1\, \eta_{\text{scale}} \frac{\text{adj}(\mathbf{H}_{kp} - E\mathbf{1}) \cdot \mathbb{A}}{\text{Det}(\mathbf{H}_{kp} - E\mathbf{1})}. \tag{9.48}$$

Let us constrain $\mathbf{H}_{kp}$ be a 2×2 matrix, with eigenenergies given by E_c and E_v. Then

$$\mathbb{F}(x) = \frac{ik_1}{\pi}\, \mathbb{A}\, (E_c - E_v) \int dk\, \frac{e^{ikx} \eta_{\text{scale}}}{(k^2 - k_1^2)(k^2 - k_2^2) \cdots (k^2 - k_N^2)}, \tag{9.49}$$

where $k_1, k_2, \ldots, k_N$ are the solution wavevectors for the determinant equation $\text{Det}(\mathbf{H}_{kp} - E\mathbf{1}) = 0$. The Hamiltonian $\mathbf{H}_{kp}$ can include terms in higher order k (such as $k, k^2, k^3, \ldots$), and hence the determinant equation will have N roots. Hence, we choose

$$\eta_{\text{scale}} = \frac{(k^2 - k_2^2) \cdots (k^2 - k_N^2)}{(E_c - E_v)}, \tag{9.50}$$

so that the incoming wavefunction $\psi(x)$ will have the form

$$\psi(x) = \mathbb{A} e^{ik_1 x}. \tag{9.51}$$

Expression derived in Eq. (9.50) will follow for 2D and higher dimensions as well.

9.6 Conclusions

We have developed a non-asymptotic quantum scattering theory using a novel method of sources and absorbers. The Cauchy BCs that are necessary in the variational formulation of scattering are reduced to simpler Dirichlet BCs by constructing a perfect absorber (stealth region) around the scattering centers. This will also allow us to define a finite computational domain. In the conventional scattering theory, the scattering properties are determined at the asymptotic distance. This will result in the loss of evanescent mode contributions in multi-band scattering processes, and when we have absorbing scattering potentials. Introducing an absorber will facilitate us to calculate the scattering properties at a finite distance, and it will preserve the distance information between the observer and the scattering center. Once we enclose the active scattering region with the absorber, we require a carrier source that will generate plane waves in order to initiate the scattering event. We have derived the expression for the circular source term introduced in the right-hand side of the Hamiltonian equation.

We have shown that even for the simplest case of scattering from a hard circle potential, the cross-section length measured at a finite distance using our formalism, and at the asymptotic distance obtained through the analytical formulation will have significant differences in their numerical values. The carrier scattering time is inversely proportional to the cross-section length. Hence, our formalism will be of great interest in nanoscale systems, where the transport measurements are made in the proximity of the scattering centers.

We have shown that bridging our scattering theory framework using the $\boldsymbol{k} \cdot \boldsymbol{P}$ perturbation theory, with inputs from *ab-initio* electronic structure calculations provides a versatile multiscale formalism. The continuum nature of our formalism will enable the modeling of realistic devices, scaling from hundreds to thousands of atoms. Carrier scattering rates obtained through our formalism combined with the phonon scattering rates obtained through DFT calculations can accurately yield the thermoelectric properties of the nano-devices. Hence, it will be very useful to design and simulate nanoelectronics circuits using 2D materials with the theoretical guidelines provided by simulations done with this formalism.

The phonon scattering rates for MX_2 (M = Mo, W; X = S, Se) monolayers were obtained through deformation potential calculations. We observed that the WS_2 ($MoSe_2$) has the highest (lowest) phonon scattering time in the family of MX_2 monolayers. This observation is consistent with other first-principles study in the literature.

As an application of our formalism, we studied the transport properties in lateral TMDC heterostructures. Material encapsulations in TMDCs act as a short-range scattering centers. We observed the emergence of novel Fano resonances for the first time in 2D materials, when the MoS_2 is encapsulated within the WS_2 monolayer. MoS_2 encapsulation here acts as a quantum well, and forms the quasi-bound states that will interact with the propagating modes.

Lastly, we studied mobility as a function of carrier density in a family of TMDC layers with triangular encapsulations. WS_2 monolayer with MoS_2 encapsulation is observed to have the highest mobility by an order of magnitude larger than the phonon-limited mobility. Hence, these lateral TMDC heterostructures should be explored as candidates to realize high-mobility devices.

Finally, we note that the formalism developed here can be extended to study transport properties in any combination of 2D materials. We have also provided theoretical framework to setup a formalism in 3D open domain as well. This will allow us to model transport properties of devices hosted in bulk materials as well. Together, the formalism provides a scalable first-principles based quantum transport framework for simulating both 2- and 3-dimensional nano-devices.

References

[1] M-Y. Li, S-K. Su, H.-S. P. Wong, and L.-J. Li, Nature **567**, 169–170 (2019); "How 2D semiconductors could extend Moore's law."

[2] J. M. Shalf and R. Leland, Computer **48**, 14–23 (2015); "Computing beyond Moore's Law;" doi: 10.1109/MC.2015.374.

[3] F. Schwierz, Nat. Nanotechnol. **5**, 487–496 (2010); "Graphene transistors."

[4] L. Tao, E. Cinquanta, D. Chiappe, C. Grazianetti, M. Fanciulli, M. Dubey, A. Molle, and D. Akinwande, Nat. Nanotechnol. **10**, 227–231 (2015); "Silicene field-effect transistors operating at room temperature."

[5] L. Li, Y. Yu, G. Jun Ye, Q. Ge, X. Ou, H. Wu, D. Feng, X. H. Chen, and Y. Zhang, Nat. Nanotechnol. **9**, 372–377 (2014); "Black phosphorus field-effect transistors."

[6] Y. Yoon, K. Ganapathi, and S. Salahuddin, Nano Lett. **11**, 3768–3773 (2011); "How good can monolayer MoS_2 transistors be?"

[7] Y. Cui, R. Xin, Z. Yu, Y. Pan, Z-Y. Ong, X. Wei, J. Wang, H. Nan, Z. Ni, Y. Wu, T. Chen, Y. Shi, B. Wang, G. Zhang, Y-W. Zhang, and X. Wang, Adv. Mater. **27**, 5230–5234 (2015); "High-performance monolayer WS2 field-effect transistors on high-κ dielectrics."

[8] A. D. Bartolomeo, F. Urban, M. Passacantando, N. McEvoy, L. Peters, L. Iemmo, G. Luongo, F. Romeo, and F. Giubileo, Nanoscale, **11**, 1538–1548 (2019); "A WSe_2 vertical field emission transistor."

[9] J. Lee, D. Wong, J. Velasco, Jr., J. F. Rodriguez-Nieva, S. Kahn, H.-Z. Tsai, T. Taniguchi, K. Watanabe, A. Zettl, F. Wang, L. S. Levitov, and M. F. Crommie, Nat. Phys. **12**, 1032–1037 (2016); "Imaging electrostatically confined Dirac fermions in graphene quantum dots."

[10] C. Gutiérrez, L. Brown, C.-J. Kim, J. Park, and A. N. Pasupathy, Nature Phys. **12**, 1068–1076 (2016); "Klein tunneling and electron trapping in nanometre-scale graphene quantum dots;" http://dx.doi.org/10.1038/nphys3806 (2016).

[11] E. Margapoti, J. Li, O. Ceylan, M. Seifert, F. Nisic, T. Le Anh, F. Meggendorfer, C. Dragonetti, C.-A. Palma, J. V. Barth, and J. J. Finley, Adv. Mater. **27**, 1426–1431 (2015); "A 2D semiconductor–self-assembled monolayer photoswitchable diode."

[12] D. M.-Gustin, L. Cabral, M. P. Lima, J. L. F. Da Silva, E. Margapoti, and S. E. Ulloa, Phys. Rev. B. **98**, 241403(R)–1–5 (2018); "Photomodulation of transport in monolayer dichalcogenides."

[13] C. C. PriceNathan, C. Frey, D. Jariwala, and V. B. Shenoy, ACS Nano **13**, 8303–8311 (2019); "Engineering zero-dimensional quantum confinement in transition-metal dichalcogenide heterostructures."

[14] W. Kohn and L. J. Sham, Phys. Rev. **140**, A1133–A1138 (1965); "Self-consistent equations including exchange and correlation effects."

[15] C. W. Groth, M. Wimmer, A. R. Akhmerov, and X. Waintal, New J. Phys. **16**, 063065–1–39 (2014); "Kwant: a software package for quantum transport."

[16] M. L. Van de Put, M. V. Fischetti, and W. G. Vandenberghe, Comput. Phys. Commun. **244**, 156–169 (2019); "Scalable atomistic simulations of quantum electron transport using empirical pseudopotentials."

[17] J. Taylor, H. Guo, and J. Wang, Phys. Rev. B **63**, 245407–1–13 (2001); "Ab initio modeling of quantum transport properties of molecular electronic devices."

[18] M. Brandbyge, J. L. Mozos, P. Ordejón, J. Taylor, and K. Stokbro, Phys. Rev. B **65**, 1654011–1–17 (2002); "Density-functional method for nonequilibrium electron transport."

[19] S. K. Adhikari, Am. J. Phys. **54**, 362–367 (1986); "Quantum Scattering in two dimensions."

[20] A. Sommerfeld, *Partial Differential Equations in Physics* (Academic, New York, NY, 1949).

[21] Lord Rayleigh, *The theory of sound*, Vol. **2** (Macmillan and Co., Ltd., New York, 1896); Ch. 17, pp. 236–284.

[22] H. Faxen and J. P. Holtsmark, Zeits. f. Physik **45**, 307–324 (1927); "Beitrag zur Theorie des Durchganges langsamer Elektronen durch Gase."

[23] C. C.-Tannoudji, B. Diu, and F. Laloe, *Quantum Mechanics*, Vol. **2** (John Wiley & Sons Inc, New York, NY, 1991).

[24] T. Liu, W-.D Li, and W-.S. Dai, J. High Energy Phys. **2014**, 87–1–11 (2014); "Scattering theory without large-distance asymptotics."

[25] W-.D Li, and W-.S. Dai, J. Phys. A: Math. Theor. **49**, 465202–1–19 (2016); "Scattering theory without large-distance asymptotics in arbitrary dimensions."

[26] Sathwik Bharadwaj, and L. R. Ram-Mohan, J. Appl. Phys. **125**, 164306–1–11 (2019); "Electron scattering in quantum waveguides with sources and absorbers. I. Theoretical formalism."

[27] H. K. Harbury and W. Porod, J. Appl. Phys. **75**, 5142–5149 (1994); "Quantum scattering states in open two-dimensional electronic system."

[28] H. Gan, P. L. Levin, and R. Ludwig, J. Acoust. Soc. Am. **94**, 1651–1662 (1993); "Finite element formulation of acoustic scattering phenomena with absorbing boundary condition in the frequency domain."

[29] P.M. Morse, and H. Feshbach, *Methods of Theoretical Physics* (McGraw-Hill Book Company, New York, NY, 1953).

[30] L. R. Ram-Mohan, *Finite Element and Boundary Element Applications in Quantum Mechanics* (Oxford University Press Inc., New York, 2002).

[31] O. C. Zienkiewicz and Y. K. Cheung, *Finite Element Methods in Structural and Continuum Mechanics* (McGraw-Hill, New York, 1967); O. C. Zienkiewicz, *The Finite Element Method* (McGraw-Hill, New York, 1977).

[32] J. Jin, *The Finite Element Method in Electromagnetics*, 2nd edition (Wiley, New York, 2002).

[33] B. Chen, M. Lazzouni, and L. R. Ram-Mohan, Phys. Rev. B **45**, 1204–1212 (1992); "Diagonal representation for the transfer-matrix method for obtaining electronic energy levels in layered semiconductor heterostructures."

[34] J. M. Luttinger, and W. Kohn, Phys. Rev. **97**, 869–883 (1955); "Motion of electrons and holes in perturbed periodic fields."

[35] J. M. Luttinger, Phys. Rev. **102**, 1030–1041 (1956); "Quantum Theory of Cyclotron Resonance in Semiconductors: General Theory."

[36] G. Bir and G. Pikus, *Symmetry and Strain-Induced Effects in Semiconductors* (Wiley, New York, NY, 1974).

[37] M. S. Dresselhaus, G. Dresselhaus, and A. Jorio, *Group Theory Application to the Physics of Condensed Matter* (Springer, Berlin, 2008).

[38] S. Haastrup, M. Strange, M. Pandey, T. Deilmann, P. S. Schmidt, N. F. Hinsche, M. N. Gjerding, D. Torelli, P. M. Larsen, A. C. Riis-Jensen, J. Gath, K. W. Jacobsen, J. J. Mortensen, T. Olsen, and K. S. Thygesen, 2D Materials **5**, 042002–1–36 (2018); "The computational 2D materials database: High-throughput modeling and discovery of atomically thin crystals."

[39] P. O. Löwdin, J. Chem. Phys. **19**, 1396–1401 (1951); "A note on the quantum mechanical perturbation Theory."

[40] E.O.Kane, J. Phys. Chem. Solids **6**, 236–241 (1958); "The influence of exchange on the effective mass formalism."

[41] E. O. Kane, *Semiconductors and Semimetals*, vol. 1, ed. R. K. Willardson, and A. C. Beer (Academic, New York, NY, 1966). See also, E. O. Kane, *Handbook on Semiconductors*, vol. 1, ed. W. Paul (North-Holland, Amsterdam,1982) p. 193.

[42] L. R. Ram-Mohan and K.-H. Yoo, J. Phys.: Condens. Matter **18**, R901–R917 (2006); "Wavefunction engineering of layered semiconductors: theoretical foundations."

[43] M. Gell-Mann, and M. Levy, Il Nuovo Cimento **16** 705–726 (1960); "The axial vector current in beta decay."

[44] R. L. Liboff, *Kinetic Theory: Classical, Quantum, and Relativistic Descriptions* (Springer, Berlin, 2003).

[45] J. Bardeen, and W. Shockley, Phys. Rev. **80**, 72–80 (1950); "Deformation potentials and mobilities in non-polar crystals."

[46] K. Kaasbjerg, K. S. Thygesen, and K. W. Jacobsen, Phys. Rev. B **85**, 115317–1–16 (2012); "Phonon-limited mobility in n-type single-layer MoS_2 from first principles."

[47] Z. Jin, X. Li, J. T. Mullen, and K. W. Kim, Phys. Rev. B **90**, 045422–1–7 (2014); "Intrinsic transport properties of electrons and holes in monolayer transition-metal dichalcogenides."

[48] X. Li, J. T. Mullen, Z. Jin, K. M. Borysenko, M. B. Nardelli, and K. W. Kim, Phys. Rev. B **87**, 115418–1–8 (2013); "Intrinsic electrical transport properties of monolayer silicene and MoS2 from first principles."

[49] B. Radisavljevic, A. Radenovic, J. Brivio, V. Giacometti, and A. Kis, Nat. Nanotechnol. **6**, 147–150 (2011); "Single-layer MoS_2 transistors."

[50] A. Ovchinnikov, A. Allain, Y.-S. Huang, D. Dumcenco, and A. Kis, ACS Nano **8**, 8174–8181 (2014); "Electrical transport properties of single-layer WS_2."

[51] B. Radisavljevic and A. Kis, Nat. Mater. **12**, 815–820 (2013); "Mobility engineering and a metal-insulator transition in monolayer MoS_2."

[52] D. M.-Gustin, S. E. Ulloa, and V. L.-Richard, Phys. Rev. B **98**, 125301–1–7 (2018); "Electron scattering in two-dimensional semiconductors: Contrasting massive Dirac and Schrödinger behavior."

[53] H. Zeng, J. Dai, W. Yao, D. Xiao, and X. Cui, Nature Nanotech. **7**, 490–493 (2012); "Valley polarization in MoS2 monolayers by optical pumping."

[54] Y. Ye, J. Xiao, H. Wang, Z. Ye, H. Zhu, M. Zhao, Y. Wang, J. Zhao, X. Yin, and X. Zhang, Nature Nanotech. **11**, 598–602 (2016); "Electrical generation and control of the valley carriers in a monolayer transition metal dichalcogenide."

[55] A. Kormányos, V. Zólyomi, N. D. Drummond, and G. Burkard, Phys. Rev. X **4**, 011034–1–16 (2014); "Spin-Orbit Coupling, Quantum Dots, and Qubits in Monolayer Transition Metal Dichalcogenides."

[56] G.-B. Liu, H. Pang, Y. Yao, and W. Yao, New J. Phys. **16**, 105011–1–23 (2014); "Intervalley coupling by quantum dot confinement potentials in monolayer transition metal dichalcogenides."

[57] Y. Wu, Q. Tong, G.-B. Liu, H. Yu, and W. Yao, Phys. Rev. B **93**, 045313–1–15 (2016); "Spin-valley qubit in nanostructures of monolayer semiconductors: Optical control and hyperfine interaction."

[58] J. Pawlowski, D. Zebrowski, and S. Bednarek, Phys. Rev. B **97**, 155412–1–12 (2018); "Valley qubit in gated MoS_2 monolayer quantum dot."

[59] R. Pisoni, Z. Lei, P. Back, M. Eich, H. Overweg, Y. Lee, K. Watanabe, T. Taniguchi, T. Ihn, and K. Ensslin, Appl. Phys. Lett. **112**, 123101 (2018); "Gate-tunable quantum dot in a high quality single layer MoS_2 van der Waals heterostructure."

[60] O. A-Ovando, D. Mastrogiuseppe, and S. E. Ulloa, Phys. Rev. B **99**, 035107–1–11 (2019); "Lateral interfaces of transition metal dichalcogenides: A stable tunable one-dimensional physics platform."

[61] C. Huang, S. Wu, A. M. Sanchez, J. J. P. Peters, R. Beanland, J. S. Ross, P. Rivera, W. Yao, D. H. Cobden, and X. Xu, Nat. Mater. **13**, 1096–1101 (2014); "Lateral heterojunctions within monolayer MoSe2-WSe2 semiconductors."

[62] Y. Gong, J. Lin, X. Wang, G. Shi, S. Lei, Z. Lin, X. Zou, G. Ye, R. Vajtai, B. I. Yakobson, H. Terrones, M. Terrones, B. K. Tay, J. Lou, S. T. Pantelides, Z. Liu, W. Zhou, and P. M. Ajayan, Nat. Mater. **13**, 1135–1142 (2014); "Vertical and in-plane heterostructures from WS2/MoS2 monolayers."

[63] X. Duan, C. Wang, J. C. Shaw, R. Cheng, Y. Chen, H. Li, X. Wu, Y. Tang, Q. Zhang, A. Pan, J. Jiang, R. Yu, Y. Huang, and X. Duan, Nat. Nanotechnol. **9**, 1024–1030 (2014); "Lateral epitaxial growth of two-dimensional layered semiconductor heterojunctions."

[64] X.-Q. Zhang, C.-H. Lin, Y.-W. Tseng, K.-H. Huang, and Y.-H. Lee, Nano Lett. **15**, 410–415 (2015); "Synthesis of lateral heterostructures of semiconducting atomic layers."

[65] M.-Y. Li, Y. Shi, C.-C. Cheng, L.-S. Lu, Y.-C. Lin, H.-L. Tang, M.-L. Tsai, C.-W. Chu, K.-H. Wei, J.-H. He, W.-H. Chang, K. Suenaga, and L.-J. Li, Science **349**, 524–528 (2015); "NANOELECTRONICS. Epitaxial growth of a monolayer WSe2-MoS2 lateral p-n junction with an atomically sharp interface."

[66] C. Zhang, Y. Chen, J.-K. Huang, X. Wu, L.-J. Li, W. Yao, J. Tersoff, and C.-K. Shih, Nat. Commun. **7**, 10349–1–6 (2016); "Visualizing band offsets and edge states in bilayer–monolayer transition metal dichalcogenides lateral heterojunction."
[67] C. Zhang, M.-Y. Li, J. Tersoff, Y. Han, Y. Su, L.-J. Li, D. A. Muller, and C.-K. Shih, Nat. Nanotechnol. **13**, 152–158 (2018); "Strain distributions and their influence on electronic structures of WSe2–MoS2 laterally strained heterojunctions."
[68] P. K. Sahoo, S. Memaran, Y. Xin, L. Balicas, and H. R. Gutiérrez, Nature **553**, 63–67 (2018); "One-pot growth of two-dimensional lateral heterostructures via sequential edge-epitaxy."
[69] S. Xie, L. Tu, Y. Han, L. Huang, K. Kang, K. U. Lao, P. Poddar, C. Park, D. A. Muller, R. A. DiStasio, and J. Park, Science **359**, 1131–1136 (2018); "Coherent, atomically thin transition-metal dichalcogenide superlattices with engineered strain."
[70] J. Wang, Z. Li, H. Chen, G. Deng, and X. Niu, Nano-Micro Lett. **11**, 48–1–31 (2019); "Recent advances in 2D lateral heterostructures."
[71] A. Kormányos, G. Burkard, M. Gmitra, J. Fabian, V. Zólyomi, N. D. Drummond, and V. I. Fal'ko, 2D Mater. **2**, 022001–1–31 (2015); "k · p theory for two-dimensional transition metal dichalcogenide semiconductors;" See also, A. Kormányos, V. Zólyomi, N. D. Drummond, G. Burkard, and V. I. Fal'ko, Phys. Rev. B **88**, 045416–1–8 (2013); "Monolayer MoS_2: Trigonal warping, the Γ valley, and spin-orbit coupling effects."
[72] B. A. Foreman, Phys. Rev. B **56**, R12748–R12751 (1997); "Elimination of spurious solutions from eight-band kp theory."
[73] U. Fano, Phys. Rev. **124**, 1866–1878 (1961); "Effects of configuration interaction on intensities and phase shifts;" U. Fano and J. W. Cooper, *ibid* **137**, A1364–A1379 (1965); "Line Profiles in the far-UV absorption spectra of the rare gases."
[74] Sathwik Bharadwaj, and L. R. Ram-Mohan, J. Appl. Phys. **125**, 164307–1–12 (2019); "Electron scattering in quantum waveguides with sources and absorbers. II. Applications."
[75] K. Kaasbjerg, T. Low, and A.-P. Jauho, Phys. Rev. B **100**, 115409–1–12 (2019); "Electron and hole transport in disordered monolayer MoS2: Atomic vacancy induced short-range and Coulomb disorder scattering."
[76] I. R. Lapidus, Am. J. Phys. **54**, 459–461 (1986); "Quantum scattering in two dimensions."
[77] S. McAlinden and J. Shertzer, Am. J. Phys. **84**, 764–769 (2016);"Quantum scattering from cylindrical barriers."
[78] P. G. Kassebaum, C. R. Boucher, and L. R. Ram-Mohan, J. Comp. Phys. **231**, 5747–5760 (2012); "Application of group representation theory to derive Hermite interpolation polynomials on a triangle."

Part III

Wavefunction Engineering

Part III

Wavefunction Engineering

10

Wavefunction engineering of semiconductor nanostructures

In this chapter:

- We present the theoretical basis of wavefunction engineering. We show that a Lagrangian formulation for the valence bands of bulk semiconductors in the $\mathbf{k} \cdot \mathbf{P}$ model provides a direct approach to determining derivative operator ordering in the multi-band description of electronic states in semiconductors in the envelope function approximation.
- The current continuity condition is also obtained through a gauge variation on the Lagrangian. This naturally leads into a finite element approach for the discretization of Schrödinger's equation. Furthermore, by including the Poisson Lagrangian, a self-consistent treatment of the Schrödinger-Poisson band bending in arbitrarily doped structures can be calculated.
- The theory is developed for both the zinc blende and wurtzite structured compound semiconductors and their heterostructures. Calculations for quantum wells and superlattices are presented to illustrate wavefunction engineering of these structures and the control achieved in obtaining desirable wavefunction localization.
- We show that when combined with optimization methods, wavefunction engineering provides a powerful new methodology for the optimized design of optoelectronic devices.

10.1 Introduction

There have been remarkable advances over the past few decades in the development of optoelectronic devices. In particular, quantum well lasers, emitting with wavelengths from the near-UV, blue, red, the mid-IR region (2–8 μm), all the way to the far-IR ($\sim$100 μm) region of the spectrum, have been successfully demonstrated. This spectacular evolution since the development of GaAs quantum well lasers in the early 1970s has been achieved by means of technological improvements in the precise growth

Finite Elements in Action. L. Ramdas Ram-Mohan, Oxford University Press. © L. Ramdas Ram-Mohan (2026).
DOI: 10.1093/oso/9780199563487.003.0010

of structures. In the following, we are concerned with the electronic properties of the III-V and II-VI compound semiconductors and their heterostructures. [1]

The geometry of the layered structure, i.e. the layer thicknesses in the multi-quantum well structure, plays a key role in determining the effective energy bandgap of the heterostructure, which arises as a manifestation of the Heisenberg uncertainty principle and carrier localization. This insight was distilled into the concept of "bandgap engineering." By choosing different materials with suitable energy bandgaps for individual layers, bandgap engineering has provided a conceptual framework for designing heterostructures with specific optical properties. [2]

Through a combination of computational modeling and advances in the growth of materials, we can progress considerably further today in exploring the details of the carrier wavefunctions, how they are shaped by choices for materials in the individual layers, and the layer geometry. We are essentially tailoring the localization and distribution of the wavefunctions in the heterostructure, and optimizing the optical transition matrix elements, which fundamentally determine the response and operation of the optoelectronic device. In this chapter, we provide a theoretical and computational overview on these advances within the framework of what we term as *wavefunction engineering.* [3]

We have emphasized earlier [4] that the finite element method (FEM) can be employed to advantage in simulating the electronic energy band structures of semiconductor heterostructures. The FEM [3, 5] may be considered to be the discretization of the action integral, which is fundamental to all of physics through its use in the principle of stationary action. We show that the Schrödinger Lagrangian (rather than the Hamiltonian), in conjunction with the action integral provides specific advantages in that we can of course obtain the equation of motion, the Schrödinger's equation. We can advance further in this framework by employing a gauge-variational method to obtain the conserved current and the current continuity boundary conditions (BCs) at interfaces [6] in an unambiguous manner; furthermore, we can show details of the approximations made within the $\mathbf{k} \cdot \mathbf{P}$ method. We can systematically include perturbations such as the built-in strain in the heterostructure and externally applied electric and magnetic fields in a natural manner. Again, the Lagrangian may be used in a discretized action in FEM calculations to obtain wavefunctions and momentum matrix elements. In the layered structure; each layer contributes additively to the global action integral.

Finally, the effect of charge redistributions in doped heterostructures due to ionized impurity dopants and free carriers, leading to energy band-edge bending, can be accounted for using the FEM. This ubiquitous problem of the Schrödinger-Poisson self-consistency occurs in essentially all active devices, and the FEM provides a convergent, stable solution in tens of iterations [7] *for arbitrary layered heterostructures with arbitrary doping profiles.* We treat this problem in greater detail in Chap. 11.

When combined with computational optimization routines, the FEM allows us to predict the material properties and optimized geometry of structures under bias that can have the desired energy level structure necessary for laser emission at *specific* wavelengths. In this sense, wavefunction engineering has matured to a level where we can *predict* the optoelectronic properties of a given heterostructure, and furthermore provide reliable guidelines for the growth of heterostructures having specific desired outcomes in terms of their optoelectronic response. [8]

In Sec. 10.2, we describe the $\mathbf{k} \cdot \mathbf{P}$ model Lagrangian used to calculate the band structure of heterostructures. In Sec. 10.3, we show how the wavefunction and current continuity BCs are derived consistent with operator ordering. We give examples of wavefunction engineering in Sec. 10.4. We compare the present methodology with other band structure modeling in Sec. 10.5, and provide concluding remarks in Sec. 10.6.

10.2 A Lagrangian formulation of the valence band structure

For our purposes here, in the envelope function approximation (EFA), the general form of the wavefunction may be considered to be a linear combination of a finite number of band wavefunctions, indexed by n, of the form

$$\psi(r) = f_n(z) e^{ik_x x} e^{ik_y y} u_n(r) \equiv F_n(r) u_n(r). \tag{10.1}$$

A clear perspective on the formulation of the band structure arises on using the principle of stationary action in setting up the $\mathbf{k} \cdot \mathbf{P}$ model. [9] A number of issues such as derivative operator ordering are resolved in a very natural manner, leading to a number of insights into the approximations that are invoked in the modeling. The action integral that leads to the Schrödinger equation for the valence bands is given by

$$\begin{aligned} \mathbf{A} &= \int dt \int_V d^3r \, (F_n^*(r) u_n^*(r)) \left[\overleftarrow{\partial} \frac{\hbar^2}{2m_0} \overrightarrow{\partial} + (V(r) - E) \right] (F_n(r) u_n(r)) \\ &= \int dt \int d^3r \mathcal{L}. \end{aligned} \tag{10.2}$$

The directed derivatives $\overleftarrow{\partial}$ and $\overrightarrow{\partial}$ act on the functions appearing to the left and to the right, respectively. All the bands are treated within degenerate perturbation theory. In the following, the integral over time is irrelevant since we are concerned with the time-independent band structure problem.

Within the spirit of the EFA, we perform "cell-averaging" by integrating over each unit cell in the crystal. The envelope functions $F_n(r)$ are typically considered to be slowly varying functions whereas the cell-periodic, and more oscillatory, Bloch functions satisfy Schrödinger's equation with band-edge energies. We use the results:

$$\begin{aligned} &\int_{cell} d^3r u_n^*(r) u_{n'}(r) = \delta_{nn'}; \\ &\int_{cell} d^3r \left[\nabla u_n^*(r) \frac{\hbar^2}{2m_0} \nabla u_{n'}(r) + u_n^*(r) V(r) u_{n'}(r) \right] = E_n \delta_{nn'}, \end{aligned} \tag{10.3}$$

and write the momentum matrix element between Bloch states as

$$\int_{cell} d^3 r u_n^*(r) \nabla u_{n'}(r) = \frac{i}{\hbar}\mathbf{p}_{nn'} = -\int_{cell} d^3 r \nabla u_n^*(r) u_{n'}(r), \quad n \neq n'. \tag{10.4}$$

The cell-averaged action integral can then be written in terms of the envelope functions alone, and we have

$$\langle \mathbf{A} \rangle = \int dt \int d^3 r \Bigg(\nabla F_n^* \frac{\hbar^2}{2m_0} \nabla F_{n'} + \frac{i\hbar}{2m_0} \left\{ F_n^* \overleftarrow{\partial} F_{n'} \mathbf{p}_{nn'} - \mathbf{p}_{nn'} F_n^* \overrightarrow{\partial} F_{n'} \right\}$$
$$+ \delta_{nn'} F_n^* (E_n - E) F_{n'} \Bigg). \tag{10.5}$$

The integrand in large parentheses in the above equation, the Lagrangian density $\langle \mathcal{L} \rangle$, may be cast in a matrix form by separating the band index n into the valence v, and the energetically higher remote bands with an index r. Following Löwdin's perturbation theory, [10] we eliminate the remote band wavefunctions in favor of the valence bands of interest by employing the equations of motion for the remote bands. If we had been developing the bulk $\mathbf{k} \cdot \mathbf{P}$ theory, the following terms would correspond to the Kane model [1, 9] for the band structure, with terms quadratic in the derivatives ($O(k^2)$ terms) within the valence bands. Ignoring spin for the moment, we have

$$\langle \mathcal{L} \rangle_{vv'} = F_v^* \left[\overleftarrow{\partial} \frac{\hbar^2}{2m_0} \overrightarrow{\partial} + (E_v - E)\delta_{vv'} \right] F_{v'}$$
$$+ \left(\frac{\hbar^2}{m_0^2} \right) \sum_r F_v^* \overleftarrow{\partial} \mathbf{p}_{vr} \frac{1}{(E - E_r)} \mathbf{p}_{rv'} \overrightarrow{\partial} F_{v'}, \quad \{v, v' = X, Y, Z\}, \tag{10.6}$$

where the p-like valence Bloch states in the III-V and II-VI semiconductors have been denoted by their symmetry-specified form in terms of X, Y, and Z. We have neglected terms of higher order in perturbation theory and used Löwdin's approximations. The advantage of the above derivation is that it explicitly retains the order of the derivative operators, each term in the Lagrangian density is Hermitian, and there is no ambiguity about reordering of operators for symmetrization. The focus is on the bands in the immediate vicinity of the bandgaps of the III-V or II-VI materials. We now follow Kane's discussion for the evaluation of the contributions of the remote bands to the valence bands . Extensions to include the conduction bands follow schemes used earlier by explicitly subtracting the contribution of the conduction band states from the Luttinger γ-parameters. [11, 12]

10.2.1 The zinc blende semiconductors

The possible intermediate states contributing to the Γ_{15} valence states belong to the energetically higher conduction bands of symmetry Γ_1 (s-like states), Γ_{12} states of symmetry $\{2z^2 - x^2 - y^2, \sqrt{3}(x^2 - y^2)\}$, Γ_{15} states of the form $\{yz, zx, xy\}$, and the Γ_{25} states. These states are compatible with the atomic s, p, d, and f orbital contributions to the valence electronic structure of semiconductors, respectively. It has been estimated that the contribution of the f orbitals to the valence electronic bands in semiconductors is negligible and hence the Γ_{25} intermediate states will be ignored. Following the notation of Foreman [13], we define

$$\begin{aligned} \sigma &= -(1/3m_0)\sum_r^{\Gamma_1} |\langle X|p_x|u_r\rangle|^2/(E_v - E_r), \\ \pi &= -(1/3m_0)\sum_r^{\Gamma_{15}} |\langle X|p_y|u_r\rangle|^2/(E_v - E_r), \\ \delta &= -(1/6m_0)\sum_r^{\Gamma_{12}} |\langle X|p_x|u_r\rangle|^2/(E_v - E_r). \end{aligned} \tag{10.7}$$

We then determine the contributions to the individual matrix elements of $\langle \mathbf{L}\rangle_{vv'}$. The matrix of derivative operators that act on the envelope functions $\{F_{vx}, F_{vy}, F_{vz}\}$ can be shown to be of the form

$$\langle \mathcal{L}\rangle_{vv'} = \left(\frac{\hbar^2}{2m_0}\right) \times \begin{bmatrix} \left\{\begin{matrix} \overleftarrow{\partial}_x A \overrightarrow{\partial}_x + \\ \overleftarrow{\partial}_y B \overrightarrow{\partial}_y + \overleftarrow{\partial}_z B \overrightarrow{\partial}_z \\ +E_v - E \end{matrix}\right\} & \left\{\begin{matrix} \overleftarrow{\partial}_x C_1 \overrightarrow{\partial}_y - \\ \overleftarrow{\partial}_y C_2 \overrightarrow{\partial}_x \end{matrix}\right\} & \left\{\begin{matrix} \overleftarrow{\partial}_x C_1 \overrightarrow{\partial}_z - \\ \overleftarrow{\partial}_z C_2 \overrightarrow{\partial}_x \end{matrix}\right\} \\ \left\{\begin{matrix} \overleftarrow{\partial}_y C_1 \overrightarrow{\partial}_x - \\ \overleftarrow{\partial}_x C_2 \overrightarrow{\partial}_y \end{matrix}\right\} & \left\{\begin{matrix} \overleftarrow{\partial}_y A \overrightarrow{\partial}_y + \\ \overleftarrow{\partial}_z B \overrightarrow{\partial}_z + \overleftarrow{\partial}_x B \overrightarrow{\partial}_x \\ +E_v - E \end{matrix}\right\} & \left\{\begin{matrix} \overleftarrow{\partial}_y C_1 \overrightarrow{\partial}_z - \\ \overleftarrow{\partial}_z C_2 \overrightarrow{\partial}_y \end{matrix}\right\} \\ \left\{\begin{matrix} \overleftarrow{\partial}_z C_1 \overrightarrow{\partial}_x - \\ \overleftarrow{\partial}_x C_2 \overrightarrow{\partial}_z \end{matrix}\right\} & \left\{\begin{matrix} \overleftarrow{\partial}_z C_1 \overrightarrow{\partial}_y - \\ \overleftarrow{\partial}_y C_2 \overrightarrow{\partial}_z \end{matrix}\right\} & \left\{\begin{matrix} \overleftarrow{\partial}_z A \overrightarrow{\partial}_z + \\ \overleftarrow{\partial}_x B \overrightarrow{\partial}_x + \overleftarrow{\partial}_y B \overrightarrow{\partial}_y \\ +E_v - E \end{matrix}\right\} \end{bmatrix} \tag{10.8}$$

in the envelope function basis. Here $A = 1 - 6\sigma - 12\delta$, $B = 1 - 6\pi$, $C_1 = 6\delta - 6\sigma$, and $C_2 = 6\pi$. While in the *bulk* semiconductor (where the envelope functions are of the form $\exp(i\mathbf{k}\cdot\mathbf{r})$) the ordering of the derivative operators is redundant, [9, 14] and way in which the band parameters are grouped is irrelevant, it becomes crucial in the heterostructure. The interface BCs depend on the particular ordering of the differentiation operators relative to the band parameters that vary across the interface. The parameters σ, δ, and π are simply related to the usual Luttinger parameters, [14] and we have

$$\begin{aligned} \gamma_1 &= -1 + 2\sigma + 4\pi + 4\delta, \\ \gamma_2 &= \sigma - \pi + 2\delta, \\ \gamma_3 &= \sigma + \pi - \delta. \end{aligned} \tag{10.9}$$

The three parameters based on the symmetry type can be determined from experiments, as are the Luttinger parameters, so that the implementation of envelope function continuity and the probability current continuity conditions at interfaces now becomes feasible. [15, 16]

The inclusion of the electron spin and the effects of spin-orbit interaction are straightforward and we obtain a 6×6 valence Lagrangian.

The spin-orbit interaction splits the six-fold degeneracy at the zone center into a fourfold degenerate heavy-hole (hh) and light-hole (lh) bands of Γ_8 symmetry with total angular momentum $J = 3/2$, and a doubly degenerate split-off (so) band of Γ_7 symmetry. The valence Lagrangian in the $|J, m_j\rangle$ basis is given by

$$\mathcal{L}_{vv'} = \begin{bmatrix} P+Q & -S_- & R & 0 & \frac{-1}{\sqrt{2}}S_- & \sqrt{2}R \\ -\widetilde{S}_+ & P-Q & C_- & R & -\sqrt{2}Q & \Sigma_- \\ R^* & -C_+ & P-Q & -\widetilde{S}_- & \Sigma_+ & \sqrt{2}Q \\ 0 & R^* & S_+ & P+Q & -\sqrt{2}R^* & -\frac{1}{\sqrt{2}}S_+ \\ -\frac{1}{\sqrt{2}}\widetilde{S}_+ & -\sqrt{2}Q & -\widetilde{\Sigma}_- & -\sqrt{2}R & P+\Delta & \widetilde{C}_- \\ \sqrt{2}R^* & -\widetilde{\Sigma}_+ & \sqrt{2}Q & \frac{1}{\sqrt{2}}\widetilde{S}_- & -\widetilde{C}_+ & P+\Delta \end{bmatrix} \begin{matrix} |\frac{3}{2},\frac{3}{2}\rangle \\ |\frac{3}{2},\frac{1}{2}\rangle \\ |\frac{3}{2},-\frac{1}{2}\rangle \\ |\frac{3}{2},-\frac{3}{2}\rangle \\ |\frac{1}{2},\frac{1}{2}\rangle \\ |\frac{1}{2},-\frac{1}{2}\rangle \end{matrix}, \tag{10.10}$$

where, specializing to the case of [001] growth direction, we have

$$\begin{aligned} P &= \gamma_1 k_\parallel^2 + \overleftarrow{\partial}_z \gamma_1 \overrightarrow{\partial}_z - (E_v - E); \\ Q &= \gamma_2 k_\parallel^2 - 2\overleftarrow{\partial}_z \gamma_2 \overrightarrow{\partial}_z, \\ R &= -\sqrt{3}\gamma_2(k_x^2 - k_y^2) + i2\sqrt{3}\gamma_3 k_x k_y; \\ S_\pm &= \frac{i}{\sqrt{3}}\left[k_\pm C_1 \overrightarrow{\partial}_z + \overleftarrow{\partial}_z C_2 k_\pm\right], \\ \Sigma_\pm &= \frac{i}{3\sqrt{2}}\left[\overleftarrow{\partial}_z (C_2 - 2C_1)k_\pm + k_\pm(C_1 - 2C_2)\overrightarrow{\partial}_z\right]; \\ C_\pm &= \frac{-i}{\sqrt{3}}\left[k_\pm(C_1 + C_2)\overrightarrow{\partial}_z + \overleftarrow{\partial}_z(C_1 + C_2)k_\pm\right]. \end{aligned} \tag{10.11}$$

Here $\{\gamma_1, \gamma_2, \gamma_3, C_1, C_2\}$ are in units of $(\hbar^2/2m_0)$ and $k_\pm = k_x \pm ik_y$. The quantities $S_\pm, C_\pm, \Sigma_\pm$ and $\widetilde{S_\pm}, \widetilde{C_\pm}, \widetilde{\Sigma}_\pm$ differ by the exchange of the location of the directed derivatives and k-components in the expressions. *Thus, the Hermiticity of the Lagrangian can be verified with more care than is usually needed.* All the parameters introduced above are expressible in terms of the three quantities σ, δ, and π, or equivalently, in terms of the Luttinger γ parameters. In the bulk semiconductor, we have $\widetilde{S}_+ = -S_+$, $\Sigma_- = \sqrt{3/2}S_-$, and $C_\pm = 0$. Thus, the operator ordering leads to the breaking up of the usual parameters into component contributions that appear to contribute differently in layered systems as compared with bulk semiconductors. Additional perturbations, e.g. the strain energy, are additively included in the Lagrangian. [15, 16]

The Luttinger parameters were determined by comparison of the valence hh and lh bands to cyclotron experiments and used in the bulk band structure calculations. This direct determination of the γ parameters of Luttinger represented the closed theoretical area of valence bands only. Later the inclusion of the conduction band into consideration for the analysis of magneto-optical experimental data required a modification of the Luttinger scenario through subtracting the conduction band contributions to the valence band parameters as was done by Pidgeon and Brown, [12] and by Weiler. [11] The contribution of the remote higher bands was treated using Löwdin perturbation theory. [10] The elimination of the remote band envelope wavefunctions is very similar to the so-called Guyen reduction of variables and the consequent reduction of the global matrices in finite element analysis of structures.

10.2.2 The wurtzite semiconductors

We consider the bands in the immediate vicinity of the bandgap of wurtzite materials at the zone center. The valence states belong to the $\{\Gamma_{5v} : \{X, Y\} + \Gamma_{1v} : \{Z\}\}$ representations. The deeper valence bands belong to the $\Gamma_{6v} : \{X^2 - Y^2, 2XY\}$ and the $\Gamma_{1v} : \{Z\}$ representations. The nearest higher conduction bands belong to the Γ_{5c} states of $\{X, Y\}$ symmetry and the Γ_{3c} states transform as the $\{Z\}$ representation. Here we are using the notation of Kane, [9] for example.

The second-order valence band terms of Eq. (10.6) are evaluated as done by Kane for bulk zinc blende semiconductors. We define parameters for the contributions of the higher Γ_{1c} conduction bands as follows:

$$\begin{aligned}
\sigma_1 &= \frac{1}{m_0}\sum_{\Gamma_1} |\langle \Gamma_5^{\left\{ X \atop Y \right\}} | p_{X,Y} | \Gamma_1^{\{S\}} \rangle|^2 / (E_v - E_{\Gamma_1}); \\
\sigma_2 &= \frac{1}{m_0}\sum_{\Gamma_1} |\langle \Gamma_1^{\{Z\}} | p_{X,Y} | \Gamma_1^{\{S\}} \rangle|^2 / (E_v - E_{\Gamma_1}), \\
\sigma_3 &= \frac{1}{m_0}\sum_{\Gamma_1} \frac{\langle \Gamma_5^{\left\{ X \atop Y \right\}} | p_{X,Y} | \Gamma_1^{\{S\}} \rangle \langle \Gamma_1^{\{S\}} | p_{X,Y} | \Gamma_1^{\{Z\}} \rangle}{E_v - E_{\Gamma_1}}.
\end{aligned} \tag{10.12}$$

The Γ_{3c} intermediate conduction states do not contribute to the valence bands. The contribution of the $\Gamma_5\{XZ, YZ\}$ intermediate states can be parameterized by writing

$$\begin{aligned}
\delta_1 &= \frac{1}{m_0}\sum_{\Gamma_5} |\langle \Gamma_5^{\left\{ X \atop Y \right\}} | p_Z | \Gamma_5^{\left\{ XZ \atop YZ \right\}} \rangle|^2 / (E_v - E_{\Gamma_5}); \\
\delta_2 &= \frac{1}{m_0}\sum_{\Gamma_5} |\langle \Gamma_1^{\{Z\}} | p_{X,Y} | \Gamma_5^{\left\{ XZ \atop YZ \right\}} \rangle|^2 / (E_v - E_{\Gamma_5}), \\
\delta_3 &= \frac{1}{m_0}\sum_{\Gamma_5} \langle \Gamma_1^{\{Z\}} | p_{X,Y} | \Gamma_5^{\left\{ XZ \atop YZ \right\}} \rangle \langle \Gamma_5^{\left\{ XZ \atop YZ \right\}} | p_Z | \Gamma_5^{\left\{ X \atop Y \right\}} \rangle / (E_v - E_{\Gamma_5}).
\end{aligned} \tag{10.13}$$

The Γ_6 states of symmetry $\{X^2-Y^2, 2XY\}$ require an additional parameter to represent their contribution. We define

$$\lambda = \frac{1}{m_0}\sum_{\Gamma_6} |\langle \Gamma_5^{\{X,Y\}}|p|\Gamma_6^{\left\{ {X^2-Y^2 \atop 2XY} \right\}}\rangle|^2/(E_v - E_{\Gamma_6}). \tag{10.14}$$

The valence band Lagrangian for the wurtzite crystal, in the $\{X, Y, Z\}$ basis, is now given by Bir and Pikus, and by Ram-Mohan, et al. [17, 18]

$$\langle\mathcal{L}\rangle_{vv'} = \begin{bmatrix} \left\{ \begin{matrix} \overleftarrow{\partial}_x L_1 \overrightarrow{\partial}_x + \\ \overleftarrow{\partial}_y M_1 \overrightarrow{\partial}_y + \overleftarrow{\partial}_z M_2 \overrightarrow{\partial}_z \\ +(E_v - E) \end{matrix} \right\} & \left\{ \begin{matrix} \overleftarrow{\partial}_x \widetilde{N}_1 \overrightarrow{\partial}_y + \\ \overleftarrow{\partial}_y \widetilde{N}_1' \overrightarrow{\partial}_x \end{matrix} \right\} & \left\{ \begin{matrix} \overleftarrow{\partial}_x \widetilde{N}_2 \overrightarrow{\partial}_z + \\ \overleftarrow{\partial}_z \widetilde{N}_2' \overrightarrow{\partial}_x \end{matrix} \right\} \\ \left\{ \begin{matrix} \overleftarrow{\partial}_y \widetilde{N}_1 \overrightarrow{\partial}_x \\ +\overleftarrow{\partial}_x \widetilde{N}_1' \overrightarrow{\partial}_y \end{matrix} \right\} & \left\{ \begin{matrix} \overleftarrow{\partial}_x M_1 \overrightarrow{\partial}_x + \\ \overleftarrow{\partial}_y L_1 \overrightarrow{\partial}_y + \overleftarrow{\partial}_z M_2 \overrightarrow{\partial}_z \\ +(E_v - E) \end{matrix} \right\} & \left\{ \begin{matrix} \overleftarrow{\partial}_y \widetilde{N}_2 \overrightarrow{\partial}_z + \\ \overleftarrow{\partial}_z \widetilde{N}_2' \overrightarrow{\partial}_y \end{matrix} \right\} \\ \left\{ \begin{matrix} \overleftarrow{\partial}_z \widetilde{N}_2 \overrightarrow{\partial}_x + \\ \overleftarrow{\partial}_x \widetilde{N}_2' \overrightarrow{\partial}_z \end{matrix} \right\} & \left\{ \begin{matrix} \overleftarrow{\partial}_z \widetilde{N}_2 \overrightarrow{\partial}_y + \\ \overleftarrow{\partial}_y \widetilde{N}_2' \overrightarrow{\partial}_z \end{matrix} \right\} & \left\{ \begin{matrix} \overleftarrow{\partial}_x M_3 \overrightarrow{\partial}_x + \\ \overleftarrow{\partial}_y M_3 \overrightarrow{\partial}_y + \overleftarrow{\partial}_z L_2 \overrightarrow{\partial}_z \\ +(E_v - E) \end{matrix} \right\} \end{bmatrix}, \tag{10.15}$$

where the parameters L_i, M_i, and N_i are given by

$$\begin{aligned} L_1 &= 1 - 2\sigma_1 - 4\lambda; & L_2 &= 1 - 2\sigma_2; & \widetilde{N}_1 &= -2\sigma_1 + 4\lambda; \\ \widetilde{N}_1' &= -4\lambda; & \widetilde{N}_2 &= -2\sigma_3; & \widetilde{N}_2' &= -2\delta_3; \\ M_1 &= 1 - 4\lambda; & M_2 &= 1 - 2\delta_1; & M_3 &= 1 - 2\delta_2, \end{aligned} \tag{10.16}$$

in units of $\hbar^2/2m_0$. The order of the derivative operators with $\widetilde{N}_1'$ and $\widetilde{N}_2'$ terms is reversed with respect to $\widetilde{N}_1$ and $\widetilde{N}_2$ terms. In the bulk semiconductor, the operator ordering is irrelevant and we obtain the bulk parameters $N_1 = \widetilde{N}_1 + \widetilde{N}_1'$ and $N_2 = \widetilde{N}_2 + \widetilde{N}_2'$. Furthermore, as has been shown by Chuang and Chang, [19] the six-fold symmetry under rotation about the c-axis leads to the relation $(L_1 - M_1) = N_1$.

The valence band edge is not triply degenerate in wurtzite semiconductors, and the crystal field splitting leads to the shifts in the band-edge energy E_v by $\langle X|L_{(c.f.)}|X\rangle = \langle Y|L_{(c.f.)}|Y\rangle = \Delta_1$, and $\langle Z|L_{(c.f.)}|Z\rangle = 0$. The spin-orbit splitting, generated by

$$\begin{aligned} L_{so} &= \frac{\hbar}{4m_0^2c^2}\nabla V \times \boldsymbol{p}\cdot\boldsymbol{\sigma} \\ &= L_{(so)x}\sigma_x + L_{(so)y}\sigma_y + L_{(so)z}\sigma_z, \end{aligned} \tag{10.17}$$

is parameterized by the relations [19]

$$\langle X|L_{(so)z}|Y\rangle = -i\Delta_2; \qquad \langle Y|L_{(so)x}|Z\rangle = \langle Z|L_{(so)y}|X\rangle = -i\Delta_3. \tag{10.18}$$

The spinor matrix elements of the standard Pauli spin matrices $\boldsymbol{\sigma}$ are readily included in the above derivation. With these definitions in place, we consider the inclusion of spin in the Lagrangian. We make the same choice as in Chuang and Chang's book, [19] for the valence band basis states with spin. The valence band Lagrangian, including the crystal field splitting and the spin-orbit interaction, is given by

$$\mathcal{L}_{vv'} = \begin{bmatrix} P & -K^* & \widetilde{S}_- & 0 & 0 & 0 \\ -K & Q & -\widetilde{S}_+ & 0 & 0 & \Delta \\ -S_+ & S_- & R & 0 & \Delta & 0 \\ 0 & 0 & 0 & P & -K & -\widetilde{S}_+ \\ 0 & 0 & \Delta & -K^* & Q & \widetilde{S}_- \\ 0 & \Delta & 0 & S_- & -S_+ & R \end{bmatrix}, \tag{10.19}$$

where, specializing to the case of [0001] growth direction, and switching the sign of the energy to be positive for the valence bands, we have

$$\begin{aligned} P &= (\Delta_1 + \Delta_2 + E_v - E) + \alpha; & Q &= (\Delta_1 - \Delta_2 + E_v - E) + \alpha, \\ R &= \left[\overleftarrow{\partial}_z A_1 \overrightarrow{\partial}_z + A_2 k_{\parallel}^2\right] + (E_v - E); & K &= A_5 (k_x + i k_y)^2, \\ S_\pm &= \tfrac{i}{\sqrt{2}} \left[\overleftarrow{\partial}_z N_2 k_\pm - k_\pm N_2' \overrightarrow{\partial}_z\right]; & \widetilde{S}_\pm &= \tfrac{-i}{\sqrt{2}} \left[\overleftarrow{\partial}_z N_2' k_\pm - k_\pm N_2 \overrightarrow{\partial}_z\right], \\ \alpha &= \left[\overleftarrow{\partial}_z (A_1 + A_3) \overrightarrow{\partial}_z + (A_2 + A_4) k_{\parallel}^2\right]; & \Delta &= \sqrt{2}\Delta_3. \end{aligned} \tag{10.20}$$

Here the Bir-Pikus parameters [17] are identified by the relations

$$\begin{aligned} A_1 &= L_2; & A_2 &= M_3; \\ A_3 &= M_2 - L_2; & A_4 &= (L_1 + M_1)/2 - M_3; \\ A_5 &= (\widetilde{N}_1 + \widetilde{N}_1')/2; & A_6 &= (\widetilde{N}_2 + \widetilde{N}_2')/\sqrt{2}. \end{aligned} \tag{10.21}$$

The parameter A_6 does not appear in the Lagrangian and instead we have to contend with the determination of two separate parameters ($\widetilde{N}_2$ and $\widetilde{N}_2'$) in order to proceed. This requires that one of the parameters σ_3 or δ_3 (see Eqs. (10.12) and (10.13)) be known from *ab-initio* calculations. We have tentatively used $\delta_3 = \delta_1$.

Mireles and Ulloa [20] have one parameter less in their analysis. The quantities $S_\pm$ and $\widetilde{S}_\pm$ differ by the exchange of the location of the directed derivatives and k-components in the expressions.

The inversion asymmetry terms [21] in the wide-gap wurtzite semiconductors have been shown to provide an improved description of the valence bands. The valence band parameter A_7 has been expressed in the convenient form $A_7 = (-i\hbar/m_0\sqrt{2})\langle X|p_x|Z\rangle$ by Dugdale et al. [22] A detailed accounting of the various contributions from band mixing and the admixture of d-orbitals to the inversion asymmetry parameter has been given by Lew Yan Voon et al. [23] In the $\{X, Y, Z\}$ orbital basis, we have

$$\mathbf{L}_{inv} = \begin{pmatrix} 0 & 0 & i\sqrt{2}A_7 k_x \\ 0 & 0 & i\sqrt{2}A_7 k_y \\ -i\sqrt{2}A_7 k_x & -i\sqrt{2}A_7 k_y & 0 \end{pmatrix}. \tag{10.22}$$

This 3×3 valence band matrix is doubled to accommodate the spin degree of freedom and the resulting 6×6 Lagrangian is added to the valence band Lagrangian in Eq. (10.19).

10.3 Interface BCs and current continuity

The action integral is varied with respect to the envelope functions F^* and the principle of least action is invoked to derive the 6 coupled Schrödinger differential equations in the six-band zinc blende $\mathbf{k} \cdot \mathbf{P}$ model. Similar considerations hold for the wurtzite materials. In the following, we simplify the discussion of the BCs by considering just the valence bands *without including spin*, with the Lagrangian given by Eq. (10.8). In a layered heterostructure, each layer contributes a piece of the action integral. The material properties change across layer interfaces, and we attach a layer index λ to each of the parameters A, B, C_1, and C_2 appearing in Eq. (10.8). With the layer growth along the z-axis, the terms in the Lagrangian are separated into those with different numbers of directed derivatives in z. The action integral takes the form

$$\mathbf{A} = \int dt \sum_{\lambda} \int_{Z_\lambda}^{Z_{\lambda+1}} dz F_i^* \Big[\overleftarrow{\partial}_z \mathbf{K}^{(\lambda)} \overrightarrow{\partial}_z + (\overleftarrow{\partial}_z \mathbf{Q}^{(\lambda)} + \mathbf{Q}^{(\lambda)\dagger} \overrightarrow{\partial}_z) + \mathbf{R}^{(\lambda)} + (E_i^{(\lambda)} \delta_{ij} - E)\mathbf{1} \Big] F_j. \tag{10.23}$$

Here the unit matrix is denoted by $\mathbf{1}$, and the other matrix coefficients in the Lagrangian are given by

$$\mathbf{K}^{(\lambda)} = \begin{bmatrix} B^{(\lambda)} & 0 & 0 \\ 0 & B^{(\lambda)} & 0 \\ 0 & 0 & A^{(\lambda)} \end{bmatrix}, \qquad \mathbf{Q}^{(\lambda)} = \begin{bmatrix} 0 & 0 & -ik_x C_2^{(\lambda)} \\ 0 & 0 & -ik_y C_2^{(\lambda)} \\ ik_x C_1^{(\lambda)} & ik_y C_1^{(\lambda)} & 0 \end{bmatrix},$$

and

$$\mathbf{R}^{(\lambda)} = \begin{bmatrix} (A^{(\lambda)} k_x^2 + B^{(\lambda)} k_y^2) & (C_1^{(\lambda)} - C_2^{(\lambda)}) k_x k_y & 0 \\ (C_1^{(\lambda)} - C_2^{(\lambda)}) k_x k_y & (B^{(\lambda)} k_x^2 + A^{(\lambda)} k_y^2) & 0 \\ 0 & 0 & B^{(\lambda)} (k_x^2 + k_y^2) \end{bmatrix}. \tag{10.24}$$

Observe that while the individual linear derivative term $\mathbf{Q}$ is not Hermitian by itself, the combination of the two terms involving $\mathbf{Q}$ and $\mathbf{Q}^\dagger$ together is Hermitian. One of the BCs at interfaces is the continuity of the envelope wavefunction. The second BC at the interfaces is the continuity of the probability current across the interfaces, which we evaluate using a gauge-variational approach that is common in particle physics. Following Gell-Mann and Levy, [6] we substitute $F_j(z) \to e^{i\Lambda(z)} F_j(z)$ in the Lagrangian, Eq. (10.23), and develop a variation of the Lagrangian with respect to the gauge function $\Lambda(z)$. In the presence of operator ordering, the conserved current is obtained in the form

$$\mathbf{J} = \frac{1}{\hbar}\left(\frac{\delta\langle\mathbf{L}\rangle}{\delta(\partial_z\Lambda)}\right)\Bigg|_{\Lambda=0} = \frac{i}{\hbar}\Big[- F_i^*\mathbf{K}_{\mathrm{ij}}^{(\boldsymbol{\lambda})}\partial_z F_j + (\partial_z F_i^*)\mathbf{K}_{\mathrm{ij}}^{(\boldsymbol{\lambda})} F_j$$
$$- F_i^*\mathbf{Q}_{\mathrm{ij}}^{(\boldsymbol{\lambda})} F_j + F_i^*\mathbf{Q}_{\mathrm{ij}}^{(\boldsymbol{\lambda})^\dagger} F_j\Big]. \tag{10.25}$$

The continuity conditions at an interface at $z = 0$ are then given by

$$\begin{aligned} F|_{0^-} &= F|_{0^+}, \\ \mathbf{K}F' + \mathbf{Q}F|_{0^-} &= \mathbf{K}F' + \mathbf{Q}F|_{0^+}. \end{aligned} \tag{10.26}$$

10.4 Examples of wavefunction engineering of quantum heterostructures

The discretized Schrödinger equation is obtained in the FEM. [5] The physical region is split into small regions, finite elements, in each layer, and the wavefunctions are expressed in terms of interpolation polynomials with as-yet unknown coefficients in each finite element. The spatial dependence is integrated out, and the element matrices are overlaid in a global matrix, for the structure as a whole, in order to ensure that continuity conditions discussed in Sec. 10.3 are implemented. We minimize the action by varying the unknown coefficients. The resulting discretized Schrödinger matrix equation is solved by sparse matrix methods. This approach ties in the action to the numerical work and provides a systematic approach to improving the variational solution. The contributions of specific layers can be emphasized in FEM by discretizing them appropriately. We have applied this method to obtain the energy eigenvalues and wavefunctions in heterostructures, in-plane energy dispersions, and optical matrix elements. The method is also extended to obtain a self-consistent solution to the Schrödinger-Poisson problem of band bending, [7] a problem of particular interest in the context of spintronics applications where the carrier concentrations in GaMnAs, for example, [24] can be as high as $\sim 10^{20}\mathrm{cm}^{-3}$. Such self-consistency calculations are discussed further in Chap. 11. In the following, we consider examples of finite element modeling of heterostructures of common interest. The calculations make use of the band parameters given in Vurgaftman et al. [25]

In Fig. 10.1, the evolution of the design of a heterostructure for a mid-IR laser is shown. The initial type-II quantum well structure in Fig. 10.1(a) has poor conduction band ($\boldsymbol{\Psi}_n$) and valence band ($\boldsymbol{\Psi}_p$) wavefunction overlap for optical matrix elements. By introducing another layer in the well to form a step potential, as in Fig. 10.1(b), the wavefunctions are better confined and show increased overlap. With yet another layer, we obtain the so-called W-structure, shown in Fig. 10.1(c). The W-structure was grown to specifications based on wavefunction engineering, and the very first structure grown exhibited lasing in the mid-IR region, at $4\,\mu$m, of the spectrum. This is a classic example of wavefunction engineering [4, 26] through which we are able to manipulate the material properties of the structure to obtain desired wavefunctions and design a working laser to operate at expected wavelengths. The W-laser holds the present record for continuous operation up to 275 K for mid-IR lasers.

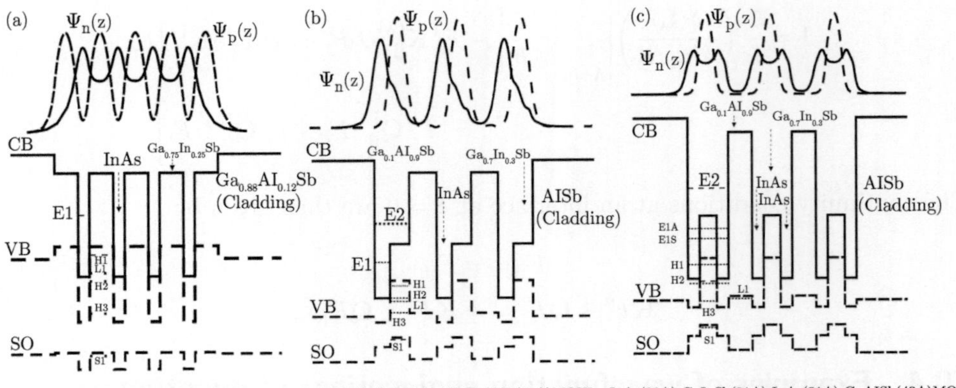

Figure 10.1 (a) The type-II heterostructure of InAs/GaInSb leads to localization of wavefunctions in different layers. This is ameliorated partially by introducing a step potential as in (b). The W-structure shown in (c) provides a substantially improved overlap (figure adapted from Ram-Mohan and Meyer [4, 26]; with permission from Elsevier).

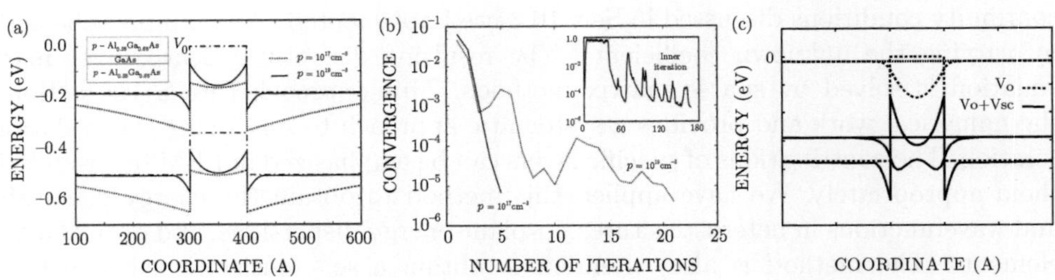

Figure 10.2 The valence band bending in a quantum well with *p*-doped barriers is shown. The FEM iterations for the Schrödinger-Poisson calculations converge to the self-consistent solution in a stable manner to the band profile shown in (a), in very few iterations as shown in (b), in the dual-loop algorithm of Ram-Mohan, Moussa, and Yoo. [7] The wavefunctions in the double well created by the self-consistent band bending are shown in (c) (after Ram-Mohan et al. [7]).

In Fig. 10.2(a), we show the band bending in a modulation-doped single quantum well, comparing the band-edge profile before and after the self-consistent calculation, with the rate of convergence for the Schrödinger-Poisson calculation shown in Fig. 10.2(b). [7] Figure 10.2(c) shows the first hh wavefunction before (dotted lines) and after (solid lines) the hole redistribution. If only the original band offset potential is considered, the wavefunction is of the simple cosine-like form. When the self-consistent potential is included, the total potential resembles that of a double quantum well, and the hh wavefunction reflects this by having two peaks. In more complex situations, the doping profile can be controlled to shape the wavefunctions differently, a situation that may be termed wavefunction engineering by modulation doping.

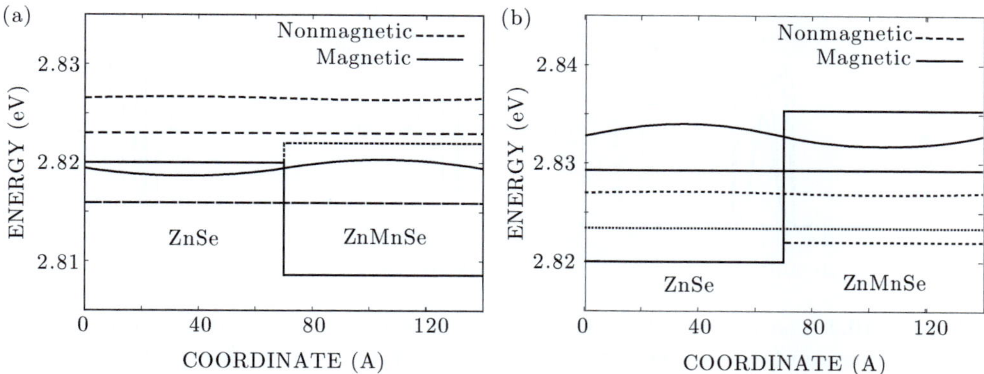

Figure 10.3 The conduction band-edge profile without (dotted lines) and with (solid lines) Mn-spin alignment in a magnetic field of 5 T for (a) the spin-down and (b) for the spin-up electron in a ZnMn/ZnMnSe superlattice is shown. The corresponding wavefunctions are shown in the "spin-superlattice" formed in the structure.

The ZnSe/$Zn_{1-x}Mn_xSe$ multilayer system was reported [27] to form a spin-superlattice. For $x \sim 0.04$, the conduction band-edge of ZnSe is almost the same as that of the diluted magnetic semiconductor (DMS) $Zn_{1-x}Mn_xSe$. In an applied magnetic field, the Mn spins are aligned, and the effective conduction band-edge of ZnMnSe for spin-up electrons is higher than that of ZnSe. It is the opposite for spin-down electrons, so that the spin-up electrons are localized in the ZnSe layer and spin-down electrons in the ZnMnSe layer. In Fig. 10.3(a), we plot the effective band-edge and the wavefunction for the lowest spin-down electron Landau level, and in Fig. 10.3(b), for the lowest spin-up level at B = 5 T.

When the Mn alignment is not considered, by neglecting the DMS Lagrangian for the DMS layer (dotted lines), the probability that the electrons reside in the ZnSe layer is 51.4% for both spin-up and spin-down electrons due to a small conduction band offset. When the Mn alignment is taken into account by including the DMS Lagrangian for the ZnMnSe layer, the probability becomes 42.1% for spin-down electrons and 60.6% for spin-up electrons. Thus, the Mn alignment through an external magnetic field alters the localization of spins in different layers in specially designed DMS superlattices. This is an example of magnetically tuned wavefunction engineering.

Figure 10.4 shows the spontaneous magnetization of Mn ions in a 100 Å $Ga_{0.95}Mn_{0.05}As/Al_{0.35}Ga_{0.65}As$ quantum well as a function of hole concentration, which can vary independently of the Mn concentration. Since the exchange interaction between Mn ions and itinerant holes increases with hole concentration, the magnetization increases with hole concentration, and saturates with a carrier density at about $3 \times 10^{19}\,cm^{-3}$. The shape of the magnetization is a reflection of the difference between spin-up and spin-down hole densities. As the hole concentration increases, higher subbands are occupied, and the shape of magnetization changes as the relative

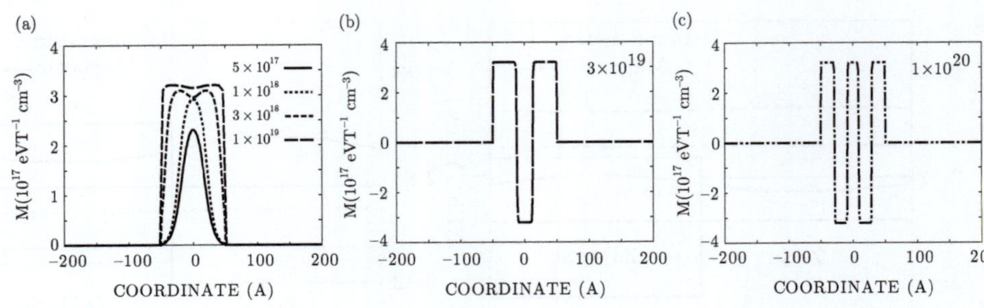

Figure 10.4 The magnetization due to Mn-spin alignment in a 100 Å GaMnAs quantum well is shown. In (a), the alteration of magnetization with increasing hole concentrations from $5\times10^{17}\,\mathrm{cm}^{-3}$ to $10^{19}\mathrm{cm}^{-3}$ is shown. With further increases in hole concentration, (b) and (c), the magnetization oscillates in orientation across the quantum well. This is due to the occupied energy levels and the corresponding wavefunctions for holes in the self-consistent potential. [24]

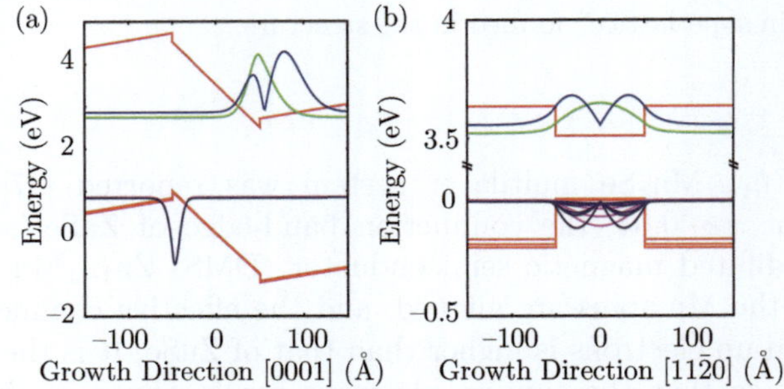

Figure 10.5 The remarkably high internal piezoelectric fields in wurtzite GaN/AlGaN structures distorts the band-edge potential in quantum wells, as shown in (a) (left), leading to a poor overlap of carrier wavefunctions. When grown along the [11$\bar{2}$0] direction, the piezoelectric fields are along the plane, leading to better characteristics for quantum well lasers. (after Ram-Mohan et al. [18] Figure reproduced with permission from Elsevier.) Strain, piezoelectric effects, and inversion asymmetry effects have been included in the calculations.

distribution of spin-up and spin-down holes changes. These Schrödinger-Poisson-DMS-self-consistent calculations were performed in the valence four-band Luttinger model using the FEM, and illustrate the power of the method in successfully accounting for such large carrier concentrations. [24]

We now consider an example of wavefunction engineering in a 100 Å wurtzite GaN/AlGaN quantum well. When the structure is grown along the c-axis the resulting polarization leads to a distortion of the quantum well profile due to the large internal piezoelectric polarization fields. The wavefunctions are then displaced in position and the wavefunction overlap between electrons and holes is substantially reduced, as shown in Fig. 10.5(a). When the growth is along the [11$\bar{2}$0] direction, the spontaneous and piezoelectric polarizations are directed in the plane (still along the c-axis, which is in the plane). The quantum well potential is no longer distorted by the internal electric field, and the wavefunction overlap is then "engineered" to be large. The calculation of the energy levels and wavefunctions proceeds by employing a rotated Lagrangian with the new z'-axis directed toward the corresponding growth direction. The coordinate transformation $\boldsymbol{x'} = \boldsymbol{Rx}$ leads to the new rotated Lagrangian given by

$$\mathbf{L}_{\{x',y',z'\}}(\boldsymbol{k'}) = \boldsymbol{R}\mathbf{L}_{\{x,y,z\}}(\boldsymbol{R}^{-1}\boldsymbol{k'})\boldsymbol{R}^{-1}. \tag{10.27}$$

The matrix elements of the transformed Lagrangian in the envelope function basis are separated, using symbolic algebra software, into terms involving powers of $k_z' = -i\partial/\partial z'$ corresponding to the new growth direction. The rotated Lagrangian is then used in the finite element calculations. [17] A discussion of the rotation of the $\mathbf{k} \cdot \mathbf{P}$ Hamiltonian was given earlier. [11, 20, 28, 29] This feature of the absence of band-edge profile distortion for the $[11\bar{2}0]$ growth direction in GaN/AlGaN quantum wells has been investigated experimentally. [30, 31]

As a final example, we consider the design of an inter-subband quantum cascade laser (QCL) structure for emitting in the terahertz (THz) region of the spectrum. Lasers in this region of the spectrum have special problems to overcome. The electronic transitions for photon emission are in the energy range 5–20 meV, so that the system has to be cooled to cryogenic temperatures, and the laser cavity has to be on the order of several hundred microns. In designing a QCL, the goal of the optimization is to have a three-level system in which the carriers transition from energy E_3 to E_2, emitting a photon. The electrons are then depleted from E_2 by a transition from E_2 to E_1 on emitting interface and layer-confined longitudinal-optic phonon. The carriers then tunnel into the next period of the cascade, arriving in the uppermost level 3 of the next stage. Recently, the operation of THz lasers has been successfully demonstrated. [32–34] Here we are concerned with the optimized design of a single period of the cascade laser.

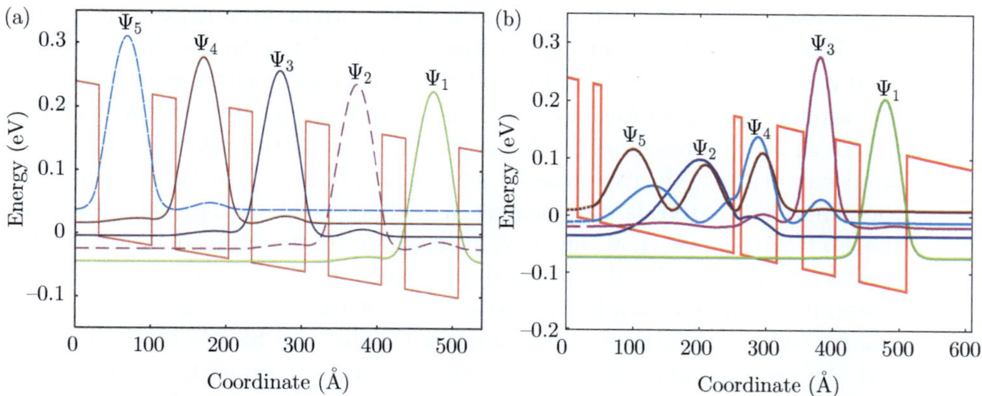

Figure 10.6 A five-well structure, representing one period of a QCL with energy level structure for emitting photons with energy 16 meV (3.9 THz) is shown before layer thickness optimization in (a). In (b) the structure after GA optimization is shown with well and barrier widths adjusted to obtain the energy levels under bias that is mentioned in the text.

In order to design such a structure, we consider a single period of a five-quantum well structure with the requirements that under an externally applied bias of say 2.5 kV/cm, the energy differences are 16 meV ($\sim$77μm or $\sim$3.9 THz) for the photonic transition, and $\sim$38 meV for the phonon emission. The calculation proceeds by coupling the one-band finite element calculations of the energy levels of the 9-layer period to a Powell or Genetic Algorithm (GA) optimizer. [35] In Fig. 10.6(a), we show the initial well widths and energy levels with no bias. Initially, the well and barrier widths are uniformly the same for each well, and the wavefunctions have been shifted by the eigenvalues for purposes of this display. This may be compared with the final structure, Fig. 10.6(b), obtained through GA optimization for the well and barrier widths in order to obtain the desired energy level structure in $GaAs/Al_{0.3}Ga_{0.7}As$. The wavefunctions have been shifted vertically to have their x-axes coincide with the energy levels for display purposes. The gain in the QCL structure per period was not optimized with respect to the rate equations, and this work is as yet preliminary [36] in this regard.

We note that the use of LO-phonon emission for depleting the carriers from E_2 to E_1 can also be optimized through the use of interface phonon modes and requiring that $(E_2 - E_1)$ lie in the set of interface mode energies. The interface modes extend over the entire structure and hence provide better overlap with the carriers. In this sense, we are employing phonon wavefunction engineering in the operation of the QCL. [36]

10.5 Comparison with other methods

We have focused mainly on the $\mathbf{k} \cdot \mathbf{P}$ model and the use of the envelope function modeling for the electronic band structure near the center of the Brillouin zone (BZ). Here we comment on the other alternatives for evaluating the electronic states in layered heterostructures so as to put their merits in perspective.

(a) In the *empirical tight binding model* (ETBM), originally put forward by Slater and Koster, [37] the electronic states are considered to be linear combinations of atomic $(s, p, d, ...)$ orbitals. The Hamiltonian's matrix elements between the atomic orbital states are not evaluated directly, but are instead introduced as free parameters to be determined by fitting the band gaps and band curvatures (effective masses) at critical points in the BZ. Depending on the number of orbitals and nearest neighbors used to represent the states, the ETBM requires that the overlap integrals be determined in terms of the measured direct and indirect band gaps and/or effective masses in the bulk material. [38] For example, the sp^3s^* basis with the second-nearest-neighbor scheme turns out to have 27 parameters for the zinc blende lattice structure, and the energies and effective masses are obtained from the diagonalization of the Hamiltonian. It has been emphasized by Lew Yan Voon and Ram-Mohan [39] that the bulk band curvatures must be reproduced well in order to extend the analysis reliably to heterostructures since the effective masses play a crucial role in determining the energy levels in quantum wells, for example. Thus, reproducing the bulk bandgaps alone is not sufficient. The bulk energy bands are nonlinear functions of the ETBM parameters, which can be fitted by trial and error or by using optimization methods, such as genetic algorithms. [40]

The lack of a direct relationship between the input parameters and the experimentally determined quantities is probably the single greatest disadvantage of the tight-binding method in making complicated band structure calculations. The strategy for employing the ETBM for heterostructures is analogous to the FEM in that we overlay atomic coupling matrices (the sub-Hamiltonians) in order to produce a global matrix for the heterostructure. The ETBM has been applied to superlattice band structure calculations. [41] A superperiodicity BC is imposed on the global matrix, and the interface BCs correspond to interpolated overlap integral parameters. It was originally thought that the momentum matrix elements for optical transitions require a refitting of the parameters for these matrix elements as well; however, it was shown by Lew Yan Voon and Ram-Mohan that one can obtain the optical matrix elements by invoking the Feynman-Hellmann theorem. We can relate the optical momentum matrix element to the derivative of the ETBM Hamiltonian with respect to the wavevector so that the same set of parameters that define the Hamiltonian can be reused in evaluating the optical matrix elements in a self-consistent manner. [39] The ETBM provides contributions from the entire BZ to the heterostructure energy levels, whereas the $\mathbf{k} \cdot \mathbf{P}$ method includes contributions only near the center of the BZ. The computational requirements for 2D and 3D modeling within the ETBM would be more demanding in our opinion than the envelope function approach with FEM. Inclusion of electric and magnetic field effects are also much more complex in the ETBM than in the procedure we have reviewed in this chapter.

(b) While the inclusion of additional bands and overlaps of higher orbitals is a possible approach to improving the tight-binding modeling of energy bands in bulk semiconductors, the *effective bond orbital model* (EBOM). Chang [42] uses spin-doubled s, p_x, p_y, p_z orbitals to generate an 8×8 Hamiltonian. A crucial difference between the EBOM and related tight-binding formulations is that the s and p orbitals are centered on the face-centered cubic lattice sites of the zinc blende crystal rather than on both of the two real atoms per lattice site. The resulting somewhat *ad hoc* formulation offers considerable computational savings in comparison with the ETBM. However, the main significance of the EBOM approach derives from the fact that the resulting secular matrix has a small-$\mathbf{k}$ expansion that exactly reproduces the form of the eight-band $\mathbf{k} \cdot \mathbf{P}$ Hamiltonian. This allows the EBOM input parameters to be readily expressed in terms of the experimentally measured parameters, such as the band gap, the split-off gap, and the zone-center mass of each band, which has not been accomplished using the more involved ETBM. In fact, the EBOM can be thought of as an extension of the $\mathbf{k} \cdot \mathbf{P}$ method to provide an approximate representation of the energy bands over the full BZ. Since short-period superlattice bands sample wavevectors throughout the BZ, we may expect the EBOM to be more accurate than the $\mathbf{k} \cdot \mathbf{P}$ model for thin-layer structures. However, the EBOM is considerably less efficient computationally than the $\mathbf{k} \cdot \mathbf{P}$ method, especially for thicker superlattices. Each lattice position must be represented in the supercell technique, i.e. no EFA is made.

(c) The influence of core electrons in keeping the valence electrons outside of the core may be represented by an effective repulsive potential in the core region. When this is added to the attractive ionic potential, the net "pseudopotential" nearly cancels at short distances. [43] The valence states are orthogonal to the core states, and the resulting band structure theory corresponds to the nearly free-electron model. In the *empirical pseudopotential model*, the crystal potential is represented by a linear superposition of atomic potentials, which are modified to obtain good fits to the experimental direct and indirect band gaps and effective masses. Further details are presented by Cohen and Chelikowsky [44] and in the reviews by Heine and Cohen. [45] *Ab-initio* approaches employ calculated band parameters (from the density-functional theory) in lieu of experimental data. Combinations of *ab-initio* and empirical methods have been developed to a high level of sophistication. [46] Extension of the pseudopotential method to heterostructures entails the construction of a supercell to assure the proper periodic BCs. With atomic potentials as the essential input, the electronic properties of the heterostructure can be determined, although the required computational effort far exceeds the demands of the $\mathbf{k} \cdot \mathbf{P}$ method. The relative merits of the $\mathbf{k} \cdot \mathbf{P}$ and pseudopotential approaches have been assessed. [47] Again, the inclusion of states from the entire BZ in the construction of heterostructure quantum states in these schemes entails heavy computational resources.

10.6 Concluding remarks

We have shown that the principle of least action can be employed to advantage in the modeling of semiconductor heterostructures by the FEM. This provides a natural way of including complex BCs into the simulation. The FEM is not a low-accuracy method, but rather it can be used to solve quantum mechanical problems with double precision accuracy at the nanoscale. This modeling approach, implemented by us over the past two decades, now allows us to explore optoelectronic properties of complex heterostructures through numerical simulations. [4, 48] The wavefunction overlap, optical matrix elements, strain-induced band splittings leading to wavefunction mixing, energy level and wavefunction manipulation through modulation doping, use of magnetic impurity layers to alter distribution of the wavefunctions, and the application of external perturbations such as electric or magnetic fields are all increasingly of interest in controlling the actual shape of the wavefunctions. These represent aspects of wavefunction engineering. The properties of the structure can now be simulated and optimized on the computer before a heterostructure is actually grown. [12, 26] The fundamental shift in paradigm represented by wavefunction engineering is leading to the exploration of new electronic mechanisms and concepts, the exploration of basic physics, and a rapid turn-around in the design-growth-characterization cycle for developing new optoelectronic devices with 2D (quantum wire) and 3D (quantum dot) confinement of carriers. We are just beginning to explore the freedom in what may be called heterostructure architecture, in the design of new optoelectronic devices in three dimensions through this combination of fundamental theory and FEM computations. Such an approach is indispensable and perhaps inevitable for modeling nanoscale systems.

References

[1] G. Bastard, *Wave Mechanics Applied to Semiconductor Heterostructures*, (Les Editions de Physique, Les Ulis, France, 1988).

[2] F. Capasso, *J. Vac. Sci. Technol.* B **1**, 457–461 (1983); "Band-gap engineering via graded gap, superlattice, and periodic doping structures: Applications to novel photodetectors and other devices;" Surf. Sci. **142**, 513–528 (1984); "New multilayer and graded gap optoelectronic and high speed devices by band gap engineering."

[3] L. R. Ram-Mohan, *Finite Element and Boundary Element Applications to Quantum Mechanics* (Oxford UP, Oxford, UK, 2002).

[4] L. R. Ram-Mohan, and J. R. Meyer, *J. Nonlin. Opt. Phys. Mater.* **4**, 191 (1995); L. R. Ram-Mohan, D. Dossa, I. Vurgaftman, and J. R. Meyer, *Handbook of Nanostructured Materials and Nanotechnology* vol **2**, edited by H. S. Nalwa (Academic Press, New York, 1999); chap 15.

[5] O. C. Zienkiewicz, *The Finite Element Method* (McGraw-Hill, New York, 1989); O. C. Zienkiewicz, and R. L. Taylor, *The Finite Element Method* (McGraw-Hill, New York, 1994); K.-J. Bathe, *Finite Element Procedures in Engineering Analysis* (Prentice-Hall, New Jersey, 1982); T. J. R. Hughes, *The Finite Element Method* (Prentice-Hall, Englewood Cliffs, 1987).

[6] M. Gell-Mann and M. Levy, *Il Nuovo Cimento* **16**, 705–726 (1960); "The axial vector current in beta decay."

[7] L. R. Ram-Mohan, J. Moussa, and K. H. Yoo, *J. Appl. Phys.* **95**, 3081–3092 (2004); "The Schrödinger–Poisson self-consistency in layered quantum semiconductor structures."

[8] I. Vurgaftman, J. R. Meyer, and L. R. Ram-Mohan, *J. Quant. Electron.* **34**, 147–156 (1998); "Mid-IR vertical-cavity surface-emitting lasers."

[9] E. O. Kane, *J. Phys. Chem. Solids* **6**, 236–241 (1958); "The influence of exchange on the effective mass formalism;" E. O. Kane, *Semiconductors and Semimetals* vol 1, edited by R. K. Willardson and A. C. Beer (Academic, New York, 1966); E. O. Kane, *Handbook on Semiconductors* vol 1, edited by W. Paul (North-Holland, Amsterdam, Holland, 1982) p. 193; J. M. Luttinger and W. Kohn, *Phys. Rev.* **97**, 869–883 (1955); "Motion of electrons and holes in perturbed periodic fields."

[10] P. O. Löwdin, *J. Chem. Phys.* **19**, 1396–1401 (1951); "A note on the quantum-mechanical perturbation theory."

[11] M. H. Weiler, *Semiconductors and Semimetals* vol **16**, edited by R. K. Willardson, and A. C. Beer (Academic, New York, 1981); p. 119.
[12] C. R. Pidgeon, and R. N. Brown, *Phys. Rev.* **146**, 575–583 (1966); "Interband magneto-absorption and Faraday rotation in InSb."
[13] B. A. Foreman, *Phys. Rev.* B **48**, 4964–4967 (1993); "Effective-mass Hamiltonian and boundary conditions for the valence bands of semiconductor microstructures." M. G. Burt, *Semicond. Sci. Technol.* Erratum: **2**, 701 (1987); "An exact formulation of the envelope function method for the determination of electronic states in semiconductor microstructures;" M. G. Burt, *J. Phys. Condens. Matter* **4**, 6651–6690 (1992); "The justification for applying the effective-mass approximation to microstructures."
[14] J. M. Luttinger, *Phys. Rev.* **102**, 1030–1041 (1956); "Quantum theory of cyclotron resonance in semiconductors: General theory."
[15] R. van Dalen, and P. N. Stavrinou, *Phys. Rev.* B **55**, 15456–15459 (1997); "Operator ordering and boundary conditions for valence-band modeling: Application to [110] heterostructures."
[16] L. R. Ram-Mohan, I. Vurgaftman, and J. R. Meyer, *Microelectron. J.* **30**, 1031–1042 (1999); "Wave function engineering of antimonide quantum-well lasers;" L. R. Ram-Mohan, *Proc. 31st Int. Symp. on Compound Semiconductors* (Seoul, Korea, 2005), (*Institute of Physics Conference Series* vol 184), edited by J.-C. Woo, H. Hasegawa, Y.-S. Kwon, T. Yao, and K.-H. Yoo (IoP, Bristol, UK, 2005); pp. 1–8.
[17] G. L. Bir, and G. E. Pikus, *Symmetry and Strain Induced Effects in Semiconductors* (Wiley, New York, 1974); G. E. Pikus, *Zh. Eksp. Teor. Fiz.* **41**, 1258 (1961); G. E. Pikus, *Sov. Phys. JETP* **14**, 898–907 (1962); "A new method of calculating the energy spectrum of carriers in semiconductors. I. Neglecting spin-orbit interaction;" Y. M. Sirenko, J. B. Jeon, K. W. Kim, M. A. Littlejohn, and M. A. Stroscio, *Phys. Rev.* B **53**, 1997–2009 (1996); "Envelope-function formalism for valence bands in wurtzite quantum wells."
[18] L. R. Ram-Mohan, A. M. Girgis, J. D. Albrecht, C. W. Litton, and T. D. Steiner, *27th International Conference on the Physics of Semiconductors* (Am. Inst. Phys., Melville, NY, 2005), edited by J. Menendez and C. G. Van de Walle, pp. 941–942; L. R. Ram-Mohan, A. M. Girgis, J. D. Albrecht, and C. W. Litton, Superlattices Microstruct. **39**, 455–477 (2006); "Wavefunction engineering of layered wurtzite semiconductors grown along arbitrary crystallographic directions."
[19] S. L. Chuang and C. S. Chang, *Appl. Phys. Lett.* **68**, 1657–1659 (1996); "Effective-mass Hamiltonian for strained wurtzite GaN and analytical solutions." *Phys. Rev.* B **54**, 2491–2504 (1996); "k . p method for strained wurtzite semiconductors."
[20] F. Mireles, and S. E. Ulloa, *Phys. Rev.* B **60**, 13659–13667 (1999); "Ordered Hamiltonian and matching conditions for heterojunctions with wurtzite symmetry: quantum wells;" *Phys. Rev.* B **62**, 2562–2572 (2000); "Strain and crystallographic orientation effects on the valence subbands of wurtzite quantum wells."

[21] G. B. Ren, Y. M. Liu, and P. Blood P, *Appl. Phys. Lett.* **74**, 1117–1119 (1999); "Valence-band structure of wurtzite GaN including the spin-orbit interaction."

[22] D. J. Dugdale, S. Brand, and R. A. Abram, *Phys. Rev.* B **61**, 12933–12938 (2000); "Direct calculation of $k \cdot p$ parameters for wurtzite AlN, GaN, and InN."

[23] L. C. Lew Yan Voon, M. Willatzen, M. Cardona, and N. E. Christensen, *Phys. Rev.* B **53**, 10703–10714 (1966); "Terms linear in k in the band structure of wurtzite-type semiconductors."

[24] S. T. Jang, K. H. Yoo, and L. R. Ram-Mohan, *Proc. 13th Int. Symp. on Physics of Semiconductors and Applications* (2006).

[25] For a compilation of band parameters, see I. Vurgaftman, J. R. Meyer, and L. R. Ram-Mohan, *J. Appl. Phys.* **89**, 5815–5875 (2001); "Band parameters for III–V compound semiconductors and their alloys." and I. Vurgaftman and J. R. Meyer, *J. Appl. Phys.* **94**, 3675–3696 (2003); "Band parameters for nitrogen-containing semiconductors."

[26] J. R. Meyer, I. Vurgaftman, R. Q. Yang, and L. R. Ram-Mohan, *Electron. Lett.* **32**, 45–46 (1996);"Type-II and type-I interband cascade lasers;" I. Vurgaftman, J. R. Meyer and L. R. Ram-Mohan, *Photonics Technol. Lett.* **9**, 170–172 (1997); "High-power/low-threshold type-II interband cascade mid-IR laser-design and modeling."

[27] N. Dai, H. Luo H, F. C. Zhang, N. Samarth, M. Dobrowolska, and J. K. Furdyna, *Phys. Rev. Lett.* **67**, 3824–3827 (1991); "Spin superlattice formation in $ZnSe/Zn_{1-x}Mn_xSe$ multilayers;" N. Dai, L. R. Ram-Mohan, H. Luo, G. L. Yang, F. C. Zhang, M. Dobrowolska and J. K. Furdyna, *Phys. Rev.* B **50**, 18153–18169 (1994); "Observation of above-barrier transitions in superlattices with small magnetically induced band offsets."

[28] C. A. Hoffman, J. R. Meyer, R. J. Wagner, F. J. Bartoli, X. Chu, J. P. Faurie, L. R. Ram-Mohan, and H. Xie, *J. Vacuum Sci. and Technol.* A **8**, 1200–1205 (1990); "Electron transport and cyclotron resonance in [211]-oriented HgTe–CdTe superlattices."

[29] K. H. Yoo, PhD Thesis: "Magneto-optical studies of HgTe/CdTe superlattices with bandgaps in the infrared" *PhD Thesis* MIT (1990).

[30] H. M. Ng, *Appl. Phys. Lett.* **80**, 4369–4371 (2002); "Molecular-beam epitaxy of $GaN/Al_xGa_{1-x}N$ multiple quantum wells on R-plane (1012) sapphire substrates;" A. Bykhovski, B. Gelmont, and M. Shur, *Appl. Phys. Lett.* **63**, 2243–2245 (1993); "Strain and charge distribution in GaN-AlN-GaN semiconductor-insulator-semiconductor structure for arbitrary growth orientation."

[31] S. Ghosh, O. Brandt, H. T. Grahn, and K. H. Ploog, *Phys. Stat. Sol. (b)* **234**, 882–886 (2002); "Electronic band structure of strained C- and M-plane GaN films investigated by polarized photoreflectance spectroscopy."

[32] S. Kumar, B. S. Williams, and Q. Hu, *Appl. Phys. Lett.* **88**, 121–123 (2006); "1.9 THz quantum-cascade lasers with one-well injector."

[33] G. Fasching, A. Benz, Zobl, A. M. Andrews, T. Roch, W. Schrenk, G. Strasser, V. Tamosiunas, and K. Unterrainer, *Physica* E **32**, 316–319 (2006).

[34] C. Worral, J. Alton, M. Houghton, S. Barbieri, H. E. Beere, D. Ritchie, and C. Sirtori, *Opt. Exp.* **14**, 171–181 (2006); "Continuous wave operation of a superlattic quantum cascade laser emitting at 2 THz."

[35] W. H. Press, S. A. Teukolsky, W. T. Vetterling, B. R. Flannery, *Numerical Recipes* (Cambridge University, Cambridge, Press, 1992).

[36] V. M. Menon, W. D. Goodhue, A. S. Karakashian, and L. R. Ram-Mohan, *J. Appl. Phys.* **88**, 5262–5267 (2000); "Phonon mediated lifetimes in intersubband terahertz lasers;" V. M. Menon, L. R. Ram-Mohan, W. D. Goodhue, A. S. Karakashian, A. Naweed, A. Gatesman, J. Waldman, *Physica* E **15**, 197–201 (2002); "Phonon engineered quantum cascade terahertz emission;" V. M. Menon, L. R. Ram-Mohan, W. D. Goodhue, A. J. Gatesman, and A. S. Karakashian, *Physica* B **316/317**, 212–215 (2002); "Role of interface phonons in quantum cascade terahertz emitters;" V. M. Menon, W. D. Goodhue, A. S. Karakashian, A. Naweed, J. Plant, L. R. Ram-Mohan, A. Gatesman, V. Badami, and J. Waldman, *Appl. Phys. Lett.* **80**, 2454–2456 (2002); "Dual-frequency quantum-cascade terahertz emitter."

[37] J. C. Slater and G. F. Koster, *Phys. Rev.* **94**, 1498–1524 (1954); "Simplified LCAO method for the periodic potential problem."

[38] D. J. Chadi, and M. L. Cohen, *Phys. Status Solidi* B **68**, 405–419 (1975); "Tight-binding calculations of the valence bands of diamond and zincblende crystals;" P. Vogl, H. P. Hjalmarson, and J. D. Dow, *J. Phys. Chem. Solids* **44**, 365–378 (1983); "A semi-empirical tight-binding theory of the electronic structure of semiconductors."

[39] L. C. Lew Yan Voon., and L. R. Ram-Mohan, *Phys. Rev.* B **47**, 15500–15508 (1993); "Tight-binding representation of the optical matrix elements: Theory and applications."

[40] G. Klimeck, R. C. Bowen, T. B. Boykin, and T. A. Cwik, *Superlattices Microstruct.* **27**, 520–524 (2000); "sp_3s^* Tight-binding parameters for transport simulations in compound semiconductors."

[41] D. L. Smith, and C. Mailhiot, *Rev. Mod. Phys.* **62**, 173–234 (1990); "Theory of semiconductor superlattice electronic structure;" J. N. Schulman and Y.-C. Chang, *Phys. Rev.* B **31**, 2056–2068 (1985); "Band mixing in semiconductor superlattices;" J. N. Schulman and Y.-C. Chang, *Phys. Rev.* B **33**, 2594–2601 (1986); "HgTe-CdTe superlattice subband dispersion." L. C. Lew Yan Voon, and L. R. Ram-Mohan, *Phys. Rev.* B **50**, 14421–14434 (1994); "Calculations of second-order nonlinear optical susceptibilities in III-V and II-VI semiconductor heterostructures."

[42] Y. C. Chang, *Phys. Rev.* B **37**, 8215–8222 (1988); "Bond-orbital models for superlattices." J. P. Loehr, *Phys. Rev.* B **50**, 5429–5434 (1994); "Improved effective-bond-orbital model for superlattices;" J. P. Loehr, *Physics of Strained Quantum Well Lasers* (Kluwer Academic, New York, 1997).

[43] J. C. Phillips, and L. Kleinman, *Phys. Rev.* **116**, 287–294 (1959); "New method for calculating wave functions in crystals and molecules;" M. H. Cohen and V. Heine, *Phys. Rev.* **122**, 1821–1826 (1961); "Cancellation of kinetic and potential energy in atoms, molecules, and solids."

[44] M. L. Cohen and J. R. Chelikowsky *Electronic Structure and Optical Properties of Semiconductors* Springer Series in Solid-State Sciences, edited by M. Cardona (Springer, Berlin, 1988), Vol. **75**.

[45] V. Heine, *The Pseudopotential Concept*, in *Solid State Physics*, edited by H. Ehrenreich, F. Seitz, and D. Turnbull, Vol. **24** (Academic, New York, 1970), p. 1; M. L. Cohen and V. Heine, *The Fitting of Pseudopotentials to Experimental Data and their Subsequent Application*, in Solid State Physics, edited by H. Ehrenreich, F. Seitz, and D. Turnbull, Vol. **24** (Academic, New York, 1970), p. 38.

[46] L.-W. Wang and A. Zunger, *in Semiconductor Nanoclusters, Studies in Surface Science and Catalysis*, Vol. **103**, edited by P. V. Kamat and D. Meisel (Elsevier Science, New York, 1996), p. 161.

[47] D. M. Wood, and A. Zunger, *Phys. Rev.* B **53**, 7949–7963 (1996); "Successes and failures of the $\mathbf{k} \cdot \mathbf{P}$ method: A direct assessment for GaAs/AlAs quantum structures;" L.-W. Wang, S.-H. Wei, T. Mattila, A. Zunger, I. Vurgaftman, and J. R. Meyer, *Phys. Rev. B* **60**, 5590–5596 (1999); "Multiband coupling and electronic structure of $(InAs)_n/(GaSb)_n$ superlattices."

[48] J. R. Meyer, C. A. Hoffman, F. J. Bartoli, L. R. Ram-Mohan, *Novel Optical Materials and Applications*, edited by I. C. Khoo, F. Simone, and C. Umeton (Wiley, New York, 1966) pp. 205–237.

11

Schrödinger-Poisson self-consistency in layered semiconductor nanostructures

In this chapter:

- We develop a self-consistent solution of the Schrödinger and Poisson equations in semiconductor heterostructures with arbitrary doping profiles and layer geometries. An algorithm for this nonlinear problem is presented in a multi-band $\mathbf{k} \cdot \mathbf{P}$ framework for the electronic band structure using the finite element method. The discretized functional integrals associated with the Schrödinger and Poisson equations are used in a variational approach. The finite element formulation allows us to evaluate functional derivatives needed to linearize Poisson's equation in a natural manner. Illustrative examples are presented using a number of heterostructures including single quantum wells, an asymmetric double quantum well, *p-i-n-i* superlattices and trilayer superlattices.

11.1 Introduction

In modulation-doped semiconductor heterostructures, the presence of discontinuities in the band edges leads to the possibility of the separation of the carriers from their parent donors or acceptors. This redistribution of the free carriers, together with the presence of ionized impurities in so-called depletion regions, gives rise to a self-consistent reconfiguration of the band edges. As is well known, this problem is numerically unstable because small changes in the potential profile lead to substantial rearrangements of free carriers. The carriers are free in the sense that they are ionized away from the impurity atoms, though they could be bound within quantum wells, for example.

Here we present a self-consistent solution of the spatial behavior of the energy band edge in arbitrary layered semiconductor nanostructures. For illustrative purposes and

Finite Elements in Action. L. Ramdas Ram-Mohan, Oxford University Press. © L. Ramdas Ram-Mohan (2026).
DOI: 10.1093/oso/9780199563487.003.0011

to describe the algorithm, we first consider the case of donor impurities and conduction band carriers in a single energy band model, and later treat the multi-band case. The carrier charge density is determined by the solution of Schrödinger's equation for the wavefunctions of the carriers in the layered quantum system. This is introduced into Poisson's equation as a source term together with the positive charge distribution of ionized donors. The resulting potential arising from the redistributed charges is obtained by solving Poisson's equation. This potential alters the initial band edge potential with flat bands, and Schrödinger's equation is solved once again for the new total potential energy. This cycle of solving the two differential equations is iterated to convergence. We do not include any polarization charges due to stress-induced piezoelectric or spontaneous polarization in the layers. Also, we do not consider structures with Type-II energy band gap alignments that have semi-metallic (zero band gap) properties, though Type-II structures with open band gaps are amenable to this analysis.

We have utilized the finite element method (FEM) for the solution of both Schrödinger's and Poisson's equations. Both of these equations are solved by employing the principle of least action. [1–3] Being a time-independent problem, this reduces to making the Lagrangian stationary to variations. In our modeling of structures with high doping, we find that the convergence is substantially faster in FEM in terms of the number of iterations and stability, even for doping concentrations as high as $10^{21}\,\mathrm{cm}^{-3}$, than conventional finite difference methods based on the shooting approach. [4]

Almost all earlier work has focused on specific structures, such as single interfaces between semiconductor materials, [5, 6] single quantum wells, bilayer superlattices, [7] or double barrier resonant tunneling structures. [8] The single symmetric quantum well with modulation doping in the barriers was treated by us earlier in a finite element framework, [1, 9] but the algorithm was specialized to that particular case.

The present method borrows from the recent work by Trellakis et al., [10] who solve the self-consistency problem for a two-dimensional distribution of carriers with finite difference methods. We specifically consider layered III-V semiconductor structures with modulation doping in terms of a finite element formulation. The finite element approach corresponds to the discretization of a functional integral over the Lagrangian density defined as the functional to be made stationary. The variational principle is invoked after the discretization of the functional is performed and the spatial dependence is integrated out. The same method is applied to both the Schrödinger and Poisson functionals. While the total functional is the sum of the Schrödinger and the Poisson parts, we vary the total functional independently with respect to the Schrödinger wavefunction and also by the electrostatic potential to obtain two coupled equations. Poisson's equation is nonlinear in the change in the potential, and the nonlinearity is accounted for by iterating over a locally linearized form. An added complication is that the electronic portion of the charge density requires the calculation of the wavefunctions for *all* in-plane wave vectors (a matrix diagonalization), which is a very time-consuming calculation. These wavefunctions are dependent on the in-plane dispersion and should be taken into account even for a one-band calculation.

Given the nonlinear nature of the calculation we separate it into two nested loops. In the outer iteration loop, the electronic charge density is calculated and updated by solving Schrödinger's equation with the latest value of the potential energy. In the inner loop, the ionized donor charge is evaluated together with the *local* feedback to the electronic charge density. Both the donor and the mobile charge densities act as source terms in the Poisson equation which is solved to obtain the change in the potential energy through the given inner iteration loop. Typically, the number of (fast) iterations in the inner loop can be $10 \sim 100$ for each outer loop, and overall convergence is achieved in $5 \sim 20$ outer loop (slower) iterations. This break-up into nested iteration loops was also advocated by Trellakis et al. [10]

Throughout the calculations, we take advantage of the finite element representation of the discretized integrals for the Schrödinger and Poisson functionals. Most of the global matrices are calculated only once in the finite element framework – an advantage in the large matrix analysis used in this iterative approach. The matrices are banded, and sparse and we exploit this in our matrix computation algorithms. We also have quadratic convergence to the minimum, in the variational sense, in the finite element approach. By using a Taylor expansion through the lowest order, *functional derivatives* are evaluated through the variation of the nodal values of the discretized functions. This too provides numerical simplifications within the finite element framework, while representing a general way of implementing functional derivatives numerically, since the values of the discretized functions at the nodes alone determine them; the interpolation polynomials are not altered.

The essential theory for obtaining electronic wavefunctions in the FEM is derived in Sec. 11.2. We describe, very briefly, the use of FEM for the determination of bound state eigenvalues and wavefunctions and for the scattering states. The algorithm for the solution of Poisson's equation is presented in Sec. 11.3. Examples of self-consistent calculations for III-V layered heterostructures are presented in Sec. 11.4.

11.2 Schrödinger's equation

11.2.1 A finite element formulation

We begin by considering Schrödinger's equation in the layered quantum heterostructure for a conduction band electron using the spherical effective mass approximation. [11] An extension to the multi-band representation of the electronic states is straightforward, and this is used in the example calculations that follow. The envelope functions satisfy the equation

$$-\frac{\hbar^2}{2}\nabla\left(\frac{1}{m^*(z)}\,\nabla\psi(\mathbf{r})\right) + V(z)\,\psi(\mathbf{r}) = E\,\psi(\mathbf{r}). \tag{11.1}$$

Here, $\psi(\mathbf{r})$ represents the wave function in the structure, where each layer is for the moment assumed to have a uniform composition and a constant effective mass. The potential energy $V(z)$ is the superposition of the potential energy due to the conduction band offsets at interfaces between layers, which may be denoted by $V_0(z)$, and $V^{ch}(z)$, which arises from the presence of ionized donors and the free charges released by them, so that

$$V(z) = V_0(z) + V^{ch}(z). \tag{11.2}$$

The changes in the potential function due to charge redistribution can be substantial, as shown in Fig. 11.1.

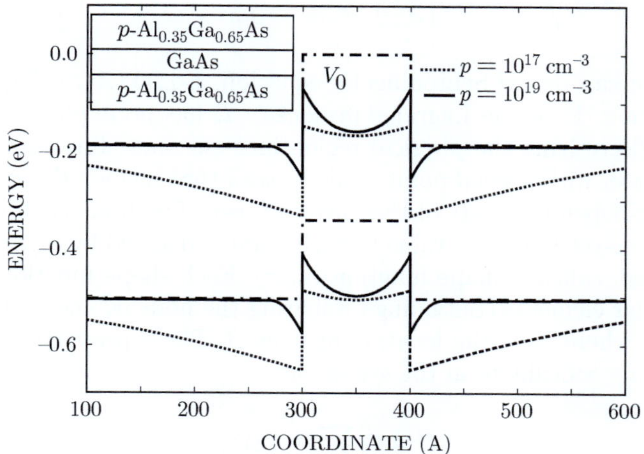

Figure 11.1 The bending of the valence band edge in a 100 Å GaAs quantum well with p-doped barriers of $Ga_{0.65}Al_{0.35}As$ (i) to 1.0×10^{17} cm^{-3} (dotted line), and (ii) to 1.0×10^{19} cm^{-3} (solid line), are shown. The dot-dashed line marked by V_0 is the potential due to band offsets before band bending. Note the larger depletion region generated in the case of lower doping.

We further assume that the electron is free in the in-plane direction, so that the envelope portion of the three-dimensional wavefunction in the effective mass approximation takes the form

$$\psi(r) = \frac{1}{L} e^{ik_x x} e^{ik_y y} f(z), \tag{11.3}$$

where L is an in-plane quantization length. Substituting Eq. (11.3) into Eq. (11.1) we obtain an equation for $f(z)$ in any given layer given by

$$-\frac{\hbar^2}{2m^*}\frac{d^2}{dz^2} f(z, k_\parallel) + \frac{\hbar^2 k_\parallel^2}{2m^*} f(z, k_\parallel) + V(z) f(z, k_\parallel) = E f(z, k_\parallel), \tag{11.4}$$

whose solution for the composite heterostructure gives us the envelope functions and the energy eigenvalues of the quantum system. Here $\mathbf{k}_\parallel = (k_x, k_y)$ represents the in-plane wave vector.

The action integral is defined by

$$\mathbf{A}[f^*, f] = \int dt\, \mathcal{A} = \int dt \int dz\, \mathcal{L}(f^*, f),$$

and

$$\mathcal{A} = \int dz f^*(z) \left[\frac{\overleftarrow{d}}{dz} \frac{\hbar^2}{2m^*} \frac{\overrightarrow{d}}{dz} + \frac{\hbar^2 k_\parallel^2}{2m^*} + V(z) - E \right] f(z), \tag{11.5}$$

from which we can derive Schrödinger's equation, Eq. (11.1), using the principle of stationary action. Here the Lagrangian density $\mathcal{L}$ has no explicit time dependence. We start by discretizing the physical region into *elements*, line segments in the one-dimensional case, with special points called *nodes* that include the endpoints. In each element, the z-dependent part of the envelope wave function, $f(z)$, is represented as the sum of its as-yet-unknown value at each node multiplied by a Lagrange interpolation polynomial, called a shape function $N_\alpha(z)$. Each shape function is nonzero only in the particular elements containing or sharing the node α, and satisfies the relation $N_\alpha(z_\beta) = \delta_{\alpha\beta}$, where z_β is the location of node β. These properties allow $f(z)$ to be expressed at any coordinate in the system as

$$f(z) = \sum_\alpha N_\alpha(z) f(z_\alpha), \tag{11.6}$$

with index α referring to the node number for the global system. If additional accuracy is desired, two degrees of freedom are used at each node by considering both $f(z_\alpha)$ and $f'(z_\alpha)$ to be nodal variables that are multiplied by Hermite interpolation polynomials for shape functions. [12] We then decompose $f(z)$ in the form

$$f(z) = \sum_\alpha N_\alpha(z) \phi_\alpha, \tag{11.7}$$

where the array elements of ϕ_α alternate between the value of the function at the node and the value of the derivative of function at the same node.

We substitute Eq. (11.7) into Eq. (11.5) and integrate out the known spatial dependence over each element. The functional integral in element *iel* is given by

$$\mathcal{A}^{iel}[f^*, f] = \phi_\alpha^* \left[A_{\alpha\beta}^{(iel)} + B_{\alpha\beta}^{(iel)} - E\, S_{\alpha\beta}^{(iel)} \right] \phi_\beta, \tag{11.8}$$

with $A^{(iel)}$ containing the kinetic energy contribution, $B^{(iel)}$ having the potential energy contributions, and $S^{(iel)}$ being the wavefunction overlap matrix. These contributions are added up to obtain the discretized form of the total functional. The individual element matrices $A^{(iel)}$, $B^{(iel)}$, and $S^{(iel)}$ are inserted into the corresponding global matrices. Appropriate care is taken to overlay the element matrices to ensure that inter-element continuity is enforced at the shared node between two elements. The resulting functional integral is given by

$$\mathcal{A}[f^*, f] = \phi^*_\alpha [A_{\alpha\beta} + B_{\alpha\beta} - E\, S_{\alpha\beta}]\phi_\beta. \tag{11.9}$$

We now invoke the (nodal) variational principle to obtain

$$\frac{\delta \mathcal{A}}{\delta \phi^*_\alpha} = 0, \tag{11.10}$$

or

$$(A_{\alpha\beta} + B_{\alpha\beta})\, \phi_\beta = E\, S_{\alpha\beta}\, \phi_\beta, \tag{11.11}$$

which is the discretized Schrödinger's equation represented here as a generalized eigenvalue problem, with eigenvectors ϕ having the nodal values and nodal derivatives of $f(z)$ as components. Here the overlap matrix $S_{\alpha\beta}$ is not diagonal because the shape functions are not orthogonal. In practice, the matrix $A_{\alpha\beta}$ is separated into terms dependent on powers of k_x and k_y. In the 1-band case

$$\begin{aligned} A_{\alpha\beta} &= \sum_{iel} A^{iel}[f^*, f] \\ &= A^{(0)}_{\alpha\beta} + k_x^2 A^{(1)}_{\alpha\beta} + k_y^2 A^{(2)}_{\alpha\beta}. \end{aligned} \tag{11.12}$$

Then the global matrices $A^{(0)}$, $A^{(1)}$, and $A^{(2)}$ are not recalculated for each value of k_x, k_y, but rather are stored in memory so that the reconstruction of A in Eq. (11.12) is very rapid. This has the added advantage that we can immediately apply the Feynman-Hellmann theorem [13] to obtain the in-plane effective mass of the carrier in the system. Given the solution of the eigenvalue problem, Eq. (11.11), and the corresponding eigenfunctions ϕ^i_β, the effective mass in the ith energy subband level is given by

$$\begin{aligned} \frac{1}{m_i^*} &= \frac{1}{\hbar^2} \left\langle f_i \,\middle|\, \frac{\partial^2 H}{\partial k_x^2} \,\middle|\, f_i \right\rangle, \\ &= \frac{2}{\hbar^2}\, \phi^i_\alpha\, A^{(1)}_{\alpha\beta}\, \phi^i_\beta. \end{aligned} \tag{11.13}$$

Again, the overlap matrix $S_{\alpha\beta}$ is computed only once. The matrix $B_{\alpha\beta}$ contains the potential energy term, and this is reevaluated as the potential energy function is updated in the iterations.

With $3 \sim 10$ quintic Hermite elements per layer, the matrix dimensions are typically $250 \sim 2500$ with a bandwidth of 6 (diagonal + supra-diagonal) leading to fairly sparse matrices.

11.2.2 Boundary and interface conditions in the FEM

For any layered semiconductor structure at hand, we now apply boundary conditions to the wavefunctions at the left and right ends of the heterostructure by specifying the values of the wavefunction or its derivative at the initial and final nodes. The boundary nodal values depend on whether we are considering bound states, free (traveling) states, or quasi-bound states.

For bound states, the wavefunction is zero at the left and right boundaries, far into the barrier regions. These boundary conditions are now built into the set of simultaneous equations, Eq. (11.11).†

For free and quasi-bound states, we have to account for the form of incoming and outgoing traveling waves. While the finite element approach gives the total wavefunction, we are concerned with boundary conditions on the incoming or outgoing wave amplitudes. This calls for a modal analysis of the wavefunction in terms of the incoming and outgoing component wavefunctions at the two ends. For such scattering problems, the energy E of the incident wave is given. We rewrite Eq. (11.5) to include additional surface terms needed to express the conditions on the incoming and outgoing probability currents at z_L and z_R.

For a wave incident from the left with amplitude a, we have

$$\begin{aligned} f(z_L) &= a\, e^{ik_L z_L} + r\, e^{-ik_L z_L}, \\ f'(z_L) &= ik_L(a\, e^{ik_L z_L} - r\, e^{-ik_L z_L}), \end{aligned} \tag{11.14}$$

where the reflected wave has amplitude r. Thus, the boundary condition on the left is given by

$$f'(z_L) + ik_L\, f(z_L) = 2ik_L\, a\, e^{ik_L z_L}. \tag{11.15}$$

Similarly, on the right side, we have

$$f'(z_R) = ik_R\, f(z_R), \tag{11.16}$$

since $f(z) = t\,\exp(ik_R z)$ in this region. These two boundary conditions are incorporated into the functional by writing

$$\begin{aligned} \mathcal{A}[f^*, f] &= \int dz\, \mathcal{L}(f^*, f) \\ &= \int dz\, f^*(z)\left[\frac{\overleftarrow{d}}{dz}\frac{\hbar^2}{2m^*}\frac{\overrightarrow{d}}{dz} + \frac{\hbar^2 k_\parallel^2}{2m^*} + V(z) - E\right] f(z) \\ &\quad - ik_R\, f^*(z) f(z)\Big|_{z_R} \\ &\quad - f^*(z)[ik_L\, f(z) - 2i\, k_L a e^{ik_L z}]\Big|_{z_L}. \end{aligned} \tag{11.17}$$

The variational principle now gives Schrödinger's equation and also the mixed boundary conditions of Eq. (11.14) required of scattering states above the barriers. The

†Since setting the boundary nodal values to zero will reduce the dimension of the matrices by two, we employ a simple trick to retain the original matrix dimension: we zero the leftmost and rightmost node's row and column corresponding to the end nodes in all the global matrices, and place a large number (say 10^3) at the diagonal entry in $A_{\alpha\beta}$ with unity for the diagonal entry in $S_{\alpha\beta}$. In effect, we have put the large matrices in a block-diagonal form such that the physically useful eigenvalues, which are of $\mathcal{O}(1)$ in eV units, will have eigenfunctions that are now zero at the ends. The two eigenfunctions with large eigenvalues are of course discarded.

discretization of the functional integral leads to global matrices analogous to the case of bound states except that now we have at hand a set of simultaneous equations with driving terms proportional to the incident amplitude a taken over to the right side.

In addition to the boundary conditions at the endpoints, we have to ensure the continuity of the wave function and of the probability current across interfaces between the layers in the heterostructure. This is implemented by the continuity of ψ'/m^* across interfaces. For the case of one degree of freedom per node, the continuity of the probability current is implicitly built in, while for two degrees of freedom per node, the element matrices need to be modified before they are overlaid to form the global matrices. For the element just to the right of the interface, we replace its right nodal derivative evaluated at the interface nodal location with $\phi_R'(z) = (m_R^*/m_L^*)\phi_L'(z)$. We can apply the interface constraint by multiplying the row and column of the element matrix iel_R appropriately by the effective mass ratio before overlaying onto the global matrix.

The bound state wavefunction is normalized such that $\int_{-\infty}^{\infty} dz|f(z)|^2 = 1$, as usual. The free state wavefunction is normalized such that the incoming wave has an amplitude of $\frac{1}{\sqrt{2\pi}}$. Now $\int_{\Omega_k} dk \int_{\Omega} dz|f(z)|^2$ is the number of states in the phase space $\Omega_k\Omega$. Alternatively, we can set the amplitude of the incoming wave to be 1 and use $\int_{\Omega_k} \frac{dk}{2\pi} \int_{\Omega} dz|f(z)|^2$ to calculate the number of states. Under these normalization conditions, $|f(z)|^2$ for a bound state has a dimension of inverse length, but is dimensionless for a free state. In Fig. 11.2, we show the lowest two heavy hole wavefunctions of the single quantum well of Fig. 11.1.

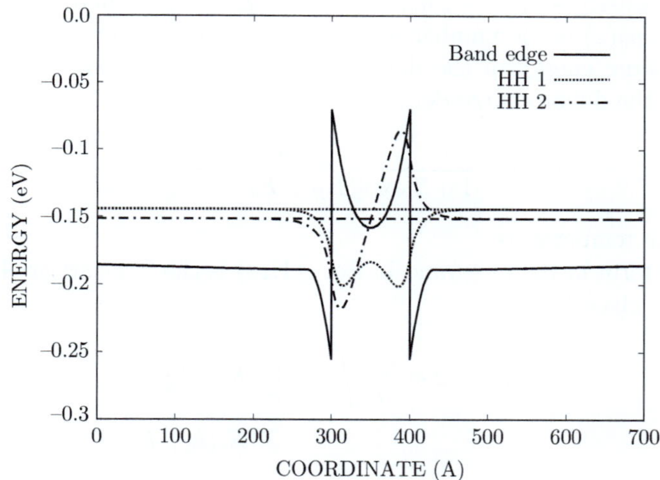

Figure 11.2 The lowest two heavy hole wavefunctions and their energy eigenvalues (horizontal lines) in a 100 Å GaAs quantum well with $Ga_{0.65}Al_{0.35}As$ barriers that are p-doped to 1.0×10^{19} cm^{-3} (Fig. 11.1) are shown. The wavefunctions and the band edge have been labeled in the legend. Note the wavefunctions which reflect the double-well nature of the quantum well under self-consistent band bending.

11.3 Poisson's equation

Now consider the Poisson equation in the self-consistent problem. [14] It is useful to multiply the potential function by $-|e|$, the electronic unit of charge, so that we directly solve for the potential energy $V^{ch}(z)$. We begin with the Poisson functional $\mathcal{P}$

$$\mathcal{P}[V] = \int dz \Bigg[\frac{1}{2} \left(\frac{dV^{ch}(z)}{dz} \right) \epsilon(z) \left(\frac{dV^{ch}(z)}{dz} \right) + 4\pi|e| \Big(\rho_d(V(z)) + \rho_n(V(z)) \Big) V^{ch}(z) \Bigg], \tag{11.18}$$

where ϵ is the dielectric function assumed to be constant in each layer, ρ is the charge density which is a function of the total potential, and the subscripts d and n refer to the ionized donor and the free electron charge densities, respectively. The potential arising from the band offsets at the heterointerfaces, and the potential arising from the spatially varying charge densities ρ_n and ρ_d, all contribute to the total potential $V(z)$. Now $\mathcal{P}$ is discretized using finite elements and is minimized with respect to $V^{ch}(z)$.

11.3.1 The source terms

The second term in the integrand in Eq. (11.18) is the ionized donor contribution to the charge density, which determines the depletion regions. At zero temperature, the donor charge density is $\rho_d(z) = e\, n_d(z)\, \theta(V(z) - E_d - E_F)$, where $\theta(x)$ is the Heaviside step function, $n_d(z)$ is the number density of the donors, $V(z)$ is the total potential, E_d is the binding energy of the donors, and E_F is the quasi-Fermi level. At finite temperature, the donor charge density is

$$\rho_d(z) = \frac{|e|\, n_d(z)}{1 + 2\, \exp[(E_F + E_d - V(z))/k_B T]}, \tag{11.19}$$

where T is the temperature.

The third term in the integrand in Eq. (11.18) is the free-carrier electronic charge density and is given by

$$\rho_n(z) = -2|e| \sum\!\!\!\!\!\!\!\int_{\nu} \int \frac{d^2 k_\parallel}{(2\pi)^2} \, |f_\nu(z, k_\parallel)|^2 \times \mathcal{F}(E_\nu(k_\parallel), E_F, T), \tag{11.20}$$

where the symbol $\sum\!\!\!\!\!\!\int$ succinctly represents the sum over discrete bound states and an integral over the continuum states, labeled by ν, associated with the energy for the motion along z. Here $f_\nu(z, k_\parallel)$ is the z component of the envelope portion of the wavefunction and $\mathcal{F}$ is the Fermi function. The wavefunction, $f_\nu(z, k_\parallel)$ and its corresponding energies, $E_\nu(k_\parallel)$, are found by solution to the Schrödinger's equation for each value of the integration variable, $k_\parallel$. We step in the integration variable, $k_\parallel$, until the additional charge becomes negligible.

Converting the integral in Eq. (11.20) directly into an integral over the density-of-states (DOS) multiplied by the Fermi factor may not be simple in general, since the dispersion relation in energy can be nonparabolic for bound states in a heterostructure. This is certainly the case for the complex valence band structure of heterostructures; it is also the case even within the simple one-band effective mass model. A simple way of evaluating this integral is to consider a series of energy intervals or bins of width $1 \sim 5$ meV. The energy spectrum is given in terms of $\{\nu, k_{\|}\}$ and the corresponding wavefunctions are known. Then the integral is evaluated in discrete cells in momentum space and stored in the corresponding energy bins. We label the DOS in the energy bin i at E_i as $\Lambda_i(z)$. The Fermi function weights the partial contributions of the integral in the various bins, and the sum is the total charge density. By staying in momentum space, and using the energy discretization, we evaluate the DOS directly. We thus have at hand the charge density in the form

$$\rho_n(z) = -|e| \sum_i \Lambda_i(z)\, \mathcal{F}(E_i, E_F, T), \tag{11.21}$$

with the accuracy depending on the level of discretization. Earlier approaches [15–18] to the evaluation of the DOS in semiconductors exploit the known energy dispersion curves over a grid of points and employ interpolation in order to perform the integrals. The determination of the Jacobian of transformation from momentum integrals to an integral over energy entails the gradient of the energy dispersion in the denominator. This requires additional care in order to evaluate it accurately. The crossing of the energy dispersion surfaces across given energy planes is required in order to obtain the energy gradients in these methods. The method used here is effective enough for the in-plane 2D phase space evaluations essential to layered semiconductors. Improving the accuracy of the DOS calculation would require us to reduce the size of the energy bins, leading to increasing time for the computation. It is clear that this issue should be revisited in order to develop a fast and numerically accurate approach to the evaluation of the DOS.

The Fermi level is determined as follows. For the modulation-doped quantum wells in which the barrier can supply enough carriers to fill the well up to Fermi level, the Fermi level for the whole heterostructure can be taken to be the Fermi level in the bulk region on either end of the heterostructure. The bulk Fermi level must be calculated using the donor and electronic charge densities. We use

$$\rho_D(z, E_f, E_D) + \rho_{n,bulk}(z, E_F) = 0 \tag{11.22}$$

to determine the Fermi level through a root-finding procedure. [19] The electronic charge density in the bulk is given at $T = 0\,\text{K}$ by

$$\rho_{n,bulk} = -|e|(2m^*(E_F - E_c)/\hbar^2)^{3/2}/3\pi^2, \tag{11.23}$$

E_c being the conduction band edge energy, and at finite temperature, we have

$$\rho_{n,\,bulk} = -\frac{|e|}{2\pi^2}\left(\frac{2m^* k_B T}{\hbar^2}\right)^{3/2} \times \int_0^\infty d\xi \,\ln\left[1 + e^{-\xi^2}\, e^{(E_F - E_c)/k_B T}\right]. \qquad (11.24)$$

The conduction band edge is assumed to be at E_c in the above expression.

If the doped barrier in a superlattice is not thick enough to supply the needed carriers, or if the well is doped, there is no easy way to predict the location of the Fermi level. In this case, the Fermi level should be found by assuming a trial Fermi level and monitoring the slope of the potential in the barrier region in the case of quantum well or watching the continuity of the slope of the potential at the interface between one period and the next period in the case of superlattice.

The solution of Eq. (11.18) entails using the charge distributions as source terms which in turn depend on the potential energy. The method of solution is given below.

11.3.2 Self-consistency iterations

The charge density is nonlinear in $V(z)$ and is linearized in each element so that the final solution may be approached iteratively starting from an initial guess for the potential energy. An overview of the method is now in order.

We have two parts to the problem. The first is a time-consuming diagonalization of the Schrödinger's equation for bound states, and solving for the wavefunctions for each free or quasi-bound state as well, using the latest potential obtained through iteration. The resulting wavefunctions are used to calculate ρ_n. The second part is the solution of the Poisson's equation for the change in the potential energy. This requires the calculation of the ionized donor charge density. This inner loop can be iterated over quickly with corresponding small corrections to the carrier densities estimated to first order. While this is a faster calculation, the ionized donor density can be a discontinuous function. Invoking superposition, we evaluate only the *changes* in the potential energy due to changes in the charges, as estimates for both ρ_n and ρ_d are updated in the inner loop iterations. Once this calculation has stabilized, we return to the slower calculation for updating solutions to Schrödinger's equation.

We now provide the notational details. Start with an initial guess for the potential energy: $V^{ch}_{j=0,\,k=0}(z)$. Here the indices j and k refer to the outer and inner loops, respectively. The function $V^{ch}_{j=0,\,k=0}(z)$ can be selected as follows. We require this potential to be zero at the edges of the layered structure under consideration where the asymptotic values of $V_0(z)$ prevail. In the absence of an external field, we also require that $V^{ch}_{0,0}(z)$ have zero derivatives at the edges. [6, 20] An initial guess of a cubic function is adequate for our purposes to account for any mismatch in the value of V_0 at the two ends. In practice, these boundary conditions will have to be modified if the modeled system is truncated and does not extend far enough into bulk regions.

For the jth iteration of the Schrödinger solver, we use the updated potential energy $V_{j,0}(z)$, and determine the charge density $\rho_n(V^{ch}_{j,0})$. The inner loop is then initiated

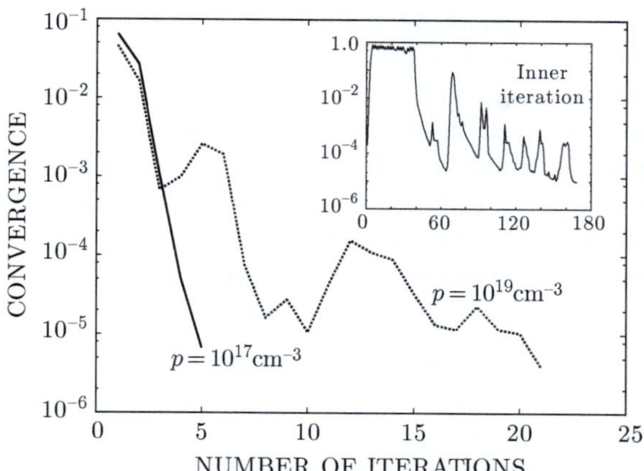

Figure 11.3 The rate of convergence for the potential function of the p-doped quantum well of Fig. 11.1 versus the number of outer (Schrödinger) iterations is shown. For a p-doping level of 1×10^{17} cm^{-3} convergence to 1×10^{-5} eV is achieved in 5 iterations, and for p-doping of 1×10^{19} cm^{-3} convergence is achieved in 21 iterations. A typical rate of convergence in the inner (Poisson) iteration is shown in the inset.

starting with the index $k = 0$. In each inner loop iteration, we evaluate the Poisson functional integral $\mathcal{P}$ minimizing it with respect to the nodal values of the potential. The resulting Poisson's equation is solved by successively working with the discretized form. Here, we solve for just the change $\delta V^{ch}_{j,k}$ in the potential energy at the $(j,k)^{th}$ iteration defined by

$$V^{ch}_{j,k+1} = V^{ch}_{j,k} + \delta V^{ch}_{j,k}. \tag{11.25}$$

We use a scaling factor λ in order to dampen oscillations in the iterative process. Here λ is a control parameter, $0 < \lambda \leq 1$, that is adjusted to accelerate convergence in the inner loop. If the changes in the charges oscillate over two consecutive iterations, λ is lowered so that a more conservative change in the potential energy is accepted. The convergence criterion for the inner loop can be numerically quantified by evaluating the vector 1-norm of the difference between the nodal values at the nodes β only, rather than integrating the modulus of the difference between these functions.

$$\frac{1}{N_g}\sum_{\beta}^{N_g} |V^{ch,\beta}_{j,k_{max}} - V^{ch,\beta}_{j,k_{max}-1}| < \eta_1, \tag{11.26}$$

where N_g is the array size of the vectors and $\eta_1 \sim 10^{-5}$, say. The parameter λ is reset to unity for each k-loop. In the initial iterations of the inner loop, the change in potential δV^{ch} can be significant enough to demand the use of λ; however, as we proceed through these iterations, we are solving for smaller and smaller quantities δV^{ch} and hence the role of λ becomes less important. The need for such damping factors is well-appreciated in the literature. [6, 21, 22] An example of the rate of convergence with the number of iterations is displayed in Fig. 11.3. The number of iterations to convergence to 1.0×10^{-5} eV in the change in the potential function is shown for a p-doped quantum well of Fig. 11.1.

Assume that we are given the final output of the inner loop as $V^{ch}_{j,k_{max}}(z)$. We re-initiate the outer loop calculation with $j \rightarrow j+1$ and $k_{max} \rightarrow 0$ and iterate until charge neutrality is achieved. Equivalently, we require that the potential energy $V^{ch}_{j+1,0}(z)$ has not changed substantially from the $V^{ch}_{j,0}(z)$ within a tolerance $\eta_2 \sim 10^{-5}$ over the entire structure. For convergence, we require

$$\frac{1}{N_g}\sum_{\beta}^{N_g} |V^{ch,\beta}_{j+1,0} - V^{ch,\beta}_{j,0}| < \eta_2. \tag{11.27}$$

On convergence, charge neutrality occurs naturally in the modeled structure extended far enough to be represented by bulk regions at the ends. Charge neutrality may not be a valid criterion if the structure is truncated before the depletion region is included in the modeled structure. However, we can still reach convergence in V^{ch} with a truncated physical region if the potential energies at the two ends are specified. Here again we may include a damping factor μ to control the fraction of δV that is included in $V^{ch}_{j+1,0}$. The outer iteration is stopped when the tolerance criterion, Eq. (11.27), is satisfied.

11.3.3 Finite element implementation

At the $(j,k)^{th}$ iteration, the Poisson functional $\mathcal{P}$ can be written in terms of

$$V^{ch}(z) = V^{ch}_{j,k}(z) + \delta V^{ch}_{j,k}(z). \tag{11.28}$$

This scalar potential is expressed in each finite element in terms of interpolation shape functions in the form

$$V^{ch}(z) = \sum_{\alpha} V^{ch}_{\alpha} N_\alpha(z), \tag{11.29}$$

and is substituted into Eq. (11.18). In the following, we suppress the dependence of ρ_n and ρ_d on the band offset potential V_0. Let us represent the change in the carrier charge density as the potential energy changes from $V^{ch}_{j,0}$ to $V^{ch}_{j,k}$ by

$$\Delta\rho_n(V^{ch}_{j,k} + \delta V^{ch}_{j,k}) = \rho_n(V^{ch}_{j,k} + \delta V^{ch}_{j,k}) - \rho_n(V^{ch}_{j,0}). \tag{11.30}$$

This can be approximated as follows. As in Eq. (11.21), we discretize the integrations in Eq. (11.20) over $\{\nu, k_{\|}\}$ into bins in the electronic total energy. Using the same DOS factors but altering the Fermi distribution function alone, we write

$$\begin{aligned} &\Delta\rho_n(V^{ch}_{j,k} + \delta V^{ch}_{j,k}) \\ &= \sum_i \Lambda_i(z)\left[\mathcal{F}(E_i + V^{ch}_{j,k} + \delta V^{ch}_{j,k} - V^{ch}_{j,0}, E_F, T) - \mathcal{F}(E_i, E_F, T)\right]. \end{aligned} \tag{11.31}$$

In other words, instead of updating E_i all over again by employing the Schrödinger solver, we account for the change in energy appearing in the Fermi function by adding the change in the potential energy to E_i. [10] The above expression is consistent with the condition $\Delta\rho_n(V^{ch}_{j,0}) = 0$, since for $k = 0$, we have already determined $\rho_n(V^{ch}_{j,0})$.

Substituting Eqs. (11.28–11.31) in Eq. (11.18), and using the expression $V_{j,k+1,\alpha} = V_{j,k,\alpha} + \delta V_{j,k,\alpha}$, we obtain

$$\begin{aligned}\mathcal{P} &= V_{j,k+1,\alpha}\frac{1}{2}\int dz\left(N'_\alpha(z)\epsilon(z)N'_\beta(z)\right)V_{j,k+1,\beta}\\ &+ V_{j,k+1,\alpha}\,4\pi|e|\int dz\,N_\alpha(z)\,\rho_d(V^{ch}_{j,k}(z) + \delta V^{\,ch}_{j,k}(z))\\ &+ V_{j,k+1,\alpha}\,4\pi|e|\int dz\,N_\alpha(z)\left(\rho_n(V^{ch}_{j,0}) + \Delta\rho_n(V^{ch}_{j,k} + \delta V^{\,ch}_{j,k})\right).\end{aligned} \tag{11.32}$$

We now write the integrals in Eq. (11.32) using a matrix notation. Define

$$M_{\alpha\beta} = \int dz\,N'_\alpha(z)\,\epsilon(z)\,N'_\beta(z). \tag{11.33}$$

The matrix $M_{\alpha\beta}$ is evaluated once and systematically reused in the iterations. While all the charge densities appearing in the integrals on the right side of Eq. (11.32) are not known, we can formally write the integrals as vector arrays with the spatial dependence integrated out.

Let us consider these integrals one by one.

(i) With $V_{j,k}(z)$ expressed in terms of the interpolation polynomials in each finite element, the integral over the donor charge density can be written as

$$4\pi|e|\int dz N_\alpha(z)\rho_d(V_{j,k}(z) + \delta V_{j,k}(z)) \equiv G_\alpha(\{V^{ch}_{j,k,\beta} + \delta V^{\,ch}_{j,k,\beta}\}), \tag{11.34}$$

where the *set* of variables $\{V^{ch}_{j,k,\beta} + \delta V^{\,ch}_{j,k,\beta}\}$ refer to the nodal values of the potential energy functions at the nodes β. A Taylor expansion of the charge density $\rho_d(V(z), z)$ with respect to $V(z) = (V_0(z) + V^{ch}(z) + \delta V^{\,ch}(z))$ entails a functional differentiation that is obviously nontrivial to evaluate. However, we can approximate $G_\alpha(\{V^{ch}_{j,k,\beta} + \delta V^{ch}_{j,k,\beta}\})$ by

$$\begin{aligned}G_\alpha(\{V^{ch}_{j,k,\beta} + \delta V^{\,ch}_{j,k,\beta}\}) &= G^0_\alpha(\{V^{ch}_{j,k,\beta}\}) + \sum_\beta \frac{\partial G_\alpha}{\partial V^{\,ch}_{j,k,\beta}}\cdot\delta V^{\,ch}_{j,k,\beta},\\ &= G^0_\alpha + \sum_\beta G^1_{\alpha\beta}\cdot\delta V^{\,ch}_{j,k,\beta}\end{aligned} \tag{11.35}$$

where $G^1_{\alpha\beta}$ in the second term is represented by

$$G^1_{\alpha\beta} = 4\pi|e| \int dz N_\alpha(z) \Big[\rho_d(V^{ch}_{j,k}(z) + [V^{ch}_{j,k,\beta} - V^{ch}_{j,k-1,\beta}]) - \rho_d(V^{ch}_{j,k}(z) - [V^{ch}_{j,k,\beta} - V^{ch}_{j,k-1,\beta}])\Big] \times \left(2(V^{ch}_{j,k,\beta} - V^{ch}_{j,k-1,\beta})\right)^{-1}.$$

Here $V^{ch}_{j,k}(z)$ and $V^{ch}_{j,k-1}(z)$ are known at every z, except in the $k = 0$ case, and ρ_d in the above expression is a known function. (For $k = 0$, this derivative is not evaluated.) The incremental difference $V^{ch}_{j,k,\beta} - V^{ch}_{j,k-1,\beta}$ in the nodal values corresponds to the change in one iteration of the inner loop and is a known, small quantity. In effect, we have approximated the functional derivative of $\rho_d(z)$ with respect to the potential energy function $V^{ch}(z)$ by the derivative with respect to the variations in the nodal variables representing V. This discretized version of the derivative may be justified by noting that the only variations that are allowed in the interpolated form of the potential energy function are the nodal values of the potential energy. Each nodal value is independently varied in order to approximate the functional variation of V^{ch} at every coordinate. Here $G^1_{\alpha\beta}$ is block-diagonal and again expressible in a sparse matrix format. (While ρ_d can be evaluated analytically, it varies abruptly with z, and its derivative becomes a highly peaked function. The numerical differentiation effectively smoothes out the result and is easier to represent in computation.)

(ii) The third term on the right side of Eq. (11.32) is

$$J_\alpha = 4\pi|e| \int dz\, N_\alpha(z)\, \rho_n(V^{ch}_{j,0}(z)). \tag{11.36}$$

Here $\rho_n(V^{ch}_{j,0}(z))$ is explicitly determined during the outer loop as soon as one steps out of the inner loop. The carrier charge density is given by Eq. (11.20), in which all the factors are known.

(iii) The last term in Eq. (11.32) is

$$\begin{aligned} 4\pi|e| \int dz\, N_\alpha(z)\, \left[\Delta\rho_n(V^{ch}_{j,k} + \delta V^{ch}_{j,k})\right] &= K_\alpha(\{V^{ch}_{j,k,\beta} + \delta V^{ch}_{j,k,\beta}\}) \\ &= K^0_\alpha + \sum_\beta \frac{\partial K_\alpha}{\partial V^{ch}_{j,k,\beta}} \cdot \delta V^{ch}_{j,k,\beta} \\ &= K^0_\alpha + \sum_\beta K^1_{\alpha\beta} \cdot \delta V^{ch}_{j,k,\beta}. \end{aligned} \tag{11.37}$$

The first term, K_α^0 is readily evaluated numerically in the $(j,\,k)^{th}$ iteration using Eq. (11.31) to obtain

$$K_\alpha^0 = 4\pi|e| \int dz\, N_\alpha(z) \sum_i \Lambda_i(z) \times \Big(\mathcal{F}(E_i + V^{ch}_{j,k}(z) - V^{ch}_{j,0}(z), E_F, T) - \mathcal{F}(E_i, E_F, T) \Big). \qquad (11.38)$$

We again approximate the functional derivative required in a Taylor expansion of $K_\alpha(V^{ch}(z) + \delta V^{ch}(z), z)$ by its discretized version, $\partial K_\alpha / \partial V^{ch}_{j,k\,\beta}$. Defining

$$\xi_i = E_i + V^{ch}_{j,k}(z) - V^{ch}_{j,0}(z), \qquad (11.39)$$

we write

$$K^1_{\alpha\beta} = 4\pi|e| \int dz\, N_\alpha(z) \sum_i \Lambda_i(z) \times \left(\frac{\partial \mathcal{F}(\xi_i)}{\partial \xi_i} \right) \left(\frac{\partial \xi_i}{\partial V^{ch}_{j,k,\beta}} \right). \qquad (11.40)$$

The derivatives are discretized and evaluated using

$$\frac{\partial \mathcal{F}(\xi_i)}{\partial \xi_i} = \frac{\mathcal{F}(E_{i+1} + V^{ch}_{j,k}(z) - V^{ch}_{j,0}(z)) - \mathcal{F}(E_i + V^{ch}_{j,k}(z) - V^{ch}_{j,0}(z))}{E_{i+1} - E_i}, \qquad (11.41)$$

and

$$\frac{\partial \xi_i}{\partial V^{ch}_{j,k,\beta}} = \frac{\partial}{\partial V^{ch}_{j,k,\beta}} \Big(E_i + V^{ch}_{j,k}(z) - V^{ch}_{j,0}(z) \Big) = \frac{\partial V^{ch}_{j,k}}{\partial V^{ch}_{j,k,\beta}} = N_\beta(z), \qquad (11.42)$$

so that we can evaluate the integral $K^1_{\alpha\beta}$ numerically.

Combining and rearranging all the terms in Eq. (11.32), the Poisson functional is given by

$$\begin{aligned} \mathcal{P}_{j,k} = {} & \frac{1}{2} V_{j,k+1,\alpha} \cdot M_{\alpha\beta} \cdot V_{j,k+1,\beta} \\ & + V_{j,k+1,\alpha} \cdot (G^0_\alpha + J_\alpha + K^0_\alpha) \\ & + V_{j,k+1,\alpha} \cdot (G^1_{\alpha\beta} + K^1_{\alpha\beta}) \cdot \delta V^{ch}_{j,k,\beta}. \end{aligned} \qquad (11.43)$$

Invoking the variational principle for the electrostatics problem, we require $\delta \mathcal{P} / \delta V_{j,k,\alpha} = 0$, which leads to

$$\left[M_{\alpha\beta} + G^1_{\alpha\beta} + K^1_{\alpha\beta} \right] \cdot \delta V^{ch}_{j,k,\beta} = - \Big(M_{\alpha\beta} \cdot V^{ch}_{j,k,\beta} + G^0_\alpha + J_\alpha + K^0_\alpha \Big). \qquad (11.44)$$

Thus at each inner loop iteration, we determine the "small" change in the potential energy δV^{ch} by solving the above set of simultaneous equations. [23] The solution at nodal locations are then used to reconstruct δV everywhere.

In the above discussion, we have purposely left the details of the heterostructure unspecified. The only features it needs to have are that it is modulation-doped, the doping profile being specified by $n_d(z)$, and the structure can have bound states, quasi-bound states, and free or traveling states. We have exploited the method of finite elements and discretized the physical region into elements in which the carrier wavefunctions and also the electrostatic potential energies are determined at nodal points. These are solved for and the complete functions reconstructed using interpolation polynomials. The method being a variational procedure, we are assured of a quadratic convergence to the minimum. By integrating out the spatial dependence, we have reduced the Schrödinger-Poisson problem into one in which we need to determine only values at nodes in the elements. Functional differentiations are reduced to nodal variational derivatives. These are concrete advantages inherent in the finite element approach to this fairly complex nonlinear problem. We now present applications of this method.

11.4 Applications

11.4.1 Multi-band $\mathbf{k} \cdot \mathbf{P}$ model and self-consistency

The approach presented in Sec. 11.3 for a single energy band is readily extended to a 6-band description of the valence heavy hole, light hole, and spin-orbit split-off bands, and also to a full 8-band representation of the conduction and valence bands used to describe the energy band structure in typical III-V or II-VI compound semiconductors. [24] The $\mathbf{k} \cdot \mathbf{P}$ band parameters for the materials in the layers are known. [25]

The required Lagrangian has been formulated elsewhere. [26–28] While the Poisson part of the problem remains the same as in the 1-band model, the calculation of the bound, quasi-bound, and traveling states requires a careful application of boundary conditions at the two ends of a heterostructure. The boundary conditions at interfaces between the layers are also more complex.

At the interfaces, we apply the current continuity conditions for the multi-band $\mathbf{k} \cdot \mathbf{P}$ model. The usual Löwdin perturbation theory [29] of degenerate bands employs the elimination of remote high-energy bands in favor of the valence and the lowest conduction bands. This "folding in" of higher energy bands [30] requires additional care in the presence of layer interfaces in heterostructures. [31, 32] With the z-axis perpendicular to the layer interfaces, the functional integral takes the general form [32–34]

$$\mathcal{A} = \int dz f_i^*(z) \Big[\overleftarrow{\partial}_z (\boldsymbol{A})_{ij} \overrightarrow{\partial_z} + \overleftarrow{\partial}_z (\boldsymbol{B}_L)_{ij} + (\boldsymbol{B}_R)_{ij} \overrightarrow{\partial}_z + (\boldsymbol{C})_{ij} - E\delta_{ij} \Big] f_j(z), \tag{11.45}$$

where the order of the derivative terms becomes important. Here the envelope functions $f_i(z)$, $i = 1, \ldots, n_b$ have components corresponding to the number of bands n_b appearing in the $\mathbf{k} \cdot \mathbf{P}$ model that is employed. The coefficients are $n_b \times n_b$ matrices. The matrices $\mathbf{B}_L$ and $\mathbf{B}_R$ are non-Hermitian, but transform into one another under

Hermitian conjugation so that the Lagrangian is Hermitian. The current continuity condition is derived from the Lagrangian through a gauge-variational approach described earlier. [28]

For bound states, again we set the wavefunctions at the boundaries to zero, while for traveling states, we specify the incoming plane waves. This requires a modal analysis to determine the wavevectors for given energy in the bulk-like regions at the two ends of the heterostructure. The $\mathbf{k} \cdot \mathbf{P}$ bulk Hamiltonian

$$H\psi(\mathbf{r}) = (Ak_z^2 + i(B_R - B_L)k_z + C)\psi(\mathbf{r}) = E\psi(\mathbf{r}) \tag{11.46}$$

is a quadratic eigenvalue equation for k_z, given E. The plane wave eigenfunctions and wavevector-eigenvalues are determined in a straightforward manner. [35, 36] This then allows us to select the traveling wave solutions for the above-barrier states and account for them in the calculation of the carrier density. With each energy band, we sum over the free carrier density by doing the finite element calculation over a discretized set of energy values. This computation is then inserted into the Poisson solver. The inclusion of strain in the layered structures generates strain-related energy-level splitting, which generates no additional significant issues.

The example structures investigated below illustrate the results from multi-band band structure computations also.

11.4.2 Further considerations

The effect of exchange and correlation [37] due to electron–electron interactions can be taken into account through numerical calculations. The full electrostatic potential energy

$$V_{ch} = -|e|\,\phi_{ch}(z) + V_{ex}(z)$$

has the electron–electron exchange and correlation effects in the local density approximation [38, 39] given by

$$V_{ex}(z) = -\left[1 + 0.0545\, r_s \ln\left(1 + \frac{11.4}{r_s}\right)\right] \times (18/\pi^2)^{1/3} R_y^*, \tag{11.47}$$

where

$$r_s = \left[\frac{4}{3}\,\pi a^{*3} n_n(z)\right]^{-1/3},$$

$$a^* = \varepsilon\hbar^2/m^* e^2,$$

$$R_y^* = m^* e^4/(2\varepsilon^2\hbar^2),$$

and $n_n(z)$ is the electron charge density. Numerical calculations which have included the effects of V_{ex} show [14, 22, 39, 40] that these effects can be neglected at least in GaAs heterostructures with doping levels of about 10^{18} cm^{-3}.

11.4.3 Results

The structures we consider include p-type modulation-doped single quantum wells, a p-type modulation-doped asymmetric double quantum well, p-i-n-i superlattices, and n-type modulation-doped trilayer superlattices. These calculations are done at zero temperature. In the final example, we studied temperature effects in an n-type modulation-doped single quantum well. In all examples, GaAs and $Al_xGa_{1-x}As$ layers are used, and input material parameters are taken from Ref. [25].

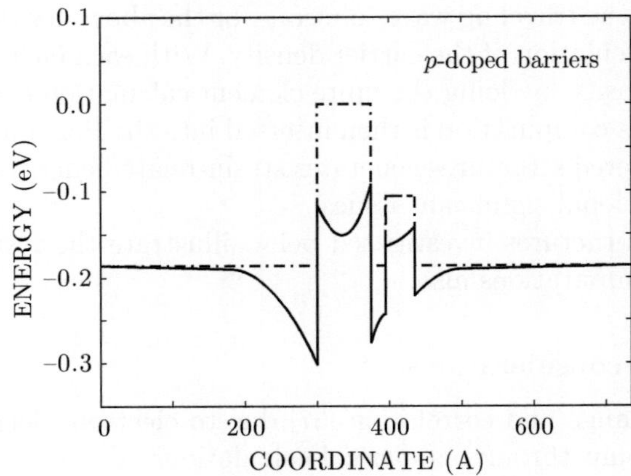

Figure 11.4 The valence band bending in a double quantum well that is asymmetric due to the well widths and due to the Al content in the well layers is shown. The wells are 75 Å and 40 Å wide separated by a 20 Å barrier, all sandwiched between wide barriers, with Al content of all barriers being $x = 0.35$. The narrower well is shallower, containing 20% Al. The end barriers are p-doped with 1×10^{18} cm^{-3} impurity concentration, while the middle barrier is p-doped with 1×10^{19} cm^{-3} impurities.

A single quantum well with *p*-type modulation doping

We calculated the self-consistent potential of single quantum wells of p-type modulation-doped $GaAs/Al_{0.35}Ga_{0.65}As$, at two p-doping concentrations, 10^{19} cm^{-3} and 10^{17} cm^{-3}. In Fig. 11.1, the degenerate heavy-hole and light-hole band edges and the spin-orbit split-off band edge are shown. V_0 is the initial potential due to band offset, and the solid and dotted lines represent the self-consistent potential for p-doping levels of 10^{19} cm^{-3} and 10^{17} cm^{-3}, respectively. As expected, the width of depletion region is shorter for higher doping; about 25 Å for $p = 10^{19}$ cm^{-3} case and over 200 Å for $p = 10^{17}$ cm^{-3} case. The band bending effectively gives rise to a symmetric double quantum well, and this is reflected in the behavior of the wavefunctions. In Fig. 11.2, we show the lowest two heavy hole states. The ground state happens to be higher in energy than the peak of the potential in the quantum well, and it clearly shows a double peak.

Figure 11.3 shows a typical convergence behavior in a semilog scale. The inner iterations converges to below 10^{-5} eV after a few hundred iterations, while the outer iterations converge after a few tens of iterations. In general, the convergence is faster if the free carrier concentration is smaller.

A *p*-type modulation-doped asymmetric double quantum well

In Fig. 11.4, we show the self-consistent potential for an asymmetric double quantum well of $Al_{0.35}Ga_{0.65}As$/GaAs(75 Å)/ $Al_{0.35}Ga_{0.65}As$(20 Å)/ $Al_{0.2}Ga_{0.8}As$(40 Å)/ $Al_{0.35}Ga_{0.65}As$. The end barriers are p-doped at 10^{18} cm^{-3}, while the center barrier is doped at $p = 10^{19}$ cm^{-3}. The left barrier, close to the wide and deep well, is depleted more than the right barrier, and the center barrier is fully depleted. Fig. 11.5 shows the first heavy-hole and the first light-hole wavefunctions. The deeply confined heavy hole is mostly confined in the wide well, while the light hole has a substantial probability in the narrow well.

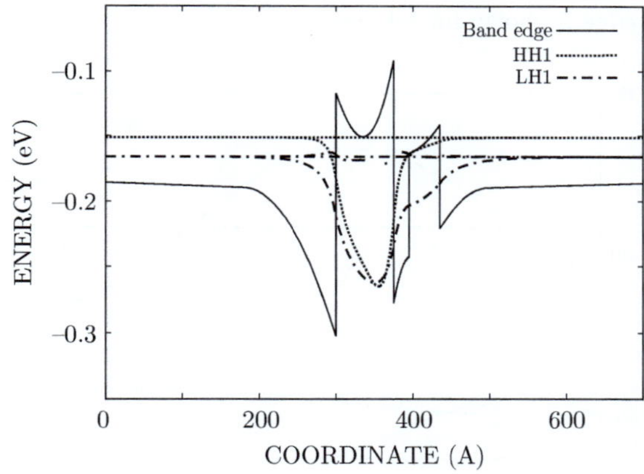

Figure 11.5 The first heavy hole and light hole wavefunctions in the asymmetric double quantum well of Fig. 11.4 are shown. Since this calculation used the 6-band $\mathbf{k} \cdot \mathbf{P}$ model including heavy hole, light hole, and spin-orbit split-off bands, the light hole wavefunction has a large light hole basis component and a small spin-orbit split-off basis component. At zero in-plane wavevector, the heavy hole components are completely decoupled from the others.

p-i-n-i superlattices

Next we consider the case of a *p-i-n-i* GaAs superlattice, each region being 50 Å. The band bending is displayed in Fig. 11.6. The dotted line is for $p = n = 10^{18}$ cm^{-3}. In this case, the Fermi level is around mid-gap and all the donors and acceptors are ionized, with the system attaining charge neutrality with no free carriers. At zero temperature, there is a wide range of the Fermi level with which all the impurities are ionized and no free carrier states are occupied, and the exact position of the Fermi

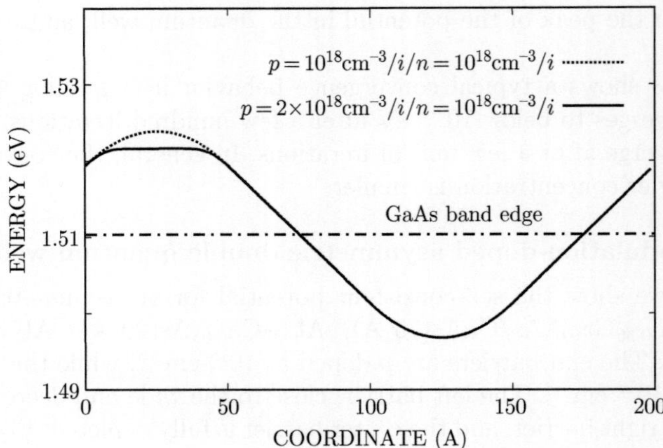

Figure 11.6 The conduction band bending in a *p-i-n-i* GaAs superlattice with each region being 50 Å is shown. (i) The dotted curve corresponds to equal n and p doping at 1.0×10^{18} cm^{-3}, while (ii) the solid curve has twice the number of acceptor doping in the *p*-layers. The original GaAs band edge is shown for reference.

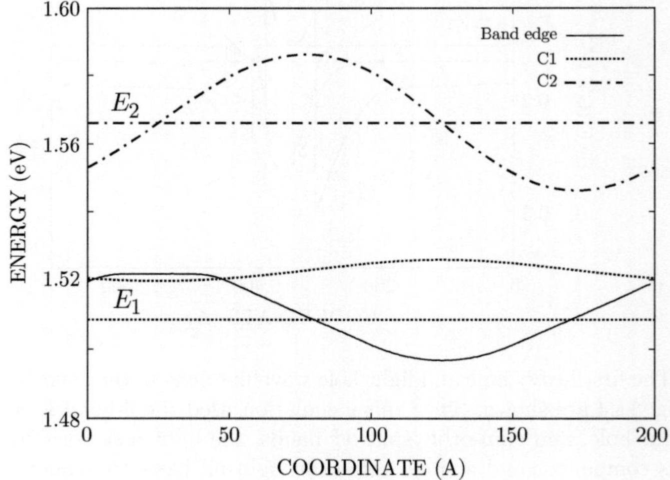

Figure 11.7 The first two zone center energy levels and wavefunctions in the *p-i-n-i* GaAs superlattice of Fig. 11.6 are shown, together with the band edge (solid curve).

level is irrelevant. The solid line in Fig. 11.6 corresponds to the case of $p = 2 \times 10^{18}$ cm^{-3} and $n = 10^{18}$ cm^{-3}. Again there are no free carriers. In order to maintain the charge neutrality, only half of the acceptors should be ionized. Only one Fermi level which is close to the valence band edge satisfies the charge neutrality.

In Fig. 11.7, we show the first two conduction eigenenergies and eigenfunctions at zero superlattice wavevector ($q = 0$) in the case of $p = 2 \times 10^{18}$ cm^{-3} and $n = 10^{18}$

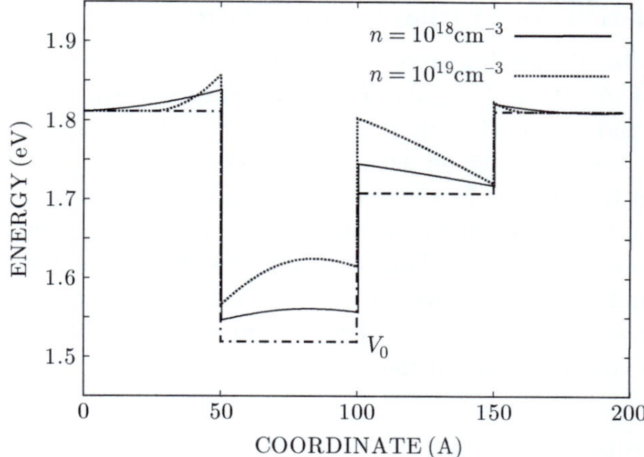

Figure 11.8 The conduction band bending in a trilayer superlattice is shown. The superlattice consists of a 50 Å GaAs quantum well followed by a 50 Å layer of $Al_{0.2}Ga_{0.80}As$ and a 100 Å barrier of $Al_{0.35}Ga_{0.65}As$. The band edge is shown for a barrier n-doped to (1) 1.0×10^{18} cm^{-3} (solid curve) and to (2) $1.0{\times}10^{18}$ cm^{-3} (dotted curve). The 100 Å barrier is shown distributed symmetrically on both sides of the stepped quantum well. For 10^{18} cm^{-3} doping, the barrier is completely depleted of unionized donors, and the Fermi level is in the quantum well.

cm^{-3}. The first conduction subband energy is below the maximum of the conduction band edge, but since the potential barrier in this p-i-n-i superlattice is not large, the wavefunction is relatively flat. The eigenenergy of the second subband energy is well above the maximum of the conduction band edge.

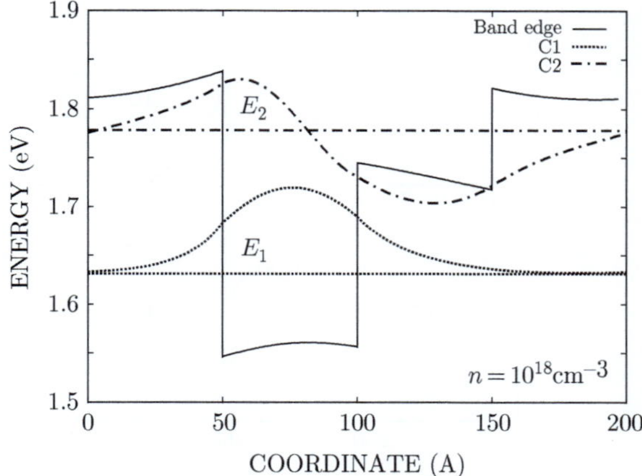

Figure 11.9 The zone center energy levels and wavefunctions for the trilayer superlattice of Fig. 11.8 are shown with the band bending for $n = 1.0 \times 10^{18}$ cm^{-3}.

n-type modulation-doped trilayer superlattices

We considered band bending under self-consistency in two trilayer GaAs(100 Å)/ $Al_{0.2}Ga_{0.8}As$(50 Å)/$Al_{0.35}Ga_{0.65}As$(50 Å) superlattices where the doped layer is $Al_{0.35}Ga_{0.65}As$ and is doped with $n = 10^{18}$ and $n = 10^{19}$ cm^{-3}, respectively. In the $n = 10^{19}$ cm^{-3} case, the barrier is partially depleted, and the Fermi level for the whole structure is the bulk Fermi level of the barrier region. On the other hand, in the $n = 10^{18}$ cm^{-3} case, the barrier is fully depleted, and the Fermi level is determined to be at 1.666 eV. Figure 11.8 shows the self-consistent potential of the two trilayer superlattices, and Fig. 11.9 shows the first two conduction subband wavefunctions at $q = 0$ for the $n = 10^{18}$ cm^{-3} case. Unlike the quantum well, the wavefunctions of the superlattice do not go to zero in the barrier region.

Effect of finite temperature

For a modulation-doped ($n = 10^{18}$ cm^{-3}), the variation of the self-consistent potential with temperature in a GaAs/$Al_{0.35}Ga_{0.65}As$ single quantum well structure is shown in Fig. 11.10. As the temperature increases from 0 K to 100 K and to 300 K, the bandgap of both the well and barrier regions decreases, which accounts for most of the displacement along the energy axis. Beside this, the most significant effect is that the transition between the depleted and neutral barrier regions becomes smoother at higher temperature.

11.5 Concluding remarks

We have shown that the finite element approach to the Schrödinger-Poisson self-consistency calculations provides a controlled means of obtaining convergence in a

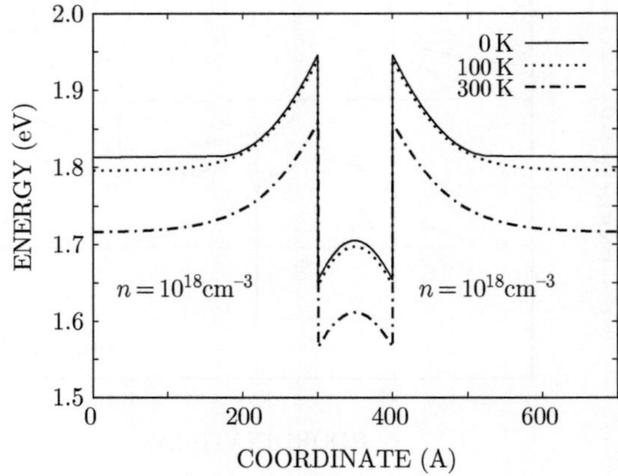

Figure 11.10 The variation of the conduction band edge with temperature in a 100 Å GaAs quantum well structure. The reduction in the band gap with temperature has been included, and the presence of free carriers above the band edge has been accounted for.

fairly general manner. This has been illustrated in this chapter by several examples which bring out the nuances in the procedure.

The formulation presented here allows the generalization to numerically evaluate functional derivatives in the "weak" sense as a variational approach in which only the nodal values of the functions are varied in evaluating the derivative. This clearly has further applications in the context of field theoretic calculations of physical quantities.

The calculation of the self-consistent potential is a ubiquitous problem that occurs in every active quantum device. For example, the first step in multi-band modeling of tunneling effects in layered structures is the determination of the band bending of the conduction and valence band edges under modulation doping. This includes the effect of strain and external electrically applied bias, and the effect of nonparabolicity in the in-plane energy band dispersion. The next stage is the computation of the multi-band tunneling coefficient, which is then included in a tunneling current calculation. Again, in quantum well lasers, the energies of the actual energy levels employed in the lasing depend on the self-consistent quantum well band profile. The problem is of central interest in layered spintronic semiconductor materials, where the ferromagnetic behavior of Mn-doped layers is mediated by the free carriers interacting with the Mn spins through the exchange interaction.

References

[1] L. R. Ram-Mohan, *Finite Element and Boundary Element Applications to Quantum Mechanics* (Oxford University Press, Oxford, 2002).

[2] O. C. Zienkiewicz, *The Finite Element Method* (McGraw-Hill, New York, 1989); O. C. Zienkiewicz and R. L. Taylor, *The Finite Element Method* (McGraw-Hill, New York, 1994).

[3] A Galerkin, and B. G. Galerkin, Vestn. Inz. Tech. Vol. 19, 897–908 (1915); "Ryady v nekotorykh voprosakh uprugogo ravnovesiya sterzhnei i plastin;" [English translation from Russian: *Series solution of some problems of elastic equilibrium of rods and plates.*] The Galerkin formulation employing the method of residuals within an FEM framework is an alternate approach to the solution of these differential equations; however, since both the Schrödinger and the Poisson differential equations can be derived by applying variational principles to appropriate functionals the Galerkin and the variational approaches are equivalent.

[4] B. C. Duncan, "Selfconsistent analysis of energy levels in modulation doped heterostructures" (Senior Thesis, Worcester Polytechnic Institute, 1990). A transfer matrix method was used for Schrödinger's equation, and discretization with the "shooting method" for the Poisson problem – this procedure was observed to be unstable for high impurity concentrations $\geq 10^{18}\,\mathrm{cm}^{-3}$.

[5] T. Ando, A. B. Fowler, and F. Stern, Rev. Mod. Phys. **54**, 437–672 (1982); "Electronic properties of two-dimensional systems."

[6] F. Stern, J. Comput. Phys. **6**, 56–67 (1970); "Iteration methods for calculating self-consistent fields in semiconductor inversion layers."

[7] I. Suemune, J. Appl. Phys. **67**, 2364–2369 (1990); "Doping in a superlattice structure: Improved hole activation in wide-gap II-VI materials."

[8] H. Mizuta and T. Tanoue, *The Physics and Applications of Resonant Tunneling Diodes* (Cambridge University Press, Cambridge, 1995).

[9] D. J. Rostcheck, "Selfconsistent calculations of electronic energy levels in quantum heterostructures" (Senior Thesis, Worcester Polytechnic Institute, 1992). The FEM was employed for both the Schrödinger and the Poisson equations, with quintic Hermite interpolation polynomials. The calculations were limited to a single modulation doped quantum well.

[10] A. Trellakis, A. T. Galick, A. Pacelli, and U. Ravaioli, J. Appl. Phys. **81**, 7880–7884 (1997); "Iteration scheme for the solution of the two-dimensional Schrödinger-Poisson equations in quantum structures."

[11] G. Bastard, *Wave Mechanics Applied to Semiconductor Heterostructures* (Les Editions de Physique, Les Ulis, France, 1988).

[12] Ref. 1, *op. cit.*, Chapter III.

[13] R. P. Feynman, Phys. Rev. **56**, 340–343 (1939); "Forces in molecules." H. Hellmann, in *Einführung in die Quantumchemie* (Deuticke, Leipzig, 1937), p. 285. For an application in the context of heterostructure bandstructure, see: L. C. Lew Yan Voon and L. R. Ram-Mohan, Phys. Rev. B **47**, 15500–15508 (1993); "Tight-binding representation of the optical matrix elements: Theory and applications."

[14] T. Ando, J. Phys. Soc. Jpn. **51**, 3893–3899 (1982); "Self-consistent results for a GaAs/$Al_xGa_{1-x}As$ heterojunciton. I. Subband structure and light-scattering spectra" *ibid.*, **51**, 3900–3907 (1982); "II. Low temperature mobility."

[15] G. Gilat and L. J. Raubenheimer, Phys. Rev. **144**, 390–395 (1966); "Accurate numerical method for calculating frequency-distribution functions in solids." Erratum Phys. Rev. **147**, 670 (1966).

[16] G. Lehmann and M. Taut, Phys. Stat. Sol. (b) **54**, 469–477 (1972); "On the numerical calculation of the density of states and related properties."

[17] M. S. Methfessel, M. H. Boon, and F. M. Mueller, J. Phys. C **16**, L949–L964 (1983); "Analytic-quadratic method of calculating the density of states."

[18] G. Wiesenekker, G. te Velde, and E. J. Baerends, J. Phys. C **21**, 4263–4283 (1988); "Analytic quadratic integration over the two-dimensional Brillouin zone."

[19] R. P. Brent, *Algorithms for Minimization Without Derivatives*, Chapters 3, 4 (Prentice Hall, Englewood Cliffs, NJ, 1973).

[20] F. Stern, and W. E. Howard, Phys. Rev. **163**, 816–835 (1967); "Properties of semiconductor surface inversion layers in the electric quantum limit."

[21] S. E. Laux, D. J. Frank, and F. Stern, Surf. Sci. **196**, 101–106 (1988); "Quasi-one-dimensional electron states in a split-gate GaAs/AlGaAs heterostructure."

[22] K. Inoue, H. Sakaki, J. Yoshino, and T. Hotta, J. Appl. Phys. **58**, 4277–4281 (1985); "Self-consistent calculation of electronic states in AlGaAs/GaAs/AlGaAs selectively doped double-heterojunction systems under electric fields."

[23] The simultaneous equations are solved numerically using the upper triangular-lower triangular (LU)-matrix decomposition method of Gauss elimination for a sparse matrix representation of the matrices.

[24] J. M. Luttinger and W. Kohn, Phys. Rev. 97, 869–883 (1955); "Motion of Electrons and Holes in Perturbed Periodic Fields."

[25] For a recent compilation of band parameters, see I. Vurgaftman, J. R. Meyer and L. R. Ram-Mohan, J. Appl. Phys. **89**, 5815–5875 (2001); "Band parameters for III–V compound semiconductors and their alloys."

[26] L. R. Ram-Mohan and J. R. Meyer, J. Nonlin. Opt. Phys. Mater. **4**, 191–243 (1995); "Multiband finite element modeling of wavefunction-engineered electro-optical devices."

[27] L. R. Ram-Mohan, D. Dossa, I. Vurgaftman, and J. R. Meyer, in *Handbook of Nanostructured Materials and Nanotechnology*, Vol. 2, Ed. H. S. Nalwa, (Academic Press, New York, 1999), Chap. 15.

[28] L. R. Ram-Mohan, I. Vurgaftman, and J. R. Meyer, Microelectron. J. **30**, 1031–1042 (1999); "Wave function engineering of antimonide quantum-well lasers."

[29] P. O. Löwdin, J. Chem. Phys. 19, 1396–1401 (1951); "A note on the quantum-mechanical perturbation theory."

[30] E. O. Kane, J. Phys. Chem. Solids **6**, 236–241 (1958); "The influence of exchange on the effective mass formalism;" E. O. Kane, in *Semiconductors and Semimetals*, Vol. 1, edited by R. K. Willardson and A. C. Beer (Academic, NY, 1966); E.O. Kane, in *Handbook on Semiconductors*, Vol. 1 p. 193, edited by W. Paul (North-Holland, Amsterdam, 1982).

[31] M. G. Burt, Semicond. Sci. Technol. **2**, 460–? ? (1987); "An exact formulation of the envelope function method for the determination of electronic states in semiconductor microstructures." Erratum **2**, 701 (1987); J. Phys. Condens. Matter 4, 6651 (1992).

[32] B. A. Foreman, Phys. Rev. B **48**, 4964–4967 (1993); "Effective-mass Hamiltonian and boundary conditions for the valence bands of semiconductor microstructures."

[33] R van Dalen and P. N. Stavrinou, Semicond. Sci. Technol. **13**, 11–17 (1998); "General rules for constructing valence band effective mass Hamiltonians with correct operator order for heterostructures with arbitrary orientations."

[34] L. R. Ram-Mohan, unpublished lecture notes (1999).

[35] D. L. Smith and C. Mailhiot, Rev. Mod. Phys. **62**, 173–234 (1990); "Theory of semiconductor superlattice electronic structure."

[36] B. Chen, M. Lazzouni, and L. R. Ram-Mohan, Phys. Rev. B **45**, 1204–1212 (1992); "Diagonal representation for the transfer-matrix method for obtaining electronic energy levels in layered semiconductor heterostructures."

[37] For the many-body theory of exchange and correlation effects see, for example, S. Raimes, *The Wave Mechanics of Electrons in Metals* (North-Holland, Amsterdam, 1961); A. L. Fetter and J. D. Walecka, *Quantum Statistical Mechanics of Many Particle Systems* (McGraw-Hill, New York, 1971); G. D. Mahan, *Many*

Particle Physics, 2nd edition (Plenum Press, New York, 1990); G. D. Mahan and K. R. Subbaswamy, *Local Density Theory of Polarizability* (Plenum Press, New York, 1990).

[38] O. Gunnarson and B. I. Lundquist, Phys. Rev. B **13**, 4274–4298 (1976); "Exchange and correlation in atoms, molecules, and solids by the spin-density-functional formalism."

[39] T. Ando, and S. Mori, J. Phys. Soc. Jpn. **47**, 1518–1527 (1979); "Electronic properties of a semiconductor superlattice. I. Self-consistent calculation of subband structure and optical spectra."

[40] F. Stern and S. Das Sarma, Phys. Rev. B **30**, 840–848 (1984); "Electron energy levels in GaAs-$Ga_{1-x}Al_xAs$ heterojunctions."

Part IV

Steady-state Current Distributions

12

The extraordinary magneto-resistance effect in metal–semiconductor structures

In this chapter:

- We study the extraordinary magnetoresistance (EMR) effect which was discovered in 2000 in semiconductor/metal hybrid structures. The remarkable feature was that in a modified van der Pauw disk of InSb with a concentric embedded Au disk, the measured magnetoresistance (MR) was $10^7\%$. (Hence, the use of the appellation, "extraordinary.")
- Using the finite element method (FEM), the room temperature EMR can be calculated with no adjustable parameters as a function of the applied magnetic field and the size/geometry of the inhomogeneity. The finite element results are nearly identical to exact analytic results and are in excellent agreement with the corresponding experimental measurements on circular geometry for the structure. With the FEM, we can now model an EMR device of any shape.
- With the FEM, several important properties of the composite InSb/Au system, such as the field dependence of the current flow and of the potential on the disk periphery can be mapped out in great detail.
- It is found that both the EMR and the output voltage depend sensitively on the placement and size of the current and voltage ports.
- The results for 2D EMR for other geometries besides the circular configuration and for 3D metal–semiconductor structures are presented.

12.1 Introduction

It has been shown experimentally [1–3] that semiconductor thin films with metallic inclusions display extraordinary magnetoresistance (EMR) at room temperature, with enhancements as high as $100\%-750,000\%$ at magnetic fields ranging from 0.05 to 4 T. The magnetoresistance (MR) is defined as $\mathrm{MR} = [R(H) - R(0)]/R(0)$, where

Finite Elements in Action. L. Ramdas Ram-Mohan, Oxford University Press. © L. Ramdas Ram-Mohan (2026).
DOI: 10.1093/oso/9780199563487.003.0012

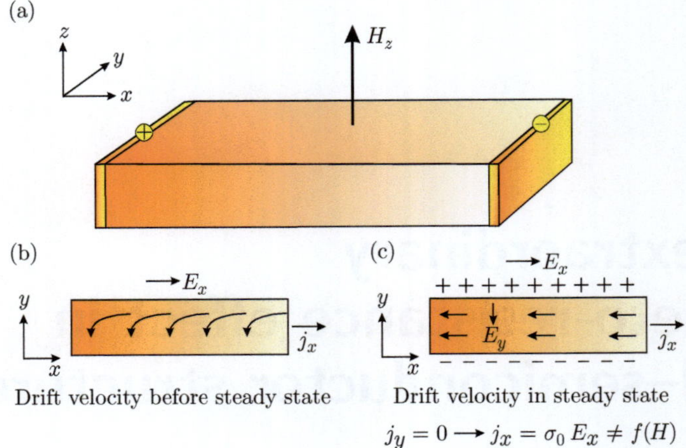

Figure 12.1 The Hall configuration: the electron trajectories in a rectangular sample, shown in (a), bend in a magnetic field so that negative charges accumulate at the edge of the xy plane, as shown in (b). In the steady state, shown in (c), this leads to an electric field E_y, the Hall field, perpendicular to the longitudinal direction in which the current j_x is flowing. This does not give a magnetic field-dependent conductivity.

$R(H)$ is the resistance at finite field H. Theoretical estimates suggest that EMR can reach $10^7\%$. The experiments were initially performed on a composite van der Pauw disk of a semiconductor matrix with an embedded metallic circular inhomogeneity that was concentric with the semiconductor disk. A similar enhancement has been reported [4] for a rectangular semiconductor wafer with a metallic shunt on one side. While the rectangular geometry with four contacts can be shown to be derivable from the circular geometry by a conformal mapping, [5] the use of a metallic shunt had not been considered earlier. Magnetic materials and artificially layered metals exhibit so-called giant magnetoresistance (GMR) and manganite perovskites show colossal magnetoresistance (CMR). However, patterned nonmagnetic InSb shows a much larger geometrically enhanced extraordinary MR even at room temperature.

In Fig. 12.1(a), the typical Hall bar configuration is shown. In an applied magnetic field, the electrons have a circular trajectory around the lines of the magnetic field, as displayed in Fig. 12.1(b). As soon as the current begins to flow, the space charge accumulation on one side gives rise to a (Hall) electric field E_y, which is measured through the voltage difference across the Hall bar, the Hall voltage. If we assume only one type of carrier with a δ-function velocity distribution (or the average drift velocity), the force on the carriers from the Hall field cancels the Lorentz force and the direct current j_x continues to remain the same, as indicated in Fig. 12.1(c). There is then no magnetic field dependence of the resistance in this case, i.e. the MR, $\Delta\rho/\rho_0 = 0$, where ρ is the resistivity and ρ_0 is the zero-field resistivity. To introduce the concepts of conductivity and magnetoresistance, consider a metal with electrons free to move about in the volume of the metal. Let there be n electrons per unit volume described by a

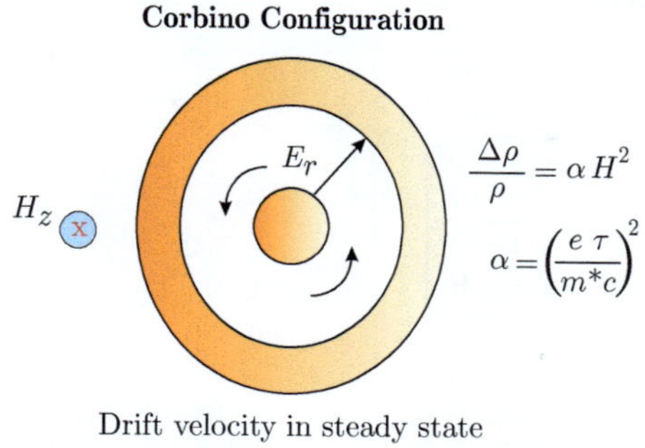

Corbino Configuration

H_z E_r

$$\frac{\Delta\rho}{\rho} = \alpha H^2$$

$$\alpha = \left(\frac{e\,\tau}{m^* c}\right)^2$$

Drift velocity in steady state

Figure 12.2 In the Corbino geometry, the electrodes are concentric, with a radial current flowing in the absence of a magnetic field. With a magnetic field present, the electron drift velocity is in the tangential (to the circular region) direction. This leads to a magnetic field-dependent conductivity that is geometry-dependent. The magnetoresistance $\Delta\rho/\rho_0$ is proportional to H^2 at low field.

Fermi-Dirac (or Maxwellian) velocity distribution function and let $\mathbf{v}_{avg}$ be the average velocity of the electrons. The equation of motion for the electrons in an external electric field $\mathbf{E}$ and magnetic field $\mathbf{H}$ is

$$m\dot{\mathbf{v}} + m^*\frac{\mathbf{v}}{\tau} = -|e|\Big[\mathbf{E} + \mathbf{v} \times \mathbf{H}\Big]. \tag{12.1}$$

Here, the effective mass of the carriers is denoted by m^* and the collisions that limit the current from increasing indefinitely in the presence of external fields are accounted by a collision time τ. In the steady state, the first term on the left side in Eq. (12.2) vanishes, and the velocity components can be isolated and related to the components of the electric field. A simple calculation leads to

$$\begin{pmatrix} v_x \\ v_y \end{pmatrix} = -\frac{\mu}{1+\mu^2 H^2}\begin{pmatrix} 1 & -\mu H \\ \mu H & 1 \end{pmatrix}\begin{pmatrix} E_x \\ E_y \end{pmatrix}, \tag{12.2}$$

where $\mu = -|e|\tau/m^*$ is the mobility of the electrons. The current density

$$\mathbf{j} = -n|e|\mathbf{v}, \tag{12.3}$$

where n is the carrier density. The current in the device is expressed in terms of a current density using $I = \mathbf{j}\cdot\mathbf{A}$, A being the cross-sectional area through which the current flows. Using Eqs. (12.3) and (12.2), we arrive at the relation

$$\begin{pmatrix} j_x \\ j_y \end{pmatrix} = -\frac{\sigma_0}{1+\mu^2 H^2}\begin{pmatrix} 1 & -\mu H \\ \mu H & 1 \end{pmatrix}\begin{pmatrix} E_x \\ E_y \end{pmatrix}. \tag{12.4}$$

Thus the field-current identity relation $j_i = \sigma_{ij}E_j$ has a modified conductivity tensor now containing a magnetic field dependence and also off-diagonal elements. These off-diagonal terms account for the circular motion of the electrons around the lines of the magnetic field.

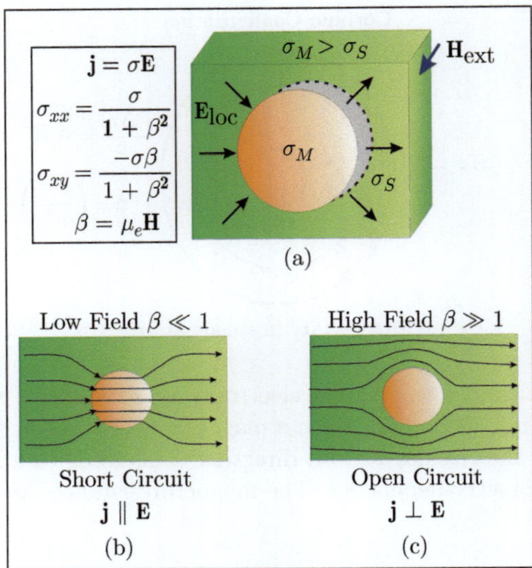

Figure 12.3 (a) The extraordinary magnetoresistance (EMR) is manifested by the effect on the boundary conditions at the metal–semiconductor interface as a function of the magnetic field H. (b) At low magnetic fields ($\beta \ll 1$), the current density j is parallel to the electric field E and the metal acts as a short circuit. (c) At high fields, ($\beta \gg 1$), the system acts as an open circuit.

In the Corbino disk, illustrated in Fig. 12.2, we have two concentric contacts with the current entering at $r = 0$ and exiting at the periphery. In the presence of a magnetic field perpendicular to the Corbino disk, the electron trajectories follow circular paths and the resistance is a function of the magnetic field. Moreover, because the conducting electrodes are in this case equipotential surfaces, no space charge accumulates on them and no Hall voltage is developed. Since there is no Hall field to produce a force that competes with the Lorentz force, there can be a large MR, which in this case is $\Delta\rho(H)/\rho_0 = (\mu H)^2$, where μ is the carrier mobility, and H is the applied magnetic field. The geometric differences in the standard Hall geometry and the Corbino geometry yield a significant field dependence of the resistance in the latter. We show here how such geometric effects can be generally exploited to enhance MR. An additional feature that we wish to employ for obtaining enhanced magnetoresistance is the following.

In the presence of metallic inhomogeneities, the narrow-gap semiconductors show marked enhancement of the MR. Because of their small carrier masses, the narrow gap high mobility semiconductors such as InSb and HgCdTe are the preferred materials to consider.

Let us suppose that we have a semiconducting slab with a cylindrical metal (Au) embedded in it, as shown in Fig. 12.3(a). Suppose that the conductivities of the semiconductor and the metal are σ_S and σ_M, respectively, in the absence of a magnetic field. In low magnetic fields, the current flowing through the material is focused into metallic regions with the metal acting as a short circuit; the current density $\mathbf{j}$ is parallel to the local electric field $\mathbf{E}_{\mathbf{loc}}$ as indicated in Fig. 12.3(b). Note that for $\sigma_M \gg \sigma_S$, the surface of the metal is essentially an equipotential. Thus $\mathbf{E}_{\mathbf{loc}}$ is normal to the interface between the metal and semiconductor. At finite magnetic field, the current deflection due to the Lorentz force results in a directional difference between $\mathbf{j}$ and $\mathbf{E}_{\mathbf{loc}}$, the angle between them being the Hall angle. For sufficiently high fields, the Hall angle approaches 90° in which case $\mathbf{j}$ is parallel to the semiconductor–metal interface and the current is deflected around the metal which acts like an open circuit as indicated in Fig. 12.3(c). The transition of the metal from a short circuit at low field to an open circuit at high field gives rise to the very large MR or EMR which has been reported by Solin et al. [1]

Under steady-state conditions, the problem of determining the current and the field in the inhomogeneous semiconductor reduces to the solution of Laplace's equation for the electrostatic potential. For some simple structures, this problem can be solved analytically. [6] In general, however, the location and material properties of the inhomogeneities can be altered and the semiconducting material can be shaped to enhance the MR. In order to have this freedom to explore the geometrical enhancement of the MR in the device and to be able to consider semiconductor films and metallic inclusions/shunts of arbitrary shape, we require a numerical approach to the simulation of the enhanced MR. In this chapter, we show that the FEM is ideally suited to this modeling and demonstrate that the modeling agrees very well with the experiments with no adjustable parameters.

12.2 Finite element analysis of the modified van der Pauw disk

For the sake of a simple presentation, and for comparison with experiments, we will first consider a circular disk of InSb of thickness t containing a concentric disk of Au, with four contacts, or ports as we refer to them, attached symmetrically to the periphery of the semiconductor (see Fig. 12.4). This configuration is called the modified van der Pauw geometry and was first studied analytically by Wolfe and Stillman. [7] The disk of InSb has a radius a while the Au inclusion has a radius b, with the ratio of the radii denoted by $\alpha = b/a$. A current enters port 1 and exits via port 2. We use "contact" and "port" interchangeably. The response of the semiconducting disk is measured by the voltage difference between ports 3 and 4. The widths of the ports have been labeled as Δ_i. The disk geometry reduces the problem to a two-dimensional (2D) one.

The conductivity at zero magnetic field is given by

$$\sigma(0) = \frac{ne^2\tau}{m^*} = ne\mu \tag{12.5}$$

as noted earlier.

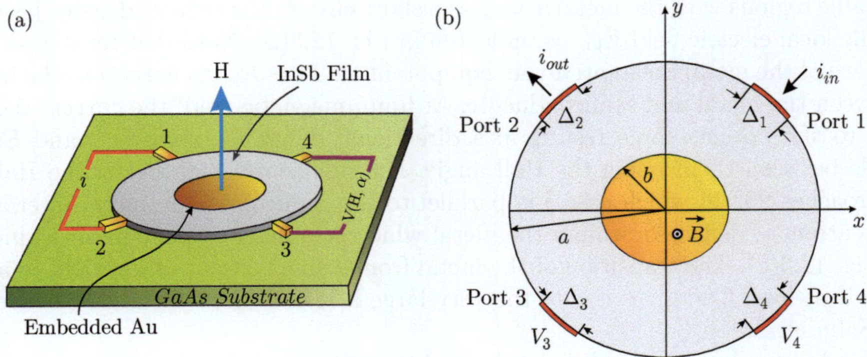

Figure 12.4 (a) Circular van der Pauw geometry showing the symmetrically placed contacts and a concentric Au inclusion in a circular InSb sample. (b) The geometry of the structure and the ports (contacts) are shown.

In the steady state, the charge conservation relation or the continuity equation reduces to

$$\frac{\partial}{\partial x_i} j_i = 0. \tag{12.6}$$

With $E_i = -\partial_i \phi$, the electrostatic potential $\phi(x, y)$ satisfies the Laplace-like differential equation

$$\partial_i[\sigma_{ij}\partial_j\phi(x, y)] = 0. \tag{12.7}$$

The boundary conditions at the outer edge at radius $r = a$ are as follows.

(a) At port 1:

$$(\mathbf{j} \cdot \hat{n}) = -J_{in} = -\frac{I_{in}}{\Delta_1 t}. \tag{12.8}$$

(b) At port 2:

$$(\mathbf{j} \cdot \hat{n}) = +J_{out} = \frac{I_{out}}{\Delta_2 t}. \tag{12.9}$$

Here, $I_{in} = I_{out}$, from current conservation. The quantities j_{in} and j_{out} are positive and their signs have been entered explicitly in the above.

(c) At ports 3, 4: we have $\phi = V_3$ and $\phi = V_4$, respectively, where these potentials have to be determined by the calculations. The entire width of each port is considered to be at the same voltage, as the typical contacts are metallic in nature.

(d) Along the rest of the semiconductor disk's edge, we set $\mathbf{j} = 0$ along the periphery.

(e) Finally, along the metal–semiconductor interface at $r = b$ the normal currents are equal. Hence, we have

$$[\sigma_{ij}^{(1)} \partial_j \phi(r = b)](\hat{n}_r)_i = [\sigma_{ij}^{(2)} \partial_j \phi(r = b)](\hat{n}_r)_i \tag{12.10}$$

or

$$(\mathbf{j} \cdot \hat{n}_r)|_{r=b^-} = (\mathbf{j} \cdot \hat{n}_r)|_{r=b^+}. \tag{12.11}$$

Here, the material index on the conductivity tensor refers to the semiconductor (index = 1) and to the metal (index = 2).

Rather than directly solving Eq. (12.7), we begin by setting up the action integral that gives rise to it. This is done in order that we may employ the FEM. [8] The action integral is given by

$$\begin{aligned} A =& \frac{1}{2} \iint dx\, dy\, [\partial_i \phi(x, y)]\, \sigma_{ij}\, [\partial_j \phi(x, y)] \\ & - \int_{\Delta_1} dl[\phi(x, y)]|_{\Delta_1} j_{in} + \int_{\Delta_2} dl[\phi(x, y)]|_{\Delta_2} j_{out}. \end{aligned} \tag{12.12}$$

The double integral in Eq. (12.12) is just the electrostatic energy in the system. It is instructive to apply the principle of least action to the above equation in order to verify that the variation of the action with respect to the potential function indeed

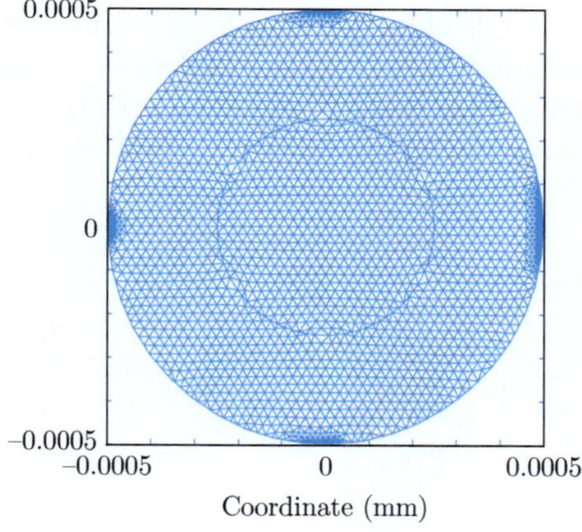

Figure 12.5 The discretization of the physical region for $\alpha = b/a = \frac{8}{16}$. The mesh has been made finer in the vicinity of the ports (contacts). The algebraic integer method was used for the mesh generation.

reproduces Eq. (12.7). Setting the variation of A with respect to the potential function $\phi(x, y)$ to zero, we obtain

$$\delta A = 0 = -\iint dx\, dy\, \delta\phi \{\partial_i \sigma_{ij}[\partial_j \phi(x, y)]\} + \left[\int_{\Gamma(r=a)} dl\, \delta\phi\, \hat{n} \cdot [\sigma_{ij}(\partial_j \phi)] - \int_{\Delta_1} dl\, \delta\phi\, j_{in} + \int_{\Delta_2} dl\, \delta\phi\, j_{out}\right]. \tag{12.13}$$

An integration by parts has been performed in order to obtain the terms in the square brackets. The variations $\delta\phi$ are arbitrary. We therefore choose them as follows.

1. Choose variations such that $\delta\phi = 0$ on the periphery, $r = a$. We then obtain the differential equation, Eq. (12.6), from the double integral in Eq. (12.12).
2. Now let $\delta\phi = 0$ inside the physical region *and also along the periphery* Δ_i *except at the input port along* Δ_1 where it is chosen to be 1. Then $\delta A = 0$ requires
$$\int_{\Delta_1} dl(j_{in} + \hat{n} \cdot \mathbf{j}) = 0.$$
In other words, we have $(\hat{n} \cdot \mathbf{j})_{\Delta_1} = -j_{in}$. This is just Eq. (12.7).
3. Next, choose $\delta\phi = 1$ along Δ_2 and zero elsewhere. Then
$$\int_{\Delta_2} dl(-j_{out} + \hat{n} \cdot \mathbf{j}) = 0,$$
implying that $(\hat{n} \cdot \mathbf{j})_{\Delta_2} = j_{out}$, as in Eq. (12.8).
4. Finally, put $\delta\phi = 0$ in the interior and along Δ_1 and Δ_2. Then $\delta A = 0$ requires that $(\hat{n} \cdot \mathbf{j})_{r=a} = 0$ on the rest of the circular periphery of the semiconductor.

We have shown that our starting action integral with the surface terms for the currents, through the principle of least action, leads to the original differential equation with its boundary conditions including the "derivative" boundary conditions on the input and output currents. This procedure is similar to that discussed in Courant and Hilbert [9] for derivative boundary conditions. We now employ the above action in numerical modeling.

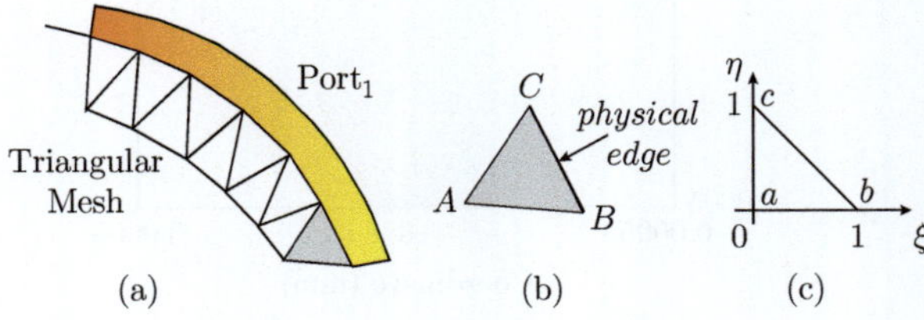

Figure 12.6 (a) One edge-triangle from the triangulated physical region near port 1 is highlighted. (b) The selected triangle ABC is shown. (c) The edge-triangle is mapped onto a standard right triangle for computation of the action integrals.

In the FEM, we begin by discretizing the action integral itself. We break up the physical region into triangles, or elements, in each of which the physics of the problem holds. This discretization is performed using an unstructured triangular mesh that is generated by the so-called algebraic integer method. [10] The result of this meshing is shown in Fig. 12.5. In each of the triangular elements, we represent the function $\phi(x, y)$ in terms of local interpolation polynomials $N_i(x, y)$. Each triangle has either three nodes, located at the three vertices of the triangle, or six nodes located at the midpoints of the sides of the triangle and at the vertices. Let

$$\phi(x, y) = \sum_i \phi_i N_i(x, y), \tag{12.14}$$

where N_i are unity at node i. The interpolation polynomials are linear for the three-nodal triangle, and quadratic polynomials in x and y in the case of a six-nodal triangle. Using the above functional form in the action integral, we integrate out the spatial dependence of the potential function and express the action in each element in the form

$$A^{(ielem)} = \frac{1}{2}\phi_i M_{ij}^{(ielem)} \phi_j. \tag{12.15}$$

Next, for continuity at the nodes, we add all the contributions from each element by setting the nodal values to be the same for all triangles having a common node. This amounts to an overlay of each element matrix $M_{ij}^{(ielem)}$ into a global matrix such that we add the nodal contributions from all triangles having that node in common. We then obtain

$$A = \frac{1}{2}\phi_\alpha M_{\alpha\beta}\phi_\beta - \int_{\Delta_1} dl[\phi(x, y)]|_{\Delta_1} j_{in} + \int_{\Delta_2} dl[\phi(x, y)]|_{\Delta_2} j_{out}. \tag{12.16}$$

The integrals in Eq. (12.16), that may be designated as "surface" terms, need further elaboration. If we assume a constant linear current density across the port, the surface terms are readily evaluated as follows.

In Fig. 12.6(a), we show the edge of the physical region along port 1. We have to integrate along the port in order to evaluate the first contour integral in Eq. (12.16). Consider one of the triangles with one edge coinciding with the contour at port 1. The physical edge is along the side BC of the triangle ABC of Fig. 12.6(b). We map this on to a "standard" right-angled triangle abc of Fig. 12.6(c). If side BC corresponds to side ab of the standard triangle, we have

$$\int_B^C dl\, \phi(x, y) j_{in} = j_{in} \int_a^b d\xi \frac{dl}{d\xi} \sum_i \phi_i N_i(\xi, o) = \frac{1}{2} l(\phi_C + \phi_B) j_{in}. \tag{12.17}$$

Here, l is the length of the edge BC. The same result is obtained if the side BC corresponds to the side ac of the standard triangle. In the case where the side BC

corresponds to the hypotenuse *bc* of the standard triangle, we obtain

$$\int_B^C dl\,\phi(x,y)j_{in} = j_{in}\int_b^c d\lambda \frac{dl}{d\lambda}\sum_i \phi_i N_i(\xi, \eta = 1-\xi). \tag{12.18}$$

Here,

$$d\lambda = \sqrt{(d\xi)^2 + (d\eta)^2} = d\xi\sqrt{1 + \left(\frac{d}{d\xi}(1-\xi)\right)^2} = \sqrt{2}d\xi$$

and

$$\frac{dl}{d\lambda} = \frac{l}{\sqrt{2}},$$

so that again Eq. (12.18) reduces to the right side of Eq. (12.17). The above calculations can be performed for quadratic interpolation functions when these are employed in representing the potential over each triangle. We thus have an expression for the discretized action given by

$$\begin{aligned} A = \frac{1}{2}\phi_\alpha M_{\alpha\beta}\phi_\beta - j_{in}&\left[\sum_{\substack{edge\\ sides\\ on\,\Delta_1}} l_i\left\{\frac{1}{6}\phi_1^{(i)} + \frac{2}{3}\phi_2^{(i)} + \frac{1}{6}\phi_3^{(i)}\right\}\right] \\ + j_{out}&\left[\sum_{\substack{edge\\ sides\\ on\,\Delta_2}} l_i\left\{\frac{1}{6}\phi_1^{(i)} + \frac{2}{3}\phi_2^{(i)} + \frac{1}{6}\phi_3^{(i)}\right\}\right]. \end{aligned} \tag{12.19}$$

Now our variational principle is implemented by varying the discretized action A of Eq. (12.19) with respect to the nodal variables ϕ_α. We obtain

$$\frac{\delta A}{\delta\phi_\alpha} = 0 = M_{\alpha\beta}\phi_\beta - C_i j_{in}\delta_{i\alpha} + C_j j_{out}\delta_{j\alpha}. \tag{12.20}$$

Here C_i and C_j are constants determined by evaluating the surface terms as described above. We then have a set of simultaneous equations for the nodal variables ϕ_α.

Due to the connectivity of the triangular mesh, the resulting coefficient matrix is sparsely occupied. We first perform a bandwidth reduction of the matrix and then decompose it into the standard **LU** form for Gauss elimination. [11]

We also equate all the nodal values for nodes appearing in ports 3 and 4 in order to define a unique potential at the ports over the lengths Δ_3 and Δ_4. Since no absolute potential values are set in the problem, we assign one of the ports to have zero potential with respect to which all other potentials are measured. The solution of the simultaneous equations now provides us with a unique solution.

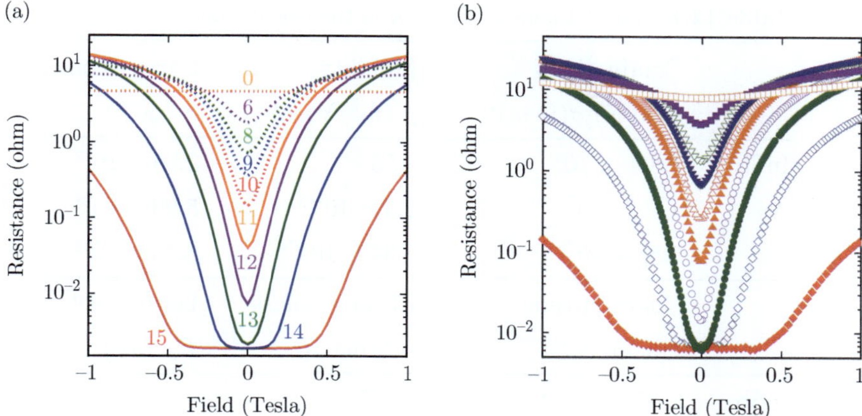

Figure 12.7 A plot of the resistance as a function of the magnetic field for $\alpha' = 16(b/a)$ ranging from 0 to 15, as obtained from (a) finite element method calculations and (b) directly from experiments. The symbols in (b) correspond to the following values of α' : $0 - \square, 6 - \blacksquare, 8 - \triangledown, 9 - \blacktriangledown, 10 - \vartriangle, 11 - \blacktriangle, 12 - \circ, 13 - \bullet, 14 - \Diamond, 15 - \blacklozenge$.

12.3 Comparison with experimental results

The principle quantity of interest is the field and geometry dependence of the effective resistance $R_{mn}(H,\alpha) = [\Delta V_{mn}(H,\alpha)]/i$, where $\Delta V_{mn}(H,\alpha) = V_m(H,\alpha) - V_n(H,\alpha)$, n and m define the voltage ports, i is a constant current, and as before $\alpha = b/a$ [see Figs. 12.4(a) and 12.4(b)]. Once the effective resistance is known, the EMR can be readily determined from

$$EMR_{mn}(H,\alpha) = \left(\frac{R_{mn}(H,\alpha) - R_{mn}(0,\alpha)}{R_{mn}(0,\alpha)} \right). \tag{12.21}$$

The effective resistance $R_{43}(H,\alpha)$ (see Fig. 12.4) is plotted in Fig. 12.7(a) using the parameters for InSb and Au specified in Table 12.1 (with values for Cu shown for reference) corresponding to experimental measurements to be described below. Thus, the influence of the metallic inhomogeneity is determined by varying the radius of the inner metallic region to change the ratio α. Note that for $\alpha > 13/16$, $R_{43}(H,\alpha)$ is very small and field-independent up to an onset field above which it increases very rapidly. This diode-like behavior may offer the opportunity for employing such constructs as a magnetic switch. The physical origin of this diode effect is understood: For a sufficiently large conducting inhomogeneity, deflected current will only flow in the correspondingly small annular ring of semiconductor when the field exceeds a critical value. Below that value, the current is completely shunted by the inhomogeneity, and its path through the semiconductor from the input to the output ports provides a negligible contribution to the resistance.

The calculations of $R_{43}(H,\alpha)$ described above are compared with the corresponding experimental results shown in Fig. 12.7(b). Those experimental results were obtained from a composite van der Pauw disk of InSb with fourfold symmetric 80

Table 12.1 Material parameters used in the calculations.

	Conductivity (300 K) $\sigma(\Omega\ \mathrm{m})^{-1}$	Mobility μ (m^2/Vsec)	Carrier conc. n (m^{-3})
InSb	1.856×10^4	4.55	2.55×10^{22}
Au	4.52×10^7	5.3×10^{-3}	5.90×10^{28}
Cu	5.88×10^7	3.34×10^{-3}	8.45×10^{28}
	Collision time $\tau(sec)$	**Effective mass** m^*/m_0	**Fermi level** E_F (eV)
InSb	3.87×10^{-13}	0.015	
Au	3.0×10^{-14}	1.0	5.51
Cu	1.9×10^{-14}	1.0	7.0

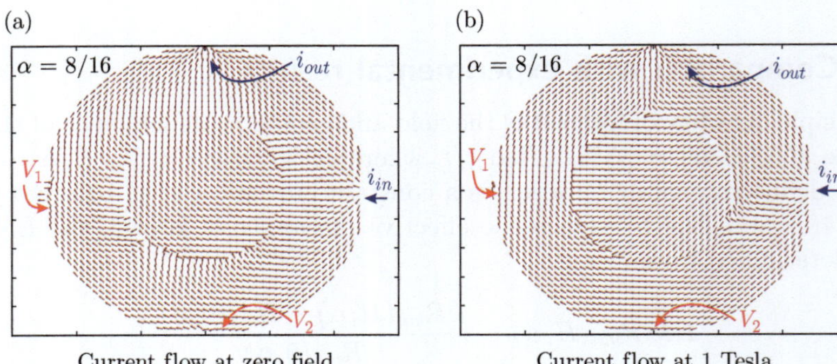

Figure 12.8 The current flow in the van der Pauw geometry for a circular InSb wafer with a concentric metallic inhomogeneity and $\alpha' = 16(b/a) = 8$, (a) at $H = 0$ and (b) at $H = 1$ T. The lengths of the arrows are not to scale.

wide current and voltage ports (i.e. $\Delta_i = 8°$, $i = 1 - 4$) and an embedded Au inhomogeneity which has been fabricated using methods described in detail elsewhere. [1] The agreement between experiment and the finite element calculation is remarkable in view of the fact that no adjustable parameters were employed in the calculation. The slight shift in the relative values of the abscissae in Figs. 12.7(a) and 12.7(b) is probably due to a finite contact resistance between the metal inhomogeneity and the semiconductor which has not been included in the calculation. Moreover, one can notice that the calculated effective resistance for $\alpha = 0$ is totally field-independent, whereas the corresponding experimental result shows a slight field-dependence. This difference results from the fact that the physical contribution to the effective resistance from the field dependence of the intrinsic parameters such as the mobility and

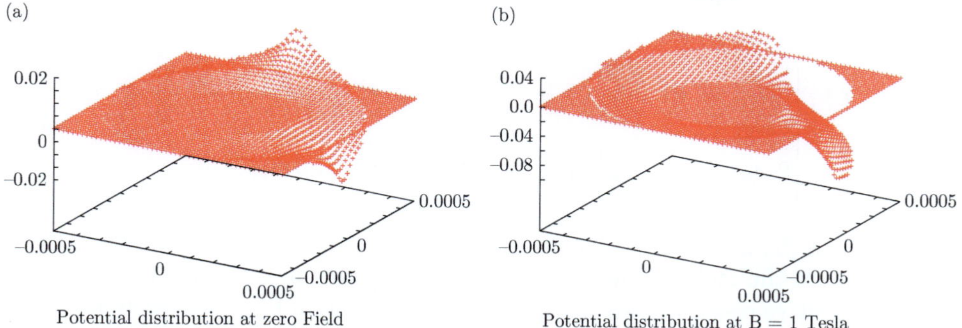

Figure 12.9 The potential distribution in the van der Pauw geometry for a steady-state current through ports 1 and 2 (a) in the absence of an applied magnetic field and (b) at $H = 1\mathrm{T}$.

carrier concentration is small but finite and has not been included in the calculated results.

12.4 Comparison with analytical calculations

For the highly symmetric centered van der Pauw structure shown in Fig. 12.4, one can analytically compute the magnetotransport properties as has been described in detail for the low-field region ($H < 0.1\mathrm{T}$) elsewhere. [6] It is useful to compare the analytic results in both the low and high field range ($0 \leqslant H \leqslant 1\,\mathrm{T}$) with the FEM discussed above. The analytic solution for the resistance of the centered van der Pauw structure shown in Fig. 12.4 with four identical ports of width Δ is [6]

$$\begin{aligned}
&\Delta V_{34}(\alpha,\gamma,\eta,\beta,\sigma_0,\phi,t,\theta) = \\
&\qquad \frac{1+\beta^2}{\sigma_0}\frac{i}{\pi t\Delta}\sum_{n=1}^{\infty}\frac{1}{n^2}\frac{1}{J^2+K^2}\Big\{[(JU-KW)(1-\alpha^{2n}\gamma) \\
&\quad -\alpha^{2n}\eta(KU+JW)]\cos(n\theta) + [(KU+JW) \\
&\qquad \times(1-\alpha^{2n}\gamma)+\alpha^{2n}\eta(JU-KW)]\sin(n\theta)\Big\},
\end{aligned} \tag{12.22}$$

where

$$\begin{aligned}
J &= 1+\alpha^{2n}\gamma+\beta\alpha^{2n}\eta, \\
K &= \beta+\alpha^{2n}\eta-\beta\alpha^{2n}\gamma,
\end{aligned} \tag{12.23}$$

with

$$\begin{aligned}
U &= \sin(n\pi/2+n\Delta/2)-\sin(n\pi/2-n\Delta/2)-2\sin(n\Delta/2), \\
W &= \cos(n\pi/2-n\Delta/2)-\cos(n\pi/2+n\Delta/2),
\end{aligned} \tag{12.24}$$

and

$$\gamma = \left[(\omega_0^2 - \omega^2) + (\omega_0\beta_0 - \omega\beta)^2\right] / \left[(\omega_0 + \omega)^2 + (\omega_0\beta_0 - \omega\beta)^2\right],$$
$$\eta = \left[2\omega(\omega_0\beta_0 - \omega\beta)2\right] / \left[(\omega_0 + \omega)^2 + (\omega_0\beta_0 - \omega\beta)^2\right], \tag{12.25}$$

while $\beta_0 = \mu_0 H$, $\omega = \sigma/(1+\beta^2)$, $\omega_0 = \sigma_0/(1+\beta_0^2)$, and t is the thickness of the disk, μ_0 and σ_0 are the mobility and conductivity of the metal, respectively.

The effective resistance for the configuration shown in Fig. 12.4 calculated directly from Eqs. (12.20)–(12.22) using no adjustable parameters can be compared with the corresponding finite element analysis, and the agreement is excellent as expected. While the two methods yield nearly identical results, the FEM is not restricted to highly symmetric structures such as that shown in Fig. 12.4.

It is clear from the results that the finite element method is able to accurately reproduce the experimental results for $R_{mn}(H, \alpha)$. Indeed, given the solution for the potential $V_m(H, \alpha)$ we can compute not only the EMR (see Eq. (12.20)) but also a number of other interesting properties of the modified van der Pauw disk. In Fig. 12.8, we show the flow lines of the current. The arrows indicate the direction of current flow at $H = 0$ and at $H = 1$T. The lengths of the arrows are not to scale in the figure. The effect of the applied magnetic field on the current deflection at the interface between the semiconductor and the inhomogeneity can be readily seen from the figure. The large perturbation to the potential at the current ports caused by the applied field as shown in Fig. 12.9 for the two cases $H = 0$ and $H = 1$T for a value of $\alpha = 0.5$ indicates that the EMR will be very sensitive to the position and width of the voltage ports. Indeed, for the modified van der Pauw structure addressed here, the output voltage, $\Delta V_{mn}(H, \alpha)$ decreases as the EMR is increased by the selection of the voltage and current port locations. The current through the structure provides the driving terms in the set of simultaneous equations obtained from the variational principle applied

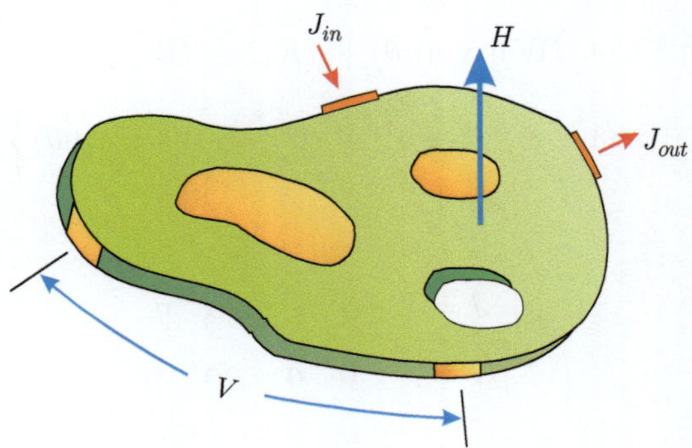

Figure 12.10 The four-contact van der Pauw geometry for an arbitrarily shaped InSb thin film with distributed metallic inclusions of arbitrary shape and with a hole in it of arbitrary shape.

to the discretized action. Here we see that the finite element approach is able to capture the important dependence of the response of the structure to the geometrical placement and location of the current ports.

We have demonstrated an advantage of the action integral formulation of the finite element analysis in its ability to apply the current (derivative) boundary conditions very directly in specific sections of the boundary, i.e. at the current ports, while allowing the values of the potential at other segments of the periphery to be self-consistently determined from the variation of the discretized action integral. The values of the potential at all the nodes located just at the voltage-measuring ports are equated to each other since the voltage is the same across the leads. This is implemented by "folding in," or adding together, of columns of the global matrix generated in the FEM that correspond to the nodes at the voltage ports. Such details of the boundary conditions can be accounted for with ease within the framework of the FEM.

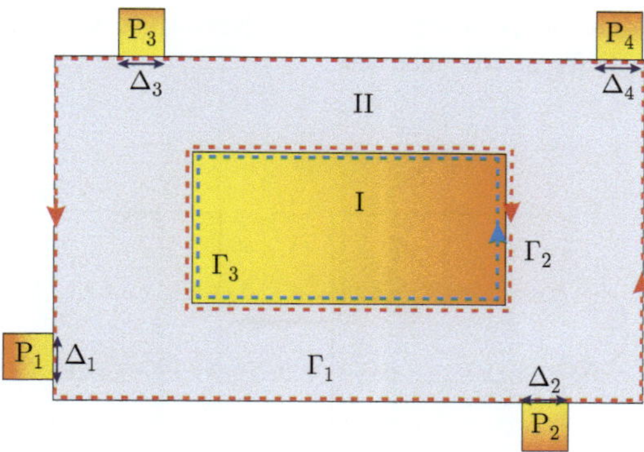

Figure 12.11 A semiconductor wafer with a rectangular metallic inclusion is shown. Contacts P_1, ..., P_4 correspond to two voltage probes P_3 and P_4, with current I coming in at say P_1 and leaving the structure at P_2. The current density entering the device is taken to be $I/(\Delta_1 t)$, where t is the thickness of the wafer and Δ_i is the width of the ith contact. The metal and semiconductor are labeled by roman numerals.

Finally, note that the use of a bandwidth reducer and sparse matrix LU solvers substantially reduces the computer time for the calculations. With 6000 nodal points in the typical unstructured mesh used in the calculations, we have a 6000×6000 (sparse) matrix that is solved for over 60 values of the magnetic field. The global matrix is evaluated afresh each time and calculations for a single curve, for a given value of α, are completed very quickly on desktop computers.

Because they are nonmagnetic and work at room temperature, EMR devices can be used in applications where typical magnetic sensors are not suitable. Furthermore, their performance continues to be impressive down to the nanoscale. Unlike traditional magnetic recording sensor technologies, such as GMR and tunneling magnetoresistance (TMR), where device resistance is determined by spin-dependent scattering, in EMR structures the magnetoresistance is modulated by utilizing the

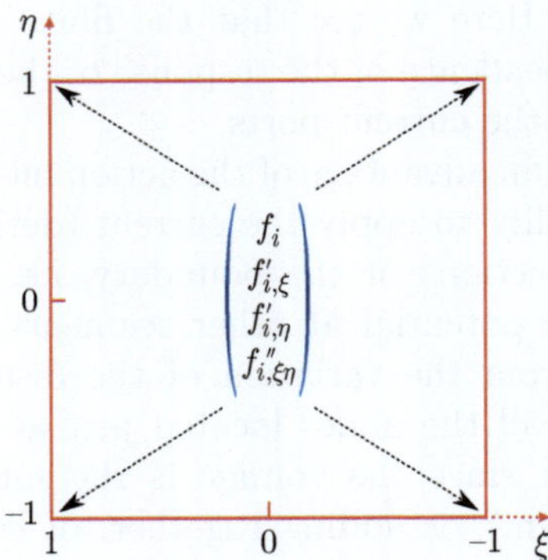

Figure 12.12 A 2D reference finite element is shown with four degrees of freedom at nodes at the four corners corresponding to the value of the function, its first derivatives with respect to ξ, η, and a second (cross) derivative. The polynomial interpolation within the element is performed using the values of the function and its derivatives at the nodes. See Ref. [12]. This scheme is extended to 3D for a cube element.

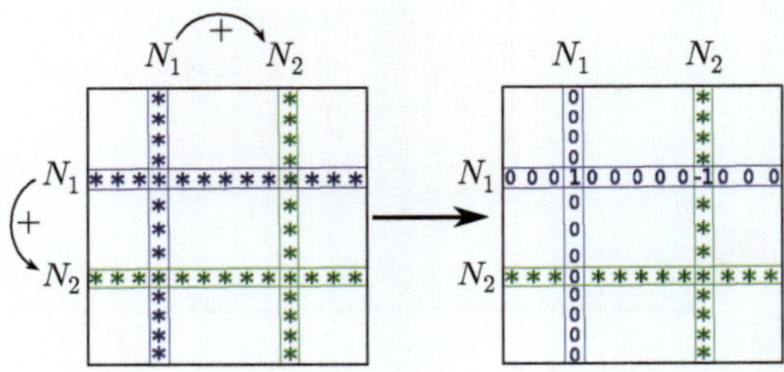

Figure 12.13 Matrix Manipulation for Setting Two Nodal Values to be the Same. Matrix entries are rearranged as discussed in the text. This is done for all nodes on P_4 to make it an equipotential contact.

Lorentz force to steer an electron current away from the high-conduction metallic regions. The carrier velocity has a nonzero Hall angle with respect to the electric field which continues to be directed normal to the essentially equipotential metal–semiconductor interface. The carrier path being diverted away from the metal into the semiconductor, which has a factor $\sim 10^3$ larger resistivity, then gives rise to the large MR.

The true flexibility and the power of the FEM comes into its own when the physical semiconductor film is not only of arbitrary shape but contains both filled (conducting) and empty (insulating) embedded inhomogeneities of arbitrary shape as shown schematically in Fig. 12.10. Given that the EMR discussed here is geometrically enhanced, in the FEM we now have the means of designing and exploring geometrically more complex heterostructures for additional improvements in the performance of EMR-based magnetic devices *before* they are fabricated for electronic applications.

12.5 More complex geometries for EMR

We now wish to consider the promise of very high MR in the metal–semiconductor structures by designing new schemes that could substantially enhance the EMR effect, and by modeling realistic 2D structures that could be fabricated using Au/InSb.

We have also considered 3D structures in the following and find very encouraging results. Here we develop the theory for such analysis, and demonstrate that geometrical enhancement of MR can be increased considerably with no more effort than used in making devices employed in earlier experiments with simple shunt configurations. We provide estimates for devices of mesoscopic dimensions, keeping in mind the recent technological advances in material fabrication today. The mesoscopic structures we consider are on the order of 1 μm, so that transport is mostly dominated by diffusive effects rather than by ballistic carriers.

12.6 Theoretical considerations in 3D

In the presence of a magnetic field, the magneto-conductivity is given in terms of $\vec{\beta} = \mu \vec{H}$, where μ is the carrier mobility and H is the magnetic field. In 3D, we have

$$\boldsymbol{\sigma} = \frac{\sigma_0}{1+\beta_x^2+\beta_y^2+\beta_z^2}\begin{pmatrix} (1+\beta_x^2) & (-\beta_z+\beta_y\beta_x) & (\beta_y+\beta_z\beta_x) \\ (\beta_z+\beta_y\beta_x) & (1+\beta_y^2) & (-\beta_x+\beta_y\beta_z) \\ (-\beta_y+\beta_z\beta_x) & (\beta_x+\beta_y\beta_z) & (1+\beta_z^2) \end{pmatrix} \tag{12.26}$$

which reduces in 2D, with $\vec{H} = \hat{z}H$ and $\beta_z = \mu H$, to

$$\boldsymbol{\sigma} = \frac{\sigma_0}{1+\beta_z^2}\begin{pmatrix} 1 & -\beta_z \\ \beta_z & 1 \end{pmatrix} \tag{12.27}$$

with only the x, y-components for the conductivity tensor. Here, the intrinsic conductivity σ_0 is the conductivity in the absence of a magnetic field.

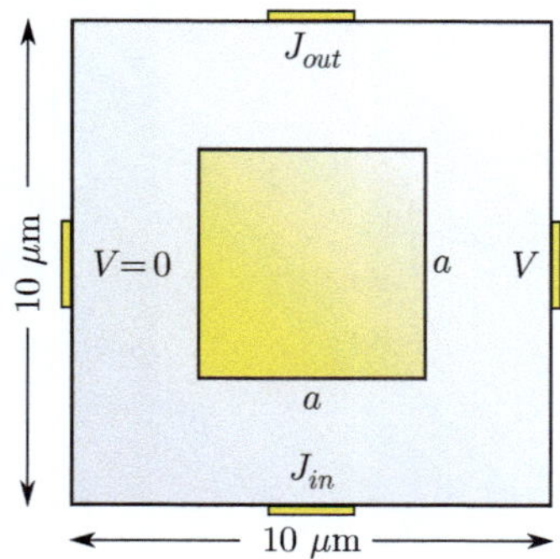

Figure 12.14 A schematic for a 10 μm square with contacts centered on all sides.

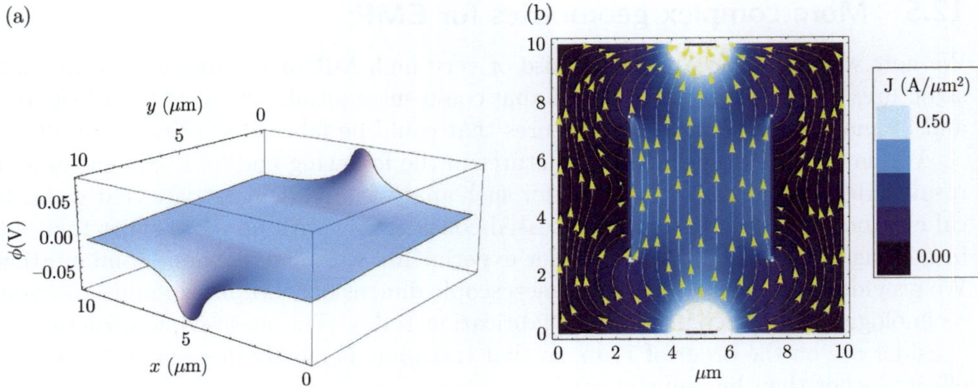

Figure 12.15 The potential (a) and the plot of the current (b) for a 10 μm square with centered contacts with a 5 μm metallic region and an applied field of $H = 0$ T are shown.

12.6.1 The action integral and Hermite finite elements

In Refs. [13, 14], we showed that a finite element approach [8, 12, 15] to the calculation of the MR in simple 2D structures provides remarkable congruence with experimental results. Only linear interpolation polynomials were used in these calculations. Here, we display the details of the theoretical development of the calculations for more complex geometries in 2D, keeping in mind that we will be investigating examples from 2D as well as 3D. We will also employ $\mathcal{C}_{(1)}$-continuous functions which provide significant advantages in terms of accuracy, and also in terms of explicitly applying current continuity conditions at internal metal–semiconductor interfaces and derivative boundary conditions along the periphery.

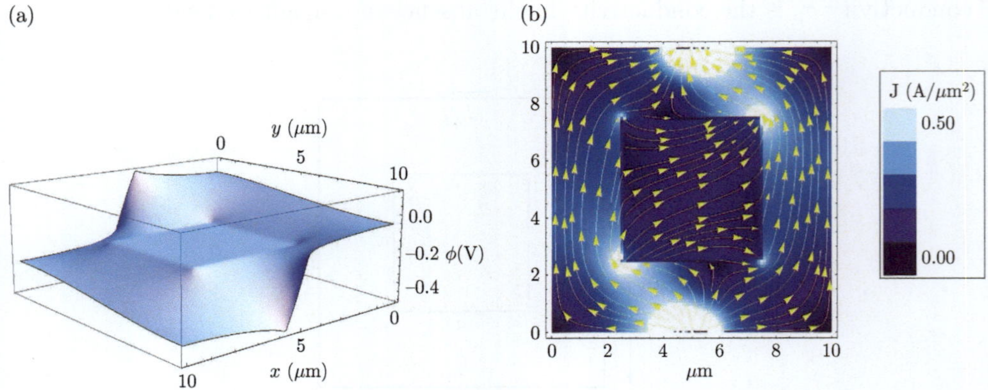

Figure 12.16 The potential (a) and the plot of the current (b) for a 10 μm square with centered contacts with a 5 μm metallic region and an applied field of $H = 1$ T are shown.

The results presented here for specific geometries are scalable to smaller dimensions down to the onset of ballistic effects in the transport. We have verified this explicitly through numerical calculations.

In the present case of steady-state conditions, the equation of continuity leads to

$$\nabla \cdot \mathbf{J} = 0 = \nabla \cdot (\boldsymbol{\sigma}\mathbf{E}), \tag{12.28}$$

or equivalently,

$$-\nabla \cdot \boldsymbol{\sigma} \cdot \nabla\phi(\mathbf{r}) = 0, \tag{12.29}$$

where the electric field $\mathbf{E}$ is expressed in terms of a scalar potential ϕ. The use of variational methods provides fast, stable convergence in the calculations, and we cast the problem using the principle of stationary action. The action integral from which this equation is derivable for Dirichlet boundary conditions is

$$A_0 = \int_0^T dt \sum_\alpha \left[\int_{\Omega_\alpha} d\mathbf{r}\, \frac{1}{2}\Big(\partial_i\phi(\mathbf{r})\, \sigma_{ij}^{(\alpha)}\, \partial_j\phi(\mathbf{r}) \Big) \right]. \tag{12.30}$$

The sum over α is to account for the actions in different regions Ω_α with their different conductivities. In the steady state under consideration here, the integration over time is trivial. Let us consider a typical 4-probe system for measuring the MR in the structure as in Fig. 12.11. The presence of current boundary conditions at two of the contacts, corresponding to derivative boundary conditions, requires a modification of the above action in order to ensure that the equation of motion can be derived consistently. We suppose that the steady current comes in at contact P_1, say, and leaves the structure at contact P_2. The additional terms that are needed can be identified by analytically attempting to obtain the equation of motion as follows. A variation of A_0 with respect to the potential function ϕ together with the usual integration by parts leads to

$$\delta_\phi(A_0/T) = 0 = \sum_\alpha \left[\int_{S_\alpha} d\mathbf{r}\; \delta\phi\Big(-\nabla \cdot \boldsymbol{\sigma}^{(\alpha)} \cdot \nabla\phi(\mathbf{r}) \Big) \right]$$
$$+ \sum_\nu \int_{\Gamma_\nu} d\ell\, \delta\phi\; \hat{n} \cdot \Big(\boldsymbol{\sigma}^{(\nu)} \cdot \nabla\phi(\mathbf{r}) \Big) \Big]. \tag{12.31}$$

Here ν corresponds to the various contours at the peripheries of the various regions and $\hat{n}$ is the normal to the counter-clockwise boundary paths in 2D. It is clear that if we had Dirichlet boundary conditions specifying the potential everywhere along the external periphery, the second term in Eq. (12.31), which we refer to as the surface term in both 2D and 3D, would vanish since ϕ is then fixed on the boundary. We note that (i) the requirement of continuity of the current across the metal–semiconductor interface always ensures that the integrals along Γ_2 and Γ_3 cancel (see Fig. 12.11); (ii) The potential at, say, P_3 is set to zero to give a reference potential, hence the boundary integral across Δ_3 is zero ($\delta\phi$ is zero there since ϕ is zero there and is therefore fixed in value); (iii) Our boundary conditions are not of the Dirichlet type along the outer periphery at the current contacts so that the portions of Γ_1 corresponding to $\Delta_{1,2}$

require special consideration. Using the relation $\mathbf{J} = -\sigma \nabla \phi(\mathbf{r})$, we can identify the integrand of the surface term in square brackets in terms of the current there. Since no current comes in or escapes along Γ_1 except at the contacts P_1 and P_2, we can set the contour integral to zero everywhere except over Δ_1 and Δ_2. The potential at contact P_4 is determined by the solution there as discussed below. (iv) The two surface terms at $\Delta_{1,2}$ are such that $\delta\phi$ are arbitrary there, and $-\sigma \nabla \phi \cdot \hat{n}$ is nonzero. Since these two surface terms cannot be set to zero, the equation of motion, Eq. (12.29), does not hold. This is remedied, as discussed in Courant and Hilbert, [9, 13] by adding two additional terms to the action that ensure that these surface terms are canceled out. Writing the new action, we have

$$A/T = \sum_{\alpha} \int d\mathbf{r} \, \frac{1}{2} \Big(\partial_i \phi(\mathbf{r}) \, \sigma_{ij}^{(\alpha)} \, \partial_j \phi(\mathbf{r}) \Big) - \int_{\Delta_1} d\ell \, \phi(x, y)|_{\Delta_1} J_{\text{in}} + \int_{\Delta_2} d\ell \, \phi(x, y)|_{\Delta_2} J_{\text{out}}, \tag{12.32}$$

with the current boundary conditions incorporated into the action. We note that while the current I_{in} must equal I_{out}, the width of the contacts $\Delta_{1,2}$ and the thickness of the semiconductor wafer determine the current densities $J_{1,2}$. The same considerations apply to a 3D geometry with the extension of the above expressions to metallic inclusions in a semiconductor volume.

We now evaluate the action directly by discretization of the physical space using the FEM, as discussed in the following.

12.7 FEM with $\mathcal{C}_{(1)}$-continuous elements

In the FEM, the physical domain is discretized into elements. In each of the elements, the variational principle holds. The potential function is represented as a polynomial multiplied by coefficients representing the value of the potential at special points in the element called nodes. On integrating out the spatial dependence, the action reduces to a bilinear expression in the as-yet unknown interpolation coefficients, which are known as the nodal variables. The principle of stationary action is invoked by varying A/T with respect to the nodal variables, which then leads to a system of simultaneous equations that represent the discretized equation of motion. [12]

The simultaneous equations are solved to obtain the potentials at the nodes, and the potential everywhere is reconstructed using the original interpolation polynomial in each element. This allows us to also obtain the spatial variation of the current density in great detail, and we then determine the MR for a range of values of the magnetic field H.

Conforming to the practical choice of rectangular device geometry, we consider finite elements of the same shape. In 2D, consider a standard square element with nodes at $\xi = \pm 1, \eta = \pm 1$. A given rectangular element can be linearly mapped into the standard element, so that the interpolation polynomials can be defined on the standard element for convenience. Each of the four nodes at the four corners of the element are associated with the values of the potential and its derivatives

$\{\phi^{(in)}, \phi_{,\xi}^{(in)\prime}, \phi_{,\eta}^{(in)\prime}, \phi_{,\xi\eta}^{(in)\prime\prime}\}$, where $in = 1, \ldots, 4$, for the four nodes. Thus there are 4 degrees of freedom (DoFs) at each of the four nodes of the element. This is shown in Fig. 12.12. The corresponding $\mathcal{C}_1$-continuous (Hermite) interpolation polynomials are given in Ref. [12]. We represent the potential function over a given element as

$$\phi(x, y) = \sum_{\nu} \phi_\nu N_\nu(x, y), \tag{12.33}$$

with the sum running over the full set of 16 DoFs for the element. The action is calculated over each element, and the spatial variables are integrated out. The resulting expression is bilinear in the nodal variables and can be cast in a matrix form. The element matrices are then overlaid to account for the continuity of the solution over the individual materials of the composite, keeping account of the interface boundary conditions. In summary, the discretized action obtained from Eq. (12.32) can be written as

$$A/T \doteq \frac{1}{2}\phi_\alpha M_{\alpha\beta}\phi_\beta - \phi_\alpha[\delta_{\alpha\mathcal{T}_1} R^{\text{in}}_{\mathcal{T}_1}]J_{\text{in}} + \phi_\alpha[\delta_{\alpha\mathcal{T}_2} R^{\text{out}}_{\mathcal{T}_2}]J_{\text{out}} \tag{12.34}$$

where the surface integrals in Eq. (12.32) are designated by the nodal values multiplied by integration of shape functions only over the current contacts in the last two terms.

12.7.1 Boundary conditions for Hermite elements

In order to simplify our discussion, we limit ourselves to a 2D example. The extension of the following to 3D is straightforward. The boundary conditions for the potential and its derivatives are readily implemented within the finite element scheme.

(a) The continuity of the potential across the metal–semiconductor interface can be enforced by setting the nodal values

$$\phi^I_{in} = \phi^{II}_{in}. \tag{12.35}$$

The continuity of the normal current across the interface requires

$$\hat{n}_i \sigma^{(I)}_{ij} \partial_j \phi^{(I)}(x, y) = \hat{n}_i \sigma^{(II)}_{ij} \partial_j \phi^{(II)}(x, y),$$

so that at each of the nodes common to the paths $\Gamma_{2,3}$ where, for example, $\hat{n}_i = \hat{y}$, we use the relation

$$\begin{pmatrix} 1 & 0 & 0 & 0 \\ 0 & 1 & 0 & 0 \\ 0 & \sigma^{(I)}_{yx} & \sigma^{(I)}_{yy} & o \\ 0 & 0 & 0 & 1 \end{pmatrix} \begin{pmatrix} \phi^{(I)} \\ \phi^{(I)\prime}_x \\ \phi^{(I)\prime}_y \\ \phi^{(I)\prime\prime}_{xy} \end{pmatrix} = \begin{pmatrix} 1 & 0 & 0 & 0 \\ 0 & 1 & 0 & 0 \\ 0 & \sigma^{(II)}_{yx} & \sigma^{(II)}_{yy} & o \\ 0 & 0 & 0 & 1 \end{pmatrix} \begin{pmatrix} \phi^{(II)} \\ \phi^{(II)\prime}_x \\ \phi^{(II)\prime}_y \\ \phi^{(II)\prime\prime}_{xy} \end{pmatrix}; \tag{12.36}$$

a similar relation holds for the current continuity of J_x across the interfaces with constant y. Thus, the first-derivative DoFs are reduced appropriately to enforce the current continuity.

(b) No current enters or leaves the device on the outer boundary Γ_2 other than at the current contacts. We therefore require that

$$J_n = 0 = \sigma_{nx}\frac{\partial \phi}{\partial x} + \sigma_{ny}\frac{\partial \phi}{\partial y}, \tag{12.37}$$

except at the current contacts. This again allows us to reduce the nodal derivative DoFs by one at every node on the external boundary. The variables $\phi^{(in)}$ and $\phi_{,xy}^{(in)\,\prime\prime}$ at the boundary nodes are not preassigned any values since they have no conditions on them.

(c) At the voltage contact P_3, the potential at the nodes is set to zero, while the normal current is eliminated as in the boundary condition (*b*) above. At the voltage contact P_4, the potential is not determined, but the normal current is again eliminated since no current leaves the system at P_4. In order to make the contact an equipotential, we ensure that the potential variables ϕ^{in} are equated to one another as follows.

Suppose there are only two nodes N_1 and N_2 on P_4, where $N_{1,2}$ refer to node numbers associated with a global node numbering for the entire structure. The discretized action leads to the matrix $M_{\alpha\beta}$ as in Eq. (12.32). The effect of equating the potentials $\phi_{N_1} = \phi_{N_2}$ is to add the row and column corresponding to N_1 onto the row and column for N_2. After the transfer, the row and column of index N_1 is then zeroed out, and the matrix entry (N_1, N_1) is set to 1, while matrix entry (N_1, N_2) is set to -1. This procedure is illustrated in Fig. 12.13.

As for the other DoFs for these two contacts, we treat them the same as the outer boundary since we do not want current going in or out. Therefore $\phi'_{,x}$ and $\phi'_{,y}$ follow the condition given in (*a*) above, and $\phi''_{,xy}$ is left floating.

The overlay of the calculations for the element matrices, consistent with the above element and interface boundary conditions, leads to the discretized action given by a global matrix M together with vectors representing the surface terms at the current contacts. We have

$$A/T \doteq \frac{1}{2}\phi_\alpha M_{\alpha\beta}\phi_\beta - \phi_\alpha[\delta_{\alpha\tau_1}R^{\mathrm{in}}_{\tau_1}]J_{\mathrm{in}} + \phi_\alpha[\delta_{\alpha\tau_2}R^{\mathrm{out}}_{\tau_2}]J_{\mathrm{out}} \tag{12.38}$$

with the surface integral evaluated explicitly using the shape functions mentioned earlier. The nodal values for the potential labeled by ϕ_α, and their values at the current contacts are limited to the nodes labeled by τ_1 and τ_2 that are located there. The principle of stationary action is implemented by varying the above discretized action with respect to ϕ_α and thereby obtaining the matrix equation that represents the original differential equation. We solve the matrix equation

$$M_{\alpha\beta}\phi_\beta = [\delta_{\alpha\tau_1}R^{\mathrm{in}}_{\tau_1}]J_{\mathrm{in}} - [\delta_{\alpha\tau_2}R^{\mathrm{out}}_{\tau_2}]J_{\mathrm{out}} \tag{12.39}$$

for the potential at the nodes over the entire domain.

12.7.2 A square metallic region embedded in a semiconductor

For a 10 μm square, we consider the contacts that are centered on all sides as shown in Fig. 12.14. The metallic inclusion is embedded in the structure, i.e. the interface between the metal and semiconductor is normal to the plane of the semiconductor. For a 5 μm metal region, Fig. 12.15 and Fig. 12.16 show plots of the potential and the current for $H = 0\,\mathrm{T}$ and $H = 1\,\mathrm{T}$, respectively. Note that the current is represented by vectors which show the direction, and by a color gradient which represents its magnitude. We see that for zero field, the current is normal to the surface of the metal, and for higher fields, the current curves around the metal and concentrates in the semiconductor region, causing a change in resistance. Figure 12.17 shows a plot of the MR versus the applied magnetic field. Note that this is a log-log plot and the dashed lines represent the negative portion of the magnetic field for each case. For an inner square width of 8 μm for $H = 1\,\mathrm{T}$ and $H = -1\,\mathrm{T}$, we see an MR on the order of almost $10^7\%$. The variation of MR with the filling factor is displayed in Fig. 12.18. We also completed calculations for a 1 μm square with the same filling factors. We obtained the same results for MR as the 10 μm square. This shows that the effect is scale-invariant.

12.7.3 Multiple metallic regions embedded in a semiconductor wafer

Because of the flexibility of FEM, we can easily add multiple metal regions to the semiconductor. Fig. 12.19 shows a 10 μm semiconductor with two embedded metal regions. The length and width of these regions can be adjusted as well as the positions of the contacts. As in Fig. 12.19, the current contacts are centered along x. Fig. 12.20 and Fig. 12.21 show plots of the current and potential for $H = 0$ and 1 T. In these plots, we are using metal regions of width $a = 2.5\,\mu$m, height $b = 5\,\mu$m, separated by a distance of $d = 2.5\,\mu$m. Fig. 12.22 shows the MR versus magnetic field for $a = 4\,\mu$m, $b = 8\,\mu$m, and d varying. The case where $d = 0$ corresponds to the case above with a 10 μm-semiconductor square with centered contacts and an embedded metal square of width 8 μm. As we increase the separation width d, the MR increases. Therefore, using the same amount of metal, we can enhance the effect further by separation of the metal regions. This pushes the MR to over $10^7\%$ at $\pm 1\,\mathrm{T}$.

12.8 The magnetoresistance in 3D

We consider a rectangular geometry for the semiconductor as shown in Fig. 12.11, with a single embedded square metal region. The contacts can be placed anywhere on the edges of the device, and the overall size as well as the size and position of the metal region can vary. First, a 10 μm semiconductor square is considered for various dimensions of the embedded metal square. This is followed with discussion of a semiconductor square with two embedded metal regions as well as a 3D semiconductor cube with a metallic cube embedded in it. The interface between the metal and semiconductor is assumed to be Ohmic.

Metallic cube embedded in a 3D semiconductor cube

Figure 12.23 shows an example of a 3D semiconductor cube. The results shown in Fig. 12.23 and Fig. 12.24 have an embedded metal region of 5 μm. We have a magnetic field only along z, with current contacts on the left and right faces and voltage contacts on the front and back. This case is analogous to the 2D case, where there is one square

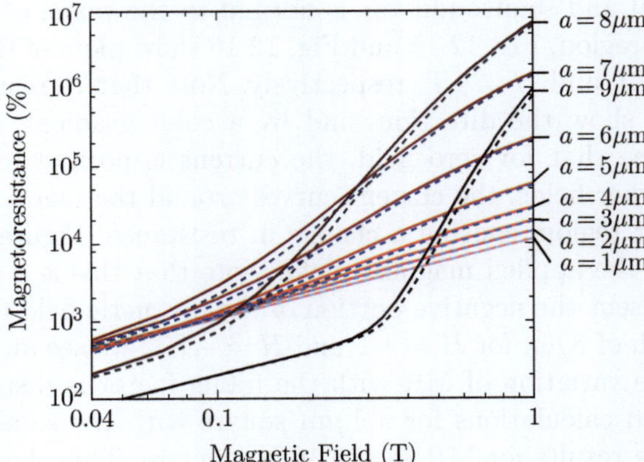

Figure 12.17 Plots of the magnetoresistance versus the magnetic field for a 10 μm square with contacts centered on all sides for different square metallic regions of side a. The dashed lines represent the MR for negative values of the magnetic field.

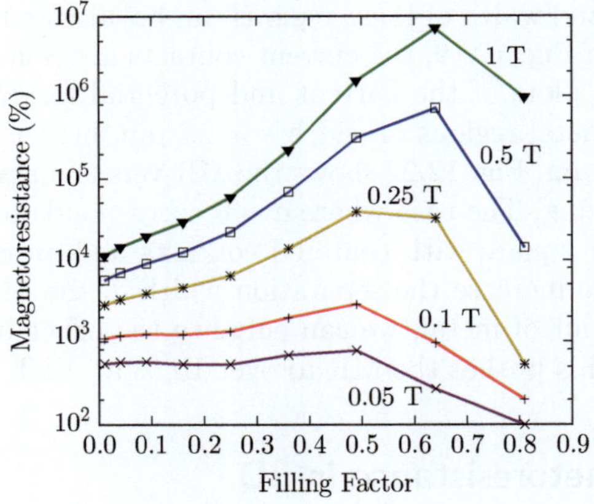

Figure 12.18 Plot of filling factor versus magnetoresistance for various values of the magnetic field for a 10 μm square with centered contacts.

metal region with contacts centered on all sides. Three-dimensional vector field plots for the current are shown in Fig. 12.23 for $H = 0\,\mathrm{T}$ and $H = 1\,\mathrm{T}$. We see a fairly large MR, as seen in Fig. 12.24. If we compare this to the 2D case with contacts centered on all sides and an inner square of 5 μm, the MR is on the same order of magnitude.

We are at present anticipating substantial improvements in the computational time for 3D calculations so that we may fully explore the variation of MR as the direction of the magnetic field is changed. The results will be reported elsewhere.

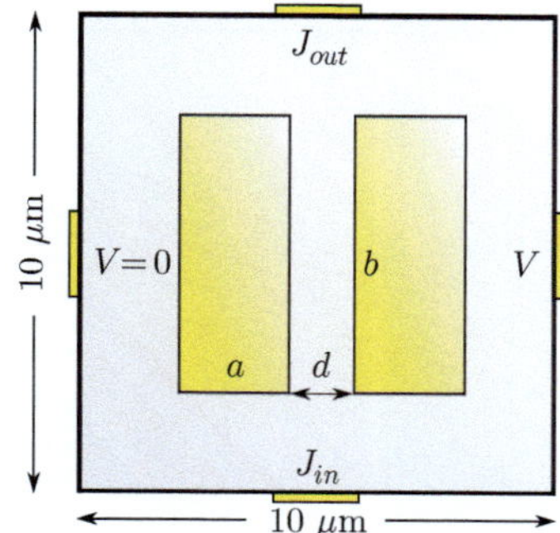

Figure 12.19 A schematic for a 10 μm square with two metal regions and current contacts along x.

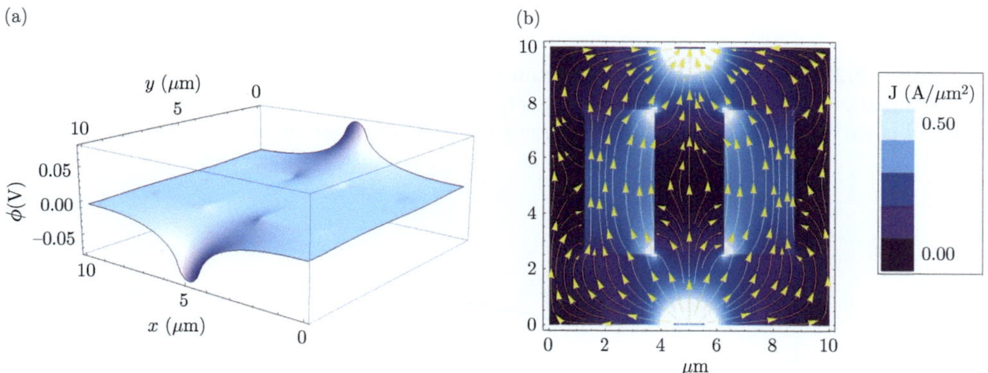

Figure 12.20 The potential (a) and the plot of the current (b) for a 10 μm square with centered contacts along x with two metal regions and an applied field of $H = 0\,\mathrm{T}$ is shown.

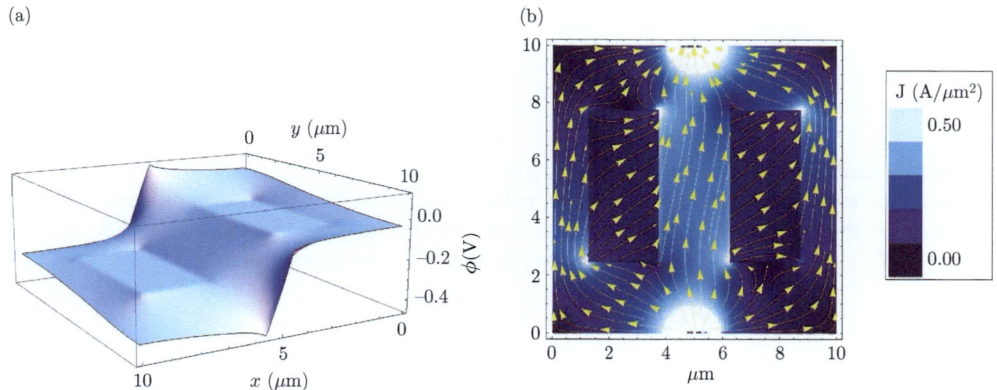

Figure 12.21 The potential (a) and the plot of the current (b) for a 10 μm square with centered contacts along x with two metal regions and an applied field of $H = 1\,\mathrm{T}$ is shown.

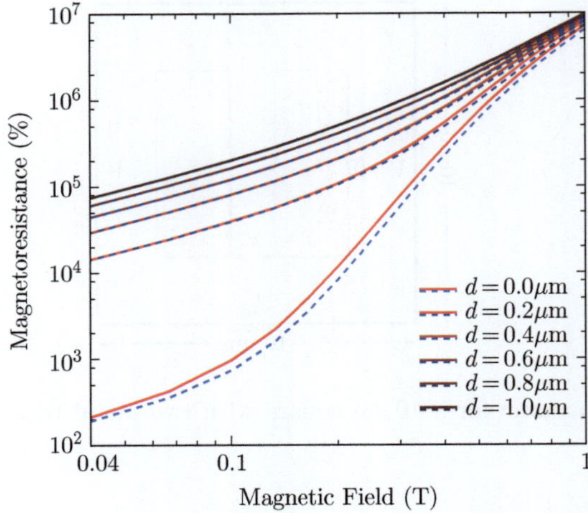

Figure 12.22 Plots of the magnetoresistance versus the magnetic field for a $10\,\mu$m square with two metal regions and current contacts centered along x. For this case, $a = 4\,\mu$m, $b = 8\,\mu$m, and d is varying. The dashed lines represent the MR for negative values of the magnetic field.

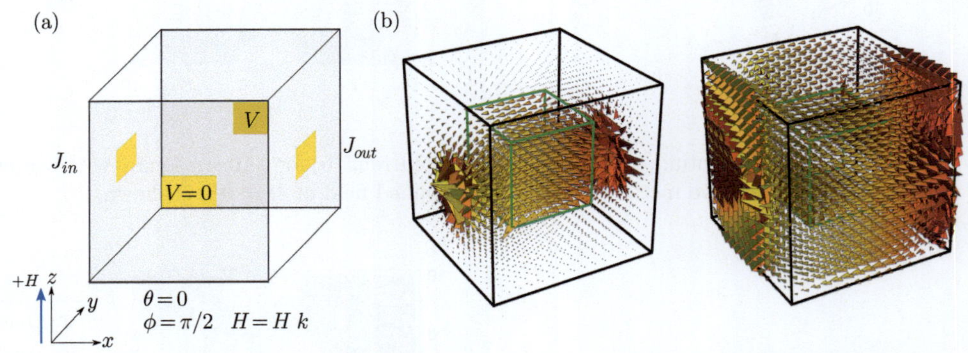

Figure 12.23 (a) A schematic for a $10\,\mu$m cube with $\vec{H}$ along z only is shown. The semiconductor cube has an embedded metal cube which is centered within it. Current plots are shown for the cube for (a) $H = 0\,$T; note that the current goes through the metal cube, and for (b) $H = 1\,$T, the current effectively avoids the metal by flowing around it.

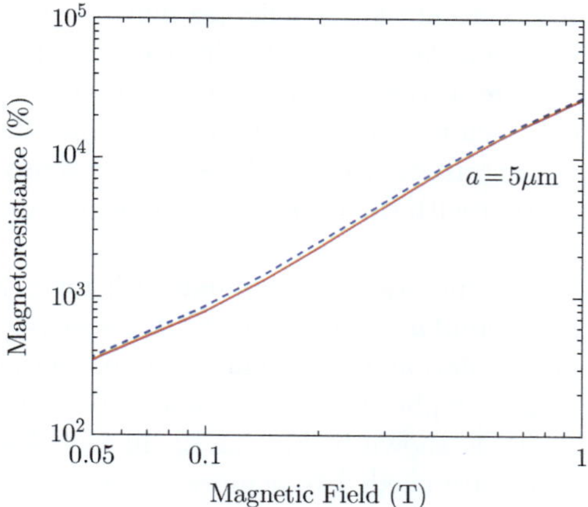

Figure 12.24 Plot of magnetoresistance versus magnetic field for the cube for an inner cube width of 5 μm.

12.9 Concluding remarks

Here we have shown that: the EMR effect can be optimized through changes in the geometry of the semiconductor/metal hybrid structure. The sensitivity of the device is based on intrinsic contributions from physical properties such as carrier mobility and energy band structure. [16] However, there is also a geometric contribution to the MR, which can play an even more important role. The geometric contribution can include the size and shape of the metallic regions and the device as a whole, the number of metallic regions, and even the orientation of the current and potential contacts. [16] We can see this especially in the geometry of the square region with centered contacts, where the highest MR is obtained for an inner metal width of 8 μm at fields of $H = \pm 1$ T and also for the 10 μm square with two metal regions and contacts centered along x for 1 T. However, we see an even greater MR in the 10 μm square with two metal regions and contacts centered along x for 1 T. If we compare these two cases, when we have the same amount of metal we can achieve a greater MR when the separation width is larger. Even for lower values of the field, we see in most cases a very large MR (10^4% to 10^5% for two regions at ± 40 mT), which means that EMR read-heads would be more sensitive than the currently used TMR read-heads. The 3D model also shows some promising results for other applications. For the cube which is analogous to the 2D case with centered contacts, we estimate an MR which is on the same order of magnitude for an inner metal width of 5 μm.

The use of the FEM produces highly accurate results, especially when employing Hermite interpolation polynomials. FEM is advantageous because the method represents the direct discretization of the action integral and permits us to directly apply derivative boundary conditions for the current. It is also a very flexible method which transcends geometrical issues, so many possibilities for the geometry-dependent EMR calculations are feasible.

The results presented here are for specific geometries. We have verified explicitly through numerical calculations that the MR is invariant under scaling of the structure as a whole as long as we are in the diffusive regime for conduction.

With 3D structures, it should be possible to determine the direction of the magnetic field as well as its magnitude using such devices as sensors. This aspect is being explored at present, together with the possibility of more complex structures for higher sensitivity.

The use of Hermite interpolation polynomials [12] for this purpose allows us to implement the derivative boundary conditions at interfaces very much more accurately than with Lagrange interpolation polynomials. All potential function and current boundary conditions can be explicitly implemented with Hermite interpolation polynomials, given their $\mathcal{C}_1$ DoFs, shown in Fig. 12.12. In fact, fewer finite elements, and therefore smaller matrices, are needed to achieve convergence as compared with earlier treatments using Lagrange interpolation for finite elements. The development of using Hermite finite elements substantially increases the accuracy in our simulations. A similar observation has been made for quantum mechanical calculations earlier where eigenvalues were obtained to double precision, and the convergence aspects of using Hermite elements have been treated in detail. [12]

We summarize here the work on EMR by other research groups. Through a series of papers, Grundler's group has investigated the influence of width-to-length ratio, mobility, interface, and contact resistance between metal and semiconductor regions, and carrier density in 2D Au-InSb structures. [17, 18] The placement of the contacts and its influence on the MR was also considered by them. [19, 20] Hewett and Kusmartsev considered a multi-branched metallic region within a circular semiconductor to investigate enhancement of EMR. [21] Bulgadaev and Kusmartsev [22] used dual transformation theory to derive MR for inhomogeneous structures that show linear variation of MR. Suh et al., [23] considered 100μm disks with embedded Au regions of 0–70μm to reconfirm experimentally the enhancement of MR. El-Ahmar and Pozniak [24] have reviewed the status as of 2010. It is evident that much remains to be done to explore the geometric MR effects.

We have considered here the more interesting case of 3D MSH structures which substantially increases the design space for EMR through the placement of voltage and current contacts, and the shape and location of the metallic inclusions. This first extension to 3D calculations in this report is limited to cubic geometry. The response of the structure depends on the direction of the magnetic field affecting its MR. This suggests the use of 3D structures in applications where the direction as well as the magnitude of the magnetic field is critical.

References

[1] S. A. Solin, Tineke Thio, D. R. Hines, and J. J. Hermans, Science **289**, 1530–1532 (2000); "Enhanced room-temperature geometric magnetoresistance in inhomogeneous narrow-gap semiconductors." Also see: S. A. Solin, Sci. Am. **291**, 70–77 (2004); "Magnetic field nanosensors."

[2] S. A. Solin, Tineke Thio, D. R. Hines, J. J. Heremans, and T. Zhou, *Proceedings of the 25th International Conference on the Physics of Semiconductors*, Osaka, Japan, edited by N. Miura (Springer, Berlin, 2001), pp. 1771–1774.

[3] S. Solin, D. R. Hines, A. C. H. Rowe, J. S. Tsai, Y. A. Pashkin, S. J. Chung, N. Goel, and M. B. Santos, Appl. Phys. Lett. **80**, 4012–4014 (2002); "Non-magnetic semiconductors as read-head sensors for ultrahigh-density magnetic recording."

[4] T. Zhou, D. R. Hines, and S. A. Solin, Appl. Phys. Lett. **78**, 667–669 (2001); "Extraordinary magnetoresistance in externally shunted van der Pauw plates."

[5] R. S. Popovic, *Hall Effect Devices* (Adam Hilger, Bristol, 1991).

[6] T. Zhou, S. A. Solin, and D. R. Hines, J. Magn. Magn. Mater. **220–230**, 1976–1977 (2001); "Extraordinary magnetoresistance of a semiconductor-metal composite van der Pauw disk."

[7] C. M. Wolfe, and G. E. Stillman, Appl. Phys. Lett. **18**, 205–208 (1971); "Anomalously high "mobility" in semiconductors;" C. M. Wolfe, G. E. Stillman, and J. A. Rossi, J. Electrochem. Soc. **119**, 250–255 (1972); "High apparent mobility in inhomogeneous semiconductors;" C. M. Wolfe and G. E. Stillman, in *Semiconductors and Semi metals*, edited by R. K. Willardson and A. C. Beer (Academic, New York, 1975), Vol. 10, p. 175.

[8] O. C. Zienkiewicz and R. L. Taylor, *The Finite Element Method*, 4th ed. (McGraw-Hill, New York, 1994).

[9] R. Courant and D. Hilbert, *Methods of Mathematical Physics* (Interscience Publishers, New York, 1953).

[10] J. M. Sullivan, in *CAD/CAM Robotics and Factories of the Future*, edited by Birendra Prasad (Springer, Berlin, 1988), Vol. I.

[11] W. H. Press, S. A. Teukolsky, W. T. Vetterling, and B. P. Flannery, *Numerical Recipies in C* (Cambridge University Press, Cambridge, UK, 1992).

[12] L. R. Ram-Mohan, *Finite Element and Boundary Element Applications to Quantum Mechanics* (Oxford University Press, Oxford, UK, 2002).

[13] J. Moussa, L. R. Ram-Mohan, J. Sullivan, T. Zhou, D. R. Hines, and S. A. Solin, Phys. Rev. B **64**, 184410–1–8 (2001); "Finite-element modeling of extraordinary magnetoresistance in thin film semiconductors with metallic inclusions."

[14] J. Moussa, L. R. Ram-Mohan, A. C. H. Rowe, and S. A. Solin, J. Appl. Phys. **94**, 1110–1114 (2003); "Response of an extraordinary magnetoresistance read head to a magnetic bit."

[15] T. J. R. Hughes, *The Finite Element Method* (Prentice-Hall, Englewood Cliffs, NJ, 1987).

[16] S. A. Solin and L. R. Ram-Mohan, Geometry-driven magnetoresistance, in *Handbook of Magnetism and Advanced Magnetic Materials*, edited by H. Kronmüller and S. Parkin, Volume **5**: Spintronics and Magnetoelectronics (John Wiley & Sons, NY, 2007); pp. 1–21.

[17] M. Hoener, O. Kronenwerth, C. Heyn, D. Grundler, and M. Holz, J. Appl. Phys. **99**, 036102–1–3 (2006); "Geometry enhanced MR of narrow Au/InAs structures incorporating a two-D electron system."

[18] M. Holz, O. Kronenwerth, and D. Grundler, Appl. Phys. Lett. **83**, 3344–3346 (2003); “Optimization of semiconductor–metal hybrid structures for application in magnetic-field sensors and read heads.”

[19] M. Holz, O. Kronenwerth, and D. Grundler, Phys. Rev. B **67**, 195312–1–10 (2003); “Magnetoresistance of semiconductor-metal hybrid structures: The effects of material parameters and contact resistance.”

[20] M. Holz, O. Kronenwerth, and D. Grundler, Appl. Phys. Lett. **86**, 072513–1–3 (2005); “Enhanced sensitivity due to current redistribution in the Hall effect of semiconductor-metal hybrid structures.”

[21] T. H. Hewett and F. V. Kusmartsev, Phys. Rev. B **82**, 212404–1–4 (2010); “Geometrically enhanced extraordinary magnetoresistance in semiconductor-metal hybrids.”

[22] S.A. Bulgadaev, and F.V. Kusmartsev, Phys. Lett. A **342**, 188–195 (2005); “Large linear magnetoresistivity in strongly inhomogeneous planar and layered systems.”

[23] J. Suh, W. Kim, J. Chang, S.H. Han, and E. K. Kim, J. Korean Phys. Soc. **55**, 577–580 (2009); “Magnetoresistance of a polycrystalline InSb disk with an embedded Au core.”

[24] S. El-Ahmar and A. A. Pozniak, J. Phys. D: Appl. Phys. **48**, 5101–1–8 (2015); “Modeling the planar configuration of extraordinary magnetoresistance”; doi:10.1088/0022-3727/48/20/205101.

13

Read-head design based on the EMR effect

In this chapter:

- We show that experimental observation of the extraordinary magnetoresistance (EMR) observed at room temperature in *nonmagnetic* semiconducting materials containing metallic inhomogeneities can be used in suitably constructed read-heads for magnetic storage devices.
- Here, we see that such read-heads are much simpler in design, and allow for higher sensitivity than is observed using magnetic layered structures that employ the phenomenon of giant magnetoresistance.
- We calculate, with no adjustable material parameters, the room-temperature response of an EMR read-head design, using finite element analysis, as a function of the position of the magnetic bits on a hard-drive relative to the read-head. The scaling property of the EMR bodes well for increasing the storage density to 1–2 Tbit/sq. in. [1] in the near future.

13.1 Introduction

Experiments on semiconductor thin films with metallic inclusions display extraordinary magnetoresistance (EMR) at room temperature. [2, 1] The remarkable observations of EMR as high as 100%–750000% at magnetic fields ranging from 0.05 to 4 T, suggest that this type of a magnetoresistance (MR) device can be used as sensitive detectors of magnetic fields. The MR is defined as $\mathrm{MR} = [R(H) - R(0)]/R(0)$, where $R(H)$ is the resistance at finite field H. The experiments were initially performed on a composite van der Pauw disk of a semiconductor matrix with an embedded metallic circular inhomogeneity that was concentric with the semiconductor disk (see Chap. 12). A similar enhancement has been reported [3] for a rectangular semiconductor wafer with an external rectangular metallic shunt on it. While the rectangular

Finite Elements in Action. L. Ramdas Ram-Mohan, Oxford University Press. © L. Ramdas Ram-Mohan (2026).
DOI: 10.1093/oso/9780199563487.003.0013

geometry with four contacts can be shown to be derivable from the circular geometry by a conformal mapping, [4] the use of a metallic shunt had not been considered before. Most recently, [5] a mesoscopic nonmagnetic magnetoresistive read-head sensor was fabricated from a narrowband-gap Si-doped InSb quantum well, that exhibited a current sensitivity of $147\,\Omega/\mathrm{T}$ at a relevant field of $0.05\,\mathrm{T}$ and a bias of 0.27 T. This sensor has a conservatively estimated areal density of 116 Gb/sq. in. [1] With this room-temperature sensitivity, the sensor is not subject to magnetic noise that limits conventional sensors to areal densities of order 100 Gb/sq. in. [1] This technological breakthrough may soon allow ultra-high-density recording at areal densities of order 1 Tb/sq. in. [1]

The phenomenon of EMR arises from the orbital rather than the spin degrees of freedom of the charged carriers and is due to the magnetic-field-induced deflection of the current from an internal (external) conducting metallic inhomogeneity embedded in (attached to) the semiconductor. The presence of the metal provides an alternate path of low resistivity at low magnetic fields that is denied the carriers at higher fields due to this field-induced deflection of the current. The electric field is normal to the metallic surface while the magnetic field deflects the current. For low effective mass of carriers and high mobility, the Hall angle between the electric field and the current reaches $90°$ at fairly low fields. It is then clear that the geometry of the metallic inclusion(s) or shunt(s) is crucial in determining the magnitude of the change in resistance with applied field.

Here we show how the modeling of the magnetoresistive response of an EMR read-head to the external magnetic field of a magnetic bit can be performed. The read-head consists of a semiconducting region with four leads attached so that a current may be passed through it while measuring the probe voltage across two of the leads. A metallic shunt is attached to the semiconducting region in order to implement the enhancement of the magnetoresistance. The magnetic bit is a small, magnetized region on the surface of a magnetic storage device. We shall consider it to be of rectangular shape with the magnetic field being a maximum at the center and falling off rapidly just at the edges of the rectangle. We then calculate the change in the probe voltage as the magnetic bit moves across the read-head. We assume that this motion is slow enough that at any position of the magnetic bit, we have quasi-equilibrium and a steady-state current through the read-head.

As shown in Chap. 12, under steady-state conditions, the problem of determining the current and the field through the physical structure reduces to the solution of Laplace's equation for the electrostatic potential. We have two issues to take into account: (i) the location in the semiconducting host, the shape, and the material properties of the inhomogeneities or shunts in the read-head, and (ii) the spatially inhomogeneous magnetic field of the magnetic bit affecting the resistance in the read-head structure that depends on the position of the bit, with the magnetic field being effective just below the magnetic bit. The magnetic field of a bit of dimension $a \times b$ is directed out of the plane of the read-head and is approximated by the (arbitrarily chosen) function

$$\mathbf{H}(x,y) = \hat{z} h_0 e^{-[(x-x_0)/(a/2)]^{12}} e^{-[(y-y_0)/(b/2)]^{12}}, \tag{13.1}$$

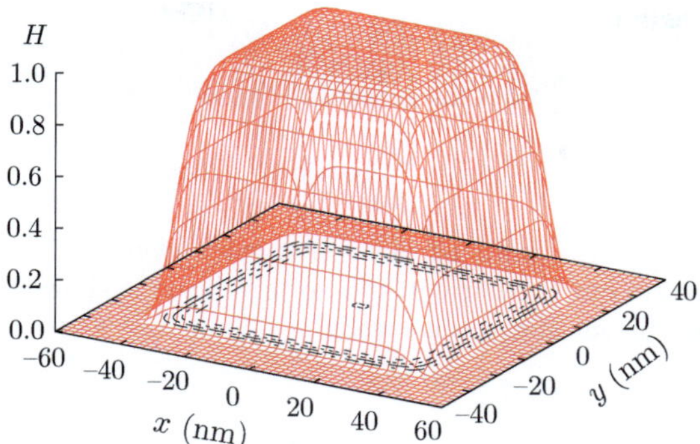

Figure 13.1 Field close to the surface of a magnetic bit has a rapid fall off away from the magnetized region. The spatial dependence of the magnetic field is approximated by Eq. (13.1). The figure shows the field of a magnetic bit of size 80 nm $\times$ 50 nm. Note the flat spatial dependence is approximated by $\simeq \exp(-x^{12})$ in order to obtain a constant field over a given region. This is different from a Gaussian-function which would fall off much faster.

which shows a rapid fall off at the edges of the magnetic bit while providing a large area over which the field is essentially constant in the interior (see Fig. 13.1). This function may be contrasted with a Gaussian in which the downturn of the field begins much closer to the center of the bit. This arbitrarily chosen function, $\exp(-[(x-x_0)/(a/2)]^{12})$, is selected only to provide an analytic expression for the spatially varying field. Note the power of x in the exponent!

In practice, the field at the read-head sensor depends on a number of complex factors such as shielding, fly height, recording conditions, demagnetization effects, etc., and various other approximations have been employed to represent it. [6] To best illustrate the efficacy of the finite element method (FEM) it is convenient and most transparent to use an analytic field expression as given in Eq. (13.1). However, we could, in principle, use any appropriate analytic expression in the FEM, although a more faithful representation of the bit field is not expected to yield qualitatively different results.

We have shown in Chap. 12 that the FEM is a natural choice for modeling the EMR in inhomogeneous semiconductor structures with metallic regions of complex geometry. [7] The complex boundary conditions along the periphery, which include the current leads, the voltage probes, and the semiconductor or metal edges, and the current continuity conditions at the semiconductor–metal interfaces, are all readily accommodated in the FEM. [7] The striking agreement between theoretical computational simulations based on the FEM on the one hand and the experimental results on a van der Pauw disk geometry on the other, provide the assurance needed in making the extension to the treatment of inhomogeneous applied magnetic fields on EMR structures of much more complex geometry. Here, we apply the FEM for the calculation of the MR in an InSb bar, as shown in Fig. 13.2, and a read-head structure (Fig. 13.3) as the magnetized bit moves across the semiconducting portion of each structure.

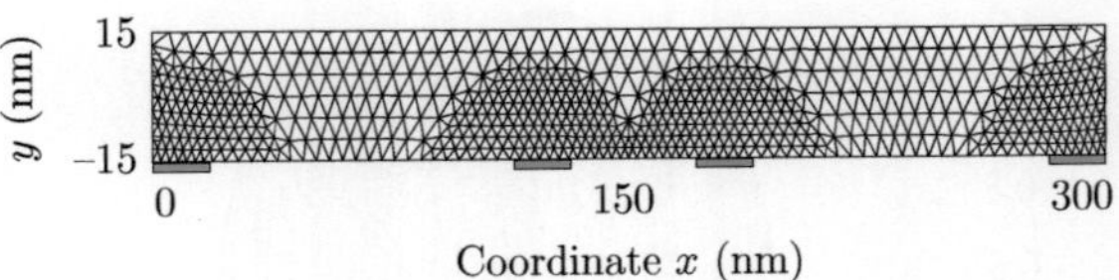

Figure 13.2 Breakup of the InSb bar into triangular elements used in the finite element analysis with the location of the current and voltage leads is shown. The mesh is adaptively improved around the current and voltage leads.

13.2 Read-head structure and FEM

The constitutive relation between the current **j** and the electric field **E** in the presence of an external magnetic field **H** along the z axis is given by

$$j_i = \sigma_{ij}(H)E_j, \tag{13.2}$$

where

$$\sigma_{ij}(H) = \frac{\sigma(0)}{1+\beta^2}\begin{bmatrix} 1 & -\beta \\ \beta & 1 \end{bmatrix}, \quad \beta = \mu H, \tag{13.3}$$

with μ being the mobility of the carriers. The conductivity is given by

$$\sigma(0) = \frac{ne^2\tau}{m^*} = ne\mu, \tag{13.4}$$

where n is the carrier density, m^* is the effective mass, e is the electronic charge, and τ is the momentum relaxation time.

In the steady state, we have $\partial_i j_i = 0$. With $E_i = -\partial_i \phi$, the electrostatic potential $\phi(x, y)$ satisfies the Laplace's differential equation

$$\partial_i[\sigma_{ij}\partial_j\phi(x, y)] = 0, \tag{13.5}$$

which we solve for the structures at hand. Note that σ_{ij} has off-diagonal elements in the presence of a magnetic field. The read-head consists of a semiconducting bar with four leads attached to it. Depending on whether we are considering the IVVI configuration or the VIVI configuration, the leads are designated as incoming or outgoing current leads and as the voltage probes. For the present theoretical development, let us suppose we are considering the IVVI configuration with the outer leads (1,4) carrying the current and the inner leads (2,3) measuring the voltage difference. The boundary conditions at the outer edges and at the leads have been discussed at length in Ref. [7]. We impose the current continuity condition at the metal–semiconductor interface. In the steady state, with our classical model, issues of electron continuity (wavefunction, etc.) do not occur. We set up the "action" integral† so that we may employ the FEM. [8] The action integral is given by

$$A = \frac{1}{2}\iint dxdy\,(\partial_i\phi(x,y))\sigma_{ij}(\partial_j\phi(x,y))$$
$$-\int_{\Delta_1} dl\,[\phi(x,y)]|_{\Delta_1} j_{in} + \int_{\Delta_4} dl\,[\phi(x,y)]|_{\Delta_4} j_{out}. \quad (13.6)$$

The double integral in Eq. (13.6) is just the electrostatic energy in the system. We have shown earlier in Chap.12 that our starting action integral with the surface terms for the currents, through the principle of least action, leads to the original differential equations with its boundary conditions including the "derivative" boundary conditions on the input and output currents. We now employ the above action in numerical modeling.

In the FEM, we begin by discretizing the action integral itself. We break up the physical region into triangles, or elements, in each of which the equation of motion (Laplace's equation) holds. This discretization is performed using an unstructured triangular mesh that is generated by the so-called algebraic integer method. [9] The result of this meshing is shown for a semiconductor bar in Fig. 13.2, and in Fig. 13.3

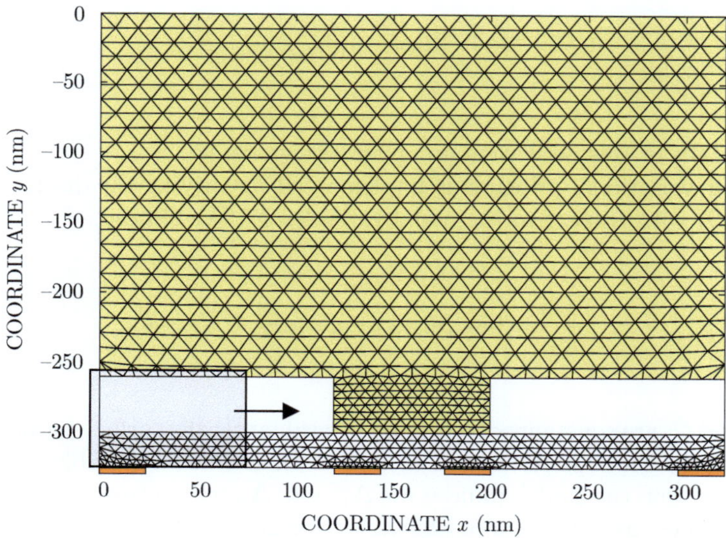

Figure 13.3 Finite element mesh for modeling the read-head structure, consisting of a rectangular bar of InSb with a metallic T-shaped shunt is shown. The four contact leads are located as shown, and the path of an 80 nm × 50 nm rectangular magnetized bit across the semiconductor-shunt region is shown.

†Strictly speaking, the action is the integral over time of the Lagrangian; however, it is convenient to label the present time-independent integral as the action, and call the variational minimization as an application of the principle of stationary action.

for a read-head. In each of the triangular elements, we represent the function $\phi(x, y)$ in terms of local interpolation polynomials $N_i(x, y)$, where N_i are unity at node i. Using this functional form in the action integral, we integrate out the spatial dependence and express the action *in each element* (i.e. Eq. (13.6) limited to the elemental area) in the form

$$A^{(i_{elem})} = \frac{1}{2}\phi_i M_{ij}^{(i_{elem})}\phi_j. \tag{13.7}$$

The input parameters for the mobility and the carrier concentrations for the semiconductor and the metallic regions are shown in Table I of Chap. 12. Next, for continuity at the nodes, we add all the contributions from each element by setting the nodal values to be the same for all triangles having a common node. This amounts to an overlay of each element matrix $M_{ij}^{(i_{elem})}$ into a global matrix such that we add the nodal contributions from all triangles having that node in common. We then obtain

$$\begin{aligned} A = \frac{1}{2}\phi_\alpha M_{\alpha\beta}\phi_\beta - \int_{\Delta_1} dl[\phi(x,y)]|_{\Delta_1} j_{in} \\ + \int_{\Delta_4} dl[\phi(x,y)]|_{\Delta_4} j_{out}. \end{aligned} \tag{13.8}$$

Further details of this derivation are given elsewhere. [7]

Now our variational principle is implemented by varying the discretized action A of Eq. (13.9) with respect to the nodal variables ϕ_α. We obtain

$$\frac{\delta A}{\delta\phi_\alpha} = 0 = M_{\alpha\beta}\phi_\beta - C_i j_{in}\delta_{i\alpha} + C_j j_{out}\delta_{j\alpha}. \tag{13.9}$$

Here, C_i and C_j are constants determined by evaluating the surface terms in Eq. (13.9). We then have a set of simultaneous equations for the nodal variables ϕ_α. The indices i and j are over the nodal indices on Δ_1 and Δ_4, respectively. Since no absolute potential values are set in the problem, we assign one of the ports V_3, say, to have zero potential with respect to which all other potentials are measured.

Due to the connectivity of the triangular mesh, the resulting coefficient matrix is sparsely occupied, and this sparsity is exploited in the computation. We first perform a bandwidth reduction of the matrix and then decompose it into the standard product of a lower triangular matrix multiplied by an upper triangular matrix (**LU**) form for Gauss elimination. [10] The solution of the simultaneous equations now provides us with a unique potential function everywhere in the physical region and the current can be calculated at every point in the structure.

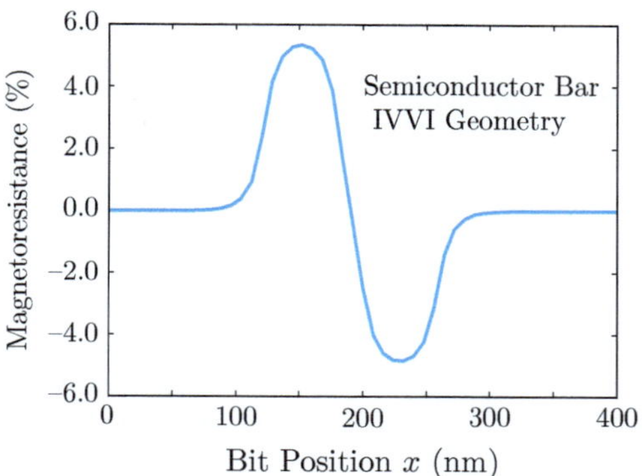

Figure 13.4 Calculated magnetoresistance $[R(H) - R(0)]/R(0)$ in an InSb bar as a magnetic bit of dimensions 80 nm $\times$ 50 nm and a magnetic field of 0.05 T moves longitudinally across the top of the bar. The semiconductor bar does not have any metallic inclusions or shunts. The external leads are 20 nm thick, with the current leads centered at ± 150 nm, and the voltage leads centered at ± 30 nm. The leads are in the IVVI configuration.

13.3 Results of calculations

The general form of the MR for the IVVI port configuration is determined from the field and geometry-dependence of the effective resistance

$$R_{23}(H, x) = [\Delta V_{23}(H, x)]/I \tag{13.10}$$

where $\Delta V_{23}(H, x) = V_2(H, x) - V_3(H, x)$, and 2 and 3 label the voltage ports, I is a constant current. The coordinate x represents the position of the magnetic bit as it moves over the read-head. An appropriate change of indices provides the expressions for the VIVI configuration.

13.3.1 MR in a semiconductor bar without a shunt

We first calculate the MR as a function of the location "x" of the magnetic bit over a semiconductor bar for dimension 320 nm $\times$ 30 nm, which does not have a metallic shunt. In the following, the dimensions of the bit are 80 nm $\times$ 50 nm, and the field is assumed to vary spatially according to Eq. (13.1). The parameter h_0 defining the field strength at the center of the bit is taken to be 0.05 T, a value that is typical close to the surface of magnetic bits in storage devices. As the center of the bit moves longitudinally along the rectangular semiconducting bar (see Fig. 13.2), the magnetoresistive response varies with the bit position as displayed in Fig. 13.4 for the IVVI configuration of the attached leads. In this configuration, the incoming current enters on the leftmost lead and exits at the rightmost lead. As the magnetic bit passes over the bar from left to right, the MR continues to be zero until the field from the bit begins to influence the region near the left voltage probe. Let us suppose that the orientation of the bit field is such that the current in the head beneath the bit is locally deflected toward the left voltage probe under the influence of the Lorentz force. Note

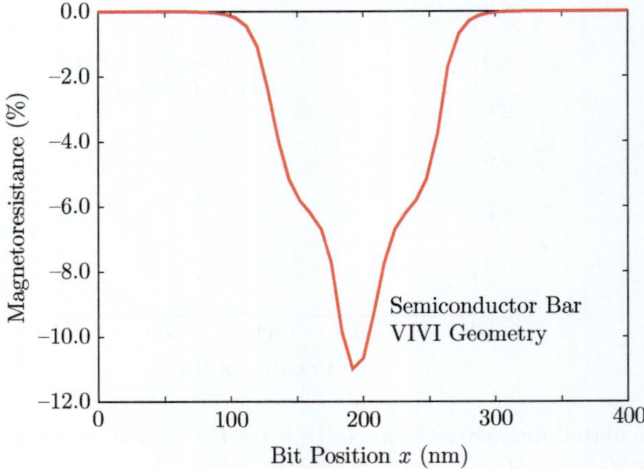

Figure 13.5 Calculated magnetoresistance in an InSb bar as a magnetic bit of dimensions 80 nm × 50 nm and a magnetic field of 0.05 T moves longitudinally across the top of the bar. The external leads are 20 nm thick, centered at ±150 and ±30 nm. The leads are in the VIVI configuration, and the semiconductor bar does not have any metallic inclusions or shunts.

that current flows into the voltage probes. The left voltage probe self-consistently raises its voltage while there is (as yet) no change in the voltage measured at the right-hand voltage probe. Thus, the voltage difference between the probes increases leading to the increase in the resistance (and the MR) as observed in Fig. 13.4. As the bit moves further to the right it begins to affect the voltage at the right-hand voltage probe for the same reasons as described above. The measured resistance thus begins to decrease, and the MR crosses through zero when the bit is centrally located along the bar. Moving the bit further to the right reduces its effect on the left-hand voltage probe but deflects more carriers away from the second voltage probe. The consequent rise in the voltage at this probe reduces the voltage difference between the probes below that measured in the absence of the bit. Thus, the MR contains a dip whose minimum corresponds to the point at which the bit is located above the right-hand voltage probe (see Fig. 13.4). Once the bit moves beyond the right-hand voltage probe, its influence is diminished and the resistance returns to its zero field value, i.e. the MR returns to zero. A similar argument explains the dip shown in Fig. 13.5 for the MR in the VIVI configuration. Note, however, that the MR is symmetric with respect to the central location of the bit in this case, whereas it is antisymmetric in the IWI configuration.

13.3.2 MR in the semiconductor read-head structure

Now consider the MR in the structure that is composed of a semiconductor bar of the same dimensions as in the above example and a T-shaped shunt attached across the middle two leads. Figures 13.6 (a) and (b) show, respectively, the calculated magnetoresistive response for the IVVI and the VIVI configurations of this device,

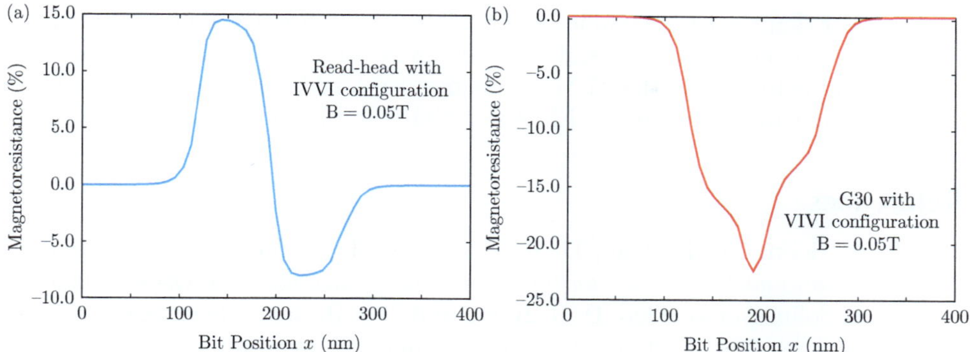

Figure 13.6 Calculated magnetoresistance in the read-head structure as a function of the position of a 80 nm × 50 nm magnetic bit with a magnetic field of 0.05 T are shown. The external leads are 20 nm thick, with the leads centered at ±150 and ±30 nm. In (a) the leads are in the IVVI configuration; in (b) the IVIV configuration of the leads is used.

which are qualitatively similar to the results obtained on the shunt-less InSb bar. However, the presence of the metallic shunt gives rise to an increase in the calculated MR of both cases by a factor of approximately 2.5. This is qualitatively consistent with recent measurements of the effect of the shunt in nanoscopic externally shunted plates of similar geometry to that considered here but in a uniform magnetic field and with a shunt that spans both the current and voltage leads. Nanoscopic devices of this type show a factor of 3.7 increase in MR from 9% to 33% in the IVVI shunted versus unshunted configuration, respectively, at 0.05T. This gives us some confidence in the predictions of the calculation, and indicates that EMR (which by definition cannot exist without the presence of a metal/semiconductor interface) is scalable down to mesoscopic device sizes. Note, also, that since the actual MR in current commercial state-of-the-art read-head sensors is of order 5%–10%, the nanoscopic devices alluded to above and the calculations addressed here are clearly technologically relevant.

13.4 Concluding remarks

An interesting feature of these calculations is that the VIVI lead configuration appears to be superior to the IVVI configuration. Not only is the response of the MR to the moving bit symmetric, but also it is significantly larger (23% compared with under 15%) in the VIVI lead configuration. This translates into a superior power signal-to-noise ratio for the VIVI lead configuration, and thus better sensor performance. [5] In terms of the existing sensor technology, the modeled EMR read-head, which has an unshielded areal density of 160 Gb/sq. in., [1] has a factor of 2–3 higher MR than typical giant MR sensors that operate at about an order of magnitude lower areal density. [11] Other advantages of EMR-type sensors for noise characteristics, fabrication costs, response time, etc., have been addressed in detail elsewhere. [5]

It is clear that the FEM is sufficiently flexible to solve nontrivial problems involving inhomogeneous magnetic fields and complex device structures. Thus, the next step in our program is to employ the FEM techniques developed here to optimize read-head design in order to achieve higher EMR sensitivity and higher bit densities.

References

[1] S. A. Solin, Tineke Thio, D. R. Hines, and J. J. Hermans, *Proceedings 25th International Conference on the Physics of Semiconductors*, Osaka (2000).

[2] S. A. Solin, Tineke Thio, D. R. Hines, and J. J. Hermans, Science (Washington, DC, U.S.) **289**, 1530–1532 (2000); "Enhanced room-temperature geometric magnetoresistance in inhomogeneous narrow-gap semiconductors."

[3] T. Zhou, D. R. Hines, and S. A. Solin, Appl. Phys. Lett. **78**, 667–669 (2001); "Extraordinary magnetoresistance in externally shunted van der Pauw plates."

[4] R. S. Popovic, *Hall Effect Devices* (Hilger, Bristol, 1991).

[5] S. A. Solin, D. R. Hines, A. C. H. Rowe, J. S. Tsai, Yu. A. Pashkin, S. J. Chung, N. Goel, and M. B. Santos, Appl. Phys. Lett. **80**, 4012–4014 (2002); "Nonmagnetic semiconductors as read-head sensors for ultra-high-density magnetic recording."

[6] R. E. Jones, Jr., C. D. Mee, and C. Tsang, in *Magnetic Recording Technology*, 2nd edition, edited by C. D. Mee, and E. D. Daniel, (McGraw-Hill, New York, 1996), pp. 6.1–6.102.

[7] J. Moussa, L. R. Ram-Mohan, J. Sullivan, T. Zhou, D. R. Hines, and S. A. Solin, Phys. Rev. B **64**, 184410–1–8 (2001); "Finite-element modeling of extraordinary magnetoresistance in thin film semiconductors with metallic inclusions."

[8] O. C. Zienkiewicz, and R. L. Taylor, *The Finite Element Method*, 4th edition (McGraw-Hill, New York, 1994); T. J. R. Hughes, *The Finite Element Method* (Prentice-Hall, Englewood Cliffs, NJ, 1987). L. R. Ram-Mohan, *Finite Element and Boundary Element Applications to Quantum Mechanics* (Oxford University Press, Oxford, U.K., 2002).

[9] J. M. Sullivan, in *CAD/CAM Robotics and Factories of the Future*, edited by B. Prasad (Springer, Berlin, 1988), Vol. I.

[10] W. H. Press, S. A. Teukolsky, W. T. Vetterling, and B. P. Flannery, *Numerical Recipes in C* (Cambridge University Press, Cambridge, U.K., 1992).

[11] IBM Techfax No. 7107 http://www.storage.ibm.com/hdd/micro/datasheet.pdf

Part V

Electrodynamics

14

Fields in electromagnetic waveguides

In this chapter:

- We solve Maxwell's electromagnetic field equations for waveguides using the finite element method with Hermite interpolation polynomials (HFEM). The electromagnetic fields, being vectors, represent significant challenges in their solution in complex geometries.
- We show that the HFEM approach yields better accuracy and smoother representations of fields than the vector finite elements that are currently in vogue for waveguide calculations. The C_1-continuity of the interpolation polynomials normal to the edges of the triangle elements and the C_2-continuity tangentially along the sides gives interpolated global fields that are not "pixelated."
- The results from Hermite elements are without any spurious solutions that used to plague Lagrange finite elements, even though the C_1-continuous Hermite polynomials are scalar in nature.
- We present solutions for propagating modes in homogeneous and dielectrically loaded inhomogeneous waveguides. We directly derive the dispersion relations for the modes in the loaded waveguides, and investigate their behavior as the ratio of dielectric constants is varied both theoretically and numerically.
- We list the limitations of the vector FEM in the context of multiscale calculations that couple electrodynamics with quantum mechanics, for example. These are not present in the method proposed here.
- The scalar Hermite interpolation polynomials are shown to provide a robust, accurate, and efficient means of solving Maxwell's equations in a variety of media, potentially offering computationally inexpensive means of designing devices for optoelectronics of increasing complexity.
- The method presented here uses FEM for a rectangular waveguide geometry. Clearly, using FEM we can then tackle a waveguide with any geometry.

Finite Elements in Action. L. Ramdas Ram-Mohan, Oxford University Press. © L. Ramdas Ram-Mohan (2026).
DOI: 10.1093/oso/9780199563487.003.0014

14.1 Introduction

Acoustic waveguides in the form of musical instruments, such as flutes, have been used since prehistoric times. The idea of guiding electromagnetic waves along conducting rods has been of interest for more than a century. [1] During the 1940s, the need to design radars for detection of aircraft and warships provided a new impetus for the analysis of waveguides. Schwinger [2, 3] was responsible for the theoretical framework for designing waveguides with complex shapes and embedded dielectrics; it was his development of variational techniques in this context that prepared him to solve issues in relativistic quantum field theory. With the advent of digital computation during the 20th century, it became possible to consider problems with waveguide geometries and characteristics that had no closed-form analytical solutions. In particular, the numerical simulation of the behavior of electromagnetic waves at microwave frequencies attracted much interest from their use in transmitters and receivers of radio waves.

Obtaining high-quality variational solutions for electromagnetic fields using triangular finite elements has been the topic of intense investigation for decades. The approach offers detailed real-space information appropriate for the analysis of fields not only in open domains but in closed regions of complicated structure, such as the dielectric, magnetic, and semiconducting layers encountered in radio-frequency integrated circuits. The impact of predictive capability with this approach in electromagnetic simulation has been profound for problems ranging from antenna design to on-chip signal propagation. Typically, the on-chip electromagnetic solvers are invoked for modeling ports and guiding structures, the device solvers are parameterized models, and the passives are lumped elements. While this design philosophy has been very successful at lower frequencies and for large devices, the eventual addition of full-wave electromagnetic solvers to fully integrated chip layout and device design modeling tools is a foregone conclusion.

This leads to the critical issue of obtaining field calculations that provide the spatial resolution adequate for multiscale problems, such as field propagation into and out of active electronic devices that are much smaller than the wavelengths of the propagating signals they manipulate. For RF circuits operating at sub-millimeter wave frequencies, the designs of transitions and interactions of signals at the transistor level are intensive engineering exercises to obtain the critical matching conditions that make useful circuits possible at these demanding frequencies. [4] Given the geometric complexities and the material loss components that are relevant with increasing frequency above 300 GHz, most design cycles require extensive back-fitting to measurement and redesign. This iterative process is very expensive in time and fabrication costs compared with design methods at lower frequencies, where the full-wave modeling can be restricted to port transitions and guiding structures of minimal loss. Therefore, a solver methodology that can obtain high spatial resolution without poor computational scaling and discretization error behaviors is a key element of integrating full-wave electromagnetics when small (10s of nanometers) active devices interact with metals and dielectrics that are far from perfect conductors or lossless.

The research literature has converged, after a period of wide exploration, on a consensus approach to triangular finite element basis polynomials. To date, the most favorable trade-off from a numerical standpoint has been to sacrifice polynomial degrees of freedom (DoFs) for the representation of either the normal or tangential fields in exchange for compliance with Nedelec conditions. The mainstay of this approach is the use of Nedelec-compliant vector finite elements (VFEM). [5–10] The most widely implemented element basis functions share a common attribute that tangential field boundary conditions are explicit at triangle boundaries and that normal field representations are one polynomial order less than that of the tangential fields. These conditions and the mixed-order field representations are well-documented and effort has been made to suppress potentially nonphysical solutions through the VFEM basis choices. [9]

The disadvantage of mixed-order polynomial approaches is the inherent imbalance in discretization error with increasing mesh density. While every discretization method introduces errors with arbitrary mesh scaling, h-convergence is achieved in a well-behaved finite element calculation, and it would be expected that increasing mesh density where solutions change rapidly should provide much better real-space functions until extremely dense conditions prevail. However, in mixed-order elements, mesh refinement toward a dense grid of equilateral triangles (in 2D) can decrease the overall quality of the field representation in polynomials because increasing portions of real space are described by lower-order polynomials since field components with projections normal to the triangle boundaries occur ubiquitously.

In the limit of a very dense mesh, the entire solution can be no better than the lowest-order description because the inter-element boundary regions dominate over the vanishing triangle interior. What is worse is that the as-written basis functions produce ambiguous vector fields at triangle vertices in the sense that approaching a point in space that is shared as a common corner node of several triangles produces fields that are unique to each triangle, creating an over-determined basis representation for the fields at vertices. (See Fig. 14.1.) This occurs because the basis decomposes the field there as projections orthogonal to adjacent edges that are not shared among all the triangles sharing the node. Therefore, an extremely dense mesh results in at best a constant value description if additional numerical techniques are not deployed to mitigate the problem of multiple definitions of the vector field at shared vertices. In addition, when such fields calculated using VFEM are employed in further applications, such as determining electron trajectories in accelerators or in high-power vacuum-tube design, these regions around vertices inject uncertainties in the charged particle trajectories.

Given the inherent side effects of vector basis element discretization, it is not surprising that most of the work on VFEM solution enhancement and refinement has focused on the construction of higher-order polynomial basis functions to increase spatial resolution rather than dense meshing (p-convergence). Hierarchical approaches are used in practice and have been published in great detail. [11, 12] In practice, the difficulty with this route to better spatial resolution is that mesh refinement, or the discretization of disparately sized physical regions, often requires the mating of finite elements with different basis orders. The enormous advantages to finding a nodal basis that scales well with meshing refinement make this an attractive topic to re-engage. The spectral pollution that arises from node-based basis functions has been analyzed elegantly from a mathematical point of view for traditional choices of polynomials. [13]

The present chapter analyzes, for comparable DoFs, the impact of choosing an alternative nodal basis formed from Hermite interpolation polynomials *versus* the use of vector finite elements. The advantages of the Hermite finite element method (HFEM) are shown for canonical waveguide problems and compared to published treatments. The nodal representation with function and derivative continuity (the core of the Hermite approach) results in unambiguous field descriptions at shared nodes among triangles and treats the field components with uniform polynomial degree (not mixed-order.) The built-in access to derivative quantities should offer ready access to quantities that are useful for sensitivity analysis. [14] This treatment results in global functions without severe coarsening which will be suitable for analyzing interactions with small features, such as those found in high-frequency transistor circuits. In an on-chip calculation, including transitions to co-planar waveguide and micro-strip waveguide structures, some thin film layers will set a mesh size that is incompatible with the overall structure size and propagating signal wavelength. At present, these treatments in circuit design software are dominated by finite-difference time-domain methods. [15]

The vector basis functions can be constructed in VFEM to systematically eliminate spurious solutions by casting them into the null space of the curl operator. [6, 9, 16] These solutions correspond to the zero-frequency (static) solutions in the EM problems. For matrix dimensions of 10^3 in typical EM calculations, nearly 20%–30% of

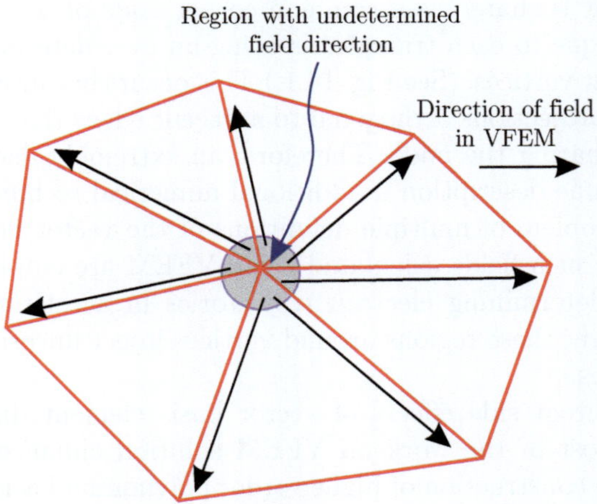

Figure 14.1 We show the vector fields at vertices in the edge-elements in VFEM. These are directed along the edges, leading to a region of ill-defined direction for the field at each vertex. Thus, more the number of vertices the larger is the region of poorly defined fields.

the solutions belong to this class and are thrown away, being unusable solutions. [9] Carrying this overhead in the calculation is computationally expensive when scaling to sophisticated structures.

In large-scale computations where the method of domain decomposition is often employed, and all solutions for each sub-domain are calculated so that the solutions for the entire domain can be constructed from those of the sub-domains. The zero-frequency modes of a sub-domain are also needed as Fourier components in order to construct the global solutions. However, it is not simple to discriminate between acceptable zero-frequency solutions and the pollution of the null space with spurious solutions.

We propose the use of the Hermite basis functions because they employ spatial derivative DoFs that directly coincide with the operators in the curl, and are completely consistent with EM theory. We show that the approach yields better accuracy, with a more physical (smoother) representation of fields, than from VFEM. This is particularly effective in 2D. This method does not generate the spurious solutions that plagued node-based Lagrange FEM encountered earlier in the 1970s, even though the $\mathcal{C}_1$-continuous Hermite polynomials are also scalar in nature. Our alternative set of polynomial basis functions, the scalar fifth-order Hermite interpolation polynomials for the numerical calculations of EM fields removes all the above difficulties. These polynomials are associated with DoFs that include both function value and spatial derivatives up to second order as given in Appendix B. Recently, Kassebaum, Boucher, and Ram-Mohan (KBR) [17] have shown how to derive these polynomials using group representation theory, giving a comparison with the earlier basis functions occurring in the literature, [18–22] which use the same derivative DoFs but are distinct from existing sets of polynomials.

In our approach, each in-plane component of the field is represented by scalar Hermite shape functions. In this work, we use H-fields, but the method works equally well with E-fields. These functions ensure tangential continuity along shared sides of triangles, thereby eliminating spurious solutions. The $\mathcal{C}_{(1)}$-continuity perpendicular to the sides of the triangular elements leads to smoother reconstruction of solutions. This representation guarantees consistency in the field direction at the vertices of triangles. These properties allow for more accurate solutions of electrodynamics problems with faster h-convergence because of the higher derivative shape functions. *We show here that the scalar HFEM yields four orders of magnitude higher accuracy with fewer elements than those needed in the presently prevalent methods for waveguides.* [23] Also, for straight waveguides, it is easy to show that $\nabla \cdot \mathbf{E} = 0$ is automatically satisfied, ensuring that there are no spurious solutions in our two-dimensional (2D) calculations.

In this chapter, we show that the use of Hermite interpolation polynomials, which are fifth-order polynomials in two dimensions, do not give rise to spurious solutions when representing field components as scalars in electromagnetic calculations. This is in contrast with the earlier use of finite elements with Lagrange interpolation polynomials. Hermite polynomials have been employed earlier in electromagnetics as will be

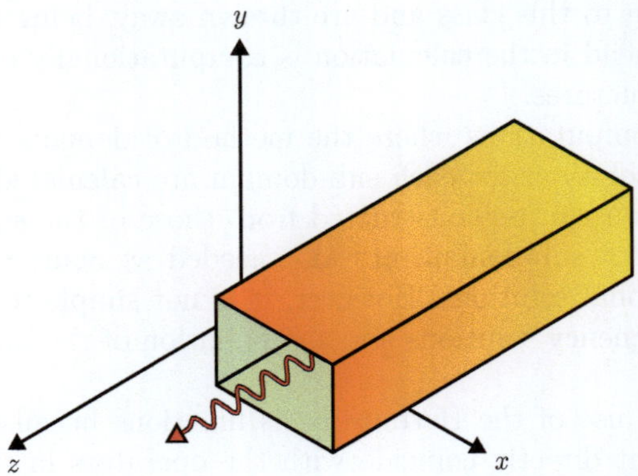

Figure 14.2 A homogeneous wave guide with a rectangular cross-section $a \times b$ is shown. The propagating wave moves along the z-axis. [24]

discussed below. The results for the eigenfrequencies in a hollow rectangular metallic waveguide are 4–5 orders of magnitude more accurate than hierarchical VFEM for the same total number of DOFs, or the global number of unknown interpolation parameters. We are thus able to benchmark the present method and quantify its superior qualities.

The full spectrum of propagating modes in a partially filled rectangular waveguide is obtained with HFEM and compared with analytic solutions. We plot the eigenfrequencies of the different modes which shows the existence of cut-offs in the propagating mode frequencies that agree perfectly with the analytic solutions. The magnetic fields for various modes are plotted to show the continuous character of the solutions in HFEM.

The capturing of eigenstates in the higher dielectric region as dielectric contrast increases is analogous to the evolution of the electronic states in an asymmetric quantum well from above-barrier states to states bound in the quantum well as the barrier energy in the well region is lowered. We plot the changes in the eigenvalues as the dielectric ratio is raised in the inhomogeneous waveguide.

14.2 HFEM formulation of the conducting waveguide

Calculations for the frequencies and vector fields for the propagating modes in a rectangular waveguide are standard results discussed in textbooks. The purpose of this section is to compare these quantities as calculated using HFEM and VFEM with exact analytic solutions. The VFEM techniques are established in the literature and the exact analytic solutions are provided in Sec. 14.3 for completeness. The HFEM has been carefully derived in the literature, [24] so we focus here on the details specific to the waveguide application because our focus is demonstrating superior numerical performance of HFEM.

We begin with Maxwell's equations expressed in MKS units,

$$\nabla \cdot \mathbf{D} = \rho, \tag{14.1}$$

$$\nabla \times \mathbf{H} - \frac{\partial \mathbf{D}}{\partial t} = \mathbf{J}, \tag{14.2}$$

$$\nabla \times \mathbf{E} + \frac{\partial \mathbf{B}}{\partial t} = 0, \tag{14.3}$$

$$\nabla \cdot \mathbf{B} = 0. \tag{14.4}$$

In the above, we express the electric displacement $\mathbf{D}$ and the magnetic induction $\mathbf{B}$ in terms of the electric and magnetic fields $\mathbf{E}$ and $\mathbf{H}$,

$$\mathbf{D} = \epsilon \mathbf{E}; \quad \mathbf{B} = \mu \mathbf{H}. \tag{14.5}$$

If the medium is isotropic, ϵ and μ are scalar quantities, rather than second-rank tensors. Let us define the dimensionless quantities ϵ_r and μ_r so that

$$\epsilon = \epsilon_r \epsilon_0, \quad \mu = \mu_r \mu_0 \tag{14.6}$$

with ϵ_0 and μ_0 being the permittivity and permeability of free space, respectively.

We assume that the dielectric regions of the waveguide are charge-free and current-free. We also assume that the fields are time-harmonic so that $\mathbf{H}(\mathbf{r}, t) = \mathbf{H}(\mathbf{r}) \exp(-i\omega t)$ and similarly for $\mathbf{E}$.

We can separate out the electric and magnetic field vectors into components that are parallel to the $\hat{z}$-axis and in the plane of the cross-section of the waveguide. We write

$$\mathbf{E} = \mathbf{E}_t + E_z \hat{\mathbf{k}}, \qquad \mathbf{H} = \mathbf{H}_t + H_z \hat{\mathbf{k}}, \tag{14.7}$$

with

$$\mathbf{E}_t = (\hat{\mathbf{k}} \times \mathbf{E}) \times \hat{\mathbf{k}}, \quad \mathbf{E}_z = \hat{\mathbf{k}} E_z, \tag{14.8}$$

and similar expressions for the components of $\mathbf{H}$. In the following, we introduce the transverse differential operator ∇_t, defined by

$$\nabla_t = \hat{\mathbf{i}} \partial_x + \hat{\mathbf{j}} \partial_y. \tag{14.9}$$

Using the plane-wave representation of fields, we can express the curl-Maxwell equations, Eqs. (14.2, 14.3), in the form

$$\hat{\mathbf{k}} \cdot (\nabla_t \times \mathbf{E}_t) = i\omega\mu \, H_z, \tag{14.10}$$

and

$$\hat{\mathbf{k}} \times [\nabla_t \times \mathbf{E}_z + \nabla_z \times \mathbf{E}_t] = i\omega\mu \, \hat{\mathbf{k}} \times \mathbf{H}_t,$$

or

$$(\nabla_t E_z) - ik_z \mathbf{E}_t = i\omega\mu \, \hat{\mathbf{k}} \times \mathbf{H}_t. \tag{14.11}$$

In a similar manner, we have

$$\hat{\mathbf{k}} \cdot (\nabla_t \times \mathbf{H}_t) = -i\omega \, \epsilon \, E_z \tag{14.12}$$

and

$$\hat{\mathbf{k}} \times [\nabla_t \times \mathbf{H}_z + \nabla_z \times \mathbf{H}_t] = i\omega\,\epsilon\,\hat{\mathbf{k}} \times \mathbf{E}_t \tag{14.13}$$

or

$$\nabla_t H_z - ik_z \mathbf{H}_t = -i\omega\,\epsilon\,\hat{\mathbf{k}} \times \mathbf{E}_t. \tag{14.14}$$

Eliminating $\mathbf{H}_t$ from Eq. (14.11) using Eq. (14.14), and substituting for $\mathbf{E}_t$ in Eq. (14.14) using Eq. (14.11) we obtain

$$\begin{aligned}(k_z^2 - \omega^2\mu\epsilon)\mathbf{E}_t &= -ik_z\nabla_t E_z + i\omega\mu\,\hat{\mathbf{k}} \times (\nabla_t H_z),\\ (k_z^2 - \omega^2\mu\epsilon)\mathbf{H}_t &= -ik_z\nabla_t H_z - i\omega\epsilon\hat{\mathbf{k}} \times (\nabla_t E_z),\end{aligned} \tag{14.15}$$

showing that the transverse fields can be solved for if the longitudinal components E_z, H_z are known.

Eliminating $\mathbf{E}_t, \mathbf{H}_t$ from Eq. (14.15) by applying a curl operator throughout and using Eqs. (14.10, 14.12) to simplify terms, we note that

$$(\nabla_t^2 - k_z^2 + \omega^2\mu\,\epsilon)\left\{\begin{array}{c} E_z \\ H_z \end{array}\right\} = 0. \tag{14.16}$$

We observe that Maxwell's equations reduce to the curl-curl form of the wave equation,

$$\nabla \times \left(\frac{1}{\epsilon_r}\nabla \times \mathbf{H}\right) - k_0^2\mu_r\mathbf{H} = 0, \tag{14.17}$$

which only includes the spatial dependence of the **H**-field. Alternatively, the curl-curl form may be derived in terms of the dual electric field by replacing $\mathbf{H} \to \mathbf{E}$ and interchanging $\epsilon_r \leftrightarrow \mu_r$ in Eq. (14.17). The wave equation for **E** is then given by

$$\nabla \times \left(\frac{1}{\mu_r}\nabla \times \mathbf{E}\right) - k_0^2\epsilon_r\mathbf{E} = 0. \tag{14.18}$$

Next, we study Eq. (14.18) and Eq. (14.17), observing that either equation may be used to set up the action integral.

In order to define the action, we begin by multiplying the differential equation, Eq. (14.17), by $\delta\mathbf{H}^*$ and integrating over the physical domain. We employ the vector identity

$$\begin{aligned}\nabla \cdot (\mathbf{P} \times \mathbf{R}) &= \left[\epsilon_{ijk}\,(\partial_i P_j)\,R_k - P_j\epsilon_{jik}\partial_i R_k\right]\\ &= (\nabla \times \mathbf{P}) \cdot \mathbf{R} - \mathbf{P} \cdot (\nabla \times \mathbf{R})\,.\end{aligned} \tag{14.19}$$

Now let $\mathbf{R} = \alpha\nabla \times \mathbf{Q}$. Then from Eq. (14.19),

$$\nabla \cdot (\mathbf{P} \times (\alpha\nabla \times \mathbf{Q})) = (\nabla \times \mathbf{P}) \cdot (\alpha\nabla \times \mathbf{Q}) - \mathbf{P} \cdot (\nabla \times (\alpha\nabla \times \mathbf{Q}))\,. \tag{14.20}$$

This leads to the integrals

$$\int_V d^3r\, \nabla \cdot [\mathbf{P} \times (\alpha\nabla \times \mathbf{Q})] = \int_V d^3r\, (\nabla \times \mathbf{P}) \cdot (\alpha\nabla \times \mathbf{Q}) - \int_V d^3r\, \mathbf{P} \cdot [\nabla \times (\alpha\nabla \times \mathbf{Q})]. \quad (14.21)$$

The left side can be reduced to a surface integral by Gauss's theorem, so we have

$$\int_V d^3r\, \mathbf{P} \cdot [\nabla \times (\alpha\nabla \times \mathbf{Q})] = \int_V d^3r\, (\nabla \times \mathbf{P}) \cdot (\alpha\nabla \times \mathbf{Q}) - \oint_S ds\, \hat{\mathbf{n}} \cdot [\mathbf{P} \times (\alpha\nabla \times \mathbf{Q})]. \quad (14.22)$$

We can now set $\mathbf{Q} = \mathbf{H}$, $\mathbf{P} = \delta\mathbf{H}^*$, and $\alpha = \epsilon_r^{-1}$ to obtain

$$\int_V d^3r\, \delta\mathbf{H}^* \cdot \left[\nabla \times \frac{1}{\epsilon_r}(\nabla \times \mathbf{H})\right] = \int_V d^3r\, (\nabla \times \delta\mathbf{H}^*) \cdot \frac{1}{\epsilon_r}(\nabla \times \mathbf{H}) - \oint_S ds\, \hat{\mathbf{n}} \cdot \left[\delta\mathbf{H}^* \times \frac{1}{\epsilon_r}(\nabla \times \mathbf{H})\right]. \quad (14.23)$$

Since

$$\mathbf{P} \cdot (\mathbf{Q} \times \mathbf{R}) = \mathbf{Q} \cdot (\mathbf{R} \times math\mathbf{P}) = \mathbf{R} \cdot (\mathbf{P} \times \mathbf{Q}), \quad (14.24)$$

the integrand of the surface term may be rewritten as

$$\hat{\mathbf{n}} \cdot \left[\delta\mathbf{H}^* \times \frac{1}{\epsilon_r}(\nabla \times \mathbf{H})\right] = -\delta\mathbf{H}^* \cdot \left[\hat{\mathbf{n}} \times \frac{1}{\epsilon_r}(\nabla \times \mathbf{H})\right]. \quad (14.25)$$

For time-harmonic fields, the relations between $\mathbf{E}$ and $\mathbf{H}$ is given by

$$\mathbf{H} = -\frac{i}{\mu\omega}\nabla \times \mathbf{E}, \quad (14.26a)$$

$$\mathbf{E} = \frac{i}{\epsilon\,\omega}\nabla \times \mathbf{H}. \quad (14.26b)$$

Therefore, $\mathbf{E}$ is proportional to $\epsilon_r^{-1}(\nabla \times \mathbf{H})$. Assuming the waveguide to be enclosed by a perfectly conducting material, one of the boundary conditions is that $\hat{\mathbf{n}} \times \mathbf{E} = 0$. Thus, the surface term in Eq. (14.23) is exactly zero.

Note that it is possible to work instead with the electric field formulation of Eq. (14.18). In this case, the surface term takes the form

$$\oint_S ds\, \hat{\mathbf{n}} \cdot \left[\delta\mathbf{E}^* \times \frac{1}{\mu_r}(\nabla \times \mathbf{E})\right]. \quad (14.27)$$

We see that for the perfectly conducting boundary, the surface term arising from an integration by parts vanishes as proven in Eq. (14.24).

We thus see that the integrated form of Eq. (14.17) is expressible as

$$\delta \int_V d^3r \, \frac{1}{2} \left[(\nabla \times \mathbf{H}^*) \cdot \frac{1}{\epsilon_r} (\nabla \times \mathbf{H}) - k_0^2 \mu_r \mathbf{H}^* \cdot \mathbf{H} \right] = 0. \tag{14.28}$$

Equation (14.28) is now interpreted as the functional variation of the action integral,† such that we may write it as

$$\delta \mathcal{A}/T = 0. \tag{14.29}$$

The principle of stationary action then identifies the action to be

$$\mathcal{A}/T = \frac{1}{2} \int_V d^3r \left[(\nabla \times \mathbf{H}^*) \cdot \frac{1}{\epsilon_r} (\nabla \times \mathbf{H}) - k_0^2 \mu_r \mathbf{H}^* \cdot \mathbf{H} \right]. \tag{14.30}$$

Similarly, using the electric field formulation of Maxwell's equations yields

$$\mathcal{A}/T = \frac{1}{2} \int_V d^3r \left[(\nabla \times \mathbf{E}^*) \cdot \frac{1}{\mu_r} (\nabla \times \mathbf{E}) - k_0^2 \epsilon_r \mathbf{E}^* \cdot \mathbf{E} \right]. \tag{14.31}$$

To formulate the Lagrangian density for our calculation, we can also start from the usual definition of the Lagrangian density for electromagnetic fields,

$$\begin{aligned} \mathcal{L} &= \frac{1}{2} (\mathbf{E} \cdot \mathbf{D}^* - \mathbf{B} \cdot \mathbf{H}^*) \\ &= \frac{1}{2} (\epsilon \mathbf{E} \cdot \mathbf{E}^* - \mu \mathbf{H} \cdot \mathbf{H}^*). \end{aligned} \tag{14.32}$$

From Eq. (14.26a) and Eq. (14.26b), we can easily get back Eq. (14.30) and Eq. (14.31) since

$$\mathcal{A}/T = \int d^3r \, \mathcal{L}. \tag{14.33}$$

For example, on substitution of $\mathbf{E}$ from Eq. (14.26b), we have

$$\begin{aligned} \mathcal{A}/T &= \frac{1}{2} \int d^3r \left[\frac{1}{\omega^2} (\nabla \times \mathbf{H}) \frac{1}{\epsilon} (\nabla \times \mathbf{H}^*) - \mu \mathbf{H} \cdot \mathbf{H}^* \right] \\ &= \frac{1}{2} \frac{c^2 \mu_0}{\omega^2} \int d^3r \left[(\nabla \times \mathbf{H}) \frac{1}{\epsilon_r} (\nabla \times \mathbf{H}^*) - \mu_r k_0^2 \mathbf{H} \cdot \mathbf{H}^* \right] \end{aligned}$$

where we used $c = 1/\sqrt{\mu_0 \epsilon_0}$ and $c = \omega/k_0$.

The extra factor in front of the action integral is irrelevant in eigenvalue problems, and we are left with similar forms of the action integral in both derivations. Unlike previous HFEM results which focused on specialized treatments of the action applicable to wave propagation through periodic systems (photonic crystals), the waveguide problem requires the discretization of the transverse fields of propagating

†Usually the action is the time integral of the Lagrangian. Here the Lagrangian is effectively independent of time since a harmonic solution in time is assumed. Hence $\mathcal{A}/T$ is appropriate.

states traveling along the waveguide axis. As with any other waveguide formulation, the action integral can be cast explicitly in terms of the transverse field components ($\mathbf{H}_t = H_x\hat{x} + H_y\hat{y}$). After manipulation, the transverse field form of the action is [25]

$$\begin{aligned}\mathcal{A}/T = \frac{1}{2}\int_{\Omega} d\Omega \bigg[&(\nabla_t \times \mathbf{H}_t^*) \cdot \frac{1}{\epsilon_r}(\nabla_t \times \mathbf{H}_t) \\ &- \mathbf{H}_t^* \cdot \left(k_0^2\mu_r - \frac{k_z^2}{\epsilon_r}\right)\mathbf{H}_t \\ &+ \frac{1}{\mu_r}\nabla_t \cdot (\mu_r \mathbf{H}_t)\, \nabla_t \cdot \left(\frac{1}{\epsilon_r}\mathbf{H}_t^*\right)\bigg] \\ - \oint_{\Gamma} d\Gamma &\left(\frac{1}{\epsilon_r\mu_r}\nabla_t \cdot (\mu_r \mathbf{H}_t)\, H_n^*\right), \end{aligned} \tag{14.34}$$

where Ω is the domain of the transverse cross-section, Γ indicates the domain boundary and H_n is the field projection normal to the boundary. Equation (14.34) is the integral that we discretize when performing the FEM calculations. The advantage of this expression for the action integral is that the problem can be formulated entirely in terms of the transverse **H**-field components as described by Jin. [25]

After discretizing the action integral, the FEM was employed on the 2D waveguide cross-section using a set of Hermite interpolation polynomials. The Hermite elements exhibit C_1 continuity throughout the finite element mesh. The components of the EM fields in this work are represented using basis functions within each triangular element. For a given Cartesian component of **H**, say $f(x, y)$, the interpolated value in terms of the 2D Hermite basis functions ϕ_i is given by

$$f(x, y) = \sum_{i=1}^{18} f_i \phi_i(x, y), \tag{14.35}$$

where the f_i are DoFs assigned to the function value and its derivatives at the vertices of the triangle. This allows the enforcement of either derivative continuity or EM boundary conditions depending upon the material composition of adjacent triangles. We have provided the basis for an equilateral triangle because that is the optimization goal of most mesh refinement and it is also the rationale for the KBR group theoretical development of the basis. The basis functions ϕ_i for a reference triangle and the numbering sequence for the assignment of function and derivative values at the nodes have been published elsewhere [17, 24] and reproduced in Table B.6 in Appendix B.

14.2.1 BCs for the rectangular waveguide revisited

Before discussing the solution to the differential equation governing each component, it is prudent to discuss the boundary conditions governing a rectangular conducting waveguide with sides parallel to the x- and y-axes. We know in this case that the electric field can only have a normal component and the magnetic field can only have a tangential component at the boundary, so that

$$\hat{\mathbf{n}} \cdot \mathbf{H} = 0, \tag{14.36}$$

$$\hat{\mathbf{n}} \times \mathbf{E} = 0. \tag{14.37}$$

Expanding Eq. (14.36), we have

$$n_x H_x + n_y H_y = 0. \tag{14.38}$$

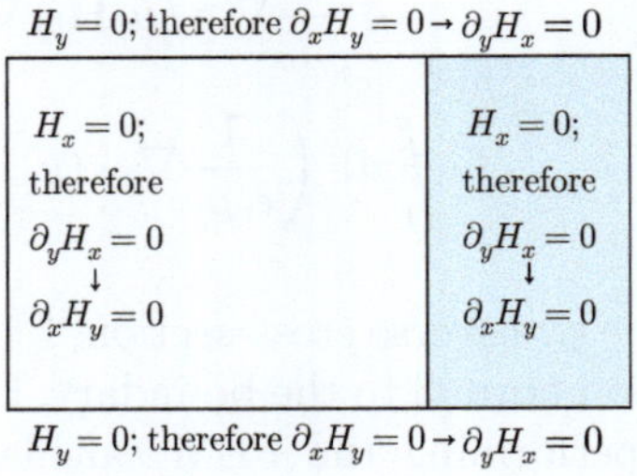

$H_y = 0$; therefore $\partial_x H_y = 0 \rightarrow \partial_y H_x = 0$

$H_x = 0$; therefore $\partial_y H_x = 0 \downarrow \partial_x H_y = 0$

$H_x = 0$; therefore $\partial_y H_x = 0 \downarrow \partial_x H_y = 0$

$H_y = 0$; therefore $\partial_x H_y = 0 \rightarrow \partial_y H_x = 0$

Figure 14.3 The boundary conditions for the electromagnetic fields in a homogeneous waveguide are shown.

For the left and right boundaries, which are parallel to the y-axis, Eq. (14.38) simplifies to

$$H_x = 0, \quad \text{for } x \in \{0, d\}. \tag{14.39}$$

For the top and bottom boundaries, we get

$$H_y = 0, \quad \text{for } y \in \{0, h\}. \tag{14.40}$$

In addition, Eq. (14.37) may be used to generate derivative boundary conditions at the edges. Recall that the differential form of the Maxwell-Ampère equation reads

$$\nabla \times \mathbf{H} - \frac{\partial \mathbf{D}}{\partial t} = \mathbf{J}. \tag{14.2}$$

Assuming that dielectric properties do not vary over time and there is no current density at the boundary, we have

$$\nabla \times \mathbf{H} = \epsilon \frac{\partial \mathbf{E}}{\partial t}. \tag{14.41}$$

Assuming that the electric and magnetic fields have time-harmonic forms, Eq. (14.41) simplifies to

$$\nabla \times \mathbf{H} = -i\omega\epsilon \mathbf{E}. \tag{14.42}$$

Solving Eq. (14.42) for $\mathbf{E}$ and substituting into Eq. (14.37) yields

$$\hat{n} \times \left(-\frac{1}{i\omega\epsilon} \nabla \times \mathbf{H} \right) = 0. \tag{14.43}$$

Expand the cross product to obtain

$$\begin{pmatrix} n_y \left(\partial_x H_y - \partial_y H_x\right) \\ -n_x \left(\partial_x H_y - \partial_y H_x\right) \\ n_x \left(\partial_z H_x - \partial_x H_z\right) - n_y \left(\partial_y H_z - \partial_z H_y\right) \end{pmatrix} = \begin{pmatrix} 0 \\ 0 \\ 0 \end{pmatrix}. \tag{14.44}$$

So the first and second vector components of the cross product each yield

$$\partial_x H_y = \partial_y H_x. \tag{14.45}$$

On the left and right boundaries, we have already shown that $H_x = 0$. Since these edges are parallel to the y-axis and $H_x = 0$ along these edges, it follows that $\partial_y H_x = 0$. From Eq. (14.45), we then obtain a derivative condition on H_y:

$$\partial_x H_y = 0, \quad \text{for } x \in \{0, d\}. \tag{14.46}$$

Similarly, on the top and bottom edges, the derivative condition is

$$\partial_y H_x = 0, \quad \text{for } y \in \{0, h\}. \tag{14.47}$$

Together, Eqs. (14.39), (14.40), (14.46), and (14.47) constitute all the boundary conditions for a conducting waveguide of width d and height h. The results are summarized in Fig. 14.3 for the homogeneous waveguide.

14.2.2 The $\nabla \cdot \mathbf{E} = 0$ condition in 2D waveguides

We show in the following that in 2D the divergence condition $\nabla \mathbf{E} = 0$ is automatically satisfied in calculations for the fields in the cross-section of the waveguide. For a waveguide with perfectly conducting surfaces, boundary conditions are given by

$$E_{/\!/} = 0; \quad H_n = 0. \tag{14.48}$$

We are interested in monochromatic waves that propagate down the waveguide, so we take the form for $\mathbf{E}$ and $\mathbf{B}$ as

$$\mathbf{E}(x, y, z, t) = \mathbf{E}(x, y)e^{i(kx-wt)}, \tag{14.49}$$

$$\mathbf{H}(x, y, z, t) = \mathbf{H}(x, y)e^{i(kx-wt)}. \tag{14.50}$$

For these waves to satisfy the Maxwell equations, the conditions are

$$\frac{\partial E_y}{\partial x} - \frac{\partial E_x}{\partial y} = i\omega\mu_r\mu_0 H_z; \quad \frac{\partial \mu_r\mu_0 H_y}{\partial x} - \frac{\partial \mu_r\mu_0 H_x}{\partial y} = -\frac{i\omega}{c^2} E_z, \tag{14.51}$$

$$\frac{\partial E_x}{\partial y} - ikE_y = i\omega\mu_r\mu_0 H_x; \quad \frac{\partial \mu_r\mu_0 H_z}{\partial y} - ik\mu_r\mu_0 H_y = -\frac{i\omega}{c^2} E_x, \tag{14.52}$$

$$ikE_x - \frac{\partial E_z}{\partial x} = i\omega\mu_r\mu_0 H_y; \quad ik\mu_r\mu_0 H_x - \frac{\partial \mu_r\mu_0 H_z}{\partial x} = -\frac{i\omega}{c^2} E_y. \tag{14.53}$$

Using the above relations, we can express

$$E_x = \frac{i}{(\omega/c)^2 - k^2}\Big(k\frac{\partial E_z}{\partial x} + \omega\frac{\partial \mu_r \mu_0 H_z}{\partial y}\Big), \tag{14.54}$$

$$E_y = \frac{i}{(\omega/c)^2 - k^2}\Big(k\frac{\partial E_z}{\partial y} - \omega\frac{\partial \mu_r \mu_0 H_z}{\partial x}\Big), \tag{14.55}$$

$$\mu_r\mu_0 H_x = \frac{i}{(\omega/c)^2 - k^2}\Big(k\frac{\partial \mu_r \mu_0 H_z}{\partial x} - \frac{\omega}{c^2}\frac{\partial E_z}{\partial y}\Big), \tag{14.56}$$

$$\mu_r\mu_0 H_y = \frac{i}{(\omega/c)^2 - k^2}\Big(k\frac{\partial \mu_r \mu_0 H_z}{\partial y} + \frac{\omega}{c^2}\frac{\partial E_z}{\partial x}\Big). \tag{14.57}$$

Hence, if we solve for E_z and H_z, we determine all other components. This simplification does not exist in full three-dimensional (3D) calculations. Numerically, we have to solve for wave equations of the form

$$\left[\frac{\partial^2}{\partial x^2} + \frac{\partial^2}{\partial y^2} + (\omega/c)^2 - k^2\right]\left\{\begin{array}{c} E_z \\ \mu_r\mu_0 H_z \end{array}\right\} = 0. \tag{14.58}$$

For TE modes, $E_z = 0$, and for TM modes $H_z = 0$. Therefore, it suffices to solve for either E_Z or H_z. If we take divergence of the field in Eq. (14.49), we obtain

$$\nabla \cdot \mathbf{E}(x, y, z, t) = \frac{\partial E_x}{\partial x} + \frac{\partial E_y}{\partial y} + ikE_z. \tag{14.59}$$

Substituting Eq. (14.54) and (14.55) for E_x and E_y, we obtain (for simplicity let us consider TM waves with $H_z = 0$)

$$\nabla \cdot \mathbf{E}(x, y, z, t) = \frac{ik}{(\omega/c)^2 - k^2}\Big(\frac{\partial^2 E_z}{\partial x^2} + \frac{\partial^2 E_y}{\partial y^2}\Big) + ikE_z. \tag{14.60}$$

From Eq. (14.59), we obtain

$$\nabla \cdot \mathbf{E}(x, z, y, t) = \frac{ik}{(\omega/c)^2 - k^2}\left[k^2 - (\omega/c)^2\right]E_z + ikE_z = 0. \tag{14.61}$$

Hence, the zero divergence condition is built-in for the generic form of the solution in Eq. (14.49).

14.2.3 Matrix representation of fields and their derivatives

From Maxwell's equations, we have the divergence condition,

$$\nabla \cdot (\mu_r \mathbf{H}) = 0, \tag{14.62}$$

which makes it possible to remove one of the three vector components from the action integral by expressing it in terms of the remaining two. Since the fields are known to have a sinusoidal variation in the z-direction, it is natural to eliminate the z-component of the field from the action. First, we define the transverse magnetic field,

$\mathbf{H}_t$, as $H_x\hat{\mathbf{i}} + H_y\hat{\mathbf{j}} + 0\hat{\mathbf{k}}$. We also introduce the transverse gradient operator, ∇_t, as $\partial_x\hat{\mathbf{i}} + \partial_y\hat{\mathbf{j}} + 0\hat{\mathbf{k}}$. In the following, we express the curl of the magnetic field in a matrix form,

$$\nabla \times \mathbf{H} = (\nabla \times \mathbf{H})_x \begin{pmatrix} 1 \\ 0 \\ 0 \end{pmatrix} + (\nabla \times \mathbf{H})_y \begin{pmatrix} 0 \\ 1 \\ 0 \end{pmatrix} + (\nabla \times \mathbf{H})_z \begin{pmatrix} 0 \\ 0 \\ 1 \end{pmatrix}. \tag{14.63}$$

In this format, we express the curl as

$$\nabla \times \mathbf{H} = \begin{pmatrix} \partial_y H_z - \partial_z H_y \\ \partial_z H_x - \partial_x H_z \\ \partial_x H_y - \partial_y H_x \end{pmatrix}. \tag{14.64}$$

Similarly, the matrix form of the transverse curl operator is

$$\nabla_t \times \mathbf{H}_t = \begin{pmatrix} 0 \\ 0 \\ \partial_x H_y - \partial_y H_x \end{pmatrix}. \tag{14.65}$$

It is also helpful to write out the divergence condition in terms of the transverse gradient operator,

$$\nabla_t \cdot \mu_r \mathbf{H}_t + \partial_z \mu_r H_z = 0, \tag{14.66}$$

which can be rewritten for time-harmonic waves propagating in the z-direction as

$$\nabla_t \cdot \mu_r \mathbf{H}_t = -ik_z \mu_r H_z. \tag{14.67}$$

We seek a means of expressing $\nabla \times \mathbf{H}$ in terms of $\nabla_t \times \mathbf{H}_t$, eliminating the axial component of the magnetic field from the action integral prior to discretization. Such a method is described in detail by Jin. [25, pp. 257–259] We note for the following that

$$k_0^2 \mu_r \mathbf{H}^* \cdot \mathbf{H} = k_0^2 \mu_r \mathbf{H}_t^* \cdot \mathbf{H}_t + k_0^2 \mu_r |H_z|^2. \tag{14.68}$$

Jin's formulation of the action

The following analysis serves to verify Jin's claim [25, p. 257] that the integral in Eq. (14.30) may be rewritten in the form

$$\mathcal{A} = \frac{1}{2} \int_V d^3 r \left[(\nabla_t \times \mathbf{H}_t^*) \cdot \frac{1}{\epsilon_r} (\nabla_t \times \mathbf{H}_t) - \mathbf{H}_t^* \cdot k_0^2 \mu_r \mathbf{H}_t - k_0^2 \mu_r |H_z|^2 \right.$$
$$\left. + (\nabla_t H_z - \partial_z \mathbf{H}_t)^* \cdot \frac{1}{\epsilon_r} (\nabla_t H_z - \partial_z \mathbf{H}_t) \right]. \tag{14.69}$$

First, we prove that the sum of the first and fourth terms of Eq. (14.69) is equivalent to the first term of Eq. (14.30). To prove this, rewrite Eq. (14.64) as

$$\nabla \times \mathbf{H} = \begin{pmatrix} P \\ Q \\ R \end{pmatrix}, \tag{14.70}$$

where

$$\begin{aligned} P &= \partial_y H_z - \partial_z H_y, \\ Q &= \partial_z H_x - \partial_x H_z, \\ R &= \partial_x H_y - \partial_y H_x. \end{aligned} \tag{14.71}$$

For isotropic materials, we then have

$$(\nabla \times \mathbf{H}^*) \cdot \frac{1}{\epsilon_r} (\nabla \times \mathbf{H}) = \frac{1}{\epsilon_r} \left(|P|^2 + |Q|^2 + |R|^2 \right). \tag{14.72}$$

We rewrite some of the terms of Eq. (14.69) as

$$\nabla_t \times \mathbf{H}_t = \begin{pmatrix} 0 \\ 0 \\ R \end{pmatrix} \tag{14.73}$$

and

$$\nabla_t H_z - \partial_z \mathbf{H}_t = \begin{pmatrix} \partial_x H_z - \partial_z H_x \\ \partial_y H_z - \partial_z H_y \\ 0 \end{pmatrix} = \begin{pmatrix} -Q \\ P \\ 0 \end{pmatrix}. \tag{14.74}$$

Hence, from Eq. (14.73) and Eq. (14.74), we have

$$\begin{aligned} (\nabla_t \times \mathbf{H}_t^*) &\cdot \frac{1}{\epsilon_r} (\nabla_t \times \mathbf{H}_t) + (\nabla_t H_z - \partial_z \mathbf{H}_t)^* \cdot \frac{1}{\epsilon_r} (\nabla_t H_z - \partial_z \mathbf{H}_t) \\ &= \frac{1}{\epsilon_r} (R^* R + (-Q)^* (-Q) + P^* P) \\ &= \frac{1}{\epsilon_r} \left(|P|^2 + |Q|^2 + |R|^2 \right) \\ &= (\nabla \times \mathbf{H}^*) \cdot \frac{1}{\epsilon_r} (\nabla \times \mathbf{H}). \end{aligned} \tag{14.75}$$

Thus, Eq. (14.30) is equivalent to Eq. (14.69). The equality of the terms containing μ_r has already been verified via Eq. (14.68).

Elimination of H_z

Continuing from Eq. (14.69), we now seek to eliminate terms involving H_z from the action. Jin asserts that

$$\frac{1}{\epsilon_r} \nabla_t H_z \cdot \nabla_t H_z^* = \nabla_t \cdot \left(H_z^* \frac{1}{\epsilon_r} \nabla_t H_z \right) - H_z^* \nabla_t \cdot \left(\frac{1}{\epsilon_r} \nabla_t H_z \right). \tag{14.76}$$

We can verify Eq. (14.76) immediately by applying the product rule to the first term on the right-hand side. Next, we verify Jin's result that

$$\nabla_t \cdot \left(\frac{1}{\epsilon_r} \nabla_t H_z \right) = \partial_z \nabla_t \cdot \left(\frac{1}{\epsilon_r} \mathbf{H}_t \right) - k_0^2 \mu_r H_z. \tag{14.77}$$

Beginning with the double-curl formulation of Maxwell's equations from Eq. (14.17), we take the z-component of each side,

$$\left[\nabla \times \left(\frac{1}{\epsilon_r}\nabla \times \mathbf{H}\right)\right]_z = k_0^2 \mu_r H_z. \tag{14.78}$$

We write Eq. (14.78) out term-by-term by multiplying Eq. (14.64) by ϵ_r^{-1} and then taking the curl of the product, yielding

$$\begin{aligned}\left[\nabla \times \left(\frac{1}{\epsilon_r}\nabla \times \mathbf{H}\right)\right]_z &= \partial_x \left[\frac{1}{\epsilon_r}\left(\partial_z H_x - \partial_x H_z\right)\right] - \partial_y \left[\frac{1}{\epsilon_r}\left(\partial_y H_z - \partial_z H_y\right)\right] \\ &= \partial_z \left[\partial_x \left(\frac{1}{\epsilon_r} H_x\right) + \partial_y \left(\frac{1}{\epsilon_r} H_y\right)\right] - \partial_x \left(\frac{1}{\epsilon_r}\partial_x H_z\right) - \partial_y \left(\frac{1}{\epsilon_r}\partial_y H_z\right) \\ &= \partial_z \nabla_t \cdot \left(\frac{1}{\epsilon_r}\mathbf{H}_t\right) - \nabla_t \cdot \left(\frac{1}{\epsilon_r}\nabla_t H_z\right).\end{aligned} \tag{14.79}$$

Substituting Eq. (14.78) into Eq. (14.79), we find that Eq. (14.77) is indeed true.

14.2.4 Results for a homogeneous waveguide

Our first computational comparison is made for a homogeneous rectangular ($d \times h$) waveguide simulated with $\epsilon_r = 1, \mu_r = 1$, and $k_z = 1$. The cross-section dimensions $d = 20$ and $h = 10$ were used. The units of k_z and k_0 are reciprocal of the units of d and h, and the angular frequency ω can be obtained by multiplying k_o by the speed of light, c. The scale of d and h can therefore be changed suitably. There is no inherent separation of TE and TM modes in the HFEM formulation, and therefore the eigenproblem returns all physical solutions. The resulting H_z field components were calculated in post-processing where those with finite magnitudes are TE modes and those with H_z approaching the numerical noise floor are TM modes. For the degenerate TE and TM modes of the homogeneous waveguide, the eigenfunctions for the field patterns were separated in post-processing since arbitrary linear combinations of degenerate modes result from numerical diagonalization.

To generate pure TE modes during the post-processing stage, we enforce the condition that

$$\frac{\partial H_y}{\partial x} - \frac{\partial H_x}{\partial y} = 0. \tag{14.80}$$

Since the electric field is derived from the curl of the magnetic field, this expression forces the z-component of the electric field to equal zero, creating a pure TE mode.

To create a pure TM mode, we recall that H_z is obtained from the in-plane components using the divergence condition. We can force H_z to equal zero by forcing

$$\frac{\partial H_x}{\partial x} + \frac{\partial H_y}{\partial y} = 0. \tag{14.81}$$

Either Eq. (14.80) or Eq. (14.81) may be enforced after calculating the eigenfunctions by multiplying one of the in-plane components, either H_x or H_y, by a scale factor

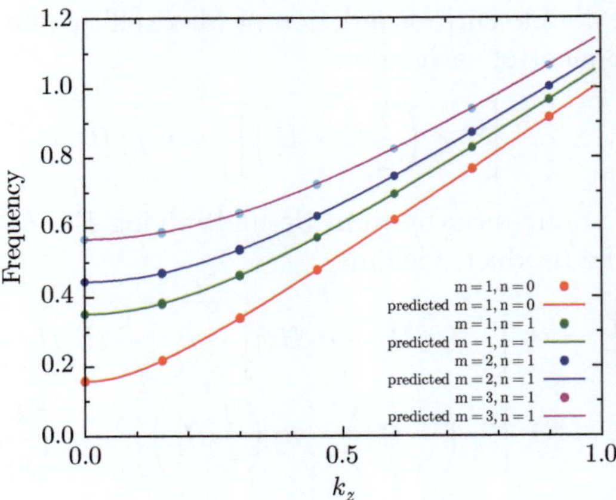

Figure 14.4 Frequency eigenvalues (ω/c) of several TE_{mn} modes of a hollow ($\epsilon_r = 1$) rectangular waveguide with 2:1 aspect ratio are shown as functions of k_z. The analytical values (curves) are compared with those calculated using HFEM (dots), and the two are essentially identical (see Table 14.1).

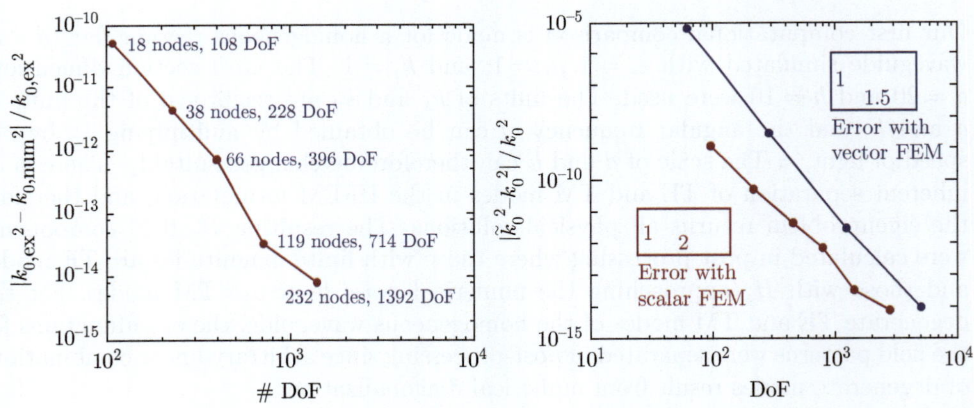

Figure 14.5 The convergence of the eigenvalue for the lowest frequency, TE_{10}-mode is displayed as the global number of degrees of freedom (DoF) is increased through mesh refinement for a homogeneous waveguide. The solid curve (red) is obtained with HFEM and compared with published results using VFEM. [23]

until one of the equations is satisfied. The analytical forms of the fields are given in Sec. 14.3.

The eigenvalues corresponding to a few propagating modes are plotted against the value of k_z in Fig. 14.4. Relative errors in the eigenvalue calculations for the homogeneous waveguide varied from 10^{-14} to 10^{-10} with approximately 2700 DoFs in the global matrix, as shown in Table 14.1. It is clear that the eigenvalues of the

Table 14.1 Numerically calculated eigenvalues versus their predicted values for the 10 lowest-energy modes of the homogeneous rectangular waveguide. The number of Hermite finite elements used was 406 with a global matrix size of 2784.

Mode	Theoretical	HFEM	Error
TE_{10}	1.024 674 011 002 72	1.024 674 011 002 69	3.0×10^{-14}
TE_{20}	1.098 696 044 010 89	1.098 696 043 949 03	6.2×10^{-11}
TE_{01}	1.098 696 044 010 89	1.098 696 044 076 11	6.5×10^{-11}
$TE_{11} + TM_{11}$	1.123 370 055 013 61	1.123 370 054 915 78	9.8×10^{-11}
$TE_{11} + TM_{11}$	1.123 370 055 013 61	1.123 370 055 117 61	1.0×10^{-10}
$TE_{21} + TM_{21}$	1.197 392 088 021 78	1.197 392 087 905 68	1.1×10^{-10}
$TE_{21} + TM_{21}$	1.197 392 088 021 78	1.197 392 088 225 77	2.0×10^{-10}
TE_{30}	1.222 066 099 024 51	1.222 066 099 118 76	9.4×10^{-11}
$TE_{31} + TM_{31}$	1.320 762 143 035 40	1.320 762 143 075 83	4.0×10^{-11}
$TE_{31} + TM_{31}$	1.320 762 143 035 40	1.320 762 143 933 56	9.0×10^{-10}

homogeneous waveguide, particularly the lowest eigenvalue, show close agreement with analytical values.

We can make a direct comparison with VFEM for this case because an analysis of the convergence of eigenvalues for hierarchical VFEM for waveguides has been published. [23] The lowest eigenvalue in the VFEM analysis has an error of 10^{-12} for approximately the same number of DoFs as compared with our lowest eigenvalue (see Table 14.1).

A few comments are in order:

1. From Table 14.1, it is clear that the eigenvalues of the homogeneous waveguide, particularly the lowest eigenvalue, show close agreement with analytical values. We note that Lee et al. [23] have made an analysis of the convergence of eigenvalues for hierarchical vector finite elements for waveguides. The lowest eigenvalue reported by them has an error of 10^{-12} for approximately the same number of DoFs as compared with our lowest eigenvalue which has double-precision accuracy as seen in Table 14.1.

   The lowest frequency propagating mode is the TE_{10} mode. The TE_{mn} and TM_{mn} modes with $m, n \neq 0$ are degenerate.
2. It should be noted that the Arnoldi/Lanczos algorithm employed in the ARPACK diagonalization software [26] leads to outputs for eigenvectors with arbitrary phase factors assigned to them. For non-degenerate eigenvalues, we resolve these phases for the eigenvectors by plotting the magnitude of each field component multiplied by the sign of the real part of the component.

3. The diagonalizer delivers a linear combination of degenerate TM and TE modes. In order to resolve the eigenfunctions into distinct TM and TE modes, it is necessary to rescale one of the in-plane magnetic field components before using the in-plane components to calculate H_z and the electric field components. The need to rescale one of these components is a consequence of the fact that the in-plane field components are used to construct the global matrix, without explicitly setting H_z or E_z to zero for TM and TE modes, respectively. While for every TM mode, there exists a TE mode with the same frequency, the resulting degeneracy cannot be resolved by a simple perturbation of the waveguide cross-sectional dimensions or global matrix elements.
4. Accidental degeneracies, which involve different modes having the same frequency, may occur, especially if one dimension of the waveguide cross-section is commensurate with the other. These accidental degeneracies may be removed by introducing a small perturbation in the global matrix, and do not require any additional post-processing.
5. We note that the transverse nature of the modes is demanded in vector finite element analysis for every element. Here we impose the transversality condition at the end of the calculation for the eigenstates.

We now determine the convergence properties of the HFEM solutions. In Fig. 14.5(a), we show the convergence of the eigenfrequency to its analytically determined value in a homogeneous waveguide. For the lowest frequency TE_{10}-mode, as the global number of DoFs is increased through mesh refinement, the accuracy improves steadily until 2400 DOF when the curve in red (lower curve) obtained with HFEM reaches down to 10^{-14}. The HFEM delivers an accuracy of 10^{-9} with just 96 DoFs (8 nodes). These results are extraordinary in terms of how quickly the frequency of the lowest mode is determined accurately. In the same figure, we have overlaid the VFEM data [23] in the blue curve (upper curve). At the low end of mesh refinement, with $\sim$100 DoFs, the hierarchical VFEM employing a comparable order quintic polynomial basis has an error of $\simeq 10^{-5}$ for a hollow rectangular waveguide. With further mesh refinement leading to $\sim$5000 DoFs, the VFEM has an error comparable to our HFEM formulation with $\sim$2400 DoFs.

We note that the error for the VFEM result in the article by Lee et al. [23] could be overestimated by 50% given the use of a structured mesh that was systematically less refined in the wider dimension of the waveguide. This effect would be greatest for the lowest DOF case (the least refined mesh). Given the many orders of magnitude spanned in this error analysis, we consider a factor of two overestimation of the error for the coarse mesh to be a small effect that does not change the order of magnitude comparison of the figure.

We can also consider the error in the eigenvalues of the higher states. The eigenvalues of solutions above the ground state have errors approximately three to four orders of magnitude higher but converge at the same rate as the ground state. This indicates that the derivative continuity of the fields is allowing for a high-quality variational solution even at modest discretization levels, which is the essence of FEM.

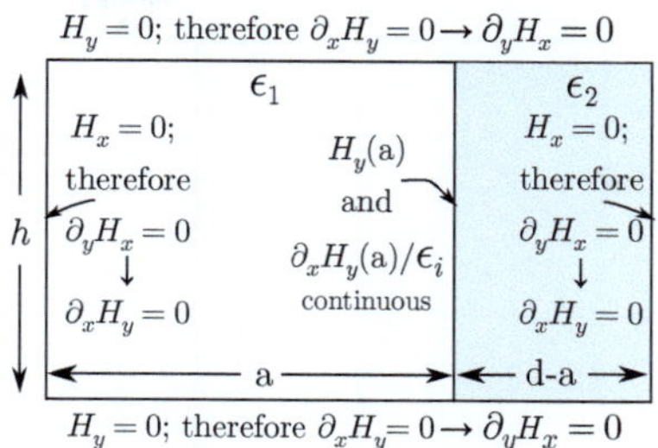

Figure 14.6 The boundary conditions and dimensions for a partially filled perfectly electrically conducting (PEC) waveguide are shown. These boundary conditions were implemented in the finite element simulations.

14.2.5 HFEM results for the inhomogeneous waveguide

We now consider a partially filled waveguide with geometry and boundary conditions as shown in Fig. 14.6. This is an especially attractive test case for HFEM because it addresses a canonical problem for which VFEM was created, namely, the presence of spurious solutions in other scalar FEM formulations. In addition, this particular waveguide configuration shows the efficacy of using HFEM to resolve spatially varying fields of both sinusoidal and exponential (sinh and cosh) dependencies. Fields of both types have analytic solutions as shown below, and therefore we can compare with analytic solutions. These field concepts are technologically important to capture with the HFEM technique because they appear in the design of slow-wave structures, on-chip waveguides, and dielectrically-loaded leaky-wave antennas. It was our expectation that HFEM would perform extremely well in resolving both types of eigenmodes because these issues are readily encountered when using scalar HFEM in the solution of quantum mechanical wavefunctions. The combination of a sine and a hyperbolic sine field solution is analogous to an asymmetric quantum well. When $(k_z^2 + (n\pi/h)^2)/\epsilon_1 < k_o^2$, this is equivalent to the potential energy of an electron in such a well being higher than the electron energy leading to an exponentially falling solution in the "barrier region" which is analogous to the dielectric with permittivity ϵ_1. When $k_o^2 > (k_z^2 + (n\pi/h)^2)/\epsilon_2$, sinusoidal solutions are obtained, corresponding to the confined state solutions in a quantum well. Similar analogies have been drawn previously. Not surprisingly, the symmetric potential well problem in 1D quantum mechanics has been compared with the electromagnetic confinement in a dielectric slab waveguide surrounded by air, [27] as this amounts to the same problem set.

We plot fields of both sinusoidal and exponential spatial variation in Fig. 14.7(a). The predicted longitudinal section electric (LSE, or TE to $\hat{x}$) eigenvalues are shown as functions of ϵ_2/ϵ_1 in Fig. 14.7(b). The quantum well analogy suggests that as ϵ_2 is increased, more modes are captured by the higher dielectric region leading to the sinusoidal behavior in the larger dielectric region and an exponential decay into the lower dielectric region, as shown schematically in Fig. 14.7(a).

The predicted eigenvalues which correspond to longitudinal section magnetic (LSM, or TM to $\hat{x}$) modes are also shown as functions of ϵ_2/ϵ_1. Note that every

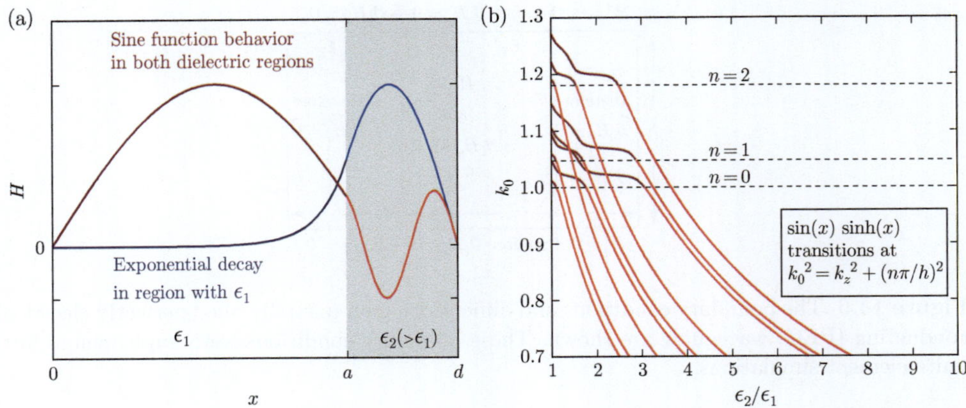

Figure 14.7 Variation of the mode frequency as the dielectric ratio is increased is shown in an inhomogeneous dielectrically loaded waveguide. (a) The waveguide sustains sine/sine and sine/sinh type solutions for H_x. (The solutions for H_y are presented in Fig. 14.8.) (b) The predicted eigenvalues for H_x are shown as functions of ϵ_2/ϵ_1. Note that every solution undergoes a transition from the sine-like regime to the hyperbolic regime, at a certain threshold value of the permittivity as the dielectric ratio is varied over 1–10.

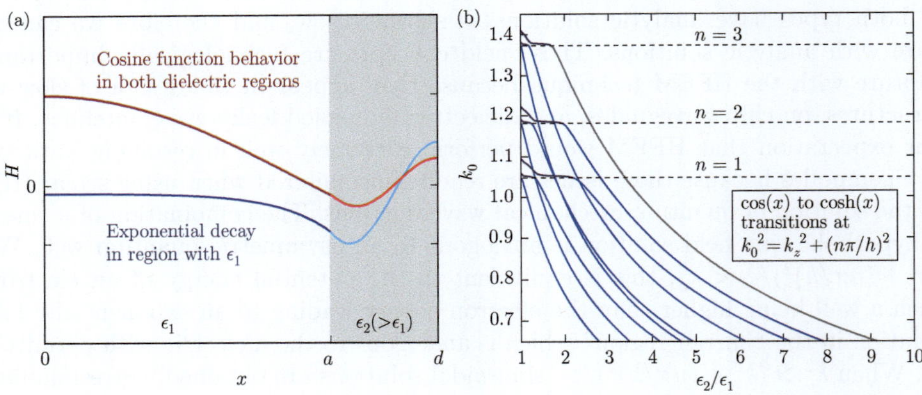

Figure 14.8 Variation of the mode frequency as the dielectric ratio is increased is shown in an inhomogeneous waveguide. (a) The waveguide sustains cosine/cosine and cosine/cosh type solutions for H_y. (b) The predicted eigenvalues for H_y are shown as functions of ϵ_2/ϵ_1. Note that every solution undergoes a transition from the cosine-like regime to the hyperbolic regime, at a certain threshold value of the permittivity as the dielectric ratio is varied over 1–10.

solution undergoes a transition from the cosine-like regime to the hyperbolic regime, as shown by the color change in the plot of each eigenvalue, at a certain threshold value of the permittivity. These threshold values depend on the value of k_z and the frequency of the eigenfunctions in the y-direction. This is shown in Fig. 14.8.

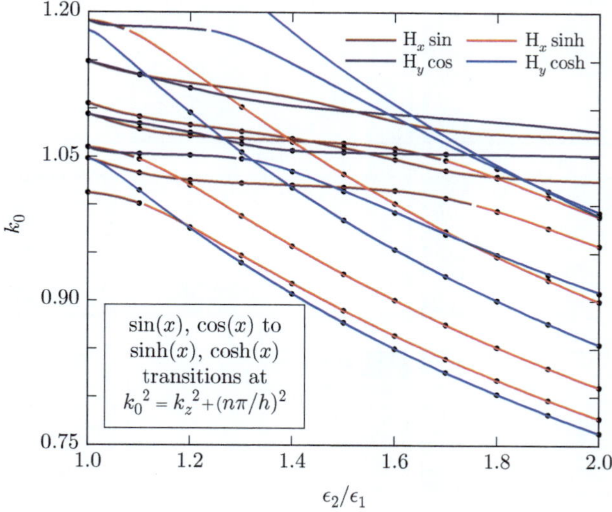

Figure 14.9 Eigenvalues of several propagating modes obtained from analytic dispersion relations, are plotted along with their HFEM computed values (points) as functions of the dielectric ratio ϵ_2/ϵ_1, which is here limited from 1.0 to 2.0. the sine (cosine)-like solutions evolve into their hyperbolic form in the lower dielectric region as the dielectric ratio increases.

The analytically determined eigenvalues were compared to the results found using the HFEM. Eigenvalues of several propagating modes obtained from analytic dispersion relations, are plotted along with their computed values (points) as functions of the dielectric ratio ϵ_2/ϵ_1 in Fig. 14.9. The agreement between the theoretically determined eigenvalues and those calculated using HFEM is excellent. The first 10 eigenvectors in the inhomogeneous waveguide, with a dielectric constant $\epsilon_2 = 1.5$, are shown in Figs. 14.9. The eigenmodes for a dielectric constant $\epsilon_2 = 2.0$ lead to more of the modes being confined in the higher dielectric region as compared with the case where $\epsilon_2 = 1.5$.

The HFEM correctly solves for the modes supported by partially dielectric-loaded waveguides (both slow- and fast-wave regimes; both LSE and LSM modes). Of particular importance is that the new formulation can solve for these various dielectric-loaded modes without spurious solutions.

14.3 Analytic solutions for homogeneous and inhomogeneous waveguides

In this section, we compare FEM calculations of waveguide eigenmodes to analytic solutions for both homogeneous and partially filled rectangular waveguides. The waveguide solutions appear in many standard texts with the exception of the hyperbolic forms present at frequencies below the standard cut-off in inhomogeneous case. These solutions provide a rigorous test of our numerical method, and given their

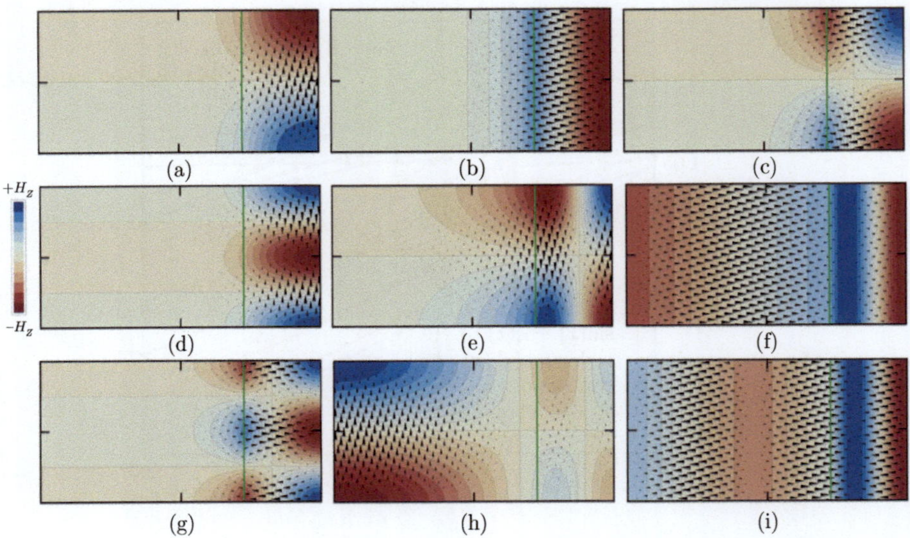

Figure 14.10 Field patterns for the nine lowest frequency LSE modes an inhomogeneous 2:1 aspect ratio PEC waveguide are plotted in (a)–(i) in increasing order. The dielectric ratio was $\epsilon_2/\epsilon_1 = 1.5$ (region 1 is left of the vertical line, region 2 to the right) and $k_z = 1.0$. The arrows represent the in-plane H-fields, while the shading shows the H_z-field pattern.

sparse mention in the literature and the absence of an elaboration of their analytic forms, we have provided them below in order to reproduce the results of this work without ambiguity. Our forms follow the classic text of Harrington, [28] as he treats both cases.

In the case of the homogeneous waveguide with ϵ_r and μ_r such that $\epsilon_1 = \epsilon_2 = 1$ and $\mu_1 = \mu_2 = 1$, the propagating transverse magnetic (TM to $\hat{z}$) and transverse electric (TE to $\hat{z}$) modes have frequencies given by

$$\frac{\omega^2}{c^2} = k_o^2 = \left(\frac{m\pi}{d}\right)^2 + \left(\frac{n\pi}{h}\right)^2 + k_z^2, \tag{14.82}$$

and the TM fields are given by

$$\begin{aligned} H_x(x,y) &= C_{mn}\frac{n\pi}{h}\sin\left(\frac{m\pi}{d}x\right)\cos\left(\frac{n\pi}{h}y\right)e^{ik_z z}, \\ H_y(x,y) &= -C_{mn}\frac{m\pi}{d}\cos\left(\frac{m\pi}{d}x\right)\sin\left(\frac{n\pi}{h}y\right)e^{ik_z z}, \\ H_z(x,y) &= 0, \end{aligned} \tag{14.83}$$

where C_{mn} is an arbitrary complex constant setting amplitude and phase. Similarly, the propagating TE modes have fields given by

$$
\begin{aligned}
H_x &= \overline{C}_{mn}\left(\frac{m\pi}{d}\right) k_z \sin\left(\frac{m\pi}{d}x\right)\cos\left(\frac{n\pi}{h}y\right)e^{ik_z z},\\
H_y &= \overline{C}_{mn}\left(\frac{n\pi}{h}\right) k_z \cos\left(\frac{m\pi}{d}x\right)\sin\left(\frac{n\pi}{h}y\right)e^{ik_z z},\\
H_z &= i\overline{C}_{mn}\left(\left(\frac{m\pi}{d}\right)^2+\left(\frac{n\pi}{h}\right)^2\right)\\
&\qquad\times\cos\left(\frac{m\pi}{d}x\right)\cos\left(\frac{n\pi}{h}y\right)e^{ik_z z},
\end{aligned}
\tag{14.84}
$$

where $\overline{C}_{mn}$ is also an arbitrary constant. Both types of modes have hard cut-off frequencies for a given m, n as $k_z \to 0$.

In the case of the partially filled waveguide with $\epsilon_1 \neq \epsilon_2$ and $\mu_1 \neq \mu_2$, the propagating modes can be found by separation of variables with a decomposition along the boundary of the material interface. The longitudinal section magnetic (LSM, or TM to $\hat{x}$) and longitudinal section electric (LSE, or TE to $\hat{x}$) modes have frequencies given by

$$
\begin{aligned}
\frac{\omega^2}{c^2} = k_o^2 &= \frac{1}{\epsilon_1\mu_1}\left(k_{1,x}^2+\left(\frac{n\pi}{h}\right)^2+k_z^2\right)\\
&= \frac{1}{\epsilon_2\mu_2}\left(k_{2,x}^2+\left(\frac{n\pi}{h}\right)^2+k_z^2\right).
\end{aligned}
\tag{14.85}
$$

The LSM eigenmodes exist when $k_{1,x}$ and $k_{2,x}$ satisfy the transcendental equation obtained by applying electromagnetic boundary conditions and given by

$$
\frac{k_{1,x}}{\epsilon_1}\tan\left(k_{1,x}a\right) = -\frac{k_{2,x}}{\epsilon_2}\tan\left(k_{2,x}\left(d-a\right)\right).
\tag{14.86}
$$

The corresponding LSM fields in the regions 1 and 2 as indicated by subscripts are given by

$$
\begin{aligned}
H_x &= 0,\\
H_{y,1} &= C_n k_z \cos\left(k_{x,1}x\right)\sin\left(\frac{n\pi}{h}y\right)e^{ik_z z},\\
H_{y,2} &= \overline{C}_n k_z \cos\left(k_{x,2}(d-x)\right)\sin\left(\frac{n\pi}{h}y\right)e^{ik_z z},\\
H_{z,1} &= iC_n\left(\frac{n\pi}{h}\right)\cos\left(k_{x,1}x\right)\cos\left(\frac{n\pi}{h}y\right)e^{ik_z z},\\
H_{z,2} &= i\overline{C}_n\left(\frac{n\pi}{h}\right)\cos\left(k_{x,2}(d-x)\right)\cos\left(\frac{n\pi}{h}y\right)e^{ik_z z},
\end{aligned}
\tag{14.87}
$$

where C_n is an arbitrary constant and

$$\overline{C}_n = C_n \cos(k_{x,1}a) / \cos(k_{x,2}(d-a)). \tag{14.88}$$

The LSE eigenmodes exist when $k_{1,x}$ and $k_{2,x}$ satisfy the transcendental equation obtained by again applying electromagnetic boundary conditions and given by

$$\frac{k_{1,x}}{\mu_1} \cot(k_{1,x}a) = -\frac{k_{2,x}}{\mu_2} \cot(k_{2,x}(d-a)). \tag{14.89}$$

The corresponding LSE fields in the regions 1 and 2 as indicated by subscripts are given by

$$\begin{aligned}
H_{x,1} &= C_n \left(\left(\frac{n\pi}{h}\right)^2 + k_z^2 \right) \sin(k_{x,1}x) \cos\left(\frac{n\pi}{h}y\right) e^{ik_z z}, \\
H_{x,2} &= \overline{C}_n \left(\left(\frac{n\pi}{h}\right)^2 + k_z^2 \right) \sin(k_{x,2}(d-x)) \cos\left(\frac{n\pi}{h}y\right) e^{ik_z z}, \\
H_{y,1} &= -C_n \left(\frac{n\pi}{h}\right) k_{x,1} \cos(k_{x,1}x) \sin\left(\frac{n\pi}{h}y\right) e^{ik_z z}, \\
H_{y,2} &= \overline{C}_n \left(\frac{n\pi}{h}\right) k_{x,2} \cos(k_{x,2}(d-x)) \sin\left(\frac{n\pi}{h}y\right) e^{ik_z z}, \\
H_{z,1} &= iC_n k_z k_{x,1} \cos(k_{x,1}x) \cos\left(\frac{n\pi}{h}y\right) e^{ik_z z}, \\
H_{z,2} &= -i\overline{C}_n k_z k_{x,2} \cos(k_{x,2}(d-x)) \cos\left(\frac{n\pi}{h}y\right) e^{ik_z z},
\end{aligned} \tag{14.90}$$

where C_n is an arbitrary constant and

$$\overline{C}_n = C_n \sin(k_{x,1}a) / \sin(k_{x,2}(d-a)).$$

When the derivation of partially filled waveguide modes is introduced in Harrington, [28] a reference is made to a seminal paper by Pincherle. [29] That paper points out the existence of propagating modes that vary exponentially as well as sinusoidally in space. This sinusoidal portion will occur in the region with smaller dielectric constant. Without loss of generality, we choose $\epsilon_2 > \epsilon_1$ and re-inspect the LSM and LSE modes under the condition that $k_{1,x} \to i\kappa_{1,x}$ which leads to an exponentially decaying mode in region 1. Note that the arguments in the trigonometric functions in Eqs. (14.86) and (14.89) are positive quantities and that allowing both $k_{1,x}$ and $k_{2,x}$ to be imaginary cannot satisfy the relation, [30] so our cases capture all possibilities.

First, from Eq. (14.85) we can see that for $n > 0$, there is no hard cut-off preventing the exponentially decaying behavior in region 1 for a propagating eigenmode. As the waveguide is a closed domain, the exponential solution will take the form of hyperbolic cosine and sine functions. Under these conditions, the LSM modes are present when

$$\frac{\kappa_{1,x}}{\epsilon_1}\tanh(\kappa_{1,x}a) = \frac{k_{2,x}}{\epsilon_2}\tan(k_{2,x}(d-a)) \tag{14.91}$$

is satisfied. The corresponding fields for the LSM modes in regions 1 and 2 are given by

$$\begin{aligned}
H_x &= 0,\\
H_{y,1} &= C_n k_z \cosh(k_{x,1}x)\sin\left(\frac{n\pi}{h}y\right)e^{ik_z z},\\
H_{y,2} &= \overline{C}_n k_z \cos(k_{x,2}(d-x))\sin\left(\frac{n\pi}{h}y\right)e^{ik_z z},\\
H_{z,1} &= iC_n\left(\frac{n\pi}{h}\right)\cosh(k_{x,1}x)\cos\left(\frac{n\pi}{h}y\right)e^{ik_z z},\\
H_{z,2} &= i\overline{C}_n\left(\frac{n\pi}{h}\right)\cos(k_{x,2}(d-x))\cos\left(\frac{n\pi}{h}y\right)e^{ik_z z},
\end{aligned} \tag{14.92}$$

where C_n is an arbitrary constant and

$$\overline{C}_n = C_n\cosh(\kappa_{x,1}a)\,/\cos(k_{x,2}(d-a)).$$

Similarly, the LSE modes are present when

$$\frac{\kappa_{1,x}}{\mu_1}\coth(\kappa_{1,x}a) = -\frac{k_{2,x}}{\mu_2}\cot(k_{2,x}(d-a)) \tag{14.93}$$

is satisfied. The corresponding fields for these LSE modes in regions 1 and 2 are given by

$$\begin{aligned}
H_{x,1} &= C_n\left(\left(\frac{n\pi}{h}\right)^2 + k_z^2\right)\sinh(\kappa_{x,1}x)\cos\left(\frac{n\pi}{h}y\right)e^{ik_z z},\\
H_{x,2} &= \overline{C}_n\left(\left(\frac{n\pi}{h}\right)^2 + k_z^2\right)\sin(k_{x,2}(d-x))\cos\left(\frac{n\pi}{h}y\right)e^{ik_z z},\\
H_{y,1} &= -C_n\left(\frac{n\pi}{h}\right)\kappa_{x,1}\cosh(\kappa_{x,1}x)\sin\left(\frac{n\pi}{h}y\right)e^{ik_z z},\\
H_{y,2} &= \overline{C}_n\left(\frac{n\pi}{h}\right)k_{x,2}\cos(k_{x,2}(d-x))\sin\left(\frac{n\pi}{h}y\right)e^{ik_z z},\\
H_{z,1} &= iC_n k_z\kappa_{x,1}\cosh(\kappa_{x,1}x)\cos\left(\frac{n\pi}{h}y\right)e^{ik_z z},\\
H_{z,2} &= -i\overline{C}_n k_z k_{x,2}\cos(k_{x,2}(d-x))\cos\left(\frac{n\pi}{h}y\right)e^{ik_z z},
\end{aligned} \tag{14.94}$$

where C_n is an arbitrary constant and

$$\overline{C}_n = C_n\sinh(\kappa_{x,1}a)\,/\sin(k_{x,2}(d-a)).$$

14.4 Alternate derivation of the exact solutions for the inhomogeneous waveguide

We calculate the exact eigenvalues and eigenfunctions of an inhomogeneous waveguide with cross-section dimensions $d \times h$, having relative permittivity $\epsilon_r = \epsilon_1$ in the region $0 \leq x < a$ and $\epsilon_r = \epsilon_2$ in the region $a < x \leq d$. In the subsequent calculations, the dielectric in the left region was assumed to be air, so that $\epsilon_1 = 1$ and $\epsilon_2 \geq \epsilon_1$.

Recall that Maxwell's equations may be expressed as

$$\nabla \times \left(\frac{1}{\epsilon_r} \nabla \times \mathbf{H} \right) - k_0^2 \mu_r \mathbf{H} = 0. \tag{14.95}$$

By assuming that the change from ϵ_1 to ϵ_2 occurs over an infinitesimally thin region centered at $x = a$, one may express the inverse of the dielectric function as a step function,

$$\frac{1}{\epsilon_r} = \left(\frac{1}{\epsilon_2} - \frac{1}{\epsilon_1} \right) \theta\left(x - a\right) + \frac{1}{\epsilon_1}. \tag{14.96}$$

In order to determine the effects of varying values of ϵ_r in the dielectric region, we assume that $\mu_r = 1$ everywhere.

Note that we can rewrite the double curl as

$$\nabla \times \left(\frac{1}{\epsilon_r} \nabla \times \mathbf{H} \right) = \nabla \left(\frac{1}{\epsilon_r} \right) \times \left(\nabla \times \mathbf{H} \right) + \frac{1}{\epsilon_r} \nabla \times \left(\nabla \times \mathbf{H} \right). \tag{14.97}$$

The second term in Eq. (14.97) can be simplified by noting that

$$\nabla \times \left(\nabla \times \mathbf{H} \right) = \nabla \left(\nabla \cdot \mathbf{H} \right) - \nabla^2 \mathbf{H}, \tag{14.98}$$

and also by noting that the divergence of $\mathbf{H}$ is zero for uniform permeability. By including the type of permittivity expressed in Eq. (14.96), the gradient in Eq. (14.97) can be rewritten as

$$\nabla \left(\frac{1}{\epsilon_r} \right) = \begin{pmatrix} \delta\left(x - a\right) \Lambda \\ 0 \\ 0 \end{pmatrix}, \quad \text{where } \Lambda = \frac{1}{\epsilon_2} - \frac{1}{\epsilon_1}, \tag{14.99}$$

so that Eq. (14.17) can be rewritten as

$$\begin{pmatrix} \delta\left(x - a\right) \Lambda \\ 0 \\ 0 \end{pmatrix} \times \begin{pmatrix} \partial_y H_z - \partial_z H_y \\ \partial_z H_x - \partial_x H_z \\ \partial_x H_y - \partial_y H_x \end{pmatrix} - \frac{1}{\epsilon_r} \nabla^2 \mathbf{H} - k_0^2 \mathbf{H} = 0. \tag{14.100}$$

The cross product in Eq. (14.2) can be expressed as

$$\begin{pmatrix} \delta\left(x - a\right) \Lambda \\ 0 \\ 0 \end{pmatrix} \times \left(\nabla \times \mathbf{H} \right) = \begin{pmatrix} 0 \\ -\delta\left(x - a\right) \Lambda \left(\partial_x H_y - \partial_y H_x \right) \\ \delta\left(x - a\right) \Lambda \left(\partial_z H_x - \partial_x H_z \right) \end{pmatrix}. \tag{14.101}$$

Finally, we have three equations for the three vector components:

$$\frac{1}{\epsilon_r}\nabla^2 H_x + k_0^2 H_x = 0, \tag{14.102a}$$

$$\delta(x-a)\,\Lambda\left(\partial_x H_y - \partial_y H_x\right) + \frac{1}{\epsilon_r}\nabla^2 H_y + k_0^2 H_y = 0, \tag{14.102b}$$

$$\delta(x-a)\,\Lambda\left(\partial_z H_x - \partial_x H_z\right) - \frac{1}{\epsilon_r}\nabla^2 H_z - k_0^2 H_z = 0. \tag{14.102c}$$

We begin by considering these vector components in order to derive a dispersion relation for the eigenvalues and eigenfunctions of the inhomogeneous waveguide. The solutions to Eqs. (14.102a) and (14.102b) will be considered. Note that it is only necessary to solve the set of differential equations for H_x and H_y, not H_z, since we construct H_z and the electric field components from the in-plane components of the magnetic field. Having assumed that the magnetic field is periodic variation in the z-direction, the divergence condition may be expressed as

$$\partial_x H_x + \partial_y H_y + ik_z H_z = 0. \tag{14.103}$$

Once H_x and H_y are determined, the divergence condition may be used to calculate H_z. In this method, there is no need to derive a dispersion relation from Eq. (14.102c).

14.5 Concluding remarks

Electromagnetic devices of higher frequencies (e.g. mm-wave) and increasing complexity are being employed in a wide variety of industries. The design of modern automobiles includes electromagnetic components with sophisticated interactions both in isolation and in integrated combinations. Designing these components requires a significant amount of modeling and simulation, and these demands continue to increase with higher levels of device integration. At the nanoscale, similar circumstances are faced for optical interconnects, quantum well laser design, and in plasmonics. Again, reliable simulations are essential to ensure that each device does not affect others near it through electromagnetic cross-talk. The novel effects exhibited by metamaterials containing negative refractive index components are all simulated before being assembled in order to optimize their optical properties as desired.

Mainstream computational electromagnetic modeling relies heavily on finite element, finite difference, and spectral methodologies. Here we focus on a scalar finite element approach in which the field components are approximated using local polynomials over discrete sub-domains. We show that the use of $\mathcal{C}_1$-continuous Hermite interpolation polynomials on triangular elements provide very accurate solutions with a minimal number of elements used in the discretization. The ability to reproduce smooth variational solutions for the fields will allow a coarsely discretized full-wave Maxwell solver to seamlessly couple to other solvers for physically small features, such as small gate geometries, quantum wells and dots, or plasmonic structures which are all deeply sub-wavelength. The Hermite interpolation polynomials are equally well suited to 3D finite elements, e.g. 40 DoFs can be used to define the basis functions on tetrahedra. We have results of using a 56 DoFs tetrahedral FEM for the cubic cavity in Chap. 16.

We have shown that the Hermite interpolation polynomials on a triangular element are able to eliminate the spurious solutions that typically occur with Lagrange-type scalar shape functions. The results for the standard rectangular waveguide with and without a dielectric inhomogeneity directly demonstrate the efficacy of this method. The eigenfrequency for the lowest mode for the homogeneous waveguide agrees with the analytical result within a relative error of 10^{-15}, which is superior to hierarchical VFEM for about the same number of DoFs. [23] The Hermite FEM delivers a relative error of 10^{-9} with just 36 DoFs (6 elements). This is displayed in Fig. 14.5(a). With mesh refinement, an error of 10^{-14} is reached by both methods, with the HFEM requiring ~1200 DoFs versus the VFEM requiring ~5000 DoFs. These are significant improvements over the performance of VFEM. The dashed lines for eigenvalues 2 and 3 in Fig. 14.5(b) represent the errors in the eigenvalues of degenerate modes. In the case of the second and third modes, an accidental degeneracy exists between the TE_{20} and TE_{01} modes for a waveguide with dimensions in a 2:1 ratio. The lines corresponding to eigenvalues 4 and 5 correspond to the TE_{11} and TM_{11} modes, which will always be degenerate for the PEC homogeneous waveguide. The error increases steadily for states higher than the ground state. However, the errors in all states converge at the same rate. For the inhomogeneous waveguide, we have presented a discussion of the eigenmodes that are trigonometric in both regions at higher frequencies, that evolve into solutions that are sinusoidal in the higher dielectric region and hyperbolic in the region with the lower dielectric as the dielectric ratio ϵ_2/ϵ_1 increases. This behavior is analogous to the development of above-barrier states in quantum wells that get localized and captured into the quantum well as the well depth is increased. This analogy suggests that as ϵ_2 is increased, more modes are captured by the higher dielectric region leading to the sinusoidal behavior in the larger dielectric region and an exponential decay into the lower dielectric region. The results shown in Figs. 14.7, 14.8, and 14.9 demonstrate the congruence of the theoretical and computed mode frequencies as they evolve with varying ϵ_2/ϵ_1. The HFEM simulations directly provide the computed results with no spurious solutions. The field distributions are displayed for sample modes in the inhomogeneous waveguide in Fig. 14.10.

We conclude by noting that the Hermite triangular elements provide significant advantages in the electromagnetic modeling of complex systems that should attract a more universal usage from FEM practitioners.

References

[1] A. Sommerfeld, Ann. der Phys. Chem. **67**, 233–290 (1899); "Uber die fortplanzung elektrodynamisches wellen langes eines drahtes."

[2] J. S. Schwinger and David S. Saxon, *Discontinuities in Waveguides: Notes on Lectures by Julian Schwinger* (Taylor & Francis US, January 1968).

[3] K. A. Milton and J. Schwinger, *Electromagnetic Radiation: Variational Methods, Waveguides and Accelerators* (Springer, Berlin, Germany, 2006).

[4] K. Leong, W. R. Deal, V. Radisic, M. X. Bing, J. Uyeda, L. Samoska, A. Fung, T. Gaier, and R. Lai, IEEE Microw. Wirel. Compon. Lett. **19**, 413–415 (2009); "A 340–380 GHz integrated CB-CPW-to-waveguide transition for sub millimeter-wave MMIC packaging."

[5] J. C. Nedelec, Numerische Mathematik, **35**, 315–341 (1980); "Mixed finite elements in R3."

[6] D. Sun, J. Manges, X. Yuan, and Z. Cendes, IEEE Antennas Propag. **37**, 12–24 (1995); "Spurious modes in finite element methods."

[7] D. R. Lynch, and K. D. Paulsen, IEEE Trans. Microw. Theory Tech. **39**, 395–403 (1991); "Elimination of vector parasites in finite element maxwell solutions."

[8] J. F. Lee, D. K. Sun, and Z. J. Cendes, IEEE Trans. Magn. **27**, 4032–4035 (1991); "Tangential vector finite elements for electromagnetic field computation."

[9] A. F. Peterson, IEEE Trans. Antennas Propag., **43**, pp. 357–365 (1994); "Vector finite element formulation for scattering from two-dimensional heterogeneous bodies."

[10] J. P. Webb, Rep. Prog. Phys. **58**, 1673–1712, (1995); "Application of the finite-element method to electromagnetics and electric topics." also see, *Efficient generation of divergence-free fields for the finite element analysis of 3D cavity resonances*, IEEE Trans. Magn. **24**, 162–165 (1988).

[11] L. S. Anderson, and J. L. Volakis, IEEE Trans. Antennas Propag. **47**, 112–120 (1999); "Development and application of a novel class of hierarchical tangential vector finite elements for electromagnetics."

[12] J. P. Webb, IEEE Trans. Antennas Propag. **47**, 1244–1253 (1999); "Hierarchal vector basis functions of arbitrary order for triangular and tetrahedral finite elements."

[13] D. Boffi, P. Fernandes, L. Gastaldi, and I. Perugia, SIAM J. Numer. Anal. **36**, 1264–1290 (1999); "Computational models of electromagnetic resonators: Analysis of edge element approximation."

[14] J. P. Webb, IEEE Trans. Magn. **37**, 3600–3603 (2001); "Design sensitivities using high-order tetrahedral vector elements."

[15] H. X. Zheng, L. Y. Feng, and Q. S. Wu, IEEE Trans. Microw. Theory Tech. **58**, 128–135 (2010); "Three-dimensionally nonorthogonal alternating-direction implicit finite-difference time-domain algorithm for the full-wave analysis of microwave monolithic circuit devices."

[16] J. P. Webb, IEEE Trans. Magn. **24**, 162–165 (1988); "Efficient generation of divergence-free fields for the finite element analysis of 3D cavity resonances."

[17] P. G. Kassebaum, C. R. Boucher, and L. R. Ram-Mohan, J. Comput. Phys., **231**, 5747–5760 (2012); "Application of group representation theory to derive Hermite interpolation polynomials on a triangle."

[18] K. Bell, Int. J. Numer. Methods Eng., **1**, 101–122 (1969); "A refined triangular plate bending finite element."

[19] I. Holand and K. Bell, *Finite Element Methods in Stress Analysis* (Tapir, Trondheim, 1970).

[20] J. H. Argyris, I. Fried, and D. W. Scharpf, Imperial College of Science and Technology, University of London, Technical Note 14, 701–709 (1968); "University of London, Technical Note 14."

[21] Gouri Dhatt and Gilbert Touzot, *The Finite Element Method Displayed*, (John Wiley & Sons, April 1984).

[22] L. R. Ram-Mohan, *Finite Element and Boundary Element Applications in Quantum Mechanics* (Oxford, New York, 2002).

[23] S. C. Lee, J. F. Lee, and R. Lee, IEEE Trans. Microw. Theory Tech. **51**, 1897–1905 (2003); "Hierarchical vector finite elements for analyzing waveguiding structures."

[24] C. R. Boucher, Z. Li, C. I. Ahheng, J. D. Albrecht, and L. R. Ram-Mohan, J. Appl. Phys. **119** (14), 143106–1–10 (2016); "Hermite finite elements for high accuracy electromagnetic field calculations: A case study of homogeneous and inhomogeneous waveguides."

[25] Jianming Jin, *The Finite Element Method in Electromagnetics*, 2nd edition (Wiley, New York, 2002).

[26] R. B. Lehoucq, D. C. Sorensen, and C. Yang, *ARPACK Users Guide: Solution of Large-Scale Eigenvalue Problems with Implicitly Restarted Arnoldi Methods* (SIAM publications, Philadelphia, 1998).

[27] A. M. Portis, *Electromagnetic Fields: Sources and Media* (John Wiley, New York, 1978); pp. 508–513.

[28] R. F. Harrington, *Time-Harmonic Electromagnetic Fields* (John Wiley and Sons, New York, 2001).

[29] L. Pincherle, Phys. Rev. **66**, 118–130 (1944); "Electromagnetic waves in metal tubes filled longitudinally with two dielectrics."

[30] J. Van Bladel and T. J. Higgins, J. Appl. Phys. **22**, 329–334 (1951); "Cut-off frequency in two-dielectric layered rectangular wave guides."

15

Modeling photonic crystals with Hermite FEM

In this chapter:

- We consider the calculation of photonic band structures of photonic crystals.
- We show that EM fields can be calculated using scalar Hermite interpolation polynomials as the numerical basis functions without having to invoke edge-based vector finite elements to suppress spurious solutions or to satisfy boundary conditions. This approach offers several fundamental advantages as evidenced through band structure solutions for periodic systems and through waveguide analysis treated in the earlier chapters.
- Compared with reciprocal space (plane-wave expansion) methods for periodic systems, advantages are shown in computational costs, the ability to capture spatial complexity in the dielectric distributions, the demonstration of numerical convergence with scaling, and variational eigenfunctions free of numerical artifacts that arise from mixed-order real-space basis sets or the inherent aberrations from transforming reciprocal space solutions of finite expansions.
- The photonic band structure of a simple crystal is used as a benchmark comparison, and the ability to capture the effects of spatially complex dielectric distributions is treated using a complex pattern with highly irregular features that would stress spatial transform limits.
- This general method is applicable to a broad class of physical systems, e.g. to semiconducting lasers which require simultaneous modeling of transitions in quantum wells or dots together with EM cavity calculations, to modeling plasmonic structures in the presence of EM field emissions, and to on-chip propagation within monolithic integrated circuits.

15.1 Introduction

Maxwell's vector field equations represent significant challenges for the numerical solution of electric and magnetic (EM) fields in complex geometries. With increasing complexity in the design of EM devices, it becomes imperative to develop compact

Finite Elements in Action. L. Ramdas Ram-Mohan, Oxford University Press. © L. Ramdas Ram-Mohan (2026).
DOI: 10.1093/oso/9780199563487.003.0015

algorithms for electrodynamic calculations that will generate high accuracy solutions while demanding minimal computational resources. We solve the field equations for the prototypical examples of two-dimensional (2D) photonic crystals (PhCs) using the finite element method (FEM) with scalar Hermite interpolation polynomials (HFEM). Nodal solutions are highly compatible with domain decomposition and are therefore highly sought after for full-wave EM calculations. The mainstream approaches for EM FEM typically do not use scalar nodal basis polynomials for reasons that are summarized in this introduction.

15.1.1 Discussion of earlier approaches

Vector Finite Element Approaches (VFEM). The typical analysis of electromagnetic fields today is focused on the VFEM. [1–12] We note that while the edge-elements are able to satisfy tangential continuity across interfaces, they do not ensure the continuity of the normal components of the fields in the absence of dielectric discontinuity between elements. This leads to lack of smoothness of the fields across the elements. The kinks in the EM wavefunctions can be large compared with the quantum structures with which they interact and are therefore a source of error under conditions of highly-varying length scales.

Since the wavelength of the EM fields is typically very long compared with quantum mechanical wavefunctions, these regions of ambiguity can be large compared with objects like quantum dots or wells where the accurate interaction with the EM field is desired. Further mesh refinement simply places numerical error associated *inside* the quantum structure where its impact is even more severe.

Vector finite element method also presents challenges when attempting to model an electrodynamics problem using the scalar, $\phi(\mathbf{r}, t)$ and vector, $\mathbf{A}(\mathbf{r}, t)$, potentials together instead of the vector $\mathbf{E}$ and $\mathbf{H}$ fields upon which the curl-conforming methods are based. Further details can be seen in Chaps. 16 and 17.

Plane-wave expansions and other spectral methods: Given the periodic nature of many systems of technical importance, it is natural that reciprocal space methods are widely used. For example, the other common approach for determining dispersion relations and wavefunctions for PhCs borrows from solid state theory and employs the method of plane waves. This method has the following drawbacks:

The Hamiltonian matrix in this basis is completely filled. In published examples and available open-source codes, it is common that $O(10^6)$ plane waves are required for adequate results, resulting in very large filled matrices to be diagonalized. Capturing the spatial variation of the dielectric discontinuities and the distribution of the dielectric material requires larger and larger number (10^6–10^7) of plane waves. Finite structures such as PhC slabs have been simulated using as many as 10^8. While plane waves have the advantage of almost trivial matrix assembly, these basis dimensions lead to very large matrices that need to be evaluated and diagonalized repeatedly to map out the dispersion relation over the Brillouin zone (BZ) of the PhC.

The reconstructed fields display the "ringing" behavior at the dielectric interfaces associated with the Gibbs phenomenon [13] arising from the representation of a discontinuous dielectric function with a large, but finite, number of continuous global spectral functions. Therefore, to use the spatial wavefunctions, a numerical filtering process is often applied to remove highly oscillatory content.

Alternative methods: For external regions, boundary integral techniques have been combined with FEMs to capture the asymptotic behavior in open systems. [14] While offering surface versus volume discretization advantages, boundary integral formulations are self-consistency relations and do not enjoy the advantages of variational quadratic convergence. The use of dyadic Green's functions in EM involves derivatives of these singular functions at the boundaries making the evaluation of the integrals more complex. In other work, new methods of dimensional continuation for the evaluation of such hypersingular integrals that substantially enhance the computational accuracy have been developed recently by us. [15] However, the matrices generated in this method are dense, and are generally nonlinear in the eigenfrequencies.

Finally, meshless methods [16–18] have been used to treat a number of physics problems, including EM, and are at the opposite extreme in terms of numerical representations to the method given in this chapter where we develop a higher order of derivative continuity in spatial basis. Meshless schemes can be interpreted in many ways but often involve the use of basis functions that are not restricted to a finite volume. The resulting matrix formulations have the advantage of being fully block-diagonal but require a separate treatment of continuity boundary conditions. In these situations, the boundary conditions are enforced as a constraint which can be a penalty method or numerical flux minimization. Derivative information is rarely treated when forming eigenfunctions in these methods.

15.1.2 Using Hermite FEM for photonic crystals

All the above difficulties are absent if an appropriate node-based interpolation scheme is implemented within the FEM framework. We propose the use of HFEM because they employ spatial derivative degrees of freedom that directly coincide with the operators in the curl, and are completely consistent with EM theory. We show that the approach yields better accuracy, with a more physical (smoother) representation of fields, than from VFEM. This method does not generate spurious solutions that plague nodal based Lagrange FEM encountered earlier in the 1970s, even though the $\mathcal{C}_1$-continuous Hermite polynomials are also scalar in nature.

The Hermite elements used for PhC calculations generate sparse banded matrix equations with occupancies of $\simeq 0.2\%$ or less for the matrices, in contrast with the fully occupied matrices generated using plane-wave methods. The occupancy decreases rapidly as the matrix sizes increase with mesh refinement. Also, the $\mathcal{C}_{(1)}$-continuous FEM representation provides the field eigenfunctions without spatial distortions or pixelation.

We note that for the same number of degrees of freedom, both VFEM and HFEM will generate matrix dimensions that are equal, but the bandwidth of the banded matrices in HFEM will be nominally larger. While computational times are expected to be about the same, the HFEM generates smoother solutions.

We focus here on EM wave propagation in 2D periodic arrangements of dielectric materials. The example of a square lattice of arrayed cylindrical dielectric posts is used to compare results with those in the published literature. [19] The photonic wavefunctions and frequency dispersion surfaces are shown for several TE and TM modes. The flexibility and power of our method is demonstrated in our analysis of an additional 2D PhC made from an Escher-like dielectric structure.

In Sec. 15.2, we use the KBR Hermite polynomials to calculate the band structures of two different types of 2D PhCs. We show that the local connectivity for finite elements leads to a sparse matrix, giving the HFEM a distinct advantage over older methods of PhC simulation based on plane-wave expansions and other spectral methods. Concluding remarks are given in Sec. 15.6.

15.2 Theory

A very brief introduction to the theory is presented below to make this chapter self-contained, and to treat carefully the derivation of the action integral, which we label $\mathcal{A}$, from Maxwell's wave equation. This is done to explicitly show where the spatial derivatives from the HFEM fit directly into the framework. We consider 2D PhCs which are a periodic arrangement of materials of different dielectric properties that only permit EM waves of certain frequencies and polarizations to propagate. Their design through the geometric placement of the dielectrics provides a level of control over the dispersion relations satisfied by photons in the medium.

The first proposals for the design of PhCs by Yablonovitch [20] and by John [21] in 1987, and further investigations by Ohtaka, Sakoda, and collaborators [22–26] and by Joannopoulos and Johnson [19] have led to a full appreciation of the physics of periodic dielectrics. With the rapid increases in computing power and simulation techniques and the design and fabrication of PhCs, a wide variety of opto-electronic devices including low-loss reflecting surfaces, waveguides, filters, flat lenses, optical inter-connects and the like, have made the efficient prediction of their optical properties a high priority for physicists and optical engineers. [19, 27–31]

By assuming that the PhC contains an arbitrarily large number of unit cells, one may use Bloch's theorem [32] to decompose $\mathbf{E}$ or $\mathbf{H}$ into the product of a periodic function, or cell function, and a plane wave, or envelope function. Currently, the most popular means of computing the band structure of a PhC is to represent the cell function itself as a linear combination of plane waves.

The approach outlined in the following is to use the HFEM. The FEM may be considered to be the discretization of $\mathcal{A}$. In each of the finite elements, the fields are represented in terms of the Hermite polynomials with coefficients corresponding to values of the fields and their spatial derivatives at the vertices of the triangles. The spatial dependence is integrated out leaving the action as a bilinear function of these coefficients. The variational principle on these as-yet unknown coefficients leads to a matrix equation representing Maxwell's wave equation. These simultaneous equations are solved with the periodic boundary conditions to obtain the coefficients for each eigenvalue (frequency).

We note that calculating the bandstructure point-by-point for a dense set of values of q, with each input value of the photonic momentum q is a tedious matter for mapping out the detailed bandstructure. It has been suggested by Hussein [33] that the exact value at special points in the Brillouin zone and a few values in the middle of the bandstructure between the special points are sufficient to form an approximate basis set for other values of q in between. This reduced Bloch-mode expansion with the Davidson vector iteration algorithm for the diagonalization of the FEM matrices can be used with little loss of accuracy. This is a great help in reducing the time to map out the bandstructures over the Brillouin zone.

The same polynomials have already been shown to deliver high accuracy and reliable performance when implemented to solve quantum mechanical problems. [34] A brief derivation of $\mathcal{A}$ for the EM field in a PhC is given below.

15.2.1 The action integral for photonic crystals

We seek to calculate the band structure and vector fields in a PhC. Maxwell's equations are

$$\begin{aligned} \nabla \cdot \mathbf{D} &= \rho; \qquad & \nabla \cdot \mathbf{B} &= 0, \\ \nabla \times \mathbf{E} + \frac{\partial \mathbf{B}}{\partial t} &= 0; \qquad & \nabla \times \mathbf{H} - \frac{\partial \mathbf{D}}{\partial t} &= \mathbf{J}, \end{aligned} \tag{15.1}$$

where $\mathbf{D} = \epsilon \mathbf{E}$ and $\mathbf{B} = \mu \mathbf{H}$ are the standard relations for the electric displacement and the magnetic induction in terms of the EM fields. If the medium is isotropic, ϵ and μ are scalar quantities, while in general, they behave as symmetric second-rank tensors. We define the dimensionless relative permittivity ϵ_r and relative permeability μ_r using $\epsilon = \epsilon_r \epsilon_0$ and $\mu = \mu_r \mu_0$, with ϵ_0 and μ_0 being the permittivity and permeability of free space, respectively. We assume that the PhC is charge-free and current-free. We also assume that the fields are time-harmonic so that $\mathbf{H}(\mathbf{r}, t) = \mathbf{H}(\mathbf{r}) \exp(-i\omega t)$ and similarly for $\mathbf{E}$, where $\omega = k_0 c$ is the frequency of the EM wave. With these assumptions, Eqs. (15.1) reduce to the curl-curl form of Maxwell's wave equation,

$$\nabla \times \left(\frac{1}{\epsilon_r} \nabla \times \mathbf{H} \right) - k_0^2 \mu_r \mathbf{H} = 0, \tag{15.2}$$

which only includes the spatial dependence of the $\mathbf{H}$-field. Alternatively, the curl-curl form may be derived in terms of the dual electric field by replacing $\mathbf{H} \to \mathbf{E}$ and interchanging $\epsilon_r \leftrightarrow \mu_r$ in Eq. (15.2).

We use Eq. (15.2) to construct $\mathcal{A}$; first, we multiply it by a variation with respect to the conjugate field, $\delta \mathbf{H}^*$, and integrate to obtain [35]

$$\int_V d^3 r \, \delta \mathbf{H}^* \cdot \left[\nabla \times \left(\frac{1}{\epsilon_r} \nabla \times \mathbf{H} \right) - k_0^2 \mu_r \mathbf{H} \right] = 0. \tag{15.3}$$

Because the PhC is two-dimensional and has homogeneous properties in the third direction, we can eliminate the integration in the direction in which the crystal has homogeneous properties. We assume that the remaining two integrations are performed over a single unit cell of the PhC, and refer to this 2D domain as Ω. In this manner, Eq. (15.3) becomes

$$\int_{\Omega} d^2r\, \delta\mathbf{H}^* \cdot \left[\nabla \times \left(\frac{1}{\epsilon_r}\nabla \times \mathbf{H}\right) - k_0^2 \mu_r \mathbf{H}\right] = 0. \tag{15.4}$$

Through a vector identity,† we also know that

$$\delta\mathbf{H}^* \cdot \left[\nabla \times \left(\frac{1}{\epsilon_r}\nabla \times \mathbf{H}\right)\right] = (\nabla \times \delta\mathbf{H}^*) \cdot \left(\frac{1}{\epsilon_r}\nabla \times \mathbf{H}\right) - \nabla \cdot \left[\delta\mathbf{H}^* \times \left(\frac{1}{\epsilon_r}\nabla \times \mathbf{H}\right)\right]. \tag{15.5}$$

Upon substitution of Eq. (15.5) into Eq. (15.4) and the application of Gauss's theorem to convert the volume (area) integration of the divergence term to a line integral over the perimeter, $\partial\Omega$, of the unit cell Ω, the following form is obtained:

$$\int_{\Omega} d^2r\, \delta\mathbf{H}^* \cdot \left[\nabla \times \left(\frac{1}{\epsilon_r}\nabla \times \mathbf{H}\right)\right] = \int_{\Omega} d^2r\, (\nabla \times \delta\mathbf{H}^*) \cdot \left(\frac{1}{\epsilon_r}\nabla \times \mathbf{H}\right) - \oint_{\partial\Omega} ds\hat{\mathbf{n}} \cdot \left[\delta\mathbf{H}^* \times \left(\frac{1}{\epsilon_r}\nabla \times \mathbf{H}\right)\right]. \tag{15.6}$$

By Eq. (15.6), we see that the integral can be expressed without second-order spatial derivatives.

15.2.2 Application of Bloch's Theorem

By the Bloch-Floquet Theorem, [32, 36, 37] we can decompose the magnetic field into two terms,

$$\mathbf{H}(\mathbf{r}) = \mathbf{U}(\mathbf{r})\, e^{i\mathbf{q}\cdot\mathbf{r}}. \tag{15.7}$$

We refer to the first term, $\mathbf{U}$, as the "cell function". We call the exponential term the "envelope function". In the examples treated in this report, in which the unit cell is a rectangle of dimensions $d_x \times d_y$, the vector $\mathbf{q}$ can be expressed as

$$\mathbf{q} = \frac{\pi q_x}{d_x}\hat{\mathbf{i}} + \frac{\pi q_y}{d_y}\hat{\mathbf{j}} + 0\hat{\mathbf{k}}, \tag{15.8}$$

in which, the range of q_x and q_y values which comprise the Brillouin zone [38] are

$$-1 < q_x \leq 1, \quad -1 < q_y \leq 1. \tag{15.9}$$

Values of q_x and q_y outside of the ranges defined in Eq. (15.9) are redundant. The cell function obeys periodic boundary conditions over a single unit cell. By applying the decomposition in Eq. (15.7) to the surface term in Eq. (15.6), we express the line integral as

†Apply the identity $\nabla \cdot (\mathbf{P} \times \mathbf{R}) = (\nabla \times \mathbf{P}) \cdot \mathbf{R} - \mathbf{P} \cdot (\nabla \times \mathbf{R})$ with $\mathbf{R} = \epsilon_r^{-1}\nabla \times \mathbf{H}$ and $\mathbf{P} = \delta\mathbf{H}^*$.

$$\oint_{\Gamma} ds\, \hat{\mathbf{n}} \cdot \left[\delta\mathbf{H}^* \times \left(\frac{1}{\epsilon_r}\nabla \times \mathbf{H}\right)\right] = \delta \oint_{\Gamma} ds\, \hat{\mathbf{n}} \cdot \left[\mathbf{H}^* \times \left(\frac{1}{\epsilon_r}\nabla \times \mathbf{H}\right)\right]$$
$$= \delta \oint_{\Gamma} ds\hat{\mathbf{n}} \cdot \left[\left(\mathbf{U}^* e^{-i\mathbf{q}\cdot\mathbf{r}}\right) \times \left(\frac{1}{\epsilon_r}\nabla \times \left(\mathbf{U}e^{i\mathbf{q}\cdot\mathbf{r}}\right)\right)\right]. \tag{15.10}$$

We expand the curl term in Eq. (15.10) as

$$\nabla \times \left(\mathbf{U}e^{i\mathbf{q}\cdot\mathbf{r}}\right) = e^{i\mathbf{q}\cdot\mathbf{r}}\left(\nabla \times \mathbf{U}\right) + \left(\nabla e^{i\mathbf{q}\cdot\mathbf{r}}\right) \times \mathbf{U}$$
$$= e^{i\mathbf{q}\cdot\mathbf{r}}\left(\nabla \times \mathbf{U}\right) + e^{i\mathbf{q}\cdot\mathbf{r}}\left(i\mathbf{q} \times \mathbf{U}\right). \tag{15.11}$$

The factors of $\exp(i\mathbf{q}\cdot\mathbf{r})$ and $\exp(-i\mathbf{q}\cdot\mathbf{r})$ in Eq. (15.10) cancel each other, reducing the surface integral to

$$\delta \oint_{\Gamma} ds\, \hat{\mathbf{n}} \cdot \left[\mathbf{U}^* \times \frac{1}{\epsilon_r}\left(\nabla \times \mathbf{U} + i\mathbf{q} \times \mathbf{U}\right)\right]. \tag{15.12}$$

Because $\mathbf{U}$ and ϵ_r are both periodic, for every point on the boundary of the unit cell, there will be a point on the opposite side of the cell with the same values of $\mathbf{U}$ and ϵ_r, but with $\hat{\mathbf{n}}$ pointing in the opposite direction. Therefore, it is clear that the line integral given in Eq. (15.12) is exactly zero. We now apply this conclusion and Eq. (15.6) to Eq. (15.4) to obtain a new expression for the action integral:

$$\int_{\Omega} d^2r \left[\left(\nabla \times \delta\mathbf{H}^*\right) \cdot \frac{1}{\epsilon_r} \cdot \left(\nabla \times \mathbf{H}\right) - \delta\mathbf{H}^* \cdot k_0^2\mu_r \cdot \mathbf{H}\right] = 0, \tag{15.13}$$

which may be expressed as

$$\delta\mathcal{A} = \delta \int_{\Omega} d^2r \left[\left(\nabla \times \mathbf{H}^*\right) \cdot \frac{1}{\epsilon_r} \cdot \left(\nabla \times \mathbf{H}\right) - \mathbf{H}^* \cdot k_0^2\mu_r \cdot \mathbf{H}\right] = 0. \tag{15.14}$$

In Eq. (15.14), we see the principle of least action applied to Maxwell's vector equations. We conclude that the action integral is defined as

$$\mathcal{A} = \int_{\Omega} d^2r \left[\left(\nabla \times \mathbf{H}^*\right) \cdot \frac{1}{\epsilon_r} \cdot \left(\nabla \times \mathbf{H}\right) - \mathbf{H}^* \cdot k_0^2\mu_r \cdot \mathbf{H}\right], \tag{15.15}$$

in which the integration is performed over one unit cell of the 2D PhC. Alternatively, we may begin the derivation of the action integral by using the electric field instead of the magnetic field. In this case, we again use the Bloch-Floquet Theorem to separate the field into a cell function $\mathbf{U}$ and an envelope function, and see that the surface term arising from Gauss's theorem is still zero. The resulting action is

$$\mathcal{A} = \int_{\Omega} d^2r \left[\left(\nabla \times \mathbf{E}^*\right) \cdot \frac{1}{\mu_r} \cdot \left(\nabla \times \mathbf{E}\right) - \mathbf{E}^* \cdot k_0^2\epsilon_r \cdot \mathbf{E}\right]. \tag{15.16}$$

In the following section, we use the Bloch-Floquet theorem to simplify the action by classifying all possible solutions into two distinct types of modes.

15.2.3 Classification of TE and TM modes

We begin by decomposing each term of $\mathcal{A}$ into cell functions and envelope functions. For fields of the form in Eq. (15.7), the second term of the integrand of $\mathcal{A}$ reduces to $\mathbf{U}^* \cdot k_0^2\mu_r \cdot \mathbf{U}$. The curl term in the integrand of $\mathcal{A}$ can be simplified using the absence of propagation in the third dimension in Eq. (15.8) as

$$\nabla \times \mathbf{H} = \quad e^{i\mathbf{q}\cdot\mathbf{r}} \left[\begin{pmatrix} \partial_y U_z - \partial_z U_y \\ \partial_z U_x - \partial_x U_z \\ \partial_x U_y - \partial_y U_x \end{pmatrix} + i\pi \begin{pmatrix} q_y U_z / d_y \\ -q_x U_z / d_x \\ q_x U_y / d_x - q_y U_x / d_y \end{pmatrix} \right]. \tag{15.17}$$

Eq. (15.17) may be simplified by classifying all possible solutions into two distinct cases: (*i*) Transverse electric (TE) modes with $E_z = 0$, which from the imposition of periodicity resulting in Eq. (15.17) forces $H_x = H_y = 0$, and all other vector components are nonzero and (*ii*) Transverse magnetic (TM) modes with $H_z = 0$ which forces $E_x = E_y = 0$, $H_z = 0$, and all other vector components are nonzero. This decomposition greatly simplifies the analysis of 2D PhCs compared with general cases. Any possible field in the 2D PhC may be expressed as a combination of TEand TM modes. It is well known that frequencies absent from the eigenspectra of both modes do not propagate because they are in the "band gap."

Given the simple modal decomposition resulting from 2D periodicity, eigenvalue problems for both TE and TM modes can be posed in terms of a single scalar quantity, U_z and its spatial derivatives. For isotropic media where μ_r and ϵ_r are constant, $\mathcal{A}$ for TM modes simplifies to

$$\mathcal{A} = \int_\Omega d^2r \left[U_z^* \mathbf{A} \frac{1}{\epsilon_r} \mathbf{B} U_z - U_z^* k_0^2 \mu_r U_z \right] \tag{15.18}$$

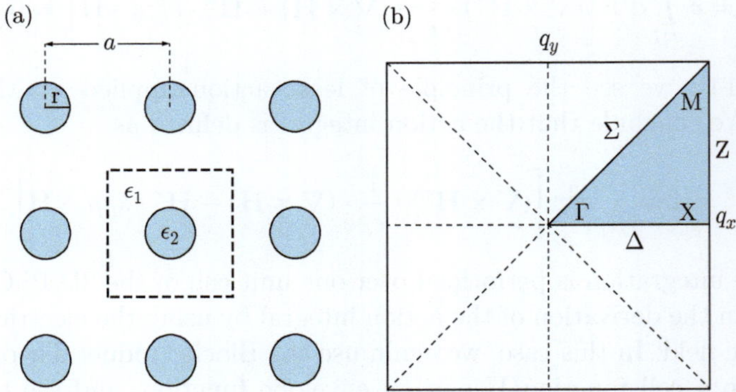

Figure 15.1 (a) An array of cylindrical dielectric posts of dielectric constant ϵ_2 arranged periodically in a medium with a dielectric constant ϵ_1 is shown. (b) The Brillouin zone for a two-dimensional photonic crystal is shown. The maximum and minimum values of the components of $\mathbf{q}$ are given by Eq. (15.9). The irreducible component of the Brillouin zone has been highlighted. The dashed lines are symmetry lines within the first Brillouin zone.

where

$$\mathbf{A} = \left[\overleftarrow{\partial_y} - \frac{\pi i q_y}{d_y}, -\overleftarrow{\partial_x} + \frac{\pi i q_x}{d_x} \right], \mathbf{B} = \begin{bmatrix} \overrightarrow{\partial_y} + \dfrac{\pi i q_y}{d_y} \\ -\overrightarrow{\partial_x} - \dfrac{\pi i q_x}{d_x} \end{bmatrix} \tag{15.19}$$

and the arrows over the derivatives denote the direction in which the derivatives operate on the quantities in Eq. (15.18). To obtain $\mathcal{A}$ for TE modes, interchange $\epsilon_r \leftrightarrow \mu_r$ in Eq. (15.18) and associate U_z with $\mathbf{E}$. Within the FEM, a variational solution is found by the eigenvalue problem $\delta\mathcal{A} = 0$ and yields the band structure of the 2D PhC.

15.2.4 The Brillouin zone

As shown in Eq. (15.9), it is only necessary to consider the eigenvalues over a finite range of $\mathbf{q}$. This range of $\mathbf{q}$ values is the first Brillouin zone; outside this zone the eigenvalues will behave periodically. Furthermore, due to reflection symmetries, it is possible to further reduce the Brillouin zone to obtain a greater density of sampling points for the same computational cost. In the first example considered, a square array of cylindrical dielectric posts, the Brillouin zone only needs to be treated over the region marked in Fig. 15.1.

The band gaps of the PhC may be determined by choosing a large number of ordered pairs (q_x, q_y) from within the irreducible Brillouin zone to determine which eigenvalues will propagate. Then the band structure in the remainder of the Brillouin zone may be determined via reflection symmetry. The entire first Brillouin zone may then be translated to adjacent zones due to the periodicity of the cell function.

In the following, we consider three examples of PhCs. The first is a square lattice of dielectric posts, for which we obtain the photonic band structure as previously reported in Joannopoulos. [38] We also identify the symmetries at various points in the dispersion relations to discuss band anticrossing and level degeneracies at special points. The eigenvector fields at various points of the bands are shown, and frequency bands over the full Brillouin zone are displayed.

The second example is that of a checkerboard lattice of dielectric regions. Here again, we provide the group-theoretic analysis, the band structure, the band surfaces over the Brillouin zone, and the eigenvector fields.

15.2.5 Group theory and photonic crystals

The eigenvector fields can be organized according to their symmetries with respect to the symmetry group of the crystal and to the group of the wavevector. An excellent exposition on the application of group representation theory to physics is provided by Dresselhaus. [39]

In the following, we follow the group-theoretic analysis of Sakoda. [23, 24, 26] The point group of the cylindrical post unit cell is C_{4v}, or the symmetry of the square. The character table of this group is shown in Table 15.1. The wavevector at the Γ-point has the full symmetry of C_{4v}. The symmetry of the Γ-point modes can be deduced by

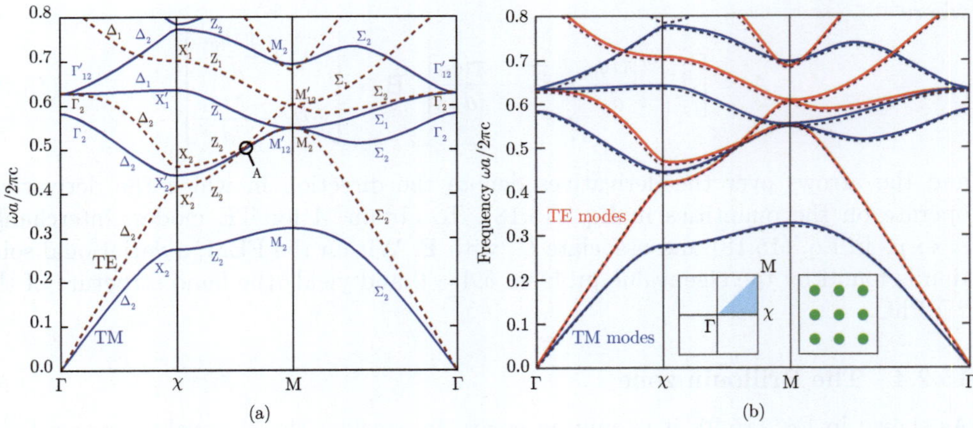

Figure 15.2 (a) Eigenvalues for the transverse electric and transverse magnetic modes as calculated using the finite element method with quintic Hermite interpolation polynomials. The point labeled as A is the location of an anticrossing site between the lowest and second-lowest TE modes. This location is shown in higher resolution in Fig. 15.3. (b) The band structure obtained using finite elements (dotted lines) is compared to the band structure given by Joannopoulos (solid lines). [38]

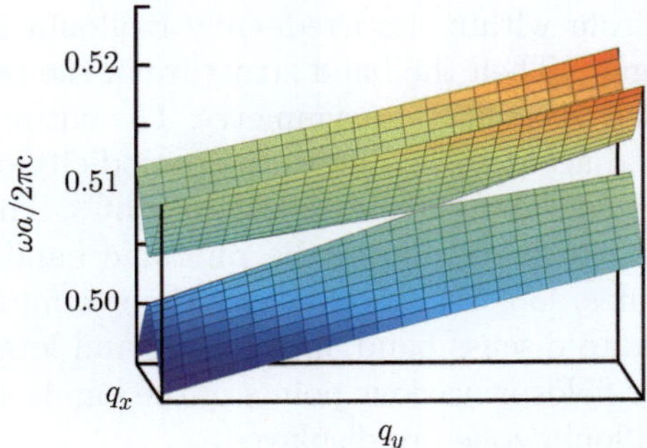

Figure 15.3 Close-up view of the anticrossing site shown at point A in Fig. 15.2(a).

inspecting the transformation properties of the eigenvectors that are transverse to the extrusion direction of the crystal. For a one-dimensional irreducible representation D_i, operation R_j in class j with character $\chi_i(R_j)$, and eigenvector field v, the eigenvector field will transform according to

$$D_i(R_j)\, v = \chi_i(R_j)\, v. \tag{15.20}$$

By inspecting the transformation of v by several $D_i(R_j)$, the character table can be used to deduce which irreducible representation the eigenvector field belongs to. As an example, consider the Γ-point mode in Fig. 15.4(a). The transverse vector field satisfies

Table 15.1 The character table for the group C_{4v}, the symmetry group of the square, is given. For the cylindrical dielectric post structure, this symmetry is exhibited at the Γ- and M-points in the Brillouin zone.

C_{4v}			E	C_2	$2C_4$	$2\sigma_v$	$2\sigma_d$
$x^2+y^2,\ z^2$	z	Γ_1	1	1	1	1	1
	R_z	Γ_2	1	1	1	–1	–1
x^2-y^2		Γ_1'	1	1	–1	1	–1
xy		Γ_2'	1	1	–1	–1	1
(xz, yz)	$(x,y),(R_x,R_y)$	Γ_{12}'	2	–2	0	0	0

Table 15.2 The character table for the group C_{2v}, the symmetry group of the rectangle, is shown. For the cylindrical dielectric post structure, this symmetry is exhibited at the X point of the Brillouin zone.

C_{2v}			E	C_2	σ_v	σ_v'
$x^2,\ y^2,\ z^2$	z	X_1	1	1	1	1
xy	R_z	X_2	1	1	–1	–1
xz	$R_y,\ x$	X_1'	1	–1	1	–1
yz	$R_x,\ y$	X_2'	1	–1	–1	1

$$D_i(C_2)v = v, \quad D_i(2C_4)v = v, \quad D_i(2\sigma_v)v = -v. \tag{15.21}$$

This mode must therefore belong to the irreducible representation with characters

$$\chi_i(C_2) = 1, \quad \chi_i(2C_4) = 1, \quad \chi_i(2\sigma_v) = -1, \tag{15.22}$$

which corresponds to the Γ_2 representation. For modes with wavevector away from the Γ-point, the symmetry of the wavevector itself must also be taken into account. The M-point has the full symmetry of C_{4v}. The X-point has the reduced symmetry group C_{2v} (the symmetry of the rectangle) with the character table given in Table 15.2. Points along Δ, Z, and Σ have the still further reduced symmetry of C_{1h} (bilateral symmetry) with the character table given in Table 15.3. Points along Z have C_{1h} symmetry due to the fact that a mirror through the line orthogonal to the q_x direction brings Z to $Z+Q$, where Q is a reciprocal lattice translation vector.

The dispersion relations for the lowest few modes of the cylindrical post labeled by their irreducible representations are shown in Fig. 15.2(a). Notice that in Fig. 15.3, there is an anticrossing site in the TE modes along Z. Since the irreducible representations form an orthogonal basis, anticrossings can only occur between modes within the same irreducible representation. Indeed, this is the case here, as the two anticrossing modes are in the Z_2 irreducible representation.

Table 15.3 The character table for the group C_{1h}, which is the bilateral symmetry group, is given. For the cylindrical dielectric post structure, this group is exhibited along the Δ and Σ lines of symmetry in the Brillouin zone.

C_{1h}			E	σ_v
x^2, y^2, z^2, xy	R_z, x, y	Δ_1	1	1
xz, yz	R_x, R_y, z	Δ_2	1	–1

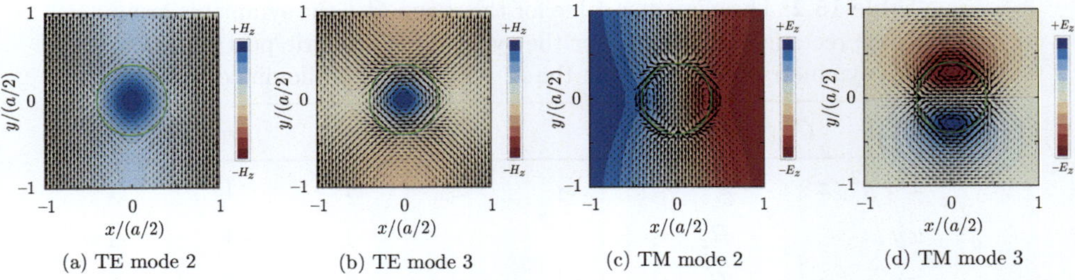

(a) TE mode 2 (b) TE mode 3 (c) TM mode 2 (d) TM mode 3

Figure 15.4 The electric and magnetic fields of the second and third modes corresponding to the Γ-point in the lattice of cylindrical posts are shown. For TE modes, the in-plane electric field is represented by vectors and the out-of-plane magnetic field is represented by the gradient background. For TM modes, the vectors represent the in-plane magnetic field and the background represents the strength of the out-of-plane electric field. Note that the first mode is not shown because the corresponding eigenvalue is zero, resulting in a trivial solution.

15.3 Eigenstates for periodic dielectric posts

The band structure for the lattice of dielectric posts was computed using a mesh of 4420 nodes, yielding a matrix size of 26520 × 26520. The mesh was refined in the region surrounding the edge of the cylindrical post. The curves shown in Fig. 15.2(a) give the behavior of the propagating frequencies of radiation at various points along the edge of the irreducible part of the Brillouin zone. The FEM reproduces a band gap in the TM modes which is also predicted by the plane-wave method. Using finite elements, it is also possible to increase the resolution close to the anticrossing site marked in Fig. 15.2(a). This is a location at which multiple eigenvalues of the same polarization (i.e. both TM or both TE) appear to touch. The close-up view of this point on the edge of the Brillouin zone is given in Fig. 15.3.

The eigenfunctions for the arrangement of cylindrical dielectric posts are shown in Figs. 15.4–15.6. Note that the point symmetries of each mode at the high-symmetry

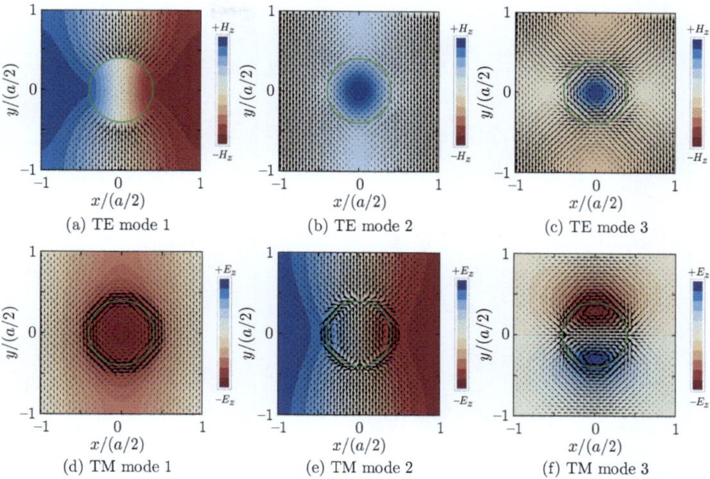

Figure 15.5 The electric and magnetic fields of the first three modes corresponding to the X-point in the lattice of cylindrical posts are shown. For TE modes, the in-plane electric field is represented by vectors and the out-of-plane magnetic field is represented by the gradient background. For TM modes, the vectors represent the in-plane magnetic field, and the background represents the strength of the out-of-plane electric field.

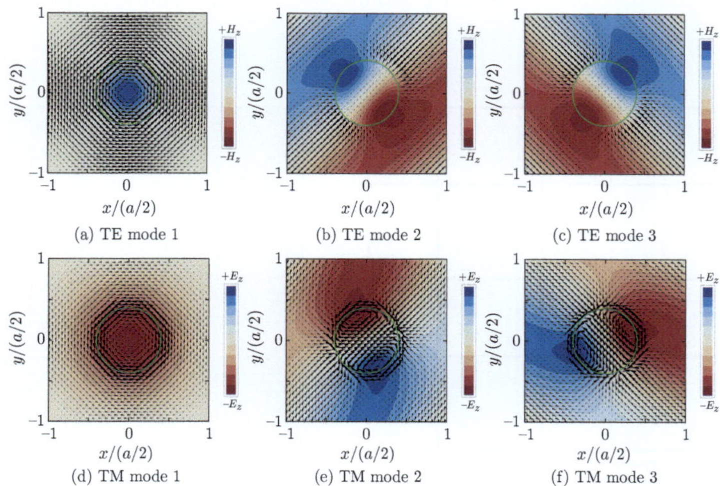

Figure 15.6 The electric and magnetic fields of the first three modes corresponding to the M-point in the lattice of cylindrical posts are shown. For TE modes, the in-plane electric field is represented by vectors and the out-of-plane magnetic field is represented by the gradient background. For TM modes, the vectors represent the in-plane magnetic field and the background represents the strength of the out-of-plane electric field.

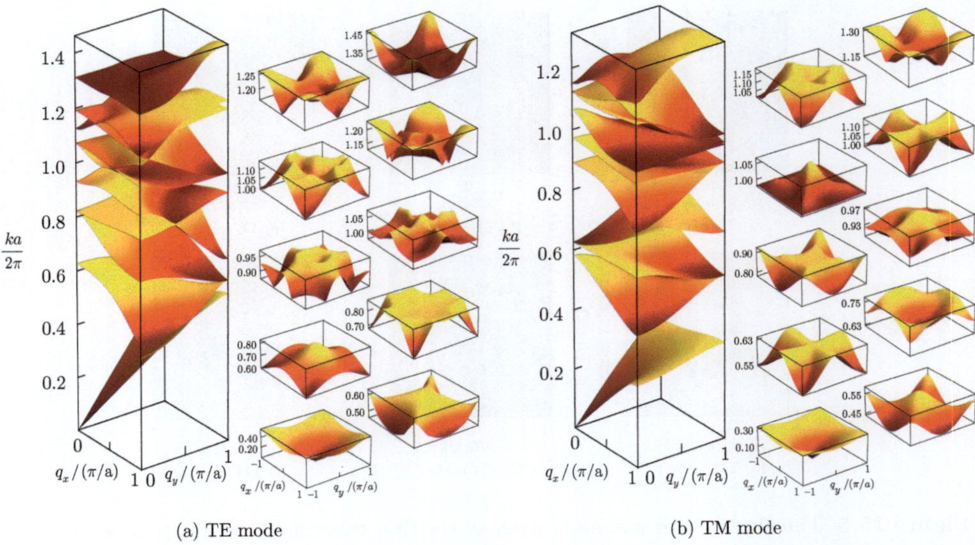

(a) TE mode

(b) TM mode

Figure 15.7 The eigenvalues of the transverse (a) electric modes and (b) magnetic modes of the periodic lattice of dielectric posts are plotted as surfaces in the first Brillouin zone. On the left side of (a) and (b), the first 10 eigenvalues are shown in the irreducible part of the Brillouin zone for TE and TM modes, respectively. On the right side, each eigenvalue has been separated from the rest and extended to the full Brillouin zone through symmetry operations.

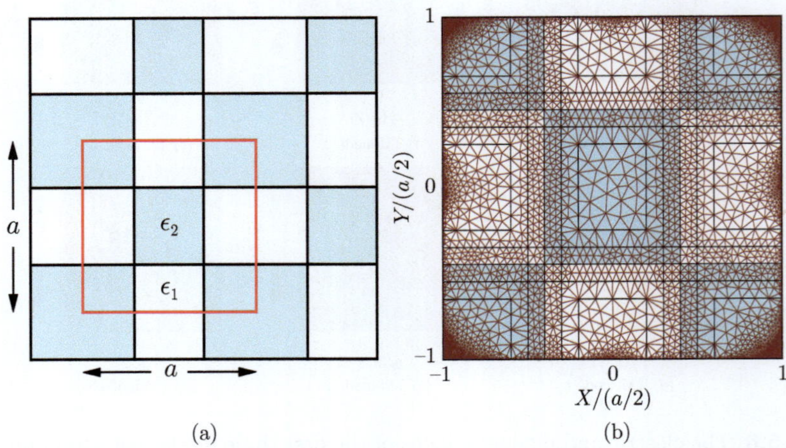

(a)

(b)

Figure 15.8 (a) A periodic checkerboard pattern with two alternating dielectric materials is shown. Note that the sizes of adjacent checkers within a single unit cell do not necessarily match. (b) A sample finite element mesh is given for the unit cell of a checkerboard lattice. Mesh refinement occurs at all of the checker boundaries.

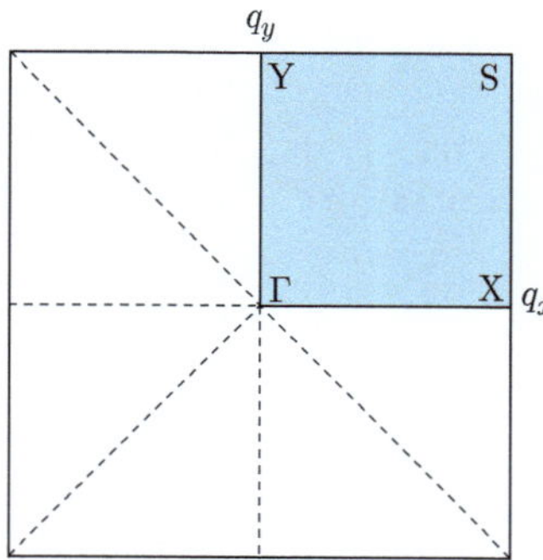

Figure 15.9 Eigenvalues for the transverse electric and transverse magnetic modes for the checkerboard arrangement.

points of Γ, X, and M can be used, along with Tables 15.1–15.3, to verify the symmetry groups shown in Fig. 15.2(a) by direct observation of the eigenvector fields.

The dispersion relations are calculated for the irreducible Brillouin zone, which is only one-eighth of the full Brillouin zone as shown in Fig. 15.1, and then their full reconstruction over the entire zone is performed. This can reduce computation time by a factor of 8. The lowest few TE and TM dispersion relations are shown in Figs. 15.7(a) and 15.7(b). These three-dimensional (3D) dispersion surfaces also provide another means of visualizing band gaps in the TE and TM modes, which are of great interest in PhC applications.

15.4 Eigenstates for a checkerboard lattice

The band structure for a checkerboard lattice was computed using a mesh of 12355 nodes, yielding a matrix size of 74130×74130. The mesh was refined in the region surrounding the edges within the checkerboard. Since the checkerboard lattice has more internal boundaries per unit cell than the cylindrical post geometry, a greater degree of mesh refinement was required, resulting in a larger global matrix than that of the lattice of cylindrical posts.

The eigenvalues are plotted over a triangular path between the Γ, X, and S points. Compared to the dielectric posts, the checkerboard shows much more activity and a denser band structure at low frequencies, but it has a smaller band gap in the TM modes. Like the cylindrical posts, this checkerboard has no TE band gap. The corresponding eigenfunctions for the lowest modes at the high-symmetry points are plotted in Figs. 15.10(a)–15.10(d).

The vectors represent the electric field in TE modes and the magnetic field in TM modes, while the shading of the background represents the intensity of the magnetic field in TE modes and the electric field in TM modes, with lighter shades corresponding to regions of greater field magnitude. Note that the eigenfunction for the lowest eigenvalue is omitted for the Γ-point for both modes of propagation.

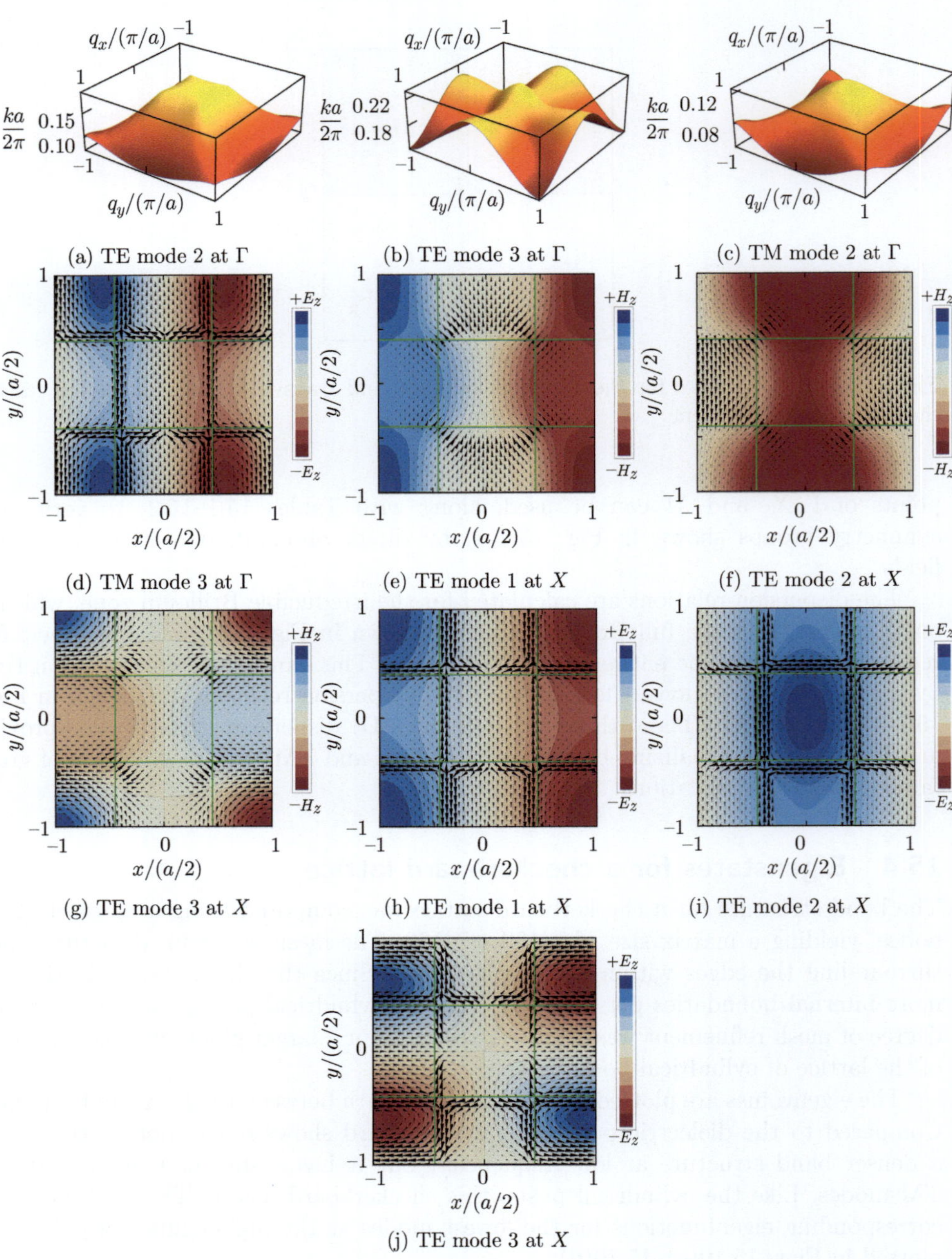

Figure 15.10 The electric and magnetic fields of the second and third modes corresponding to the Γ-point in the checkerboard lattice are shown from (a) to (d), while the fields of the first three modes corresponding to the X-point are shown from (e) to (j). For TE modes, the in-plane electric field is represented by vectors, and the out-of-plane magnetic field is represented by the gradient background. For TM modes, the vectors represent the in-plane magnetic field and the background represents the strength of the out-of-plane electric field. Note that the first mode is not shown because the corresponding eigenvalue is zero, resulting in a trivial solution.

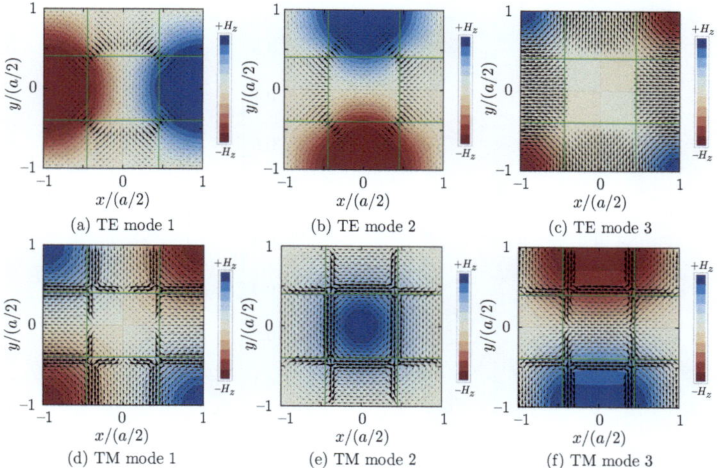

Figure 15.11 The electric and magnetic fields of the first three modes corresponding to the S-point in the checkerboard lattice are shown. For TE modes, the in-plane electric field is represented by vectors and the out-of-plane magnetic field is represented by the gradient background. For TM modes, the vectors represent the in-plane magnetic field and the background represents the strength of the out-of-plane electric field.

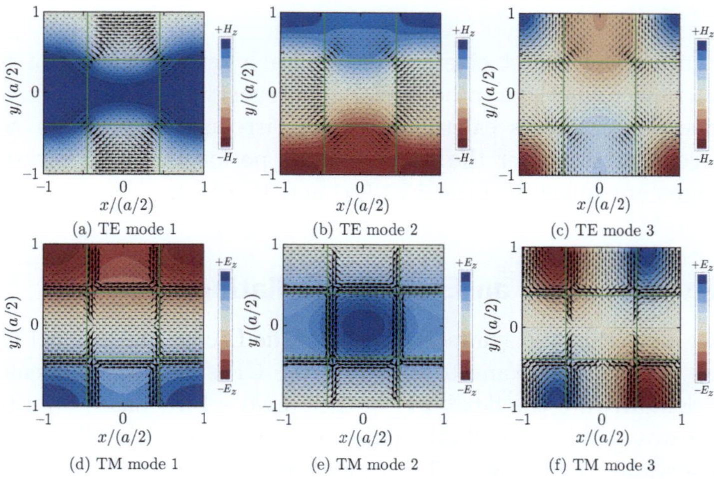

Figure 15.12 The electric and magnetic fields of the first three modes corresponding to the Y-point in the checkerboard lattice are shown. For TE modes, the in-plane electric field is represented by vectors and the out-of-plane magnetic field is represented by the gradient background. For TM modes, the vectors represent the in-plane magnetic field and the background represents the strength of the out-of-plane electric field.

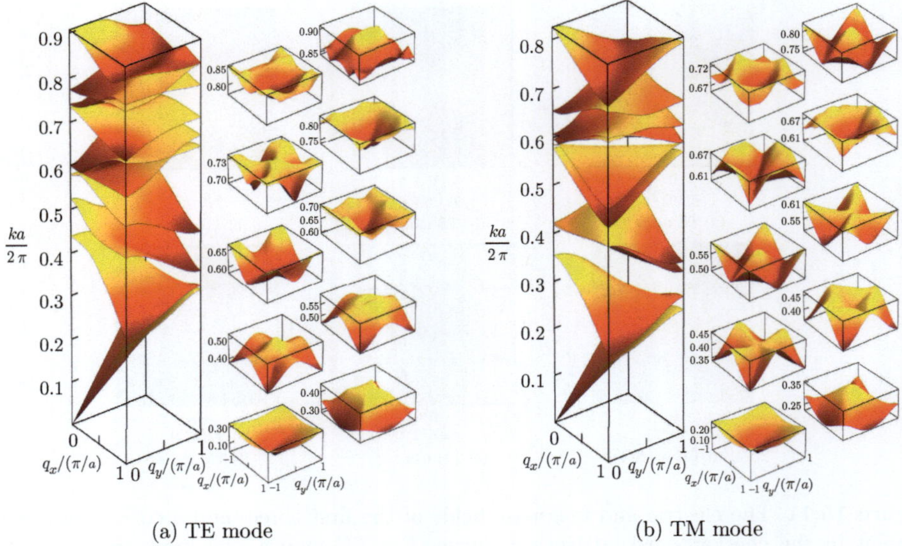

Figure 15.13 The eigenvalues of the transverse (a) electric modes and (b) magnetic modes of the checkerboard lattice of dielectric posts are plotted as surfaces in the first Brillouin zone. On the left side of (a) and (b), the first 10 eigenvalues are shown in the irreducible part of the Brillouin zone for TE and TM modes, respectively. On the right side, each eigenvalue has been separated from the rest and extended to the full Brillouin zone through symmetry operations.

This is because those lowest eigenvalues approach zero at the Γ-point, causing the corresponding eigenfunctions to be trivial (zero everywhere).

The dispersion relations calculated for the irreducible Brillouin zone and then their full reconstruction over the entire zone is performed. The lowest few TE and TM dispersions are shown in Figs. 15.13(a) and 15.13(b).

15.5 Eigenstates for an Escher tessellation

Coxeter has shown that the symmetry group of the Escher drawing is PG; [41] and the drawing was also used by Yang to illustrate the CP symmetry of weak interactions. In order to demonstrate the flexibility of the FEM, a PhC based on a tessellation by M. C. Escher was simulated and its band structure was calculated. The image used to produce the crystal was Escher's "Horsemen" as shown in Fig. 15.14. A sample mesh is given in Fig. 15.14(b).

The band structure for the Escher tessellation was computed using a mesh of 54945 nodes, yielding a matrix size of 329670 × 329670. Since the Escher unit cell does not have the same reflection symmetries as cylindrical and checkerboard unit cells, the entire Brillouin zone was tested instead of a small fraction of it.

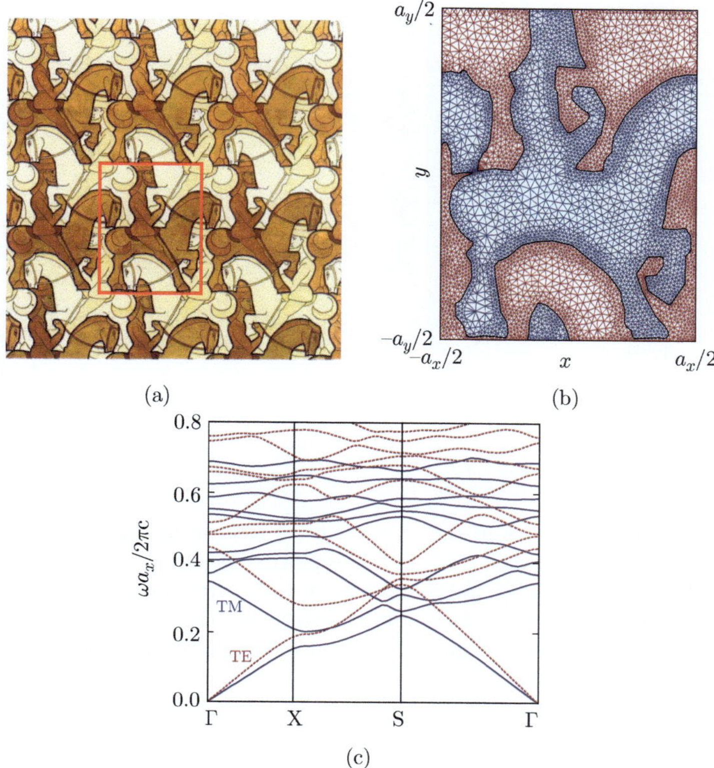

Figure 15.14 (a) The unit cell of a PhC based on "The Horsemen" by M. C. Escher is shown. The Escher tessellation was chosen to illustrate the capacity of the HFEM to calculate the band structures of complicated geometries with unique symmetry properties. [40] (b) A sample mesh is given for the unit cell of an Escher tessellation. The regions are assigned $\epsilon_r = 8.9, 1.0$ to form a 2D PhC. Most calculations used more refined meshes than shown in (b), including $54,945$ nodes for a total of $329,670$ global degrees of freedom. (c) TE (dashed) and TM (solid) eigenmodes for the associated 2D PhC are shown.

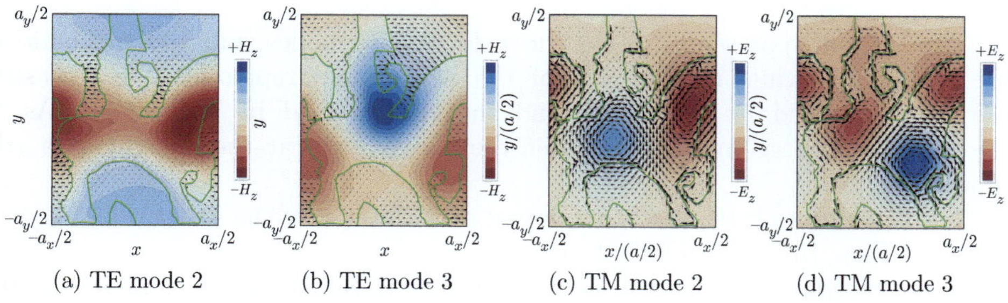

(a) TE mode 2 (b) TE mode 3 (c) TM mode 2 (d) TM mode 3

Figure 15.15 The **E** and **H** fields of the second and third modes at Γ for the Escher lattice are shown. For (a,b), $E_{x,y}$ are represented by vectors and H_z by the contours. For (c,d), $H_{x,y}$ are represented by vectors and E_z by the contours.

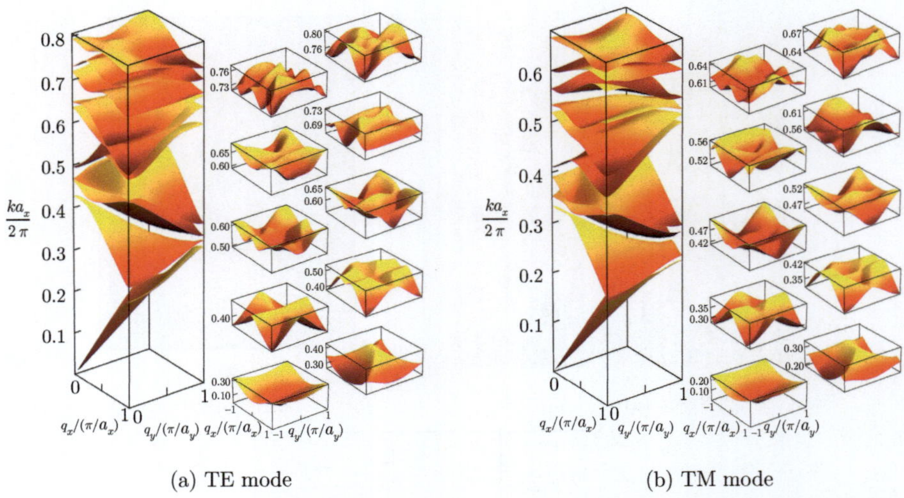

(a) TE mode (b) TM mode

Figure 15.16 The eigenvalues of the transverse (a) electric modes and (b) magnetic modes of the Escher superlattice of dielectric posts are plotted as surfaces in the first Brillouin zone. On the left side of (a) and (b), the first 10 eigenvalues are shown in the irreducible part of the Brillouin zone for TE and TM modes, respectively. On the right side, each eigenvalue has been separated from the rest and extended to the full Brillouin zone through symmetry operations.

The eigenvalues are plotted over a triangular path between the Γ, X, and S points in Fig. 15.14(c). The transverse electric modes appear to converge to a band structure similar to that of the cylindrical post, even featuring an anticrossing site in approximately the same position. However, the transverse magnetic modes fail to converge properly, even when using several tens of thousands of nodes. This may be due to the high complexity of the dielectric structure coupled with the slow error convergence of the action formulation based on $\mathbf{D}$. The corresponding eigenfunctions for the lowest modes at the high-symmetry points are plotted in Figs. 15.15(a)–15.15(d).

The vectors represent the electric field in TE modes and the magnetic field in TM modes, while the shading of the background represents the intensity of the magnetic field in TE modes and the electric field in TM modes. As with the other crystal geometries, we omit the lowest state at the Γ-point as a trivial solution.

The eigenvalues have also been plotted as surfaces over the first Brillouin zone, as shown in Figs. 15.16(a) and 15.16(b).

In conclusion, we anticipate that the use of Hermite FEM will allow the treatment of multiscale problems associated with PhCs with embedded quantum dots, defects, and the like. The spatial representation of the fields using Hermite triangular interpolation is much more economical than employing plane-wave methods for such structures allowing the deployment of more finite elements strategically in specific regions as needed. The resulting global matrices are still sparse and banded due to the local connectivity, which leads to far more compact matrices than in other schemes with the concomitant reduction in compute-time. While the transverse magnetic modes continue to converge slowly in complex geometries, the efficient calculation of the TE modes allows one to easily determine which geometries have TE band gaps. Furthermore, the extension to 3D crystals, in which a separate formulation based on **D** is no longer needed, may alleviate this problem.

15.6 Discussion

Given the periodic nature of the PhCs, it is natural that the vast majority of published works on photonic band structure calculations are analogous to the reciprocal space analysis common to the solid state physics analysis of propagating electronic states in crystals. The analogy to solid state has key flaws. In particular, the abrupt, macroscopic discontinuities of dielectric regions make it difficult to transform the dielectric function and EM states between reciprocal space and real space without artifacts. In practical use, these numerical artifacts can become sources of serious error when subsequent calculations in real space are required or the system symmetry is lowered. Interactions with sub-wavelength features, such as quantum structures that are most often at interfaces, can be very difficult to resolve if the EM field description is coarse. Other demanding examples are the computation of localized states (such as defects) and slab geometries. [42] Under conditions of a periodic slab, the predominant approach is to move away from the use of 10^6 plane waves (the 2D periodic system of a slab), plane waves corresponding to the third dimension, plus supercells corresponding to any irregularity [43, 44] toward the time domain where finite-difference time-domain calculations are now widely used for the calculation of real-space EM fields.

Vector finite element method calculations for the eigenmodes of PhCs and VFEM-time-domain calculations for waveguide and defect geometries are not common, but examples include the real-space construction of localized Wannier basis function from the perfect crystal eigenmodes to compute localized defect modes. [45] One disadvantage of such calculations is the pixelization of the resulting fields owing to the lower order of normal field continuity in vector element formalism or the spatial gridding of finite-difference time domain analysis. This mixed-order real-space description results in an EM field that is spatially coarser than the quantum mechanical features of embedded solid state structures such as quantum dots, wells, and other features common to modern semiconductor devices.

It is ultimately more desirable to obtain field patterns that have continuous spatial derivatives within dielectric layers for convenient calculation of quantum mechanical or deeply sub-wavelength interactions. The present HFEM approach provides smoothly varying EM wavefunctions using a nodal mesh description and derivative continuity, yet preserves the necessary boundary conditions and numerical constraints that have been demanded of the VFEM.

For decades, PhC analysis by reciprocal space techniques has produced accurate eigenvalue results as would be expected of a variational approach, however the HFEM offers a flexible, robust means of computing the eigenstates of a PhC with much more physical eigenfunctions at far lower computational cost. In our calculations, we note that the typical matrix dimensions for the PhC with square geometry containing cylindrical posts are on the order of 26×10^3; however, the local connectivity within the HFEM leads to a banded matrix with 0.158% occupancy. This sparsity is a demonstrable advantage over the plane-wave method. The ability to construct the field distributions from the nodal eigenvectors with no discontinuities in the reconstructed function and its derivatives is an additional benefit and allows for a high-quality description of the dual fields. Furthermore, the plane waves are global functions, and the eigenfunctions constructed using these functions have the usual errors on the order of the square root of the errors in the eigenvalues. However, in the FEM, this error can be distributed non-uniformly by emphasizing areas (or volumes) of interest through the redistribution of elements, putting more elements in those areas that are of particular interest and fewer elsewhere. The detailed agreement with the published results for the square lattice of dielectric posts shows that the HFEM provides accurate, reliable results that are derivable with banded, sparse matrices. The anticrossing of the first and second TE dispersions between the X and M points for the case of a PhC of cylindrical posts arrayed in a square lattice has been highlighted, and the degeneracies at the special points have been indicated in the displayed wavefunctions.

We have shown that the Hermite interpolation polynomials on a triangular element are able to eliminate the spurious solutions that typically occur with Lagrange-type scalar shape functions. In the course of this work, other cases were studied with comparable results to the cases given in detail here, such as the photonic band structure for a checkerboard superlattice of dielectrics. This was done because abrupt corners and interfaces gives rise to severe Gibbs phenomena in the spatial reconstruction of the EM fields. The ease with which the band structure and the wavefunctions are obtained using the HFEM framework holds great promise for obtaining physical eigenfunctions within a variational approach that are absent of many numerical artifacts that pervade reciprocal space methods.

The ability of the FEM to represent complex geometries is highlighted by considering the use of an Escher tessellation to define a unit cell of a PhC. The same example treated with reciprocal space methods would require an enormous number of Fourier components to capture the details of the geometry.

For the HFEM to be fully extensible across a wider range of PhC modeling, it will be necessary to demonstrate its applicability in three dimensions. The development of a practical, $\mathcal{C}_{(1)}$-continuous set of Hermite interpolating polynomials in three dimensions will make this method a feasible option for engineering applications. Indeed, we have already developed such polynomials in 3D, using group representation theory. As a nodal based method, the HFEM is suited for domain decomposition and can be connected with existing frameworks for treating the open domains of finite systems as well. A final feature that would be required in this context is the exploration of the time-domain evolution of solutions. The finite element time-domain techniques that are already prevalent in modeling such structures can readily incorporate the new developments we are offering here.

In conclusion, we can anticipate that the use of HFEM will allow the treatment of multiscale and multi-physics problems that require detailed spatial descriptions of EM fields. With the solutions given on the vertices of the triangle and the continuity guaranteed both for the normal and tangential derivatives at triangle interfaces, it becomes substantially simpler to mix scalar-vector field calculations involving curl operators. This a distinct advantage over reciprocal space methods as we have shown through cases designed to stress the sophistication of the spatial reconstruction of fields.

References

[1] J. C. Nedelec, Numerische Mathematik **35**, 315–341 (1980); "Mixed finite elements in R3."

[2] D. K. Sun, J. Manges, X. Yuan, and Z. J. Cendes, IEEE Trans. Antennas Propag. **37**, 12–24 (1995); "Spurious modes in finite element methods."

[3] D. R. Lynch, and K. D. Paulsen, IEEE Trans. Microw. Theory Techn. **39**, 383–394 (1991); "Origin of vector parasites in numerical Maxwell solutions."

[4] D. R. Lynch and K. D. Paulsen, IEEE Trans. Microw. Theory Techn. **39**, 395–403 (1991); "Elimination of vector parasites in finite element Maxwell solutions."

[5] J. F. Lee, D. K. Sun, and Z. J. Cendes, IEEE Trans. Microw. Theory Techn. **27**, 1262–1271 (1991); "Full-wave analysis of dielectric waveguides using tangential vector finite elements."

[6] J. F. Lee, D. K. Sun, and Z. J. Cendes, IEEE Trans. Magn. **27**, 4032–4035 (1991); "Tangential vector finite elements for EM field computation."

[7] A. Kameari, IEEE Trans. Magn. **35**, 1394–1397 (1999); "Symmetric second order edge elements for triangles and tetrahedra."

[8] A. F. Peterson, IEEE Trans. Antennas Propag. **43**, 357–365 (1994). "Vector finite element formulation for scattering from two-dimensional heterogeneous bodies."

[9] A. Ahagon and T. Kashimoto, IEEE Trans. Magn. **31**, 1753–1756 (1995). "Three-dimensional electromagnetic wave analysis using high order edge elements."

[10] C. J. Reddy, M. D. Deshpande, C. R. Cockrell, and F. B. Beck, NASA Technical Paper **3485**, Dec 1994. "Finite element method for Eigenvalue problems in electromagnetics."

[11] J. P. Webb, IEEE Trans. Magn. **24**, 162–165 (1988); "Efficient generation of divergence-free fields for the finite element analysis of 3D cavity resonances."

[12] J. P. Webb, Rep. Prog. Phys. **58**, 1673–1712 (1995); "Application of the finite-element method to eletromagnetics and electric topics."

[13] H. S. Carslaw, *Introduction to the Theory of Fourier's Series and Integrals*, 3rd edition (Dover Publications, New York, NY, 1950). R. V. Churchill and J. W. Brown, *Fourier Series and Boundary Value Problems* (McGraw-Hill, New York, NY,).
[14] O. Tuncer, Chuan Lu, N. V. Nair, B. Shanker, and L. C. Kempel, IEEE Trans. Antennas Propag. **58**, 887–889 (2010); "Further development of vector generalized finite element method and its hybridization with boundary integrals."
[15] Z. Li and L. R. Ram-Mohan, Phys. Rev. E **85**, 016706–1–14 (2012); "Taming hypersingular integrals using dimensional continuation."
[16] L. Xuan, B. Shanker, Z. Zeng, and L. Udpa, Int. J. Appl. Electromagn. Mech. **19**, 463–466 (2004); "Element-free Galerkin method in pulsed eddy currents."
[17] J. Sladek, V. Sladek, and E. Pan, Int. J. Solids Struct. **50**, 3975–3983 (2013); "Bending analyses of 1D orthorhombic quasicrystal plates."
[18] G. R. Liu and Y. T. Gu, *An Introduction to Meshfree Methods and Their Programming*, (Springer, Dorderecht, 2005).
[19] S. G. Johnson and J. D. Joannopoulos, *Photonic Crystals: The Road from Theory to Practice* (Kluwer, Massachusetts, 2002), and references therein.
[20] E. Yablonovitch, Phys. Rev. Lett. **58**, 2059–2062 (1987); "Inhibited spontaneous emission in solid-state physics and electronics."
[21] S. John, Phys. Rev. Lett. **58**, 2486–2489 (1987); "Strong localization of photons in certain disordered dielectric superlattices."
[22] K. Ohtaka and M. Inoue, Phys. Rev. B **19**, 5057–5067 (1979); "Energy band of photons and low-energy photon diffraction."
[23] K. Sakoda, Phys. Rev. B **52**, 7982–7986 (1995); "Symmetry, degeneracy, and uncoupled modes in two-dimensional photonic lattices."
[24] K. Sakoda, Phys. Rev. B **55**, 15345–15348 (1997); "Group-theoretical classification of eigenmodes in three-dimensional photonic lattices."
[25] K. Ohtaka and Y. Tanabe, J. Phys. Soc. Jpn. **65**, 2670–2684 (1996); "Photonic bands using vector spherical waves. III. Group-theoretical treatment."
[26] K. Sakoda, *Optical Properties of Photonic Crystals* (Springer, Berlin, Germany, 2001).
[27] J. B. Pendry, J. Mod. Opt. **41**, 209–229 (1994); "Photonic band structures."
[28] J. B. Pendry, Phys. Rev. Lett. **85** 3966–3969 (2000); "Negative refraction makes a perfect lens."
[29] D. R. Smith, Willie J. Padilla, D. C. Vier, S. C. Nemat-Nasser, and S. Schultz, Phys. Rev. Lett. **84**, 4184–4187 (2000); "A composite medium with simultaneously negative permeability and permittivity."
[30] H. Benisty, V. Berger, J.-M Gérard, D. Maystre, and A. Tchelnokov, *Photonic Crystals: Towards Nanoscale Photonic Devices* (Springer, Berlin, Germany, 2005).
[31] V. G. Veselago, Soviet Phys. Usp. **10**, 509–514 (1968); Usp. Fiz. Nauk **92**, 517–526 (1967); "The electrodynamics of substances with simultaneously negative values of ϵ and μ."
[32] C. Kittel, *Introduction to Solid State Physics* (John Wiley & Sons, New York, 2005).

[33] N. Mahmoud I. Hussein, Proc. R. Soc. A **465**, 2825–2848 (2009); "Reduced Bloch mode expansion for periodic media band structure calculations."

[34] L. R. Ram-Mohan, *Finite Element and Boundary Element Applications in Quantum Mechanics* (Oxford, New York, 2002).

[35] J. Jin, *The Finite Element Method in Electromagnetics*, 2nd edition (Wiley, New York, 2002).

[36] F. Bloch, Zietshrift der Physik **52**, 555–600 (1928); "Über die quantenmechanik der electronen in kristallgittern."

[37] G. Floquet, Annales de l'École Normale Supérieure **12**, 47–88 (1883); "Sur les équations différentielles linéaries à coefficients périodiques."

[38] J. D. Joannopoulos, S. G. Johnson, J. N. Winn, and R. D. Meade, *Photonic Crystals: Molding the Flow of Light, 2nd edition*, (Princeton, New Jersey, 2008), pp. 66–71. (The solid curves in Fig. 1(a) are reproduced with permission from Princeton University Press.)

[39] M. S. Dresselhaus, G. Dresselhaus, and A. Jorio, *Group Theory: Application to the Physics of Condensed Matter* (Springer, Berlin, Germany, 2008).

[40] Fig. 4(a) has been used by permission, www.mcescher.com. Copyrights for all M.C. Escher works© 2010 are with the M.C. Escher Company - the Netherlands. All rights reserved.

[41] H. S. M. Coxeter, *Introduction to Geometry* (John Wiley and Sons, New York, NY, 1969); pp. 57–58. The same Escher drawing was invoked very effectively by C. N. Yang to illustrate CP symmetry in weak interactions using color conjugacy and glide symmetry in the drawing instead of charge conjugation and parity. See also: C. N. Yang, *Elementary Particles: A Short History of Some Discoveries in Atomic Physics* (Princeton University Press, Princeton, NJ, 1961).

[42] B. Jiang, W. Zhou, W. Chen, A. Liu, and W. Zheng, J. Appl. Phys. **111**, 053103–1–6 (2012); "Improved fake mode free plane-wave expansion and three planar-slab waveguides method."

[43] S. Fan and J. D. Joannopoulos, Phys. Rev. B **65**, 235112–1–8 (2002); "Analysis of guided resonances in photonic crystal slabs."

[44] S. G. Johnson, S. Fan, P. R. Villeneuve, J. D. Joannopoulos, and L. A. Kolodziejski, Phys. Rev. B **60**, 5751–5758 (1999); "Guided modes in photonic crystal slabs."

[45] P. Sotirelis and J. D. Albrecht, Phys. Rev. B **76**, 075123–1–8 (2007); "Numerical simulation of photonic crystal defect modes using unstructured grids and Wannier functions."

16 Cavity electrodynamics and symmetries

In this chapter:

- We consider the issue of removing spurious zero-frequency solutions. In 3D, this has to b re-examined; an active approach is required to identify such modes in cavity electrodynamics and eliminating them.
- We investigate fields in an empty cubic metallic cavity and explain the level degeneracy that is larger than what is to be expected from the geometrical O_h symmetry of the cube. This behavior is identified as an example of "accidental degeneracy," and is explained in detail. We show that the inclusion of a smaller dielectric cube of relative permittivity ϵ_2 within the cubic cavity leads to the removal of this accidental degeneracy so that the eigenfields have the symmetry O_h. A further reduction of symmetry is obtained by allowing the dielectric function in the enclosed cube to have a linear dependence on z.
- The proposed method should be effective in obtaining results for scalar-vector coupled field problems such as in modeling quantum well cavity lasers and in plasmonics modeling, while allowing multiscale physical calculations.

16.1 Introduction

We showed in Chap. 15 that the present approaches for the solution of Maxwell's equations in complex geometries have limitations that can be overcome through the use of $C_{(1)}$-continuous Hermite interpolation polynomials. Our approach of calculating fields using the Hermite finite element method yields better accuracy by several orders of magnitude than comparable applications of the edge-based vector finite element method (VFEM). We note that the vector finite element that is widely used yields pixelated solutions, and ill-defined vector solutions at nodes. Our solutions have smooth representation within and across the elements, and well-defined directions for the fields at the nodes.

Finite Elements in Action. L. Ramdas Ram-Mohan, Oxford University Press. © L. Ramdas Ram-Mohan (2026).
DOI: 10.1093/oso/9780199563487.003.0016

Obtaining solutions to Maxwell's electromagnetic field equations is a challenge when we encounter the presence of confining boundaries and material interfaces. This is evidenced by the theory of waveguides, cavities, photonic crystals, and structures being developed for plasmonics. [1–4] The inclusion of inhomogeneities in waveguides were solved by Schwinger, [5] who pioneered variational methods for such problems for radar applications. More recent calculations for antenna design, optical fibers, quantum well lasers in resonant cavities, and high-frequency electronic circuit design – all are examples illustrating the need for reliable and accurate methods for their simulation. The urgency for the development of such methods lies in the ever-increasing electronic and electrodynamic component density in devices. [6] The amelioration of cross-talk between components here for electromagnetics, in some sense, parallels the effort toward adding more transistors on a single computer chip. The geometrical complexities of such systems demand a robust approach to the modeling of the electromagnetic fields. The finite element method, which is rooted in the principle of stationary action, is well established an approach toward a variational solution obtained by discretizing the physical domain, and is a natural choice to overcome the geometrical complications.

Most design cycles for EM devices require extensive comparison to experimental measurements and further redesign. This iterative process is very expensive in time, and in fabrication costs. The development of "right-the-first-time" computational schemes for obtaining solutions for EM fields with spatial resolution adequate for multiscale problems has thus been an area of active investigation for decades.† A computational scheme that can obtain high spatial resolution over multiscale systems and low discretization errors is necessary for integrating full-wave electromagnetics when small (nanometer-sized) active devices interact with metals and dielectrics that are far from ideal.

Electromagnetic fields in a physical system should satisfy the wave equation along with a divergence-free condition. For example, in the electric field formulation (E-field), we solve Maxwell's equations

$$\nabla \times \nabla \times \mathbf{E} = \epsilon \mu \omega^2 \mathbf{E}; \tag{16.1}$$

$$\nabla \cdot \epsilon \mathbf{E} = 0, \tag{16.2}$$

where ω is the eigenfrequency and ϵ and μ are the permittivity and permeability of the material. Computationally, if we simply attempt to solve the wave equation Eq. (16.1), we are not guaranteed that the obtained solutions satisfy the divergence-free condition of Eq. (16.2). Solutions with either zero frequency ($\omega = 0$) or with nonzero divergence ($\nabla \cdot \epsilon \mathbf{E} \neq 0$) that are obtained while solving Eq. (16.1) are known as the spurious (nonphysical) solutions. Such spurious solutions also corrupt the desired eigenspectra.

In two dimensional (2D) waveguides, (see Chap. 7), we circumvent this issue by solving for only one of the field components (H_z or E_z), and obtain the other two components using boundary conditions (BCs) and the divergence-free condition. [7–9]

†We suggest that a new version of "Moore's Law" for photonic and electromagnetic integrated circuits will play a role in future device developments.

However, such freedom does not exist in three dimensions (3D). In a numerical calculation, the anticipated zero-frequency solutions can have nonzero frequencies due to discretization, and will occur inter-mixed with the physical spectrum. [10, 11]

The physical region is discretized into triangles and squares in 2D, and tetrahedra and cubes in 3D; the representation of fields in each of these finite elements is in terms of so-called edge-based interpolation polynomials. Due to the nature of vector elements, field directions are ill-defined at each node, hence making VFEM unsuitable to be employed in several problems: particle trajectories in electron microscope design, vacuum tubes, and accelerators. VFEM is also ineffective in modeling multi-physics problems which will involve quantum mechanical and electrodynamic aspects in tandem, such as designing cavity lasers. In such instances, independent numerical treatments are needed to obtain the wavefunctions and EM fields. Furthermore, Nedelec-compliant edge elements [12–17] do not have normal continuity of the displacement vector **D** along element interfaces. [18] This discontinuity is unphysical, and leads to the formation of artificial charges, [19] pushing the solutions toward nonzero divergence values.

Nodal representation of field components with scalar functions and their derivatives results in an unambiguous field description at shared nodes among adjacent elements and treats the field components on a uniform footing. [20] The built-in derivative degrees of freedom (DoFs) allow us to readily calculate the additional quantities such as surface currents, while providing smoother solutions. This treatment results in global functions without severe coarsening which will be suitable for analyzing interactions with small features, such as those found in high-frequency transistor circuits.

In this chapter, we propose a method of using scalar Hermite interpolation polynomials (HFEM), which employs polynomials that correspond to the function and derivative values (i.e. the DoFs) at each node on a cube. These polynomials are constructed from the outer product of the 1D Hermite polynomials. [21] *We then have polynomials with both tangential and normal derivative continuity across the element boundary.* We note that:

The present approach yields better accuracy, with a smoother representation of fields than those obtained using VFEM. The HFEM scheme yields several orders of higher accuracy with fewer elements than those needed in the presently prevalent 3D implementations of VFEM. [22–24]

The node-based HFEM representation guarantees consistency in the direction of fields at the vertices of elements. Multiscale calculations can now be done easily, something that is not feasible with VFEM due to the lack of unique directionality for fields at shared nodes in the finite element mesh.

We impose the divergence-free condition through a constant Lagrange multiplier term introduced into the action integral, and also by explicitly requiring a zero-divergence condition at each node through the derivative DoFs available at each node in our formulation.

Any surviving $\omega \neq 0$ spurious solutions are then eliminated by identifying them using their large $|\nabla \cdot \mathbf{E}|/|\nabla \times \mathbf{E}|$ ratio. We demonstrate that this procedure does not alter or influence the accuracy of the physical solutions.

In VFEM implementations, the eigenfrequencies of spurious solutions are pushed to zero, either through the Nedelec conditions or through their removal at each iteration, as mentioned earlier. In either case, this is an expensive numerical procedure. In the literature, the first approach for eliminating spurious solutions with scalar polynomials was the penalty factor method (i. e. a Lagrange multiplier scheme). [25] However, a fixed choice of the penalty factor fails to impose the zero-divergence condition adequately for all frequencies. [26] Furthermore, the penalty term itself can introduce an additional set of spurious solutions. [27]

Within the HFEM framework, we now have the luxury of explicitly imposing a zero-divergence condition at each node while using Hermite interpolation polynomials since we have derivative DoFs there. [28–30] While this does not ensure the complete removal of the divergence in the interior of the finite element through interpolation, it reduces it substantially, especially as the size of the element is reduced. This is demonstrated explicitly below. In the calculations, we use a constant penalty factor, and impose zero-divergence at all nodes to identify the spurious solutions for elimination.

We calculate the electric fields in a metallic cubic cavity. We recognize the existence of a degeneracy of frequencies in the spectrum larger than predicted by the geometrical symmetry of the system. We attribute this to the presence of accidental degeneracy and, the system provides a rigorous test for our eigenvalue analysis. We present the additional operators that explain the additional degeneracies in the frequency spectrum beyond the degeneracies of $\{1, 2, 3\}$ allowed by the geometric symmetry, O_h. Introducing a region filled with a dielectric material inside the cavity removes this additional degeneracy, returning it to that corresponding to the geometrically recognized symmetry. The splitting in mode frequencies corresponding to such a reduction in degeneracy is predicted using group representation theory. These results are verified from the eigenvalues and explicit field distributions calculated using HFEM.

In Sec. 16.2, the HFEM formulation is discussed, along with a synopsis of the origin of spurious solutions in EM problems. This is in contrast with the earlier use of finite elements, where the use of these additional constraints have not been able to resolve the issues completely. In Sec. 16.3, we use HFEM to solve Maxwell's equations in an empty cubic cavity with conducting boundaries in order to verify the method with analytically obtained solutions. In Sec. 16.4, we discuss the presence of accidental degeneracy in the eigenvalues obtained, and identify the new operators that extend the symmetry group from O_h to a larger group which accounts for the additional degenerate states. Fields in a loaded dielectric cavity are analyzed in detail in Sec. 16.5. Group representation theory is used to determine theoretically the splitting of mode degeneracies when the dielectric is included. The surface-induced currents for various modes are displayed in figures that yield further insights into the symmetry and structure of the modes. A primer on the accidental degeneracy, and the role of symmetry in electromagnetic calculations is presented in Sec. 16.6. Concluding remarks are given in Sec. 16.7.

16.2 The HFEM formulation for the cubic cavity

16.2.1 The action integral for the cubic cavity

We wish to solve Maxwell's equations for time-harmonic fields, so that the electric field $\mathbf{E}(\mathbf{r},t) = \mathbf{E}(\mathbf{r})e^{-i\omega t}$, and similarly for the magnetic field $\mathbf{H}$. Maxwell's equations then reduce to the 3D wave equation. For the electric field,

$$\nabla \times \left[\mu_r^{-1} \nabla \times \mathbf{E}\right] = k_0^2 \, \epsilon_r \, \mathbf{E}, \tag{16.3}$$

where ϵ_r is the relative permittivity tensor, μ_r is the relative permeability tensor, and the eigenvalue $k_0 = \omega / c$. The electromagnetic wave equation can be cast into a functional $\mathcal{S}$ (action integral) of the form

$$\mathcal{S} = \int_V dV \left[\nabla \times \mathbf{E}^* \cdot \mu_r^{-1} \nabla \times \mathbf{E} - k_0^2 \, \mathbf{E}^* \cdot \epsilon_r \, \mathbf{E}\right]. \tag{16.4}$$

Additionally to minimize the divergence, we introduce a penalty term $\lambda \left|\nabla \cdot \epsilon_r \mathbf{E}\right|^2$ in the functional $\mathcal{S}$, given by

$$\mathcal{S} = \int_V dV \left[\nabla \times \mathbf{E}^* \cdot \mu_r^{-1} \nabla \times \mathbf{E} - k_0^2 \, \mathbf{E}^* \cdot \epsilon_r \, \mathbf{E} + \lambda \left|\nabla \cdot \epsilon_r \mathbf{E}\right|^2\right], \tag{16.5}$$

where λ is the Lagrange multiplier. The fields are represented by Hermite interpolation polynomials on hexahedral elements multiplied by the values of fields and their derivatives at the vertices (nodes) of each element. The integral can be discretized over the elements to obtain a matrix equation in terms of the nodal parameters. We invoke the principle of stationary action, and set the variation of $\mathcal{S}$ with respect to $\mathbf{E}^*$ equal to zero. This yields a generalized eigenvalue problem which is solved to obtain the frequencies and field distributions of the resonating modes. The magnetic fields are readily obtained from the electric fields using Maxwell's equations.

We consider a cubic cavity with perfectly conducting metallic boundaries. We assume that the dielectric regions of the cavity are charge-free and current-free. At the surface of a perfect electrical conductor, the electric and magnetic fields satisfy the BCs [1, 2, 4]

$$\hat{\mathbf{n}} \times \mathbf{E} = 0 \quad \text{and} \quad \hat{\mathbf{n}} \cdot \mathbf{H} = 0. \tag{16.6}$$

These relations give the BCs on the periphery of the cavity. When working with electric fields, the tangential components of the field are set to zero at the boundary, while the normal components are determined variationally.

16.2.2 Origin and nature of spurious solutions

Numerical solutions of Maxwell's equations are polluted with non-physical spurious solutions. These have their origin in the wave equation itself. Taking the divergence on both sides of Eq. (16.3) gives

$$k_0^2 \, \nabla \cdot \epsilon_r \, \mathbf{E} = 0. \tag{16.7}$$

This condition is satisfied when either $k_0 = 0$ or $\nabla \cdot \epsilon_r \, \mathbf{E} = 0$. The solutions can now be classified into two categories:

1. $k_0 \neq 0$, and $\nabla \cdot \epsilon_r \, \mathbf{E} = 0$,
2. $k_0 = 0$, and $\nabla \cdot \epsilon_r \, \mathbf{E} \neq 0$.

Solutions falling in the first category are the physical solutions, while those falling in the second category are spurious ones. In theory, these spurious solutions have zero frequency. However, due to discretization, the eigenvalues of the spurious modes are not computed exactly as zero, and can have numerical values comparable to those of the physical solutions. Consequently, the spurious eigenvalues cannot be easily separated from the desired eigenvalues. [11, 31] For the time-harmonic problem, spurious solutions have zero curl and a finite divergence. This manifests numerically as a very large divergence-to-curl ratio, compared to physically admissible solutions. In the following section, we show within our HFEM approach, we can increase this ratio with increase in the mesh density. We use this criterion to filter out spurious solutions from the physical admissible solutions. Examples of the spurious solutions are shown in Fig. 16.1. Note how the field distribution shows source-like behavior within the cavity, indicative of nonzero divergence.

16.2.3 The penalty method and the $\nabla \cdot \mathbf{E} = 0$ constraint

The penalty method [25] has been proposed to remove spurious solutions in nodal finite element implementations. The penalty term pushes most spurious eigenfrequencies outside the spectral range of interest. However, a fixed Lagrange multiplier does not remove all spurious modes. A low value of the penalty factor leaves behind some spurious modes, and a large value of penalty factor causes errors in the eigenvalues. One proposed algorithm is to use a different multiplier for each mode. [26] This, however, is an expensive iterative scheme. Nevertheless, a constant penalty offers an inexpensive method for removing most of the zero-frequency spurious modes, and can be further enhanced. In our calculation, we use a fixed Lagrange multiplier, $\lambda = 1$.

One key feature of spurious modes is a large divergence-to-curl ratio $|\nabla \cdot \mathbf{E}|/|\nabla \times \mathbf{E}|$. [26] There is then the possibility of identifying and removing any remaining spurious solutions based on this ratio, after we determine the eigenfields. However, the divergence-to-curl ratio for spurious and physically admissible modes can become comparable, as seen in the first and second columns of Table 16.1. It is clear that the penalty factor alone does not eliminate all the spurious solutions. There still are spurious solutions that can intermix with the physically acceptable solutions. Note that the Hermite shape functions supports a nonzero divergence value within the brick volume.

We resolve this issue by explicitly imposing the divergence-free condition at each node, using the derivative DoFs. [28, 29, 32] At the matrix level, one of the terms in the zero-divergence condition, equation $\nabla \cdot \epsilon_r \mathbf{E}$, is eliminated in favor of the other two. The procedure is demonstrated in Fig. 16.2 for the simpler case of a constant ϵ_r.

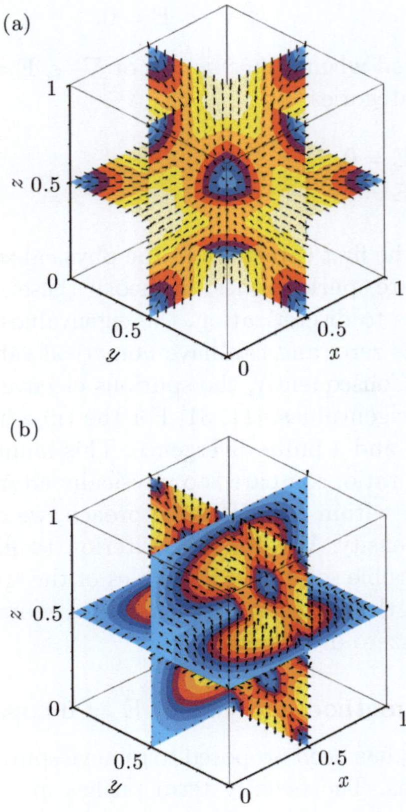

Figure 16.1 Spurious solutions in the empty cubic cavity of unit dimensions, with $|\nabla \cdot \mathbf{E}|/|\nabla \times \mathbf{E}|$ ratio equal to (a) 1783.8 and (b) 1481.7. Note the field distribution, which is indicative of the presence of sources, even though the cavity is empty. Light yellow (gray) regions correspond to amplitude antinodes, and blue (darker) regions correspond to amplitude nodes.

Applying this technique drives the divergence-to-curl ratio of physically admissible solutions even lower, and that of spurious solutions higher, as can be seen from the third and fourth columns of Table 16.1. A comparison of columns 2 and 4 in this table show the enhancement of the ratio $|\nabla \cdot \mathbf{E}|/|\nabla \times \mathbf{E}|$ for spurious solutions and a substantial reduction of this ratio for the physical solutions.

Additionally, since the divergence condition is applied at each node, the total divergence of the physically admissible solutions decreases further with mesh refinement, whereas it increases considerably for spurious solutions (see columns 4 and 6 in Table 16.1). This is manifested as the element size is reduced, and the interpolation from the nodes having the divergence-free condition into the interior of the elements is more effective, with increasing mesh density.

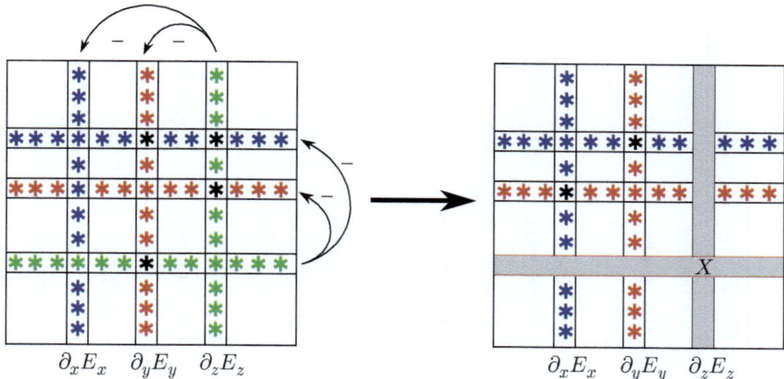

Figure 16.2 Explicitly imposing the divergence-free condition by performing global matrix row and column operations. One of the derivative DoF is eliminated in favor of the remaining two. Here $X = 1$ for the left-hand side matrix, and X is set to a large value in the right-hand side matrix, in the generalized eigenvalue problem. This choice of a large number pushes the eigenvalue of the redundant 1×1 subspace out of the spectral range of interest.

This is in contrast to VFEM, where the normal discontinuity of edge elements leads to the formation of artificial charges at element interfaces, thus increasing the total divergence of the solutions; this problem worsens with mesh refinement. [19] Note that in VFEM, the zero-frequency spurious solutions are separated by filtering out the null-space of the curl operator from the spectrum using iterative techniques, [33, 34] or by finding eigenvalues in the interior of the spectrum. This is necessary, since if the physical solution space is not normal to the null-space, the physical solutions will be polluted by null vectors.

The solutions for the lowest few states in an empty electrodynamics cavity, along with the corresponding divergence-to-curl ratios are listed in Table 16.2. The number of DoFs in this calculation is 42438. As can be seen in the divergence-to-curl values in Table 16.2, using tetrahedral elements for the breaking of the cavity, imposing the penalty factor only does not fully separate the physical solutions with spurious solutions. However, also from Table 16.2, including both the penalty factor and the divergence-free condition drastically improves the tagging of spurious solutions. We also observe less number of spurious solutions polluting the spectrum as compared to the case with cubic elements. Another feature observed is that the eigenvalues of the spurious solutions are distinct from those of the physical solutions, as opposed to the cubic element case, where the spurious solutions are degenerate with the physical ones.

16.3 Fields in an empty cubic cavity

To demonstrate our method, we first model an empty cubic cavity with conducting boundaries, with $\epsilon_r = 1$ and $\mu_r = 1$ inside. Consider a cavity of unit dimensions. The frequency of the modes ω can be obtained from the relation $k_0^2 = \omega^2/c^2$, where k_0^2 is

Table 16.1 The enhancement of the global divergence-to-curl ratio, for the lowest few eigenstates, by imposing the divergence-free condition at each node at the element matrix level are listed.

Penalty factor method only		Penalty factor with Divergence-free condition				
DoF: 8232		DoF: 8232		DoF: 52728		Solution type
k_0^2	$\lvert\nabla \cdot \mathbf{E}\rvert / \lvert\nabla \times \mathbf{E}\rvert$	k_0^2	$\lvert\nabla \cdot \mathbf{E}\rvert / \lvert\nabla \times \mathbf{E}\rvert$	k_0^2	$\lvert\nabla \cdot \mathbf{E}\rvert / \lvert\nabla \times \mathbf{E}\rvert$	
19.739334159	0.000000000	19.739221048	0.000000000	19.739209007	0.000000000	Physical
19.739334159	0.000000000	19.739221048	0.000000000	19.739209007	0.000000000	Physical
19.739334159	0.000000000	19.739221048	0.000000000	19.739209007	0.000000000	Physical
29.609001232	0.005891138	29.608831573	0.000000197	29.608813510	0.000000003	Physical
29.609001232	0.000389894	29.608831573	0.000000197	29.608813510	0.000000003	Physical
29.609001232	172.872234679	31.328579414	220.518451137	31.501166266	545.498991584	Spurious
49.358373749	0.000000000	49.349260110	0.000000000	49.348046601	0.000000000	Physical
49.358373749	0.000000000	49.349260110	0.000000000	49.348046601	0.000000000	Physical
49.358373749	0.000000000	49.349260110	0.000000000	49.348046601	0.000000000	Physical
49.358373749	0.000000000	49.349260110	0.000000000	49.348046601	0.000000000	Physical
49.358373749	0.000000000	49.349260110	0.000000000	49.348046601	0.000000000	Physical
49.358373749	0.000000000	49.349260110	0.000000000	49.348046601	0.000000000	Physical
59.228040826	0.572767636	59.218870635	0.000006213	59.217651104	0.000000056	Physical
59.228040826	0.575823907	59.218870635	0.000006195	59.217651104	0.000000080	Physical
59.228040826	0.137630780	59.218870635	0.000006005	59.217651104	0.000000112	Physical
59.228040826	0.135054548	59.218870635	0.000006204	59.217651104	0.000000094	Physical
59.228040826	0.000065277	59.218870635	0.000004082	59.217651104	0.000000043	Physical
59.228040826	0.277005817	59.218870635	0.000008226	59.217651104	0.000000129	Physical
59.228040826	1.063357739	62.432033897	167.942572387	62.873334761	321.742164741	Spurious
59.228040826	1.083589808	62.432033897	167.942572387	62.873334761	321.742164741	Spurious
59.228040826	3.580885859	62.432033897	167.942572387	62.873334761	321.742164741	Spurious

Table 16.2 The improvement of the divergence-to-curl ratio when imposing the divergence-free condition at each node is listed for the case of 5th-order HFEM with tetrahedral elements. The eigenmodes are obtained using 42438 DOFs.

Penalty factor only		Penalty factor with $\nabla \cdot \mathbf{E} = 0$		Solution
k_0^2	$\vert\nabla \cdot \mathbf{E}\vert/\vert\nabla \times \mathbf{E}\vert$	k_0^2	$\vert\nabla \cdot \mathbf{E}\vert/\vert\nabla \times \mathbf{E}\vert$	type
19.739208827	0.000000000	19.739208827	0.000000000	Physical
19.739208829	0.000000000	19.739208829	0.000000000	Physical
19.739208848	0.000000000	19.739208849	0.000000000	Physical
29.608813476	0.486030689	29.608813484	0.000000002	Physical
29.608813484	0.486018524	29.608813497	0.000000002	Physical
29.608813497	0.495455431	41.117998502	8.090436581	Spurious
49.348030093	0.000000045	49.348030229	0.000000041	Physical
49.348030963	0.000000051	49.348031125	0.000000045	Physical
49.348032032	0.000000054	49.348032161	0.000000046	Physical
49.348032074	0.000000058	49.348032241	0.000000049	Physical
49.348032134	0.000000059	49.348032302	0.000000051	Physical
59.217642131	0.274980399	59.217643847	0.000000072	Physical
59.217644056	0.242246173	59.217645045	0.000000080	Physical
59.217645004	0.230927738	59.217654541	0.000000124	Physical
59.217652841	0.276708712	59.217656977	0.000000135	Physical
59.217654456	0.287009378	59.217657319	0.000000137	Physical
59.217655419	0.276443701	59.217657643	0.000000135	Physical
59.217656808	0.635597226	N/A	N/A	Spurious
59.217657209	0.834678138	N/A	N/A	Spurious
59.217657452	0.856486293	N/A	N/A	Spurious

an eigenvalue of the matrix problem. The calculations are done using HFEM, with 8232 DoFs, within a parallel computing environment. [35–37]

From Table 16.3, it is clear that the eigenvalues of the empty cavity obtained through our scheme have very small errors, when compared to the analytical values. In Fig. 16.3, we show the convergence of the calculated frequencies to their analytical values in an empty cube of unit dimensions for both HFEM and VFEM (obtained using the software package MFEM [24]). As the global number of DoFs is increased through mesh refinement, the accuracy improves steadily. Quintic HFEM delivers an accuracy of 1 part in 10^9 with just 8232 DoFs. The second curve from bottom (in green) obtained using 5th-order VFEM shows about 10 times larger error for comparable DoFs. Even with further mesh refinement, VFEM has an error higher than our HFEM scheme. HFEM gives a higher accuracy than VFEM, even with half the number of DoFs. This reduction in required number of DoFs can lead to improvement in computation time.

The matrix bandwidth is defined as the sum of sub- and supra-diagonal arrays together with the main diagonal. For a total of 8232 DoFs (52728 DoFs), the cubic Hermite polynomials utilize a bandwidth of 16175 (105167), while the quintic Hermite interpolation polynomials occupy a comparable bandwidth of 16223 (105167). The occupancy of a matrix is defined as the percentage of nonzero entries in the matrix. While going from the cubic to quintic Hermite polynomials, there is only a nominal increase in the matrix occupancy from 0.04% to 0.125% for a matrix of dimensions 8232×8232. With an increase in DoFs to 52728, the occupancy decreases further to 0.0083% for the cubic Hermite, and is 0.025% for the quintic Hermite interpolation polynomials. We have used the Krylov-Schur algorithm as implemented in SLEPc [36] for all calculations.

We also consider the error in eigenvalues of higher frequency modes in Fig. 16.3. The errors converge at the same rate as the error in the first mode.

In Fig. 16.4, further comparison is made with readily available software, COMSOL [22] and HFSS [23], for the empty cavity calculation. Compared to HFEM, COMSOL and HFSS show poor convergence with mesh refinement. As seen from Fig. 16.4, the error in eigenvalues computed using COMSOL and HFSS converges only up to a value of 10^{-4}.

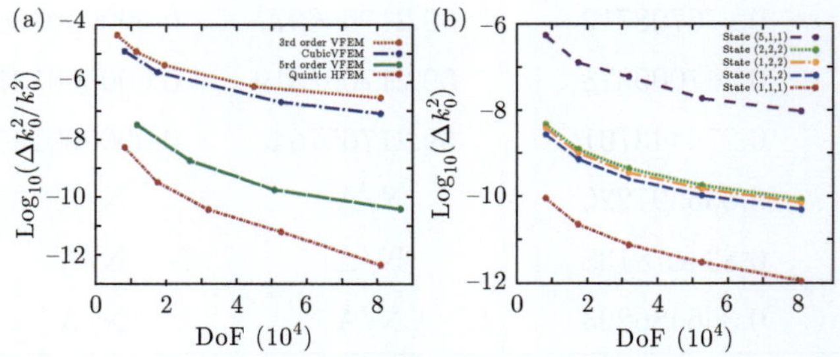

Figure 16.3 The convergence of errors in eigenvalues of (a) the first mode for the 3rd-order VFEM, 5th-order VFEM, (1D-cubic) Hermite, and (1D-quintic) Hermite interpolation polynomials, and (b) the higher frequency modes for (1D-quintic) HFEM only are shown for the case of an empty cubic cavity of unit dimensions. Using Hermite interpolation polynomials, we can reduce the error in the first mode up to 10^{-9} with just 27 elements and 8232 DoFs, with further reduction in error possible with mesh refinement. The total DoFs corresponds to the global matrix dimension.

Table 16.3 Eigenvalues for the 40 lowest frequency modes of the empty cubic cavity, calculated numerically using Hermite interpolation polynomials are listed. The eigenvalues are compared with their analytical values and the absolute errors are displayed.

Mode	Degeneracy	Analytical	Hermite FEM	Error
$(0,1,1)$	3	19.7392088021787	19.7392088021791	4.4E-13
$(1,1,1)$	2	29.6088132032680	29.6088132032692	1.2E-12
$(0,1,2)$	6	49.3480220054467	49.3480220078586	2.4E-9
$(1,1,2)$	6	59.2176264065361	59.2176264089494	2.4E-9
$(0,2,2)$	3	78.9568352087148	78.9568352135412	4.8E-9
$(1,2,2)$	6	88.8264396098042	88.8264396146369	4.8E-9
$(0,1,3)$	6	98.6960440108935	98.6960442784766	2.6E-7
$(1,1,3)$	6	108.5656484119829	108.5656486795616	2.6E-7
$(2,2,2)$	2	118.4352528130723	118.4352528203570	7.2E-9

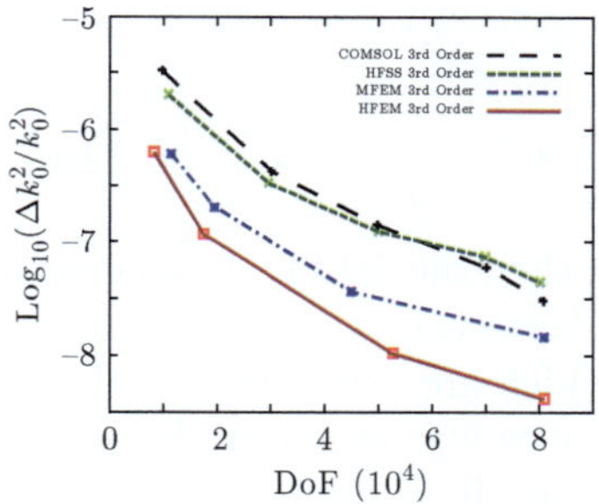

Figure 16.4 The convergence of the errors in eigenvalues of the first mode in COMSOL, HFSS, MFEM, and HFEM. The total DoFs correspond to the global matrix dimension.

16.4 Accidental degeneracies in EM cavities

Physical properties arising from the symmetry of the system can be treated efficiently using group representation theory. It has been well appreciated in quantum mechanics that the degeneracies in the energy spectrum arise from the symmetry group of the corresponding Hamiltonian. [38, 39] The degeneracy of an eigenvalue is equal to the dimensionality of the corresponding irreducible representation of the symmetry group. If we have any other additional degeneracy in the spectrum which cannot be explained by the obvious geometrical symmetry of the system, it is labeled as "accidental degeneracy." In this section, we discuss the presence of such accidental degeneracy and its removal for EM modes in a cavity. Pedagogical remarks on accidental degeneracy are presented in Sec. 16.6.

Let G be a group of order g and $\Gamma^{(i)}$ be an l_i-dimensional representation of G. For a group element R in G, its representation is given by a $l_i \times l_i$ square matrix $\Gamma^{(i)}(R)$. Then the projection operator [38] corresponding to $\Gamma^{(i)}$ is given by

$$\mathcal{P}^{(i)} = \frac{l_i}{g} \sum_R \chi^{(i)}(R) \cdot P_R, \tag{16.8}$$

where $\chi^{(i)}(R)$ is the character and P_R is the operator corresponding to the element R. The transformation of electric fields under the operation P_R is defined by

$$P_R \cdot \mathbf{E}(\boldsymbol{r}) = R \cdot \mathbf{E}\left(R^{-1} \cdot \boldsymbol{r}\right). \tag{16.9}$$

The projection operator $\mathcal{P}^{(i)}$ projects out the part of the field $\mathbf{E}$ that belongs to the representation $\Gamma^{(i)}$.

The electric or magnetic fields corresponding to a degenerate eigenfrequency will form *a set of vector basis-functions for the irreducible representations of the symmetry group.* Previously, we have derived a coefficient formula to recognize the irreducible representation corresponding to an eigenenergy, and obtain the symmetry-adapted wavefunctions in quantum dots. [40] An analogous coefficient formula exists even in the context of electric (magnetic) modes in EM cavities. Let $\{\mathbf{E}_i\}_{i=1}^n$ be the set of eigenfields for the physical system under consideration. Then the coefficient formula [40] is given by

$$c_{jk}^{(i)} = \int_V d^3r \; \mathbf{E}_j^\dagger \cdot \left(\mathcal{P}^{(i)} \mathbf{E}_k\right). \tag{16.10}$$

If the coefficient is nonzero, then the field $\mathbf{E}_k$ has a component in the ith-representation and $\mathbf{E}_j$ is a partner. Using these coefficients, we can construct the symmetry-adapted electric (magnetic) fields which are exclusively in the ith-representation.

In case of an empty cubic cavity resonator of length a with conducting boundaries. The eigenmodes supported by the cubic resonator have the eigenvalues

$$k_0^2 = \left(n_1^2 + n_2^2 + n_3^2\right) \frac{\pi^2}{a^2}, \tag{16.11}$$

and the electric field components are given by

$$\begin{aligned} E_x &= E_{0x} \cos\left(\frac{n_1 \pi x}{a}\right) \sin\left(\frac{n_2 \pi y}{a}\right) \sin\left(\frac{n_3 \pi z}{a}\right); \\ E_y &= E_{0y} \sin\left(\frac{n_1 \pi x}{a}\right) \cos\left(\frac{n_2 \pi y}{a}\right) \sin\left(\frac{n_3 \pi z}{a}\right); \\ E_z &= E_{0z} \sin\left(\frac{n_1 \pi x}{a}\right) \sin\left(\frac{n_2 \pi y}{a}\right) \cos\left(\frac{n_3 \pi z}{a}\right), \end{aligned} \tag{16.12}$$

where n_1, n_2, n_3 are nonzero integers, and E_{0x}, E_{0y}, E_{0z} are the field amplitudes in each direction.

Note that there are three kinds of degeneracies present. The first kind is due to the permutation of mode numbers. The second kind is a consequence of the divergence-free condition $\nabla \cdot \mathbf{E} = 0$. On substituting Eq. (16.12) in the divergence-free condition we

Table 16.4 Different possible even and odd combinations of eigenmodes in an empty cubic cavity, their degeneracy, and corresponding irreducible representations for the symmetry group O_h are listed. Here the indices m, n, and ℓ are nonzero integers. The column labeled "Irrep." refers to the irreducible representation.

Mode number	Degeneracy	Irrep.
$(0, 2n-1, 2n-1)$	3	T_{1u}
$(0, 2n-1, 2m)$	6	$T_{1g} \oplus T_{2g}$
$(0, 2n, 2n)$	3	$A_{2u} \oplus E_u$
$(0, 2n-1, 2m-1)$	6	$T_{1u} \oplus T_{2u}$
$(0, 2n, 2m)$	6	$A_{1u} \oplus A_{2u} \oplus 2\,E_u$
$(2n-1, 2n-1, 2n-1)$	2	E_g
$(2n, 2n, 2n)$	2	E_u
$(2n-1, 2n-1, 2m)$	6	$T_{1u} \oplus T_{2u}$
$(2n, 2n, 2m-1)$	6	$T_{1g} \oplus T_{2g}$
$(2m-1, 2n-1, 2n-1)$	6	$A_{1g} \oplus A_{2g} \oplus 2E_g$
$(2m, 2n, 2n)$	6	$A_{1u} \oplus A_{2u} \oplus 2E_u$
$(2n-1, 2m-1, 2\ell)$	12	$2\,T_{1u} \oplus 2\,T_{2u}$
$(2n, 2m, 2\ell-1)$	12	$2\,T_{1g} \oplus 2\,T_{2g}$
$(2m, 2n, 2\ell)$	12	$2\,A_{1u} \oplus 2\,A_{2u} \oplus 4\,E_u$
$(2m-1, 2n-1, 2\ell-1)$	12	$2\,A_{1g} \oplus 2\,A_{2g} \oplus 4\,E_g$

obtain the constraint $n_1\,E_{0x} + n_2\,E_{0y} + n_3\,E_{0z} = 0$. Hence, if $n_1, n_2, n_3 \neq 0$, we see that there are two independent field components; hence, for a given mode (n_1, n_2, n_3), we will have at least two degenerate field solutions. [41] The third kind occurs when two distinct sets of mode numbers give the same frequency, occurring when the following relation is satisfied:

$$n_1^2 + n_2^2 + n_3^2 = m_1^2 + m_2^2 + m_3^2, \tag{16.13}$$

with $n_i \neq m_i$, for $i = 1, 2, 3$.

We know that the cubic cavity has the geometrical symmetry of O_h. In Table 16.4, we list all different possible combinations of mode numbers and their corresponding irreducible representations from the symmetry group O_h. Here, we have accounted for only the first two kinds of degeneracies. For most of the combinations of mode numbers, we observe accidental degeneracy since they belong to two or more distinct irreducible representations. The accidental degeneracy associated with permutation of mode numbers can be rendered normal by identifying the existence of two dynamical operators $\boldsymbol{\Omega} = (\Omega_1, \Omega_2)$, given by

$$\begin{aligned}\Omega_1 &= (\partial_x^2 - \partial_y^2) \\ \Omega_2 &= (2\,\partial_z^2 - \partial_x^2 - \partial_y^2)\end{aligned} \tag{16.14}$$

which connect the degenerate field solutions. [42] The full covering group in this case is $\mathcal{G} = O_h \otimes \boldsymbol{\Omega}$.

As described in the following section, the larger symmetry group $\mathcal{G}$ of the cavity is reduced to O_h by introducing a concentric cubic dielectric inclusion inside the cavity. This inclusion removes the accidental degeneracy due to the way the fields occupy the corner regions exterior to the cubic dielectric.

16.5 Fields in a dielectrically loaded cubic cavity

We consider a cubic conducting cavity of dimensions $1 \times 1 \times 1$ mm^3. This cavity is loaded with a concentric cubic dielectric inclusion, of dimensions $0.5 \times 0.5 \times 0.5$ mm^3, and permittivity ϵ_2, as shown in Fig. 16.5. The permittivity in the rest of cavity is ϵ_1. The eigenvalues of the first few modes are tabulated for the dielectric ratios $\epsilon_2/\epsilon_1 = 1.2$ and $\epsilon_2/\epsilon_1 = 5.0$. The calculations are done with 17576 DoFs, to accurately model the dielectric function (Table 16.5).

Traditionally, while using Hermite finite elements, the discontinuity of the field across a dielectric interface is employed by equating the corresponding weighted row vectors, to account for different material properties. This is a computationally expensive and slow process. We tackle this problem by smoothing the step function behavior of the dielectric constant by a Fermi distribution function in position space for a cube of size a centered at the origin. [40]

For convenience, let us define a function

$$f(\zeta) = \left[1 + \exp\left(\frac{(\zeta^2 - a^2/4)}{\delta^2}\right)\right]^{-1}. \tag{16.15}$$

Then the dielectric function is given by

$$\epsilon(x, y, z) = \epsilon_1 - (\epsilon_1 - \epsilon_2)\Big(f(x) \times [f(y) \times f(z)\Big). \tag{16.16}$$

Here, δ is the smoothing parameter in the Fermi distribution function, which controls the steepness of the dielectric fall-off between the two regions. By decreasing the parameter δ to a small but finite value, we can approach a discontinuous dielectric function. *The most important benefit of this smoothing is that the properties of the material parameter represented in terms of the Fermi function are the same on either side of the interface at any coordinate at any frequency. Hence, there are no jump conditions to implement in the calculations.* The interface region concede a denser mesh. We note that the Fermi function smoothing may be argued to be a more physical material interface, since at an atomic level, material diffusion leads to smoother interfaces. [44]

In Figs. 16.6–16.8, we show electric field distributions for the first few modes in the loaded cavity with $\epsilon_2/\epsilon_1 = 1.2$. As shown in Fig. 16.6, the first three modes in the loaded cavity remain degenerate, and they belong to the 3D representation T_{1u} of the group O_h. The doublet in Fig. 16.7 corresponding to the mode numbers $(1, 1, 1)$ also remains degenerate; however, the two independent modes are now symmetry-adapted partners and related by a C_4 rotation. An instance of the removal of accidental degeneracy can be seen in the $(1, 1, 3), (1, 3, 1), (3, 1, 1)$ modes, which in the empty cubic cavity form a degenerate sextuplet. From Table 16.4, we see that this sextuplet decomposes into four separate irreducible representations. Figures 16.8(a,b), show the singlet modes in the loaded cavity belonging to the irreducible representations A_{1g} and A_{2g}, respectively. We note that in Fig. 16.8(b), the magnitude of the electric field has complete O_h symmetry, while the vectors flip their directions under an inversion. Similarly, Figs. 16.8(c,d) and 16.8(e,f) show symmetry-adapted partners, which belong to the 2D representations E_g.

In Fig. 16.9, the evolution of mode frequency on varying the dielectric constant ϵ_2 in the interior is shown for the lowest few modes. We observe level crossings akin to the case of bound states in a finite quantum well as the well depth is varied. [40]

In Figs. 16.10 and 16.11, we plot the surface currents on the conducting cavity. These currents ensure that the magnetic field outside the cube is identically zero. Note that the surface currents are symmetry-adapted as well, and appear in singlets, doublets, and triplets. This is in agreement with the O_h symmetry of the system.

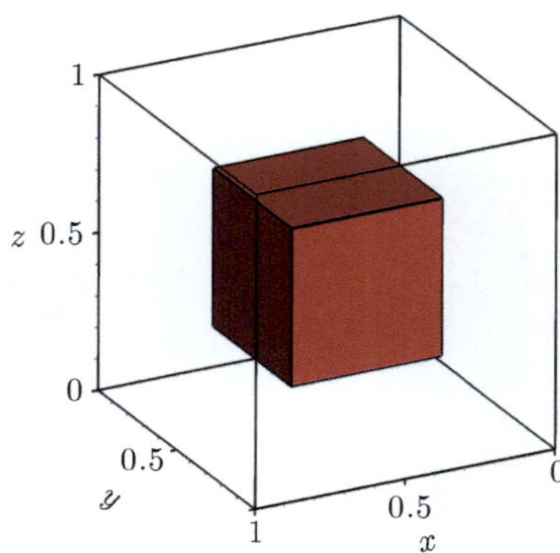

Figure 16.5 Schematics of a dielectrically loaded cavity with O_h symmetry. The external conducting cavity has dimensions $1 \times 1 \times 1$ mm^3; the cubic dielectric loading has dimensions $0.5 \times 0.5 \times 0.5$ mm^3, with dielectric constant ϵ_2. The permittivity in the rest of cavity is ϵ_1.

As a final example, we consider a linear z-dependent perturbation to the dielectric function in the interior dielectric block. This perturbation reduces the symmetry group of the loaded cavity from O_h to C_{4v}, resulting in a further reduction in mode degeneracies. In Table 16.6, we have listed the eigenfrequencies obtained with our method, and classified them into corresponding irreducible representations of the group C_{4v}. In Fig. 16.12, we show electric field distributions for the modes $(1, 1, 1)$, which are degenerate in the unperturbed loaded cavity, but are now split in frequency in the perturbed loaded cavity. As seen from Fig. 16.12, the electric field magnitudes have complete C_{4v} symmetry; the first mode belongs to the representation A_1 of C_{4v}, while the second mode belongs to the representation B_1.

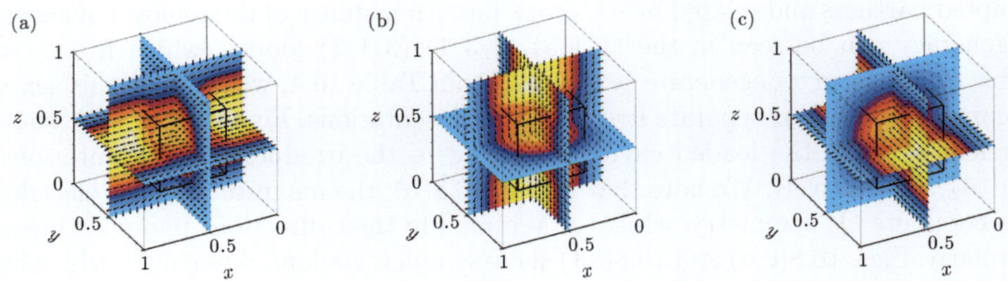

Figure 16.6 The symmetry-adapted triplet states $(0, 1, 1)$, $(1, 0, 1)$, $(1, 1, 0)$ of the dielectrically loaded cavity with an eigenvalue $k_0^2 = 18.5267$ are shown in (a), (b), and (c), respectively. The inserted cubic dielectric has dimensions $0.5 \times 0.5 \times 0.5$ mm^3, with a dielectric constant $\epsilon_2/\epsilon_1 = 1.2$. Light yellow (gray) regions correspond to amplitude antinodes, and blue (darker) regions correspond to amplitude nodes.

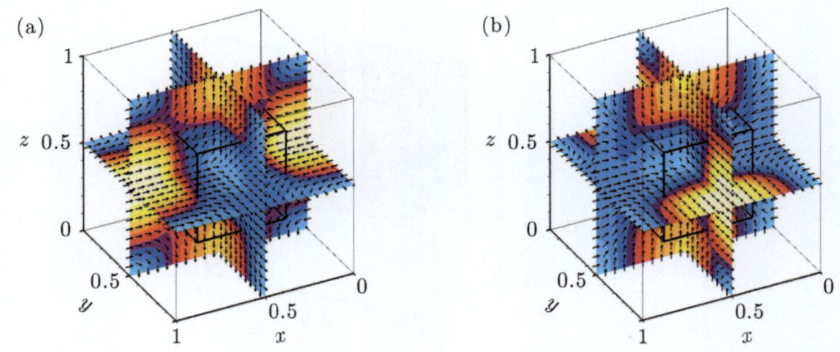

Figure 16.7 The symmetry-adapted doublet states $(1, 1, 1)$ of the dielectrically loaded cavity having eigenvalue $k_0^2 = 28.8995$ are shown in (a) and (b). The inserted cubic dielectric has dimensions $0.5 \times 0.5 \times 0.5$ mm^3, with a dielectric constant $\epsilon_2/\epsilon_1 = 1.2$. Light yellow (gray) regions correspond to amplitude antinodes, and blue (darker) regions correspond to amplitude nodes.

Quality factor: A classic benchmark in computational electromagnetics is to obtain the Q-factor in a resonant cavity with lossy walls. Here, we calculate the Q-factor in a loaded cavity, which has contributions from (a) the dissipation of energy at the cavity walls, and (b) the dielectric loss when the permittivity has both real and imaginary

Table 16.5 Numerically calculated eigenvalues for the lowest few frequency modes of the dielectrically loaded cubic cavity with 17576 total DoFs using quintic Hermite interpolation polynomials. The column labeled "Irrep." corresponds to the irreducible representation the multiplet belongs to. The conducting cavity has dimensions $1 \times 1 \times 1$ mm^3; the cubic dielectric loading has dimensions $0.5 \times 0.5 \times 0.5$ mm^3. The eigenvalues are compared against their values for the modes in the empty cavity of unit dimensions. [43]

Mode number	Irrep.	k_0^2		
		$\epsilon_2/\epsilon_1 = 5.0$	$\epsilon_2/\epsilon_1 = 1.2$	$\epsilon_2/\epsilon_1 = 1.0$
$(0, 1, 1)$		7.919791186	18.526768246	19.739208802
$(1, 0, 1)$	T_{1u}	7.919791186	18.526768246	19.739208802
$(0, 1, 1)$		7.919791186	18.526768246	19.739208802
$(1, 1, 1)$	E_g	18.202154429	28.899507681	29.608813203
		18.202154429	28.899507681	29.608813203
$(0, 1, 2), (2, 1, 0)$		21.354532231	47.387744521	49.348022007
	T_{1g}	21.354532231	47.387744521	49.348022007
$(1, 0, 2), (2, 0, 1)$		21.354532231	47.387744521	49.348022007
		23.416822776	47.404561358	49.348022007
$(0, 2, 1), (1, 2, 0)$	T_{2g}	23.416822776	47.404561358	49.348022007
		23.416822776	47.404561358	49.348022007
$(1, 1, 2), (2, 1, 1)$		35.614536102	57.284302371	59.217626408
	T_{1u}	35.614536102	57.284302371	59.217626408
$(2, 1, 1), (1, 1, 2)$		35.614536102	57.284302371	59.217626408
		38.219214013	58.327254711	59.217626408
$(1, 2, 1), (1, 2, 1)$	T_{2u}	38.219214013	58.327254711	59.217626408
		38.219214013	58.327254711	59.217626408
$(0, 2, 2)$	E_u	38.305292320	76.979818578	78.956835213
$(2, 0, 2)$		38.305292320	76.979818578	78.956835213
$(2, 2, 0)$	A_{2u}	42.790037374	77.001769904	78.956835213
		55.137384588	103.727559345	108.565648679
$(1, 1, 3)$	$2E_g$	55.137384588	103.727559345	108.565648679
$(1, 3, 1)$		77.568304705	107.596740348	108.565648679
$(3, 1, 1)$		77.568304705	107.596740348	108.565648679
	A_{1g}	67.549698454	107.431515611	108.565648679
	A_{2g}	77.460283215	107.591198480	108.565648679

Table 16.6 Lowering of symmetry, and splitting of mode degeneracy in the presence of a dielectric inclusion that has a preferential z-axis.

O_h	C_{4v}	k_0^2
T_{1u}	A_1	14.840987
	E	14.841426
		14.841426
E_g	A_1	26.303562
	B_1	26.309544
T_{1g}	A_2	39.928882
	E	39.937310
		39.937310
T_{2g}	B_2	40.430146
	E	40.431522
		40.431522
T_{1u}	A_1	51.226115
	E	51.202143
		51.202143
T_{2u}	B_1	54.625215
	E	54.631936
		54.631936
E_u	A_2	68.152735
	B_2	68.160536
A_{2u}	B_2	68.993259

parts, $\epsilon = \epsilon_r + i\epsilon_i$. For the dielectric losses, the Q-factor associated with the resonator [3, 4] is given by

$$Q_d = \frac{\iiint_V dV\ \epsilon_r\,|\mathbf{E}|^2}{\iiint_V dV\ \epsilon_i\,|\mathbf{E}|^2}. \tag{16.17}$$

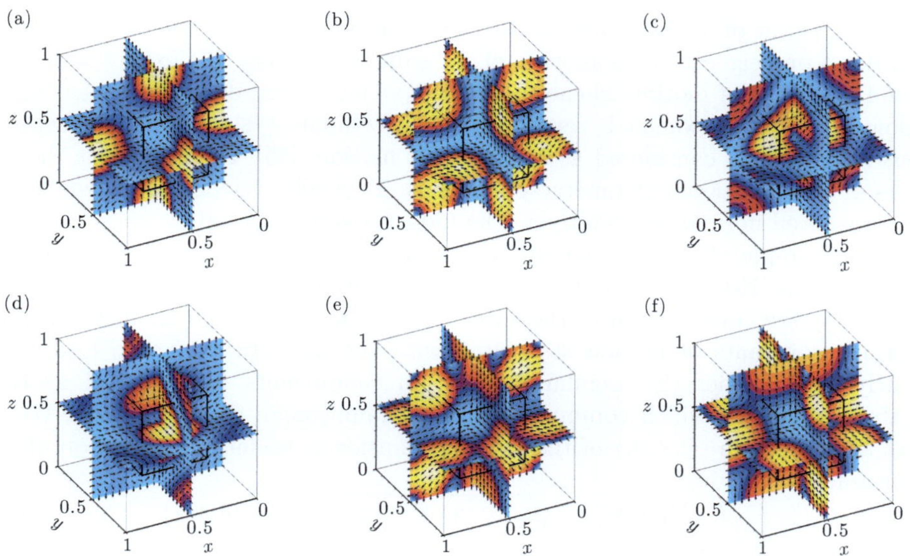

Figure 16.8 Electric fields for the symmetry-adapted modes of the dielectrically loaded cavity $(1,1,3),(1,3,1),(3,1,1)$, with eigenvalues (a) $k_0^2 = 107.4315$, (b) $k_0^2 = 107.5911$, (c, d) $k_0^2 = 103.7275$, and (e, f) $k_0^2 = 107.5967$ are shown. The cubic dielectric has dimensions $0.5 \times 0.5 \times 0.5$ mm^3, with a dielectric constant $\epsilon_2/\epsilon_1 = 1.2$. The singlets belong to the 1D representations (a) A_{1g}, and (b) A_{2g} of the group O_h. Fields in (c, d) and (e, f) are partners, and each pair form a basis for the 2D representation E. Light yellow (gray) regions correspond to amplitude antinodes, and blue (darker) regions correspond to amplitude nodes.

We note that when a constant dielectric loading fills the entire cavity, we obtain $Q_d = \epsilon_r/\epsilon_i$, irrespective of the eigenfrequency. We have verified this in the limit of full dielectric occupancy. In Table 16.7, we have shown the eigenvalues and their corresponding Q_d values in a partially loaded cubic cavity with $\epsilon_r = 2$ and $\epsilon_i = 10^{-6}$. The degeneracy spectrum again follows O_h symmetry.

When the cavity resonator has imperfect conducting walls, we define the corresponding quality factor [1] as

$$Q_c = \omega\sigma\delta_s \frac{\iiint_V dV \ \mu \left|\mathbf{H}\right|^2}{\oiint\limits_{\substack{\text{cavity}\\ \text{walls}}} dS \ \left|\mathbf{n}\times\mathbf{H}\right|^2}, \tag{16.18}$$

where σ is the conductivity of the metallic surface and δ_s is the skin depth at the resonant frequency ω. In Table 16.7, we list the Q_c values at the resonant frequencies in a partially loaded cubic cavity with gold boundary walls.

16.6 Remarks on the accidental degeneracy

Let $\tilde{G}$ be an infinitesimal transformation for the coordinate system. Given a Hamiltonian $\hat{H}$, if $[\tilde{G}, \hat{H}] = 0$, and $\tilde{G}$ does not explicitly depend on time, then we say that

$\tilde{G}$ is a constant of motion. Such constants of motion generate symmetries since they transform one eigenstate to another of the same eigenvalue. We expect to find additional constants of motion whenever we observe accidental degeneracies as explained below. If the Hamiltonian is separable in a coordinate system, then the separation constants may be considered as constants of motion. [45] These are just the generators of the additional symmetry operations. Typically, accidental degeneracies are then rendered normal by identifying the hidden covering group.

The example of the familiar hydrogen atom best illustrates the symmetry argument. In the H-atom, with its Coulomb potential having geometrical rotational symmetry, the conservation of the three components of angular momentum provide us three constants of motion associated with the 3D rotation group $O(3)$. Equivalently, we consider the angular momentum components $\{L_+, L_-, L_z\}$ as the set of three operators which commute with the Hamiltonian of the H-atom. We know that an eigenstate $|E, \ell, m\rangle$ of the H-atom transforms under the operation of ladder operators as

$$L_\pm |E, \ell, m\rangle = \sqrt{(\ell \mp m)(\ell \pm m + 1)}\, |E, \ell, m \pm 1\rangle\,. \tag{16.19}$$

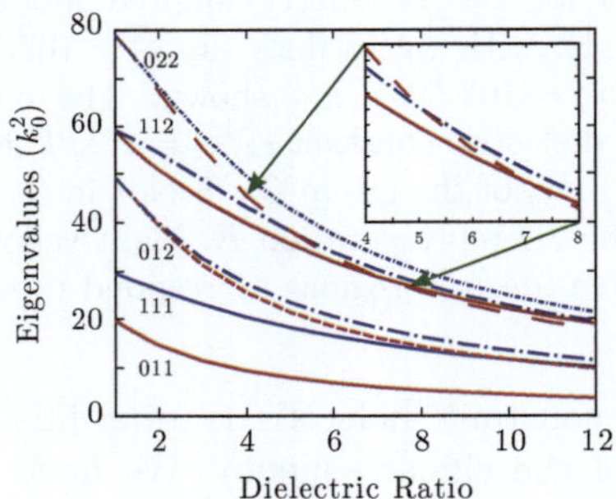

Figure 16.9 The evolution of eigenvalues as the dielectric constant in the cavity is varied from 1.0 to 12.0. Sextuplet-level degenerate modes of the empty cavity are seen to split into triplets with the inclusion of cubic dielectric loading.

Hence, angular momentum operators transform degenerate eigenstates of the same ℓ but of different azimuthal quantum numbers m into one another. For a given ℓ, we find that $(2\ell + 1)$ states are degenerate. However, the eigenstates of different allowed ℓ are also degenerate here, leading to a text-book example of accidental degeneracy. Fock [46] identified the hidden four-dimensional rotational symmetry group $O(4)$ as the true symmetry of the H-atom, which explains these additional degeneracies manifesting as the familiar $s, p, d, f, \ldots$ states being degenerate for a given principal quantum number n. We expect to find additional operators (constants of motion) which commute with the Hamiltonian and connect eigenstates of different ℓ quantum numbers. These operators are just the three components of the new conserved vector, the Runge-Lenz vector, [47] $\boldsymbol{A}$, which transform degenerate eigenstates of different ℓ

Table 16.7 We have tabulated the numerically calculated eigenvalues and corresponding Q-factors for a loaded cubic cavity with boundary walls made out of gold. The conducting cavity has dimensions $1 \times 1 \times 1$ mm^3; the cubic dielectric loading has dimensions $0.5 \times 0.5 \times 0.5$ mm^3, with dielectric ratios $\epsilon_r = 2$ and $\epsilon_i = 10^{-6}$. The conductivity of gold is taken to be $\sigma = 4.46 \times 10^7$ S/m.

k_0^2	Q_d-factor	Q_c-factor
14.64474995 $+i$ 0.00000402	3645779.29	78241.13
14.64474995 $+i$ 0.00000402	3645779.29	78241.13
14.64474995 $+i$ 0.00000402	3645779.29	78241.13
26.12581823 $+i$ 0.00000338	7718462.64	62382.25
26.12581823 $+i$ 0.00000338	7718462.64	62382.25
39.45307840 $+i$ 0.00000942	4186986.51	156821.95
39.45307840 $+i$ 0.00000942	4186986.51	156821.95
39.45307840 $+i$ 0.00000942	4186986.51	156821.95
40.02056958 $+i$ 0.00000864	4633182.94	110705.70
40.02056958 $+i$ 0.00000864	4633182.94	110705.70
40.02056958 $+i$ 0.00000864	4633182.94	110705.70
50.87611080 $+i$ 0.00000720	7061660.73	78692.17
50.87611080 $+i$ 0.00000720	7061660.73	78692.17
50.87611080 $+i$ 0.00000720	7061660.73	78692.17
54.34357813 $+i$ 0.00000540	10065812.56	86034.59
54.34357813 $+i$ 0.00000540	10065812.56	86034.59
54.34357813 $+i$ 0.00000540	10065812.56	86034.59
67.50785266 $+i$ 0.00001276	5290879.93	211112.95
67.50785266 $+i$ 0.00001276	5290879.93	211112.95
68.46687536 $+i$ 0.00001114	6146166.21	139534.39
75.30032691 $+i$ 0.00000986	7636781.09	77564.09
75.30032691 $+i$ 0.00000986	7636781.09	77564.09
75.30032691 $+i$ 0.00000986	7636781.09	77564.09

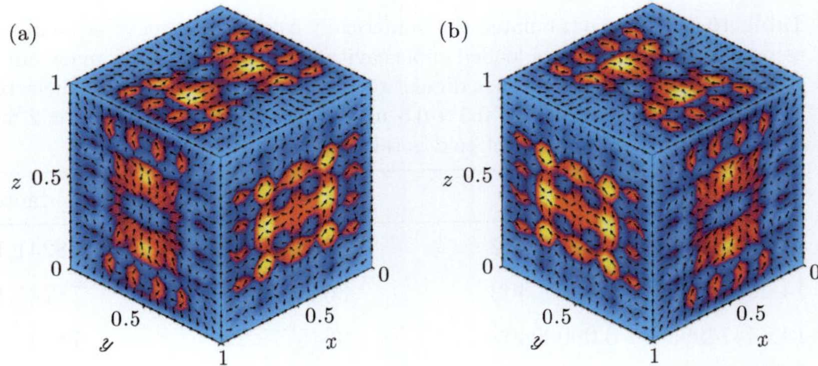

Figure 16.10 Surface currents of the doubly degenerate mode $(1, 1, 1)$ of the dielectrically loaded cavity with eigenvalue $k_0^2 = 28.8995$ are shown in (a) and (b). The cubic dielectric has dimensions $0.5 \times 0.5 \times 0.5$ mm^3, with a dielectric constant $\epsilon_2/\epsilon_1 = 1.2$. Light yellow (gray) regions correspond to current antinodes, and blue (darker) regions correspond to current nodes.

quantum numbers into one another, [48] analogous to Eq. (16.19). The components of the angular momentum $\boldsymbol{L}$ and the Runge-Lenz vector $\boldsymbol{A}$ generate the symmetry group $O(4)$. We note that even though the components of $\boldsymbol{L}$ and $\boldsymbol{A}$ commute with the

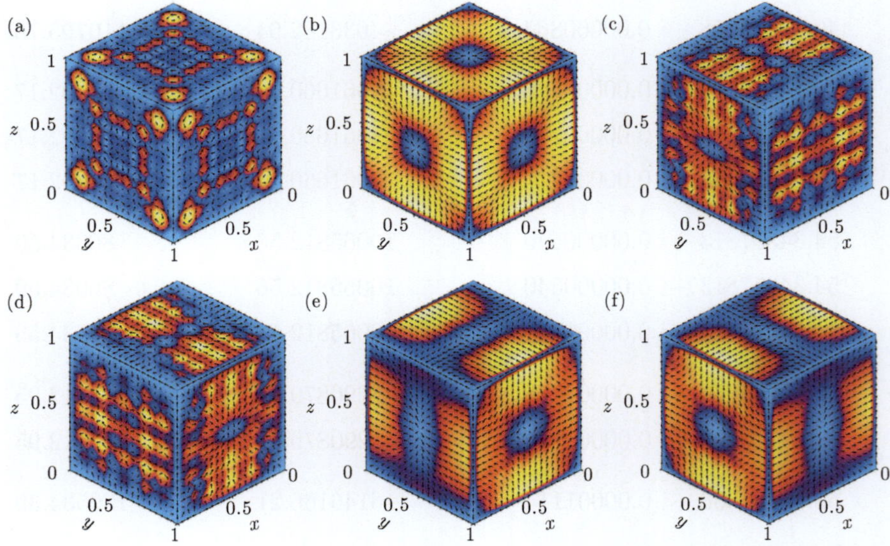

Figure 16.11 Surface currents for the symmetry-adapted modes $(1, 1, 3), (1, 3, 1), (3, 1, 1)$, of the dielectrically loaded cavity, with eigenvalues (a) $k_0^2 = 107.4315$, (b) $k_0^2 = 107.5911$, (c, d) $k_0^2 = 103.7275$, and (e, f) $k_0^2 = 107.5967$ are shown. The cubic dielectric has dimensions $0.5 \times 0.5 \times 0.5$ mm^3, with a dielectric constant $\epsilon_2/\epsilon_1 = 1.2$. Light yellow (gray) regions correspond to amplitude antinodes, and blue (darker) regions correspond to amplitude nodes.

Hamiltonian, they will not mutually commute with each other. These components are subject to kinematic constraints of the Casimir operators for the group $O(4)$. [49] Hence, the eigenstates of the Hamiltonian are represented by the complete set of commuting operators $\{H, L^2, L_z\}$. Such an analysis, based on the symmetries of a physical system, is also feasible for the EM cavities.

To further clarify the aspects of degeneracy, we briefly consider the example of a 2D empty square cavity of length a, surrounded with metal boundaries. This system has C_{4v} geometrical symmetry. The character table for different point groups are given in the texts by Dresselhaus [39] and by Tinkham [38]. We know that the eigenvalues supported in the cavity are given by

$$k_0^2 = \left(n_x^2 + n_y^2\right) \frac{\pi^2}{a^2}, \tag{16.20}$$

where n_x and n_y are nonzero integers.

In Table 16.8, we list all symmetry-adapted basis functions and irreducible representations for different combinations of (n_x, n_y) modes, derived using Eq. (16.10). The (odd, odd) or (even, even) doublet with $n_x \neq n_y$ belongs to two distinct irreducible representations. Hence, the degeneracy of these modes is not entirely explained by the symmetry group C_{4v}; therefore they exhibit accidental degeneracy. This is analogous to the situation in an infinite square quantum well, where the accidental degeneracy occurs for the eigenenergies due to the separability of the infinite well potential. [50] Such an accidental degeneracy is rendered normal, in the usual parlance, by recognizing that an additional operator $\Omega = \left(\partial_x^2 - \partial_y^2\right)$ exists, which connects the basis functions of A_1 (A_2) and B_1 (B_2) representations. Hence, the true symmetry of a $2D$ empty square cavity will be a covering group, which is a semidirect product of the geometrical symmetry group C_{4v} and a 1D compact continuous group generated by the operator $\Omega = \left(\partial_x^2 - \partial_y^2\right)$. [51] *We can remove the accidental*

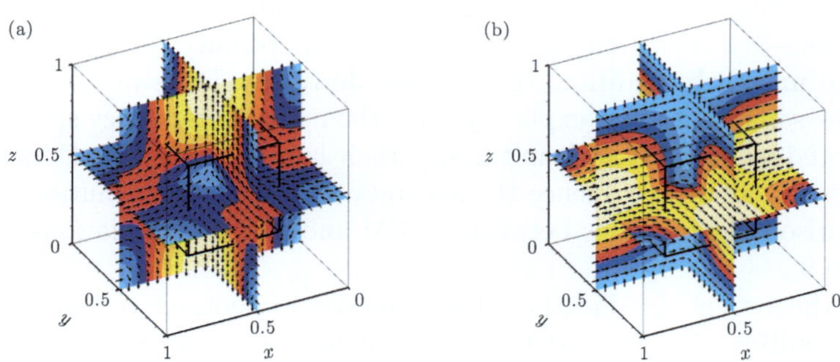

Figure 16.12 Non-degenerate modes $(1, 1, 1)$ of the perturbed loaded cavity belonging to the irreducible representations (a) A_1 with $k_0^2 = 26.303562$ and (b) B_1 with $k_0^2 = 26.309544$ are shown. Light yellow (gray) regions correspond to amplitude antinodes, and blue (darker) regions correspond to amplitude nodes.

Table 16.8 Different possible even and odd combinations of 2D eigenmodes and their corresponding irreducible representations for the symmetry group C_{4v} are shown. Here, the indices n, m are nonzero integers, and the column labeled "Irrep." refers to the irreducible representations of the multiplet.

Mode number	"Irrep."	Basis functions
$(2n-1, 2n-1)$	B_1	$\mathbf{E}_{(2n-1,2n-1)}$
$(2n, 2n)$	A_2	$\mathbf{E}_{(2n,2n)}$
$(2m-1, 2n-1)$	B_1	$\mathbf{E}_{(2m-1,2n-1)} + \mathbf{E}_{(2n-1,2m-1)}$
	A_1	$\mathbf{E}_{(2m-1,2n-1)} - \mathbf{E}_{(2n-1,2m-1)}$
$(2m, 2n)$	A_2	$\mathbf{E}_{(2m,2n)} + \mathbf{E}_{(2n,2m)}$
	B_2	$\mathbf{E}_{(2m,2n)} - \mathbf{E}_{(2n,2m)}$
$(2m-1, 2n)$	E	$\mathbf{E}_{(2m-1,2n)}, \mathbf{E}_{(2n,2m-1)}$

degeneracy in a 2D empty square cavity by introducing a concentric square dielectric inclusion. Such accidental degeneracy and its removal occurs even in rectangular cavities.

16.7 Concluding remarks

We have shown that electromagnetic simulations done with Hermite elements deliver high accuracy and smoother representation of fields. We have compared our formalism with analytical results. Fewer finite elements are needed to achieve comparable results for eigenvalue calculations.

The divergence-free constraint for the electromagnetic fields results in spurious solutions for the wave equation. Their eigenfrequencies are pushed to zero in the VFEM, either through Nedelec compliance or through their removal at each iteration. In either case, this is an expensive numerical procedure. In our approach, we imposed the divergence-free condition by adding a constant penalty term (a Lagrange multiplier, set to unity). In addition, through the derivative DoFs at each node, we have imposed the same constraint explicitly. Now the nonzero frequency spurious solutions are eliminated by identifying them through their large $|\nabla \cdot \mathbf{E}|/|\nabla \times \mathbf{E}|$ ratio. This procedure does not alter or influence the accuracy of the physical solutions. A summary and comparison of properties between VFEM and HFEM calculations are presented in Table 16.9.

Group theoretical classification of eigenmodes in photonic crystals, [53] in radiofrequency cavities, [54, 55] and in metamaterials [56] were previously presented in the literature. In this chapter, we have considered the symmetries of a metallic cubic cavity, with and without a dielectric inclusion. We have identified the origin of the high degeneracy of frequencies in a cubic cavity, and have attributed it to accidental degeneracy. We have recognized three factors contributing to the degeneracy. The operators additional to those of the symmetry group O_h have been determined. We

Table 16.9 We present here a summary, and contrast the properties between VFEM and HFEM calculations.

VFEM	HFEM
In 2D, there are no spurious solutions with edge elements.	In 2D, there are no spurious solutions using Hermite finite elements with triangles. [52, 20]
Field directions are ill-defined at all the nodes. With increasing mesh density, we have more area (2D) and volume (3D) around nodes, where the field direction is not defined. Thus, mesh refinement does not give improved results in applications.	This is a node-based FEM, and there are no issues with field directions at nodes and throughout, including for 3D hexahedral Hermite elements. Mesh refinement allows the improved identification of spurious solutions in the positive spectrum.
In 3D, all zero-frequency solutions are pushed to the null-space through Nedelec-compliant shape functions. Estimates are that for a matrix dimensions of 10^3 in typical EM calculations, there are about 20%–30% solutions to the matrix that are in this category. [11] They have to be calculated and then thrown away, being unusable solutions. Carrying this overhead in the calculation is computationally expensive when considering sophisticated structures.	In 3D, a modest penalty factor $\lambda = 1$ pushes spurious solutions to the zero-frequency sector and they do not appear in the calculated range of spectrum. Some are left over, and these affect the nonzero frequency spectrum. Now the node-based divergence condition is super-imposed on the penalty calculation; this leads to substantial improvement in the separation and identification of the spurious solutions. The key is the tag provided by $\|\nabla \cdot \mathbf{E}\|/\|\nabla \times \mathbf{E}\|$. This ratio keeps increasing for the spurious solutions, whereas it decreases substantially for physically admissible solutions as the mesh is refined.
The multiscale modeling for multi-physics systems cannot be performed: for example, modeling a vertical cavity surface emitting laser.	Being node-based, the modeling can accommodate multiscale problems for multi-physics applications.

have presented a coefficient formula which will classify, and project out the symmetry-adapted modes of the corresponding irreducible representation. The computed field distributions have been shown to be symmetry-adapted, as predicted from group theory.

The accidental degeneracy is lifted with the insertion of a concentric cubic dielectric of a smaller size. The variation of the spectrum as the ratio ϵ_2/ϵ_1 is changed has been explored. We have shown that this leads to a reordering of some of the mode frequencies.

By considering a spatially linearly varying dielectric, we can reduce the symmetry of the system further from O_h to C_{4V}. This is analogous to applying an external electric field in a semiconductor quantum dot, but with flat potential barriers outside the smaller cube.

Since this method is based on geometry discretization, we are now free to change the shape of the cavity and still obtain a high accuracy using HFEM. This method is well suited for mixed-physics applications, such as for quantum well lasers in electromagnetic cavities. This is because we have node-based finite elements with scalar shape functions.

Applications to multi-scale analysis is now feasible using the present method. This option is not open to VFEM due to the lack of directionality for fields at shared nodes in the finite element mesh. Very dense meshes lead to larger regions in which field directions are ill-defined.

It has been shown that the scalar Hermite polynomials have several fundamental advantages for obtaining the band structure of periodic systems such as photonic crystals, while compared with VFEM and other plane-wave expansion methods. [57] Advantages are observed in computational costs, the ability to capture spatial complexity in the dielectric distributions, a substantially higher numerical convergence with scaling, and in obtaining variational eigenfunctions free of numerical artifacts. We note that the method delineated in this chapter is well suited to model and design composite structures such as 3D photonic band-gap crystals, metamaterials, and topological photonic systems for applications in ultra-small optical integrated circuits.

While we have treated simple geometries in this chapter to show the accuracy and feasibility of using HFEM, it should be clear that this can be extended to any geometry using a 56-DoFs tetrahedral Hermite finite element.

The approach presented here is expected to show great promise for the simulation of electrodynamics, plasmonics, high frequency circuitry, and especially in mixed-physics problems.

Finally, we note that the importance of accidental degeneracy and its consideration is because the periodic table of elements and its structure depend on it. The progression of elements in the table with the addition of more and more electrons to the atoms is governed by the Pauli exclusion principle and his "aufbau prinzip." This then governs all of chemistry and hence all of biology. May we say that life itself depends on accidental degeneracy?

References

[1] J. D. Jackson, *Classical Electrodynamics*, 3rd edition (Wiley, New York, 1999).
[2] D. J. Griffiths, *Introduction to Electrodynamics*, 4th edition (Pearson Education, 2013).
[3] D. S. Jones, *The Theory of Electromagnetism* (The Macmillan Company, New York, 1964).
[4] R. F. Harrington, *Time-Harmonic Electromagnetic Fields* (Wiley-IEEE Press, 2001).
[5] D. S. Saxon, *Notes on Lectures by Julian Schwinger: Discontinuities in Waveguides*, (MIT Radiation Laboratory Report, Cambridge, 1945). Also see: K. A. Milton and J. Schwinger, *Electromagnetic Radiation: Variational Methods, Waveguides and Accelerators* (Springer-Verlag, Berlin, Germany, 2006).
[6] C. A. O'Donnabhain, 3rd International Conference on Computation in Electromagnetics, 348–351 (1996); "Calculation of modes in a microstrip line with no zero or spurious modes."
[7] K. Hayata, M. Koshiba, M. Eguchi, and M. Suzuki, IEEE Trans. Microw. Theory Tech. **MTT-34** 1120–1124 (1986); "Vectorial finite-element method without any spurious solutions for dielectric waveguiding problems using transverse magnetic-field component."
[8] Z. J. Cendes, and P. Silvester, IEEE Trans. Microw. Theory Tech. **MTT-18**, 1124-1131 (1971); "Numerical solution of dielectric loaded waveguides. I – Finite-element analysis."
[9] S. Ahmed, and P. Daly, IEE Proc., **116**, 1661–1664 (1969);"Finite-element methods for inhomogenous waveguides."
[10] Jian-Ming Jin, *The Finite Element Method in Electromagnetics*, 2nd edition (John Wiley & Sons, 2002).
[11] A. F. Peterson, IEEE Trans. Antennas Propag. **42**, 357–365 (1994); "Vector finite element formulation for scattering from two-dimensional heterogeneous bodies."
[12] J. C. Nedelec, Numer. Math. **35**, 315–341 (1980); "Mixed finite elements in R3."
[13] S. H. Wong, and Z. J. Cendes, IEEE Trans. Magn. **MAG-25**, 3019–3021, (1989); "Numerically stable finite-element methods for the Galerkin solution of eddy current problems."
[14] C. W. Crowley, P. P. Silvester, and H. Hurwitz, IEEE Trans. Magn. **MAG-24**, 397–400, (1988); "Covariant projection elements for 3D vector field problems."
[15] A. Bossavit and I. Mayergoyz, IEEE Trans. Magn., **MAG-25**, 2816–2821, (1989); "Simplicial finite elements for scattering problems in electromagnetism."
[16] J. F. Lee, D. K. Sun, and Z. J. Cendes, IEEE Trans. Magn. **MAG-27**, 41032–41035, (1991); "Tangential vector finite elements for electromagnetic field computation."
[17] Z. J. Cendes, IEEE Trans. Magn. **MAG-27**, 3953–3966, (1991); "Vector finite elements for electromagnetic field computation."
[18] J. P. Webb, IEEE Trans. Magn. **29**, 1460–1465 (1993); "Edge elements and what they can do for you."

[19] Gerrit Mur, IEEE Trans. Magn. **30**, 3552–3557 (1994); "Edge elements, their advantages and their disadvantages."

[20] C. R. Boucher, Z. Li, C. I. Ahheng, J. D. Albrecht, and L. R. Ram-Mohan, J. Appl. Phys. **119**, 143106–1–10 (2016); "Hermite finite elements for high accuracy electromagnetic field calculations: A case study of homogeneous and inhomogeneous waveguides."

[21] L. R. Ram-Mohan, *Finite and Boundary Element Applications in Quantum Mechanics* (Oxford University Press, New York, 2002).

[22] COMSOL Multiphysics®, version 5.2. https://www.comsol.com, COMSOL AB, Stockholm, Sweden.

[23] Ansoft. ANSYS HFSS®, 3D full-wave electromagnetic field simulation; https://www.ansys.com/products/electronics/ansys-hfss.

[24] MFEM: Modular Finite Element Methods Library, doi: 10.11578/dc.20171025.1248.

[25] B. M. A. Rahman and J. B. Davies, IEEE Trans. Microw. Theory Tech. **MTT-32**, 20–28 (1984); "Finite element analysis of optical and microwave waveguide problems."

[26] J. P. Webb, IEEE Trans. Magn. **24**, 162–165 (1988); "Efficient generation of divergence-free fields for the finite element analysis of 3D cavity resonances."

[27] J. P. Webb, IEEE Trans. Microw. Theory Tech. **MTT-33**, 635–639 (1985); "The finite-element method for finding modes of dielectric-loaded cavities."

[28] A. Konrad, IEEE Trans. Magn. **25**, 2822–2824 (1989); "A method for rendering 3D finite element vector field solutions nondivergent."

[29] C. M. Pinciuc, A. Konrad, and J. D. Lavers, IEEE Trans. Magn. **50**, 7200113 (2014); "Basis functions with divergence constraints for the finite element method."

[30] C. M. Pinciuc, *Basis Functions With Divergence Constraints For The Finite Element Method*, Doctoral dissertation, University of Toronto (2012).

[31] D. Sun, J. Manges, X. Yuan, and Z. J. Cendes, IEEE Antennas Propag. Mag. **37**, 12–24 (1995); "Spurious modes in finite-element methods."

[32] S. H. Wong, and Z. J. Cendes, IEEE Trans. Magn. **24**, 2685–2687 (1988); "Combined finite element-modal solution of three-dimensional eddy current problems."

[33] P. Arbenz and R. Geus, Appl. Numer. Math. **54**, 107–121 (2005); "Multilevel preconditioned iterative eigensolvers for Maxwell eigenvalue problems."

[34] A. Dziekonski, M. Rewienski, P. Sypek, A. Lamecki, and M. Mrozowski, Commun. Comput. Phys. **22**, 997–1014 (2017); "GPU-accelerated LOBPCG method with inexact null-space filtering for solving generalized eigenvalue problems in computational electromagnetics analysis with higher-order FEM."

[35] Satish Balay et al., *PETSc Users Manual*, Argonne National Laboratory, ANL-95/11 - Revision 3.7, 2016.

[36] V. Hernandez, J. E. Roman, and V. Vidal, ACM Trans. Math. Software, **31**, 3, 351–362 (2005); "SLEPc: A scalable and flexible toolkit for the solution of eigenvalue problems."

[37] P. R. Amestoy, I. S. Duff, J.-Y. L'Excellent, and J. Koster, SIAM J. Matrix Anal. Appl. **23**, 1, 15–41 (2006); "A fully asynchronous multifrontal solver using distributed dynamic scheduling."

[38] M. Tinkham, *Group Theory and Quantum Mechanics* (Dover Publications, New York, 2003).

[39] M. S. Dresselhaus, G. Dresselhaus, and A. Jorio, *Group Theory: Application to the Physics of Condensed Matter* (Springer, Berlin, 2008).

[40] Sathwik Bharadwaj, Siddhant Pandey, and L. R. Ram-Mohan, Phys. Rev. B **96**, 195305-1–13 (2017); "Removal of accidental degeneracy in semiconductor quantum dots."

[41] M. H. Nayfeh and M. K. Brussel, *Electricity and Magnetism* (John Wiley & Sons, New York, 1995); Ch. 16, 549–552.

[42] F. M. Fernandez, arXiv:1310.5136, (2014); "On the symmetry of the quantum-mechanical particle in a cubic box."

[43] L. R. Ram-Mohan, S. Pandey, S. Bharadwaj, and J. D. Albrecht, Bull. Am. Phys. Soc. (2019). "Electrodynamic calculations with Hermite interpolation: Role of symmetry and degeneracies of fields in a cavity."

[44] L. R. Ram-Mohan, Sathwik Bharadwaj, Siddhant Pandey, and C. M. Pierce, Trans. JASCOME, **17**, 1–7 (2017); "Modeling of 3D semiconductor quantum dots in a parallel computing environment."

[45] D. F. Greenberg, Am. J. Phys. **34**, 1101–1109 (1966); "Accidental degeneracy."

[46] V. Fock, Z. Phys. **98**, 145–154 (1935); "Zur Theorie des Wasserstoffatoms."

[47] W. Pauli, Z. Physik **36**, 336–363 (1926); "Über das Wasserstoffspectrum von Standpunkt der neuen Quantunmechanik."

[48] C. E. Burkhardt and J. J. Leventhal, Am. J. Phys. **72**, 1013–1016 (2004); "Lenz vector operators in a spherical Hydrogen atom eigenfunction."

[49] H. V. McIntosh, *Symmetry and degeneracy*, in *Group Theory and Its Applications*, pp. 75–144, edited by E. M. Loebl (Academic Press, NY, 1971).

[50] J. Shertzer and L. R. Ram-Mohan, Phys. Rev. B **41**, 9994–9999 (1990); "Removal of accidental degeneracies in semiconductor quantum wires."

[51] F. Leyvarz, A. Frank, R. Lemus, and M. V. Andrez, Am. J. Phys. **65**, 1087–1094 (1997); "Accidental degeneracy in a simple quantum system: A new symmetry group for a particle in an impenetrable square-well potential."

[52] P. G. Kassebaum, C. R. Boucher, and L. R. Ram-Mohan, J. Comp. Physics. **231**, 5747—5760 (2012); "Application of group representation theory to derive Hermite interpolation polynomials on a triangle."

[53] K. Sakoda, Phys. Rev. B **52**, 7982–7986 (1995); "Symmetry, degeneracy, and uncoupled modes in two-dimensional photonic lattices." See also, K. Sakoda, Phys. Rev. B **55**, 15345–15348 (1997); "Group-theoretical classification of eigenmodes in three-dimensional photonic lattices."

[54] S. Sakanaka, Phys. Rev. ST Accel. Beams **8**, 072002-1–11 (2005); "Classification of eigenmodes in RF cavities using the group theory."

[55] P. R. McIsaac, IEEE Trans. Microw. Theory Tech. **MTT-23**, 421–433 (1975); "Symmetry-induced modal characteristics of uniform waveguides — I: Summary of results."

[56] W. J. Padilla, Opt. Express **15**, 1639–1646 (2007); "Group theoretical description of artificial magnetic metamaterials utilized for negative index of refraction."

[57] C. R. Boucher, Zehao Li, J. D. Albrecht, and L. R. Ram-Mohan, J. Appl. Phys. **115**, 154101–1–10 (2014); "Efficient modeling of photonic crystals with local Hermite polynomials."

17

Dimensional continuation of EM singularities in structures with re-entrant geometry

In this chapter:

- We give an accurate analysis of the electromagnetic fields inside waveguides and cavities with re-entrant boundaries. This geometry is frequently present for many multiscale device applications. In earlier approaches, the singularities that exist in electromagnetic field gradients inside such waveguides and cavities significantly reduce the accuracy of the calculations. In the present novel approach, we use dimensional continuation to show that the action integral is finite when Bessel functions are used as trial functions! These singular behaviors are treated using Hermite finite elements combined with analytical regularization of the action integral. The re-entrant corners are sometimes referred to as Fichera corners.
- We explicitly show that this approach provides considerably higher accuracy as compared with other methods employed in the literature. Specific calculations are presented for three typical regions: (a) an L-shaped waveguide, (b) an L-cavity, and (c) a cubic cavity with an octant removed to form a triple-junction corner (imagine a Rubik's cube with an octant removed). The method can be adapted to evaluate very high electromagnetic fields in any complex structures displaying "hot points."
- The re-entrant geometry, together with the Laplacian operator, occurs in many fields, including fluid mechanics, diffusion, as well as electromigration of impurities in electronic devices. In all such cases, the action integral and the dimensional continuation method provides the solution.

Finite Elements in Action. L. Ramdas Ram-Mohan, Oxford University Press. © L. Ramdas Ram-Mohan (2026).
DOI: 10.1093/oso/9780199563487.003.0017

17.1 Introduction

It is illuminating to have a historical perspective to the present problem. It has been known since the experiments of Benjamin Franklin [1] in 1749 that sharp, pointed metallic objects concentrate electric field lines at their tips. A convergence of field lines and the corresponding electromagnetic field-focusing also occurs inside devices that have concave cavities where the metallic surface has a sharp negative curvature. For example, the electromagnetic field in an L-shaped waveguide (Fig. 17.1) has singular *field gradients* at the re-entrant corner. Analogs of this phenomenon can be found in heat conduction inside an L-shaped object, or in structural mechanics where it is well-known that the strains at sharp re-entrant corners can generate cracks in the material [2]. Similar issues with divergent electromagnetic field gradients at re-entrant regions and contacts also occur in a wide variety of physical systems such as in transistors based on two-dimensional (2D) materials, [3] Terahertz electronic devices, [4] surface-plasmon polaritons, [5, 6] and modern implementations of quantum computing using superconducting circuits. [7, 8] The Laplacian operator is at the core of the issue and is common to all the above examples.

The development of accurate and efficient techniques for modeling electromagnetic (EM) fields in complex geometries has been of great importance to the study of new physical phenomena, as well as to the analysis and design of novel devices. For the vast majority of available electromagnetic solvers, the problem of very high or singular fields remains an obstacle to achieving accurate results. Modeling such behavior accurately is thus of great interest not only from the point of view of fundamental considerations but also for understanding practical consequences of such geometries that occur naturally in the design of electronic circuits. Sharp corners in circuit interconnects generate very high local fields and focus electronic currents. Similarly, in semiconductor devices that have very narrow channels with rectangular metallic contacts, the inclusion of such gradient field "hot-spots" in the design considerations becomes essential. This would be particularly crucial as the device dimensions shrink to the nanoscale. Hence, precise modeling of the electric field properties in re-entrant regions is of universal importance to a wide variety of multiscale device applications.

One typical theoretical approach in treating such fields proceeds through an amelioration of the problem by rounding off of the concave corners, so that estimations for the function values near such singular points can be done. Other techniques include conformal mapping, [9] finite difference methods, [10–13] the finite element method (FEM) using vector (edge) elements, [14, 15] the FEM with special elements called super-elements, [16–18] singular function mapping, [19, 20] and the like. These methods suffer several drawbacks: the finite difference method, though simple to implement, offers limited room for improvement due to its rigidity in the structure of the computational grid. The conformal mapping approach, on the other hand, is not straightforwardly generalizable to beyond 2D problems. The vector-FEM approach faces difficulties at concave corners due to the ambiguity in the orientation of the vector fields that arises from the multiple edges attached to the node at the corner. In addition to the methods mentioned above, the method of Green's functions and boundary integrals [21–24] for potential problems has been extended to L-waveguides

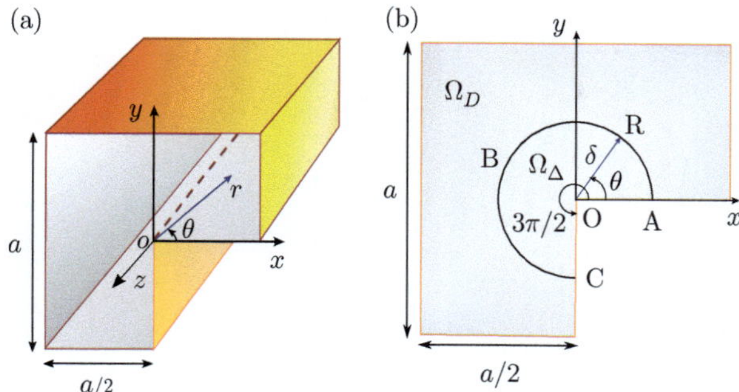

Figure 17.1 (a) The L-shaped waveguide is shown. The field gradient is singular along the dashed (red) line at the concave edge. The dimensions correspond to 1 unit for the shorter sides and 2 units for the longer sides. In (b) the circular region, denoted by Ω_Δ and enclosed by the arc OABCO, the fields are treated separately from the rest of the waveguide cross-section. The remaining area, denoted Ω_D, is treated with finite element discretization.

but requires the numerical evaluations of improper integrals at the re-entrant corners. [25–27] Webb [28] complements the standard polynomials used in FEM with singular trial functions to model the fields at sharp corners. Shu et al. [29] apply the generalized differential quadrature method to analyze waveguides with rectangular boundaries. Proekt et al. [30] use the method of overlapping patches for the fields near conducting cusps, and Ikeuchi et al. [31] employ the so-called α-interpolation scheme to model the L-waveguide.

In this chapter, we consider an approach that uses scalar interpolation polynomials in FEM for EM field calculations. The action integral is discretized using finite elements, and the spatial part of the integral is evaluated in terms of the interpolation polynomials with the as-yet unknown coefficients. The variational principle is applied to obtain the discretized Maxwell's equations.

We focus our attention first on the rectangular L-shaped waveguide as in Fig. 17.1(a); the problem is separated into a 2D geometry of the cross-section of the waveguide. The fields in a circular region around the concave corner are Bessel functions of fractional order.

Next we consider a metallic cavity created by the closing of a length of an L-waveguide at both ends, shown in Fig. 17.2(a); (we refer to this as an L-cavity). The line of the re-entrant corner has a cylindrical region where the fields behave again as Bessel functions of fractional order.

Finally, we treat the cubic cavity in which an octant of volume is removed and the remaining volume enclosed by the metal surface as shown in Fig. 17.2(b); the singular field gradients occur along the cylindrical lines and at the central spherical region, where the solutions are spherical Bessel functions. This structure, with three concave edges coming together at the center, can be referred to as the triple-junction cavity.

We show that even when the fields at re-entrant corners have *singular derivatives*, the action remains non-singular. The action integral being finite should allow the

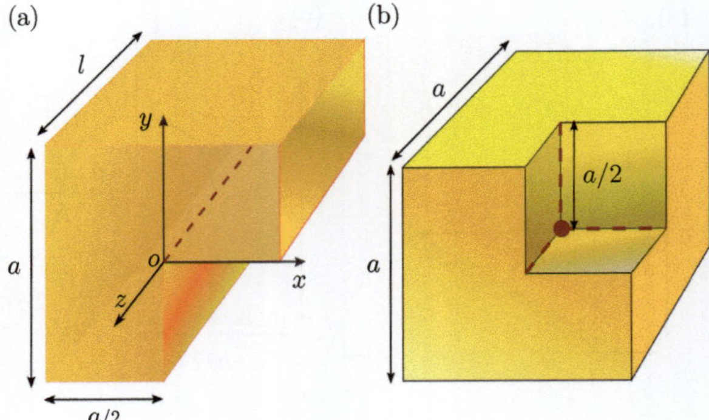

Figure 17.2 In (a) The L-shaped electromagnetic cavity (L-cavity) is shown. The field *gradient* is singular in the vicinity of the concave corner, identified by a dashed (red) line. (b) A triple-junction of re-entrant lines in a cavity with metallic surfaces is shown. The fields are very high along the interface boundaries indicated by dashed (red) lines. We refer to this structure as the triple-junction cavity.

calculations to proceed using the usual FEM approaches. However, this is not so simple for certain modes in the structures, which will be addressed in our analysis of this problem.

In Sec. 17.2, we discuss the solution for the 2D Helmholtz equation. Sec. 17.2.1 provides a detailed derivation of the action integral and regularization of singularities. The arguments are very similar to our taming of boundary integral (element) formulation of the interior or the exterior regions used to calculate the electrostatic field. In that case, we recognized the equivalence of the derivative of the Green's function under the integral sign to a loop diagram in quantum field theory, and employed dimensional continuation to isolate and cancel the singularities. Here, we are again employing dimensional continuation to investigate the nature of the singularities in the action integral. [32]

In Sec. 17.2.2, the FEM implementation is discussed. Appendix B discusses the properties and generation of the tetrahedral interpolation polynomials used here. These elements are used in benchmarking the results for an empty cubic cavity. We provide the results for the L-waveguide in Sec. 17.3.3, the 3D L-cavity in Sec. 17.3.5, and the 3D triple-junction cavity in Sec. 17.3.6. Concluding remarks are given in Sec. 17.4.

17.2 The Helmholtz equation for the L-waveguide

The vector electric and magnetic fields, collectively denoted by $\Psi = \{\mathbf{E}, \mathbf{H}\}$, satisfy the Helmholtz equation in the L-waveguide, shown in Fig. 17.1(a). The wave equation

$$\boldsymbol{\nabla}^2\Psi(\boldsymbol{r},t) + k_0^2\Psi(\boldsymbol{r},t) = 0, \tag{17.1}$$

is expressed in cylindrical coordinates in the neighborhood of the edge in the L-waveguide, as shown in Fig. 17.1(b). Here $k_0 = \omega\sqrt{\mu_0\epsilon_0}$, with $c = 1/\sqrt{\mu_0\epsilon_0}$ being the velocity of light in the empty waveguide. The field propagation in the waveguide is assumed to be of the form

$$\Psi = \Psi(r,\theta)\exp(ik_z z - i\omega t). \tag{17.2}$$

We are interested in the solutions for the transverse magnetic (TM) and transverse electric (TE) waves inside the waveguide. For these modes, knowing the solutions for E_z (TM) and H_z (TE) is sufficient to determine all other components of the electromagnetic field. The boundary conditions (BCs) at the surfaces for these two modes of solutions are

$$\Psi \equiv E_z = 0, \quad \text{for TM modes}, \tag{17.3}$$

$$\partial\Psi/\partial n \equiv \partial H_z/\partial n = 0, \quad \text{for TE modes}. \tag{17.4}$$

We focus on the region $0 \le \theta \le 3\pi/2$ and $0 \le r \le \delta$, and following the separation-of-variables method, the solution for Ψ can be written as

$$\Psi(r,\theta) = R(r)\Theta(\theta). \tag{17.5}$$

The solutions of the angular part $\Theta(\theta)$ that satisfies the BCs in Eqs. (17.3)–(17.4) are

$$\begin{aligned} \Theta(\theta) &\sim \sin(\nu_n\theta), \quad \text{for TM modes}, \\ \Theta(\theta) &\sim \cos(\nu_n\theta), \quad \text{for TE modes}, \end{aligned} \tag{17.6}$$

with $\nu_n = 2n/3$. The cutoff frequency is obtained from $k^2 = k_0^2 - k_z^2$. The solution for $R(r)$ is given by

$$R(r) = a_n J_{2n/3}(kr) + b_n Y_{2n/3}(kr). \tag{17.7}$$

Here, J and Y are Bessel functions of the first and second kind, respectively, and $n = 1, 2, 3, \ldots$. With the BCs in Eqs. (17.3)–(17.4), only the Bessel functions of the first kind survive in TM and TE solutions. Now consider the gradient of the final TM solution,

$$\begin{aligned} \nabla\Psi(r,\theta) = {} & \frac{1}{2}\hat{r}k\big(J_{(2n/3-1)}(kr) \\ & - J_{(2n/3+1)}(kr)\big)\sin(2n\theta/3) \\ & + \hat{\theta}\frac{2n}{3r}J_{2n/3}(kr)\cos(2n\theta/3). \end{aligned} \tag{17.8}$$

For $n = 1$, the first Bessel function in Eq. (17.8) becomes $J_{-1/3}(kr)$, and we have the following relation

$$J_{-1/3}(kr) = J_{1/3}(kr)\cos(\pi/3) - Y_{1/3}(kr)\sin(\pi/3). \tag{17.9}$$

Note that although $J_{2n/3}$ are finite everywhere, $Y_{1/3}$ is singular at $r = 0$, as shown in Fig. 17.3. Therefore, the gradient of the field at the origin is singular. The singular gradient at non-convex corners is the source of errors in calculating eigenmodes and eigenvalues. Next, we show the technique to eliminate this singularity with our formalism.

17.2.1 The action integral

The L-waveguide:
The action for the Helmholtz equation is

$$\mathcal{A} = \int \left[\frac{1}{2} \left(\nabla \Psi(\mathbf{r}) \right)^2 - \frac{1}{2} k^2 \Psi^2(\mathbf{r}) \right] r dr\, d\theta. \tag{17.10}$$

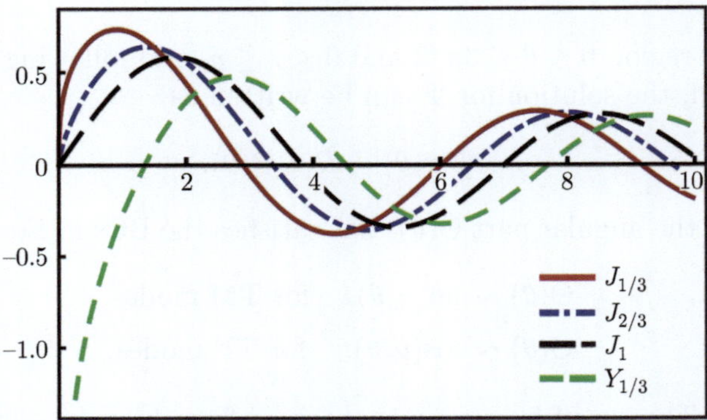

Figure 17.3 The Bessel function of the first kind J of the first few orders, and the Bessel function of the second kind $Y_{1/3}$ is plotted. Note how all the J functions are finite at the origin, while $Y_{1/3}$ is singular.

Let us first focus on a circular region Ω_δ centered at the concave corner. The rest of the physical domain Ω_D is easily treated using FEM. The action integral is then given by

$$\mathcal{A}(\epsilon \leq r \leq \delta) = \frac{1}{2} \int_0^{3\pi/2} d\theta \int_\epsilon^\delta r dr \Big[(\partial_r \Psi)^2 + (\partial_\theta \Psi / r)^2 - k^2 \Psi^2 \Big] + \frac{1}{2} \int_0^{3\pi/2} r\, d\theta\, \Psi (\partial_r \Psi) \Bigg|_{r=\epsilon}, \tag{17.11}$$

with $\epsilon \to 0$. The second integral over θ in Eq. (17.11) is needed to apply the natural BCs in order to cancel the boundary integral arising from the integration by parts. [33–35] Assuming $\Psi(r, \theta) = R(r)\Theta(\theta)$, the action becomes

$$\mathcal{A} = \frac{1}{2}\int_0^{3\pi/2} d\theta \int_\epsilon^\delta dr \Big(rR'^{\,2}(r)\Theta^2(\theta) + rR^2(r)\Theta'^{\,2}(\theta)/r^2 - k^2 rR^2(r)\Theta^2(\theta)\Big) + \frac{1}{2}\int_0^{3\pi/2} d\theta\, rR(r)R'(r)\Theta^2(\theta)\Big|_{r=\epsilon}. \tag{17.12}$$

To combine the two integrals, we rewrite

$$rRR'\Big|_{r=\epsilon} = rRR'\Big|_{r=\delta} - \int_\epsilon^\delta dr\,(rRR')'. \tag{17.13}$$

We then have

$$\mathcal{A}(\epsilon \le r \le \delta) = \frac{1}{2}\int_\epsilon^\delta dr \int_0^{3\pi/2} d\theta \Big[-\Big(k^2 rR^2(r) + rR(r)R''(r) + R(r)R'(r)\Big)\Theta^2(\theta) + R^2(r)\Theta'^{\,2}(\theta)/r\Big] + \frac{1}{2}\int_0^{3\pi/2} d\theta\Theta^2(\theta)\, rR(r)R'(r)\Big|_{r=\delta}. \tag{17.14}$$

Note that the boundary integral on the right side of Eq. (17.14) is finite, since rRR' is finite at $r = \delta$. The remaining part of the action is finite if the angular integral over θ is zero. In fact, from Eq. (17.6), we know that $\int_0^{3\pi/2} d\theta(\Theta'^2 - \nu^2\Theta^2) = 0$, so that the action integral is given by

$$\mathcal{A} = \frac{1}{2}\int_0^{3\pi/2} d\theta\Theta^2(\theta) \int_\epsilon^\delta dr\Big(-\frac{R(r)}{r}\Big)\Big[r^2R''(r) + rR'(r) + (k^2r^2 - \nu^2)R(r)\Big] + \frac{1}{2}\int_0^{3\pi/2} d\theta\Theta^2(\theta)\, rR(r)R'(r)\Big|_{r=\delta}. \tag{17.15}$$

The expression inside the square brackets is the definition of Bessel's equation. Hence, if $R(r)$ is a physical solution, then the double integral in Eq. (17.15) is effectively zero. The overall action inside the circular region $\epsilon \le r \le \delta$ is then finite.

The L-cavity:

Here we consider the action integral for the 3D L-cavity. As before in 2D, we split the physical domain into the semi-cylindrical region, Ω_Δ, around the sharp concave surface and the rest of the cavity, Ω_D. Using the separation of variables, with $\Psi = R(r)\Theta(\theta)Z(z)$, the solution for the θ-component is determined to be of the form given in Eq. (17.6). Similarly, the solutions for the z-component satisfying the BCs for

TM and TE modes also take the sinusoidal form. The action for the 3D L-cavity near the concave surface can be written as

$$\mathcal{A}(\epsilon \le r \le \delta) = \int_0^{3\pi/2} d\theta \int_0^{\ell} dz \int_\epsilon^\delta r\, dr \frac{1}{2}\Big[(\nabla^2\Psi) - k^2\Psi\Big] + \frac{1}{2}\int_0^{3\pi/2} d\theta \int_0^{\ell} dz\, r\, \Psi(\partial_r\Psi)\Big|_{r=\epsilon}. \tag{17.16}$$

Following the analysis paralleling that of the L-waveguide, we can again show that the action integral for the 3D L-cavity can also be regularized and is finite.

The triple-junction cavity:

Applying $\Psi = R(r)\Theta(\theta)\Phi(\phi)$ to the Helmholtz equation in spherical coordinates in the subdomain Ω_Δ, we obtain $\Phi(\phi) \sim e^{im\phi}$, $\Theta(\theta) \sim P_l^m(\cos(\theta))$, and $R(r) \sim j_l(kr)$, which is the spherical Bessel function. The action integral is written generally as

$$\begin{aligned}\mathcal{A} = &\frac{1}{2}\int_{\phi_1}^{\phi_2} d\phi \int_{\theta_1}^{\theta_2} d\theta \int_\epsilon^\delta dr\, r^2 \sin(\theta)\big[(\nabla\Psi)^2 - k^2\Psi^2\big] \\ &+ \frac{1}{2}\int_{\phi_1}^{\phi_2} d\phi \int_{\theta_1}^{\theta_2} d\theta \sin(\theta)\, r^2\Psi(\partial_r\Psi)\Big|_{r=\epsilon} \\ &- \frac{1}{2}\int_{\phi_1}^{\phi_2} d\phi \int_{r_1}^{r_2} dr\, r \sin(\theta)\Psi(\partial_\theta\Psi)\Big|_{\theta=\theta_2},\end{aligned} \tag{17.17}$$

where the last two integrals are added to eliminate the boundary integrals arising from an integration by parts where natural (Neumann) BCs apply. In Eq. (17.17) above, the angular integrals for $d\phi$ and $d\theta$ are done over the solid angle $(7/8)4\pi$ subtended by the triple-junction corner. The action can be rearranged as

$$\begin{aligned}\mathcal{A} = &\frac{1}{2}\int_{\phi_1}^{\phi_2} d\phi \int_{\theta_1}^{\theta_2} d\theta \int_\epsilon^\delta dr \Big\{ \sin(\theta)\Big[r^2R'^2\Theta^2\Phi^2 + R^2\Theta'^2\Phi^2 + \frac{1}{\sin^2(\theta)}R^2\Theta^2\Phi'^2 \\ &- k^2r^2R^2\Theta^2\Phi^2 - (r^2RR')'\Theta^2\Phi^2\Big] - [\Theta(\theta)\Theta'(\theta)\sin(\theta)]'R^2\Phi^2\Big\} \\ &+ (F.V)_1 + (F.V)_2,\end{aligned} \tag{17.18}$$

where

$$\begin{aligned}(F.V.)_1 &= \frac{1}{2}\int_{\phi_1}^{\phi_2} d\phi \int_{\theta_1}^{\theta_2} d\theta \sin(\theta)\Theta^2\Phi^2r^2RR'\Big|_{r=\delta}, \\ (F.V.)_2 &= -\frac{1}{2}\int_{\phi_1}^{\phi_2} d\phi \int_\epsilon^\delta dr\Phi^2R^2\Theta\Theta'\sin(\theta)\Big|_{\theta=\theta_1},\end{aligned} \tag{17.19}$$

are the finite boundary terms. Using the relations $\int_{\phi_1}^{\phi_2} \Phi'^2 d\phi = m^2 \int_{\phi_1}^{\phi_2} \Phi^2 d\phi$, and

$$(r^2RR')' = r^2R'^2 + R(r^2R')', \tag{17.20}$$

as well as the definition of associated Legendre polynomials, we obtain

$$\mathcal{A} = \frac{1}{2}\int_{\phi_1}^{\phi_2} d\phi \int_{\theta_1}^{\theta_2} d\theta \int_{r_1}^{r_2} dr R\Phi^2\Theta^2 \sin(\theta)\Big[l(l+1)R - k^2r^2R - (r^2R')'\Big] + (F.V)_1 + (F.V)_2. \quad (17.21)$$

It is clear that the solution of the integrand is the spherical Bessel function. Therefore, we have shown that if the solution for the field inside the triple-junction cavity is physical, the action integral will be finite. This result suggests that a full domain discretization with the application of the FEM is feasible.

In the following, we show that for just the modes in which the gradients of the fields are singular at the concave surfaces, the convergence to high accuracy eigenvalues is very slow. In order to extend the accuracy further, we split the physical domain into a region close to the concave surface, where analytical solutions are derived, and the rest of the physical domain where Hermite-FEM is used.

17.2.2 Finite element analysis of the L-waveguide

The finite element calculations are done with triangular elements in the 2D cross-section of the waveguide, and with tetrahedral elements for the cavity calculations in 3D. In 2D calculations, we employ $C^{(2)}$-continuous quintic Hermite polynomials for interpolation, [33, 36] with derivative degrees of freedom (DoFs) at the triangle vertices. The inter-element continuity across the sides of the triangle corresponds to $C^{(1)}$-continuity, and the tangential continuity is $C^{(2)}$. [37] These finite elements have been shown to give very high accuracy solutions. [38, 39]

The action integrals describing the fields in systems that contains sources of singularities are divided into two parts: an integral over the region surrounding the singular point, and a second integral over the remaining geometry. The action inside the region surrounding the re-entrant corner, denoted by Ω_Δ is computed analytically as discussed in Sec. 17.2.1, while the outer region, denoted by Ω_D is treated with finite element analysis. The area (volume) in Ω_D is broken into elements, within which the representation of the field is given by

$$\Psi = \sum_i a_i N_i(x, y), \quad (17.22)$$

where N_i are the interpolation polynomials, and a_i are the unknown values of the field and its derivatives at the nodes in the FEM mesh.

In 2D, we use $C^{(2)}$-continuous quintic Hermite interpolation polynomials which guarantee continuity for function values, as well as for first and second derivatives. For 3D calculations, we generate a set of polynomials for tetrahedral elements that satisfies $C^{(2)}$-continuity at the vertices of the tetrahedra, and $C^{(1)}$-continuity at the face centers.

The minimization of the action transforms the problem into an eigenvalue problem, with k^2 being the eigenvalues. Solving this eigenvalue problem gives the cutoff

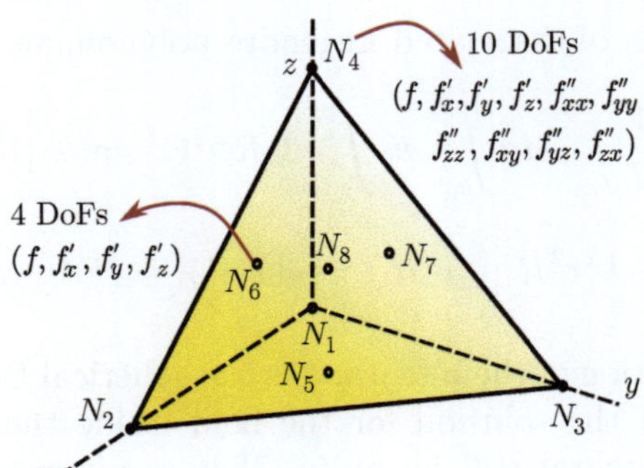

Figure 17.4 A tetrahedral element with vertices and face centers with labeled nodes is shown. The degrees of freedom (DoFs) are shown at one vertex and a node at one of the faces of the element, leading to a total of 56 DoFs for the tetrahedron.

frequency (wavevector) k and the coefficients a_i, with which the function value at any point in the geometry can be reconstructed.

17.3 Results

17.3.1 The zero divergence condition in 2D

We show in the following that in 2D, the divergence condition $\nabla \cdot \mathbf{E} = 0$ is automatically satisfied in calculations for the fields in the cross-section of the waveguide. For a waveguide with perfectly conducting surfaces, the BCs are given by

$$E_{\parallel} = 0; \quad B_n = 0. \tag{17.23}$$

We are interested in monochromatic waves that propagate down the waveguide, given by

$$\mathbf{\Psi}(x, y, z, t) = \mathbf{\Psi}(x, y)e^{i(kz-wt)}, \tag{17.24}$$

where $\mathbf{\Psi} = \{\mathbf{E}, \mathbf{B}\}$. Using Eq. (17.24) in Maxwell's equations, we get

$$E_x = \frac{i}{(\omega/c)^2 - k^2}\Big(k\frac{\partial E_z}{\partial x} + \omega\frac{\partial B_z}{\partial y}\Big), \tag{17.25}$$

$$E_y = \frac{i}{(\omega/c)^2 - k^2}\Big(k\frac{\partial E_z}{\partial y} - \omega\frac{\partial B_z}{\partial x}\Big), \tag{17.26}$$

$$B_x = \frac{i}{(\omega/c)^2 - k^2}\Big(k\frac{\partial B_z}{\partial x} - \frac{\omega}{c^2}\frac{\partial E_z}{\partial y}\Big), \tag{17.27}$$

$$B_y = \frac{i}{(\omega/c)^2 - k^2}\Big(k\frac{\partial B_z}{\partial y} + \frac{\omega}{c^2}\frac{\partial E_z}{\partial x}\Big). \tag{17.28}$$

Hence, if we solve for E_z and B_z, we determine all other components. This simplification does not exist in full 3D calculations.

We have to solve numerically the wave equations of the form

$$\left[\frac{\partial^2}{\partial x^2} + \frac{\partial^2}{\partial y^2} + (\omega/c)^2 - k^2\right]\left\{ \begin{array}{c} E_z \\ B_z \end{array} \right\} = 0. \tag{17.29}$$

For TE modes, $E_z = 0$, and for TM modes $B_z = 0$. Therefore, it suffices to solve for either E_z or B_z. Taking the divergence of Eq. (17.24), while substituting Eqs. (17.25) and (17.26) for E_x and E_y, we obtain the zero divergence condition satisfied, as asserted at the beginning.

17.3.2 Interface continuity conditions between Ω_D and Ω_Δ

At the interface between the subdomains Ω_Δ and Ω_D, a procedure has to be set up to "patch" the two regions together, so that the function value and the derivatives can be continuous across the interface. The solutions within Ω_Δ are obtained analytically. For the L-waveguide, the continuity condition for the function value is given by

$$\Psi^\Delta\Big|_{r=\delta^-} = \Psi^D\Big|_{r=\delta^+}. \tag{17.30}$$

Considering a node along the interface between Ω_D and Ω_Δ, this condition is equivalent to

$$\sum_n b_n J_{\nu_n}(k\delta_i)\sin(\nu_n\theta_i) = a_i^0, \tag{17.31}$$

where a_i^0 are the value of the field at the ith node. As before, $\nu_n = 2n/3$ refers to the nth mode, with $n =$ integer. Applying Eq. (17.31) to all the nodes at the interface, we can write the relationships between b_n, the nodal variables corresponding to the DoFs there, and a_i^0, the mode coefficients in matrix form

$$\boldsymbol{T}^0 \cdot \boldsymbol{b} = \boldsymbol{a}^0, \tag{17.32}$$

where $\boldsymbol{T}^0$ is the transformation matrix, and

$$T_{ij}^0 = J_{\nu_j}(k\delta)\sin(\nu_j\theta_i). \tag{17.33}$$

The continuity of function derivatives can be done in a similar manner. The condition for the x-derivative is given by

$$\frac{\partial\Psi^\Delta}{\partial x}\Big|_{r=\delta^-} = \frac{\partial\Psi^D}{\partial x}\Big|_{r=\delta^+}. \tag{17.34}$$

The transformation matrix for the x-derivative, $\boldsymbol{T}^x$ is then defined by

$$T_{ij}^x = \frac{k}{2}\left[J_{\nu_j-1}(k\delta_i) + J_{\nu_j+1}(k\delta)\right]\sin(\nu_j\theta_i)\cos(\theta_i) - \frac{\nu_j}{\delta}J_{\nu_j}(k\delta)\cos(\nu_j\theta_i)\sin(\theta_i). \tag{17.35}$$

Table 17.1 The eigenvalues for the first few TM and TE modes in an *L-shaped waveguide* are shown. The TE values are compared with the values from Ref. [40]. The length of the long edge of the waveguide is $a = 2\,\text{cm}$. The radius of the circular region Δ enclosing the singular point is set to $\delta = 0.05\,\text{cm}$, and 10 Bessel terms are used. The states in column 1 with asterisks have singular derivatives at the re-entrant corner.

State	Full domain FEM	Split domain			Ref. [40]
	4402 nodes	617 nodes	1374 nodes	4340 nodes	
TE1*	1.47589875453	1.47575776792	1.47562173166	1.47562166095	1.47562182408
TE2	3.53403139051	3.53403138964	3.53403136963	3.53403136842	3.53403136678
TE3	9.86960440109	9.86960440214	9.86960440109	9.86960440109	9.86960440109
TE4	9.86960440109	9.86960440216	9.86960440109	9.86960440109	9.86960440109
TE5	11.3894794163	11.3894794410	11.3894794023	11.3894794005	11.3894793979
TM1*	9.64004319852	9.64078748618	9.63964189690	9.63980531881	—
TM2	15.1974867145	15.1934450874	15.1965998327	15.1971734321	—
TM3	19.7384561952	19.7312465752	19.7378191986	19.7390439912	—
TM4	29.5204718334	29.5092709247	29.5194087661	29.5212354602	—
TM5*	31.9139164672	31.9024201699	31.9101295556	31.9125628531	—

Table 17.2 The eigenvalues k^2 for the first few TE modes in an *L-shaped waveguide* are shown in comparison with the values from Ref. [40]. The length of the long edge of the waveguide is shown in the inset of Fig. 17.5. The number of nodes used for each value of δ is held at ≈ 1300 nodes, with 10 Bessel terms used. The states in column 1 with asterisks are the ones displaying singular gradients at the re-entrant corner.

State	$\delta = 0.2$	$\delta = 0.1$	$\delta = 0.05$	Ref. [40]
TE1*	1.47646595215	1.47575212890	1.47562173166	1.47562182408
TE2	3.53403235756	3.53403175436	3.53403136963	3.53403136678
TE3	9.86960454816	9.86960441563	9.86960440109	9.86960440109
TE4	9.86960457523	9.86960443108	9.86960440109	9.86960440109
TE5	11.3899727284	11.3895223149	11.3894794023	11.3894793979

In a similar manner, we can construct the transformation matrix for the remaining derivatives,

$$T^{y}_{ij} = \frac{k}{2}\big[J_{\nu_j-1}(k\delta) + J_{\nu_j+1}(k\delta)\big]\sin(\nu_j\theta_i)\sin(\theta_i) + \frac{\nu_j}{\delta}J_{\nu_j}(k\delta)\cos(\nu_j\theta_i)\cos(\theta_i), \quad (17.36)$$

$$\begin{aligned} T^{xx}_{ij} &= \frac{k^2}{4}\big[J_{\nu_j-2}(k\delta) - 2J_{\nu_j}(k\delta) + J_{\nu_j+2}(k\delta)\big]\sin(\nu_j\theta_i)\cos^2(\theta_i) \\ &\quad - 2\nu_j\left\{\frac{k}{2\delta}\big[J_{\nu_j-1}(k\delta) - J_{\nu_j+1}(k\delta)\big] - \frac{J_{\nu_j}(k\delta)}{\delta^2}\right\}\cos(\nu_j\theta_i)\sin(\theta_i)\cos(\theta_i) \\ &\quad + \left\{\frac{k}{2}\big[J_{\nu_j-1}(k\delta) - J_{\nu_j+1}(k\delta)\big] - \frac{\nu_j^2}{\delta}J_{\nu_j}(k\delta)\right\}\frac{1}{\delta}\sin(\nu_j\theta_i)\sin^2(\theta_i), \end{aligned} \quad (17.37)$$

$$\begin{aligned} T^{xy}_{ij} &= \frac{k^2}{4}\big[J_{\nu_j-2}(k\delta) - 2J_{\nu_j}(k\delta) + J_{\nu_j+2}(k\delta)\big]\sin(\nu_j\theta_i)\sin(\theta_i)\cos(\theta_i) \\ &\quad - \left\{\frac{k}{2\delta}\big[J_{\nu_j-1}(k\delta) - J_{\nu_j+1}(k\delta)\big] - \frac{J_{\nu_j}(k\delta)}{\delta^2}\right\}\nu_j\cos(\nu_j\theta_i)\big[\sin^2(\theta_i) - \cos^2(\theta_i)\big] \\ &\quad - \left\{\frac{k}{2}\big[J_{\nu_j-1}(k\delta) - J_{\nu_j+1}(k\delta)\big] - \frac{\nu_j^2}{\delta}J_{\nu_j}(k\delta)\right\}\frac{1}{\delta}\sin(\nu_j\theta_i)\sin(\theta_i)\cos(\theta_i), \end{aligned} \quad (17.38)$$

$$\begin{aligned} T^{yy}_{ij} &= \frac{k^2}{4}\big[J_{\nu_j-2}(k\delta) - 2J_{\nu_j}(k\delta) + J_{\nu_j+2}(k\delta)\big]\sin(\nu_j\theta_i)\sin^2(\theta_i) \\ &\quad + 2\nu_j\left\{\frac{k}{2\delta}\big[J_{\nu_j-1}(k\delta) - J_{\nu_j+1}(k\delta)\big] - \frac{J_{\nu_j}(k\delta)}{\delta^2}\right\} \times \cos(\nu_j\theta_i)\sin(\theta_i)\cos(\theta_i) \\ &\quad + \left\{\frac{k}{2}\big[J_{\nu_j-1}(k\delta) - J_{\nu_j+1}(k\delta)\big] - \frac{\nu_j^2}{\delta}J_{\nu_j}(k\delta)\right\}\frac{1}{\delta}\sin(\nu_j\theta_i)\cos^2(\theta_i). \end{aligned} \quad (17.39)$$

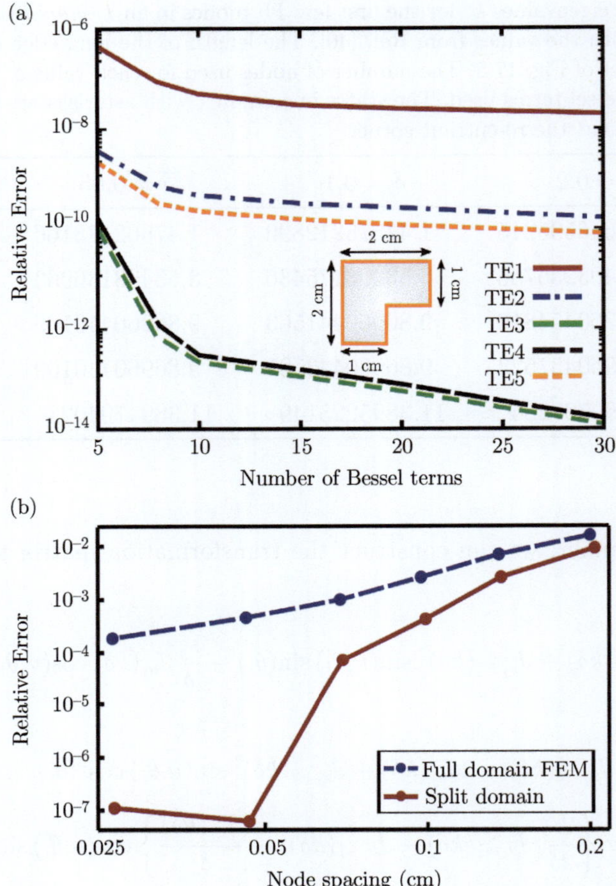

Figure 17.5 (a) The errors in eigenvalues of the L-waveguide calculated using our method relative to values provided by Ref. [40] are plotted. (b) The errors in the eigenvalue for TE1 mode calculated using full domain Hermite-FEM (dashed blue) and our split-domain method (solid red) are shown.

The $\nu_j = 2j/3$ corresponds to the order of the Bessel functions. The complete transformation matrix $\boldsymbol{T}$ can then be used to patch the solutions for the two regions at the nodes at their interfaces:

$$\begin{bmatrix} \boldsymbol{T}^0 & & & & & \\ & \boldsymbol{T}^x & & & & \\ & & \boldsymbol{T}^y & & & \\ & & & \boldsymbol{T}^{xx} & & \\ & & & & \boldsymbol{T}^{xy} & \\ & & & & & \boldsymbol{T}^{yy} \end{bmatrix} \cdot \begin{bmatrix} \boldsymbol{b}^0 \\ \boldsymbol{b}^x \\ \boldsymbol{b}^y \\ \boldsymbol{b}^{xx} \\ \boldsymbol{b}^{xy} \\ \boldsymbol{b}^{yy} \end{bmatrix} = \begin{bmatrix} \boldsymbol{a}^0 \\ \boldsymbol{a}^x \\ \boldsymbol{a}^y \\ \boldsymbol{a}^{xx} \\ \boldsymbol{a}^{xy} \\ \boldsymbol{a}^{yy} \end{bmatrix}.$$

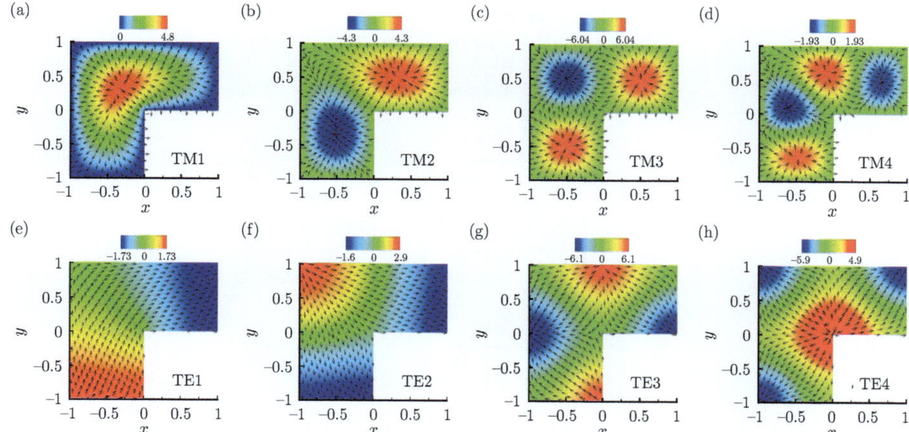

Figure 17.6 The field Ψ (shaded), along with the gradient vector field $-\nabla\Psi$ (arrows), is plotted for the first four TM modes and first four TE modes of an L-waveguide.

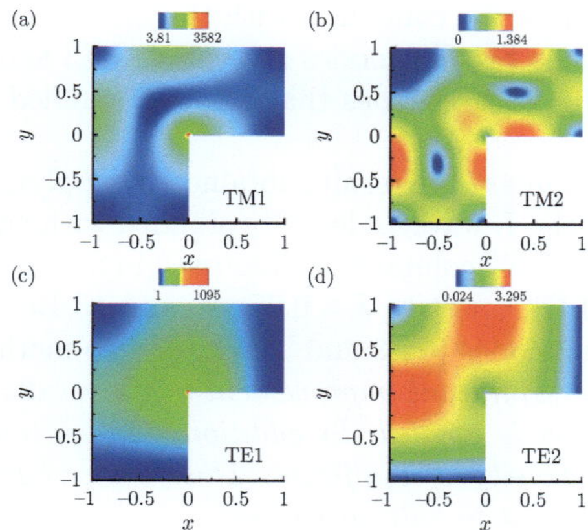

Figure 17.7 The gradient magnitudes $|\nabla\Psi|$ of the first two TM modes in an L-waveguide are shown in (a) and (b), while the gradient magnitudes $|\nabla\Psi|$ of the first two TE modes are shown in (c) and (d). Note the singularities at the re-entrant corner in (a) TM1 and (c) TE1.

The transformation matrices for TE modes are constructed by following the same steps, with a change in the θ-dependent component of Ψ from $\sin(\nu\theta)$ to $\cos(\nu\theta)$.

The interface BCs in the 3D L-cavity and also the triple-junction cavity are not included here, as they are very similar to the above considerations and are straightforward.

17.3.3 Results for the L-shaped waveguide

The eigenvalues k^2 for the first few TE and TM modes of an L-waveguide defined in Fig. 17.1(a) are calculated for $a = 2\,\text{cm}$, and the eigenvalues are presented in

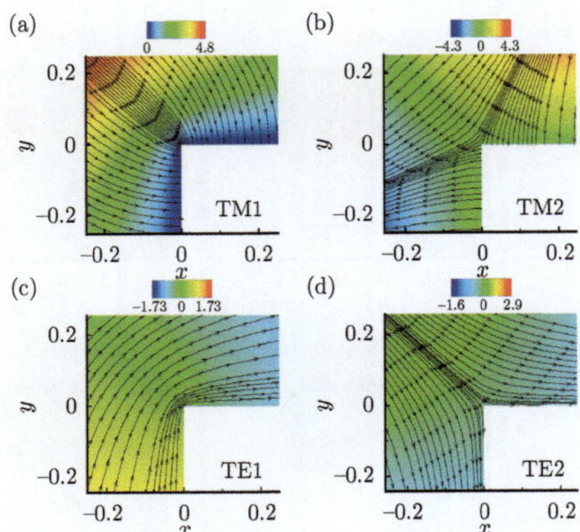

Figure 17.8 A close-up view of the tracelines of the EM field near the re-entrant corner are plotted for the first two TM modes in an L-waveguide in (a) and (b), while the tracelines for the first two TE modes are shown in (c) and (d).

Table 17.1. The TE values are compared with the 11-digit precision benchmarks provided by Ref. [40]. Of the five TE modes presented, TE3 and TE4 are a doublet with an analytical eigenvalue of π^2. Modes that have unbounded singularities are marked by asterisks.

In Table 17.1, we compare the results obtained by applying Hermite-FEM over the entire cross-section of the L-waveguide (2nd column), with values computed using the split-domain scheme ($3^{\rm rd}$–$5^{\rm th}$ columns) as discussed in Sec. 17.2.1. For the split-domain method, the radius of Ω_Δ is set to $\delta = 0.05\,\text{cm}$, and 10 Bessel terms are used. With non-singular analytical modes (TE3 and TE4), the two methods yield similar results. *However, we observe a significant improvement in the results for singular modes when the split-domain method is applied. In addition, the convergence of the eigenvalues with increasing mesh density is significant. This further authenticates our scheme of regularizing singularities at re-entrant corners.*

In Table 17.2, the TE eigenvalues computed using different values of δ are presented. The number of nodes used is ≈ 1300, with 10 Bessel terms. We see that smaller δ corresponds to better accuracy. This is because at small $k\delta$, Bessel functions of low orders are more dominant than the high-order ones. Therefore, with a fixed number of Bessel terms, the smaller radius of the singular region Ω_Δ gives more precise calculations of Ψ.

The relative error of TE eigenvalues with respect to results in Ref. [40] is plotted in Fig. 17.5(a) against the number of Bessel functions used to model Ψ inside Ω_Δ. The radius δ is fixed at 0.05 cm. After about 10 Bessel terms, the errors for singular modes such as TE1-2 flatten out and not much improvement can be made by adding more terms. This is because only the first few Bessel terms contribute significantly to Ψ at small δ, particularly the first Bessel function, $J_{2/3}$, whose derivative is singular, and are crucial to represent the singular gradients.

Of the modes calculated, it is unsurprising that modes that contain a singular gradient at the re-entrant corner have the largest error, as can be seen for TE1 in

Fig. 17.5(a). We perform a further analysis of the error in the calculation of this mode by plotting in Fig. 17.5(b) the scaling as a function of the average node spacing in the mesh. The error in TE1 due to the application of quintic Hermite-FEM to the full cross-section of the waveguide is $\mathcal{O}(h^2)$, with h being the average node spacing. The convergence is significantly improved when the split-domain (with $\delta = 0.05$) method is applied to handle the singular point. We notice that the accuracy is dramatically improved when the spacing is ≈ 0.047, which corresponds to a mesh size of ≈ 1300 elements. As seen in Fig. 17.5(b), this low error deviates from the trend of the convergence and is most likely due to a felicitous combination of the circular region Ω_Δ with the surrounding mesh that allows for accurate representation of the field pattern there. Nevertheless, the convergence of the split-domain method with this exceptional point taken out still improves to $\mathcal{O}(h^{4.5})$, which is close to the convergence usually expected in quintic Hermite-FEM when applied to geometries that do not contain sharp concave corners. This result indicates that the singularity appearing at the corner is appropriately captured by the Bessel function expansion so that the error does not propagate out of the circular patch Ω_Δ and hence does not exacerbate the calculation of the eigenvalues and the overall field distribution in the region where Hermite-FEM is used.

The fields and their gradient vectors are shown in Fig. 17.6 for the first four TM modes and first four TE modes of an L-waveguide with $a = 2\,\text{cm}$. In Fig. 17.7, the gradient magnitudes $|\nabla\Psi|$ of the first two TM and TE modes are plotted. In particular, modes TM1 and TE1, plotted in Fig. 17.7(a) and Fig. 17.7(c) respectively, have singular divergence at the re-entrant corner. In Fig. 17.8, the trace lines of $-\nabla\Psi$ near the re-entrant corner are plotted to display finer details. We can see that for singular modes (Fig. 17.8(a) and Fig. 17.8(c)), the lines "lean" toward the re-entrant corner, whereas for non-singular modes (Fig. 17.8(b) and Fig. 17.8(d)), they bend outwardly when they approach the corner.

For a further comparison with the literature, the cut-off frequency k computed for L-waveguide of size $a = 2\,\text{cm}$ using our method are presented along with published results in Table 17.3. Analytical solutions for some non-singular modes are available and are also shown. For analytical modes, the frequencies calculated using our method are exceptionally accurate. For example, the accuracy is up to 14 digits for modes TE3 and TE4. For singular modes, we have at least 8 digits of accuracy.

17.3.4 Test of the 56-DoF tetrahedron element: the 3D cubic cavity

As a confirmation of the efficacy of the new shape functions developed for tetrahedral elements, we calculated the eigenstates of the field inside an empty *cubic* electrodynamic cavity of unit length. The action integral for the electric field is given by

$$\mathcal{A} = \int_\Omega d\Omega \big[(\nabla \times \mathbf{E}^*) \cdot \mu_r^{-1} (\nabla \times \mathbf{E}) - k_0^2 \mathbf{E}^* \cdot \epsilon_r \mathbf{E} \\ + \nu \big|\nabla \cdot \epsilon_r \mathbf{E}\big|^2\big], \tag{17.40}$$

Table 17.3 Cut-off frequencies k for the first eight TM modes and first eight TE modes in an *L-shaped waveguide* of length $a = 2\,\text{cm}$ are computed using our method and compared with published results and analytical solutions. The number of nodes in the finite element mesh is 6050. The radius δ that surrounds the singular point is 0.05 cm, and 10 Bessel terms are used in the expansion. The absences of analytical solutions for certain modes are indicated by the dashes. Singular modes are marked by asterisks in column 1.

State	Our method	Ref. [16]	Ref. [11]	Ref. [25]	Ref. [26]	Analytical
TE1*	1.2147588724764	1.2149	1.2135	1.2012	1.2071	—
TE2	1.8799019567837	1.8800	1.8796	1.8516	1.8816	—
TE3	3.1415926535899	3.1423	3.1402	3.0959	3.1572	3.1415926535898
TE4	3.1415926535900	3.1423	3.1402	—	—	3.1415926535898
TE5	3.3748302769863	3.3757	3.3736	3.3314	3.4098	—
TE6*	3.5457641249569	—	3.5432	—	—	—
TE7	4.4428840857363	—	4.4410	—	—	4.4428829381583
TE8*	4.6287144481372	—	4.6219	—	—	—
TM1*	3.1048099712854	3.1054	3.1083	3.0910	3.0897	—
TM2	3.8983651580155	3.8989	3.8957	3.8964	3.9505	—
TM3	4.4428827345132	4.4438	4.4400	4.4392	4.5181	4.4428829381583
TM4	5.4333671278429	5.4375	5.4266	5.4308	5.5513	—
TM5*	5.6491529253578	5.6551	5.6477	—	—	—
TM6*	6.4400877551111	—	6.4301	—	—	—
TM7	6.7043629705046	—	6.6877	—	—	—
TM8	7.0248141010382	—	7.0091	—	—	7.0248147310407

where the last term in the action is the penalty factor that is included to help filtering spurious solutions, and ν is the Lagrange multiplier, which is set to unity. Various choices of ν were explored. None other than $\nu = 1$ was satisfactory both in terms of reducing the number the spurious solutions and in enhancing the accuracy of the eigenvalues. To further separate spurious solutions from physical solutions, we also need to impose the zero divergence condition

$$\nabla \cdot \mathbf{E} = 0 \tag{17.41}$$

at the nodes. The solutions for the lowest few states in a cavity of unit dimensions, along with the corresponding divergence-to-curl ratios are listed in Table 17.4. The number of DoFs in this calculation is 42438. As can be seen in the divergence-to-curl ratio in Table 17.4, using tetrahedral elements for the discretization of the cavity and

Table 17.4 The calculated eigenvalues for the EM fields in a simple cubic cavity of unit length are listed. The improvement of the divergence-to-curl ratio when imposing the divergence-free condition at each node is listed for the case of 5th-order HFEM with tetrahedral elements. The eigenmodes are obtained using 42438 DoFs. When a spurious mode is not present in the calculated spectrum, its entry is marked by "N/A."

Penalty factor only		Penalty factor with $\nabla \cdot \mathbf{E} = 0$		Solution
k_0^2	$\lvert\nabla \cdot \mathbf{E}\rvert / \lvert\nabla \times \mathbf{E}\rvert$	k_0^2	$\lvert\nabla \cdot \mathbf{E}\rvert / \lvert\nabla \times \mathbf{E}\rvert$	type
19.739208827	0.000000000	19.739208827	0.000000000	Physical
19.739208829	0.000000000	19.739208829	0.000000000	Physical
19.739208848	0.000000000	19.739208849	0.000000000	Physical
29.608813476	0.486030689	29.608813484	0.000000002	Physical
29.608813484	0.486018524	29.608813497	0.000000002	Physical
29.608813497	0.495455431	41.117998502	8.090436581	Spurious
49.348030093	0.000000045	49.348030229	0.000000041	Physical
49.348030963	0.000000051	49.348031125	0.000000045	Physical
49.348032032	0.000000054	49.348032161	0.000000046	Physical
49.348032074	0.000000058	49.348032241	0.000000049	Physical
49.348032134	0.000000059	49.348032302	0.000000051	Physical
59.217642131	0.274980399	59.217643847	0.000000072	Physical
59.217644056	0.242246173	59.217645045	0.000000080	Physical
59.217645004	0.230927738	59.217654541	0.000000124	Physical
59.217652841	0.276708712	59.217656977	0.000000135	Physical
59.217654456	0.287009378	59.217657319	0.000000137	Physical
59.217655419	0.276443701	59.217657643	0.000000135	Physical
59.217656808	0.635597226	N/A	N/A	Spurious
59.217657209	0.834678138	N/A	N/A	Spurious
59.217657452	0.856486293	N/A	N/A	Spurious

imposing the penalty factor only does not fully separate the physical solutions from the spurious solutions. However, also from Table 17.4, including both the penalty factor and the divergence-free condition drastically improves the tagging of spurious solutions. We can compare our calculations with published results from our research group that uses quintic Hermite polynomials over cubic (brick) elements. [41] An advantage of using tetrahedral elements over cubic elements is the flexibility in efficiently discretizing complicated geometries. *We also observe less number of spurious solutions polluting the spectrum as compared to calculations done using cubic elements. Another feature observed is that the eigenvalues of the spurious solutions are distinct from those of the physical solutions, as opposed to the cubic element case, where the spurious solutions are degenerate with the physical ones.*

17.3.5 Results for the 3D L-shaped cavity

We consider an L-cavity with dimensions $a = \ell = 2\,\text{cm}$. The fields and their gradient vectors inside the L-cavity are shown in Fig. 17.9 for the first two TM and TE modes. In Table 17.5, we report the eigenvalues for the first four TE modes and first four TM modes calculated using different numbers of DoFs. The values converge rapidly with decreasing mesh size. Among the modes shown in Table 17.5, TM4 is an analytically determinable mode, with an eigenvalue which our numerical calculation agrees with up to 11 digits.

In the EM spectrum of this L-cavity, there exist modes that have the same eigenvalues as with certain modes in the L-waveguide. These are modes TE1 and TE3 in Table 17.5. They have a constant solution along the z-direction, resulting in only the nonzero x- and y-components contributing to the eigenvalues of such modes. Therefore, the eigenvalues of these modes confirm the efficacy of our method for 3D calculations, as they can be verified by comparing with 2D L-waveguide values in Table 17.1.

In Table 17.6, the eigenvalues calculated using different numbers of Bessel functions in the series expansion are presented. We observe that using more Bessel functions leads to higher precision. This is best shown in the results for the mode TM4. Moreover, comparisons with eigenvalues of the L-waveguide in Table 17.1 show that the values for singular modes having a constant z-component field in the L-cavity converge well as the number of Bessel terms increases.

The magnitude of the gradient of the field $|\nabla\Psi|$ is plotted in Fig. 17.10. As seen in Fig. 17.10(c), the mode TE1 has a singular gradient along the re-entrant edge as expected, since the mode TE1 in the L-cavity has the same singular behavior at the corner as TE1 of the L-waveguide.

17.3.6 The triple-junction cavity

We consider a triple-junction cavity with dimensions defined as shown in Fig. 17.2(b). The dimension of the triple-junction is $a = 2\,\text{cm}$. The eigenvalues for the first four TE modes and the first four TM modes are calculated and presented in Table 17.7. The convergence of the eigenvalues with increasing number of DoFs is rapid. We also investigate how the number of *spherical* Bessel functions used in the modeling of the fields near the corner affects the convergence. These results are shown in Table 17.8; we observe a fast convergence of the eigenfrequencies with increasing number of spherical Bessel terms in the expansion.

The fields and their gradient vectors inside the cavity are shown in Fig. 17.11 for the first two TM and TE modes. The changes in field density across the space inside the cavity resemble what was observed in the case of the L-cavity or L-waveguide.

Table 17.5 The eigenvalues for the first few TM and TE modes in an *L-shaped cavity* are shown. The dimensions of the cavity are $a = \ell = 2\,\text{cm}$. The radius of the cylindrical region, Δ, surrounding the singular point is 0.05 cm, and 15 Bessel terms are used in the expansion. The modes with asterisks in column 1 have singular gradients at the re-entrant surface.

State	15903 nodes	53636 nodes	103025 nodes	Analytical
TE1*	1.4763400603088	1.4760635672800	1.47590889278907	—
TE2	2.7766826944860	2.7572368479096	2.73595505153786	—
TE3	3.5340335122259	3.5340321279768	3.53403169077342	—
TE4*	3.9437545098641	3.9434874938644	3.94330059388969	—
TM1	12.200614432968	12.160579641758	12.1448949261616	—
TM2	17.666073098432	17.665084690222	17.6648679112429	—
TM3*	19.612996329130	19.569431796915	19.5476919251284	—
TM4	22.206610008630	22.206609904146	22.2066099026140	22.2066099024510

Table 17.6 The eigenvalues for the first few TM and TE modes in an *L-shaped cavity*, computed using different number of Bessel functions in the expansion, are shown. The dimensions are $a = \ell = 2\,\text{cm}$. The radius of the cylindrical region, Δ, surrounding the singular point is 0.05 cm. The number of nodes is 103025.

State	5 Bessel terms	10 Bessel terms	15 Bessel terms	Analytical
TE1*	1.4761695513304	1.4761442865252	1.47590889278907	—
TE2	2.7708164845417	2.7600412089775	2.73595505153786	—
TE3	3.5340325942958	3.5340324431322	3.53403169077342	—
TE4*	3.9436107871005	3.9435352913820	3.94330059388969	—
TM1	12.174061275858	12.165444799643	12.1448949261616	—
TM2	17.665292776192	17.665421802698	17.6648679112429	—
TM3*	19.582390639181	19.568290219809	19.5476919251284	—
TM4	22.206609921036	22.206609906114	22.2066099026140	22.2066099024510

This is due to the similarities among these shapes; all three objects are constructed out of rectangles and cubes, therefore oscillating modes of these fundamental objects will appear in the overall solutions of the triple-junction or L-shape cavity.

The gradient magnitude of the field $|\nabla\Psi|$ is plotted in Fig. 17.12. The gradient of TM1 in Fig. 17.12(a) shows a singular point concentrated at the corner where the

Table 17.7 The eigenvalues for the first few TM and TE modes in a *triple-junction cavity* are shown. The length of the long edge of the cavity is $a = 2\,\mathrm{cm}$. The radius of the spherical region, δ, surrounding the singular point is 0.05 cm, and 15 Bessel terms are used in the expansion. Singular modes are marked by asterisks in column 1 below.

State	19181 nodes	59668 nodes	87738 nodes
TE1*	1.9420081868522	1.9412311063307	1.9404066098274
TE2*	1.9423637421950	1.9413094050296	1.9404128297105
TE3*	1.9425843727349	1.9413694927342	1.9404195746256
TE4	3.1065139543369	3.1065110021175	3.1065093092361
TM1*	10.586437408378	10.574134988391	10.572542468436
TM2*	16.617106922666	16.611616561876	16.602458621967
TM3*	16.619766176739	16.612841584585	16.604746625486
TM4	20.534179322099	20.528150845999	20.525583696207

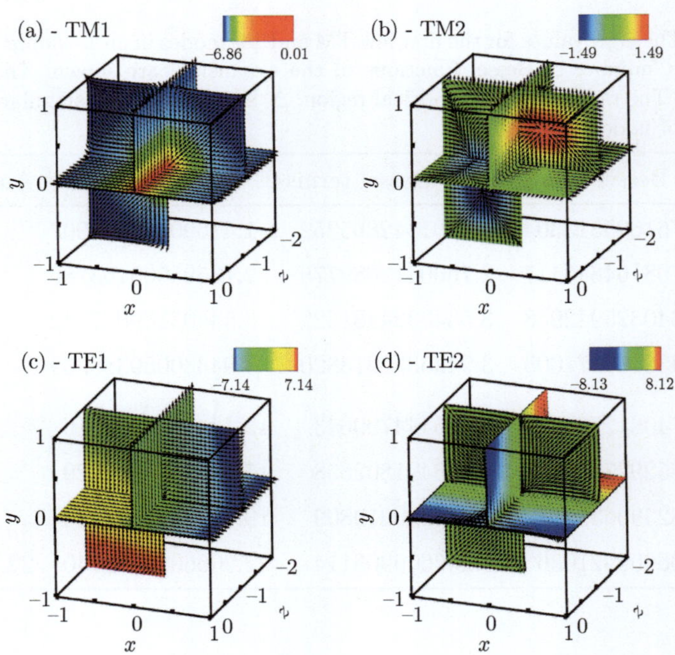

Figure 17.9 The field Ψ, along with the gradient vector field $-\nabla\Psi$, is plotted for the first two TM modes of an L-*cavity* in (a) and (b), and the first two TE modes in (c) and (d). The dimensions of the cavity are $a = \ell = 2\,\mathrm{cm}$.

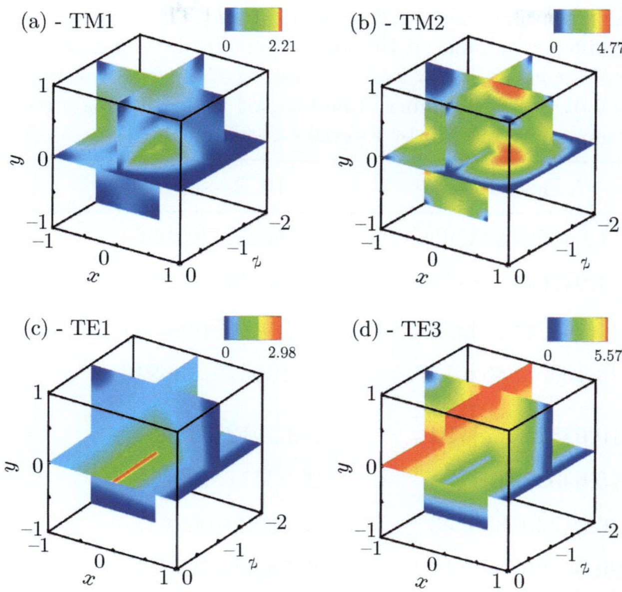

Figure 17.10 The magnitudes of the gradient $|\nabla\Psi|$ of the two TM modes TM1 and TM2 in an L-*cavity* are shown in (a) and (b), while the gradient magnitudes $|\nabla\Psi|$ of the two TE modes TE1 and TE3 are shown in (c) and (d). The dimensions of the cavity are $a = \ell = 2\,\text{cm}$. Note the singularity at the corner edge in TE1.

three junctions meet. In Figs. 17.12(c)–(e), we observe that the gradient plots for the triplet TE1, TE2, and TE3 are essentially rotations of each other; all three modes have singular gradient at one of their three junction edges.

17.4 Concluding remarks

This is the fourth in a series of publications in the area of electrodynamic modeling using Hermite interpolation [38, 39, 41]. In this chapter, we have demonstrated that electromagnetic fields in regions surrounded by peripheries with re-entrant corners can be computed through the variational method with Hermite finite elements. Such re-entrant peripheries are called Fichera corners. [42–45]

In such re-entrant geometries, the gradient of the fields are singular; despite that, the action integral is shown to be finite. This means that variational methods that employ the discretization of the action, such as the FEM can be used to calculate the electromagnetic fields. Within our Hermite element scheme, the modes with singular gradients are solved but still do not deliver the high accuracy that we seek.

The sharply varying values of electromagnetic fields near re-entrant boundaries can be effectively modeled by a series expansion of the analytical solutions satisfying the BCs over a smaller domain of circular/cylindrical/spherical symmetry. The built-in singularity of the Bessel function of fractional order that occurs in the series allows for

Table 17.8 The eigenvalues for the first few TM and TE modes in a *triple-junction cavity*, computed using different number of Bessel functions in the expansion, are shown. The length of the long edge of the cavity is $a = 2\,\text{cm}$. The radius of the spherical region, δ, surrounding the singular point is 0.05 cm. The number of nodes used in discretizing the cavity is held at 87738 nodes. Singular modes are marked by asterisks in column 1.

State	5 Bessel terms	10 Bessel terms	15 Bessel terms
TE1*	1.9465974382966	1.9445347032964	1.9404066098274
TE2*	1.9470457183565	1.9447366167374	1.9404128297105
TE3*	1.9477957543668	1.9449657362267	1.9404195746256
TE4	3.1065388646127	3.1065267936913	3.1065093092361
TM1*	10.625154771794	10.620121779434	10.572542468436
TM2*	16.640572101923	16.634715137652	16.602458621967
TM3*	16.642493850543	16.636007032682	16.604746625486
TM4	20.555997597810	20.548064157282	20.525583696207

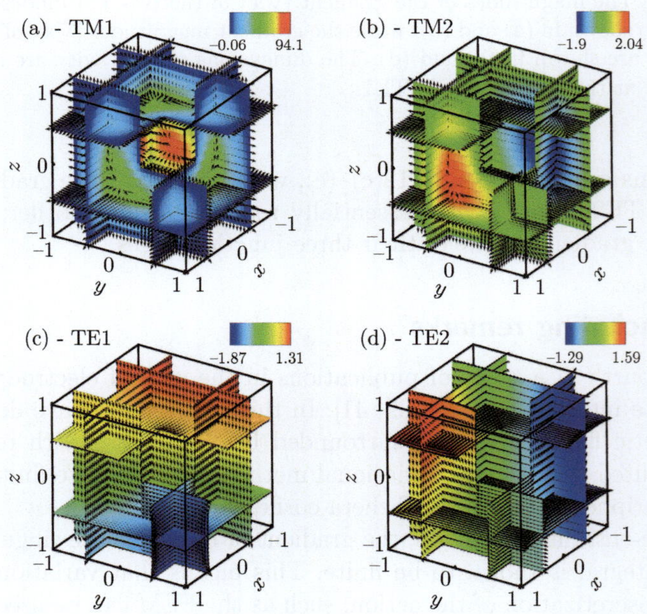

Figure 17.11 The field Ψ, along with the gradient vector field $-\nabla\Psi$, is plotted for the first two TM modes in (a) and (b), and the first two TE modes in (c) and (d) of a triple-junction cavity on planar sections. The size of the cavity is $a = 2\,\text{cm}$.

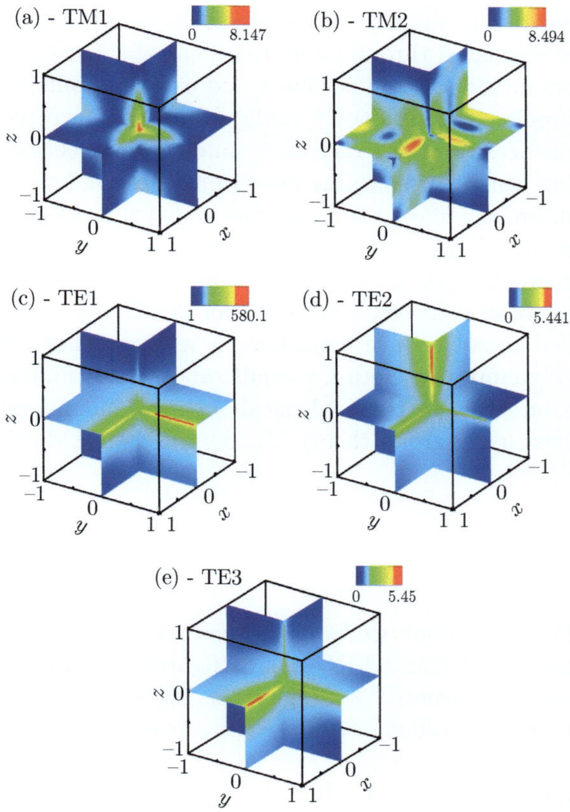

Figure 17.12 The gradient magnitudes $|\nabla\Psi|$ of the modes TM1 and TM2 in a *triple-junction cavity* are shown projected on to planar sections in (a) and (b), while the gradient of the triplet TE1, TE2, and TE3 are shown in (c), (d), and (e). The size of the cavity is $a = 2\,\text{cm}$.

accurate representation of the field and its singular gradient near the concave surface, while ensuring a finite action integral, which can then be evaluated very accurately over the rest of the domain through finite element analysis and patched with the analytic solution.

Our results for the fields and the cutoff frequencies of the 2D L-waveguide showed that very precise solutions can be obtained with no spurious modes contaminating the spectrum. In the calculations for the 3D cavities, we showed that Hermite polynomials in tetrahedral elements offer both an efficient breakup of the region and a high level of accuracy in modeling the fields. Applying the divergence-free condition at nodes together with a penalty (Lagrangian multiplier) condition in the variational scheme provides an effective way to suppress almost all of the spurious modes that appear during numerical computations. A further tag to identify the spurious solutions is the ratio $|\nabla \cdot \Psi|/|\nabla \times \Psi|$ which converges to zero very rapidly for the good solutions and is finite for the spurious ones. The results for the L-cavity and the triple-junction cavity confirm that highly-singular fields in certain modes are present along concave edges, and can be evaluated through our treatment of the action integral.

The method we introduce here is robust and can be applied to the calculations of EM fields in any region with arbitrary configuration of sharp re-entrant boundaries. In modeling the devices used in electronics, antenna engineering, or optoelectronics, singularities arising from sharp turning points or concave edges need to be accounted for, since they can greatly affect the properties of fields inside the devices. Therefore, we believe the method developed here will be crucial to the modeling and analysis of electromagnetic fields in systems of ever-increasing complexity. Moreover, since it is grounded in the analysis of the action integral for the Helmholtz equation, our method can also be applied to solve wave equations in various other areas such as thermal transport, electrodynamics of meta-materials, and elastic materials by deriving the respective action integrals in these physical settings and using the Hermite shape functions we derived using group theory [37, 33] for the FEM calculations.

References

[1] Benjamin Franklin, *Poor Richard's Almanack*, 1753 edition. Also see: http://www.ushistory.org/FRANKLIN/science/lightningrod.htm

[2] A. Carpinteri, P. Cornetti, N. Pugno, A. Sapora, and D. Taylor, Eng. Fract. Mech. **75**, 1736–1752 (2008); "A finite fracture mechanics approach to structures with sharp V-notches."

[3] J. Y. Park, J. Cho, and S. C. Jun, J. Mech. Sci. Technol. **32**, 539–547 (2018); "Review of contact-resistance analysis in nano-material."

[4] Y. H. Ren, X. G. Wang, and C. J. Xiao, J. Appl. Phys. **132**, 023101 (2022); "A terahertz signal enhancement implemented by subwavelength metallic grooves."

[5] D. K. Gramotnev, and S. I. Bozhevolnyi, Nat. Photon. **4**, 83–91 (2010); "Plasmonics beyond the diffraction limit."

[6] Z. Gao, H. Xu, F. Gao, Y. Zhang, Y. Luo, and B. Zhang, Phys. Rev. Appl. **9**, 044019–1–8 (2018); "Surface-wave pulse routing around sharp right angles."

[7] J. Wenner, R. Barends, R. C. Bialczak, Yu Chen, J. Kelly, Erik Lucero, Matteo Mariantoni, A. Megrant, P. J. J. O'Malley, D. Sank, A. Vainsencher, H. Wang, T. C. White, Y. Yin, J. Zhao, A. N. Cleland, and John M. Martinis, Appl. Phys. Lett. **99**, 113513–1–3 (2011); "Surface loss simulations of superconducting coplanar waveguide resonators."

[8] D. N. Pham, W. Fan, M. G. Scheer, and H. E. Türeci, Phys. Rev. A **107**, 053704 (2023); "Flux-based three-dimensional electrodynamic modeling approach to superconducting circuits and materials."

[9] W. P. Calixto, B. Alvarenga, J. C. da Mota, L. da Cunha Brito, M. Wu, A. J. Alves, L. M. Neto, and C. F. R. L. Antunes, Mathematical Problems in Engineering Volume 2010, Article ID 742039, 19 pages (Hindawi Publishing Corporation, 2010). *Electromagnetic Problems Solving by Conformal Mapping: A Mathematical Operator for Optimization.*

[10] M. J. Beaubien, and A. Wexler, Comput. J. **14**, 263–269 (1969). "Iterative, finite difference solution of interior eigenvalues and eigenfunctions of Laplace's operators."

[11] J.-M. Guan, and C.-C. Su, IEEE Trans. Microw. Theory Tech. **43**, 374–382 (1995); "Analysis of metallic waveguides with rectangular boundaries by using the finite-difference method and the simultaneous iteration with the Chebyshev acceleration."

[12] G. R. Hadley, J. Lightwave Technol. **20**, 1219–1231 (2002). "High-accuracy finite-difference equations for dielectric waveguide analysis II: Dielectric corners."

[13] Y.-P. Chiou, Y.-C. Chiang, C.-H. Lai, C.-H. Du, and H.-C. Chang, J. Lightwave Technol. **27**, 2077–2085 (2009). "Finite-difference modeling of dielectric waveguides with corners and slanted facets."

[14] D. Boffi, P. Fernandes, L. Gastaldi, and I. Perugia, SIAM J. Numer. Anal. **36**, 1264–1290 (1999); "Computational models of electromagnetic resonators: analysis of edge element approximation."

[15] D. Boffi, M. Farina, and L. Gastaldi, Computers and Structures **79**, 1089–1096 (2001); "On the approximation of Maxwell's eigenproblem in general 2D domains."

[16] B. Schiff, and Z. Yosibash, IEEE Trans. Microw. Theory Tech. **48**, 214–220 (2000): "Eigenvalues for waveguides containing re-entrant corners by a finite-element method with superelements."

[17] B. Schiff, J. Comput. Phys. **76**, 233–242 (1988); "Finite element eigenvalues for the Laplacian over an L-shaped domain."

[18] B. Schiff, IEEE Trans. Microw. Theory Tech. **39**, 1034–1039 (1991); "Eigenvalues for ridged and other waveguides containing corners of angle $3\pi/2$ or 2π by the finite element method."

[19] P. Grisvard, *Singularities in Boundary Value Problems* (Springer-Verlag, Berlin, 1992).

[20] V. A. Kozlov, V. G. Maz'ya, and J. Rossmann, *Spectral Problems Associated with Corner Singularities of Solutions to Elliptic Equations* (American Mathematical Society, Providence, Rhode Island, 2001).

[21] R. E. Collin, *Field Theory of Guided Waves* (IEEE Press, Piscataway, NJ, 1990).

[22] D. S. Jones, *The Theory of Electromagnetism* (Macmillan, New York, 1964); chapter 5, pp. 240–289.

[23] J. D. Jackson, *Classical Electrodynamics* (Wiley, New York, NY, 3rd edition, 1999).

[24] Jianming Jin, *The Finite Element Method in Electromagnetics* (Wiley, New York, 2nd edition, 2002); chapter 7, pp. 233–271.

[25] M. Swaminathan, E. Arvas, T. K. Sarkar, and A. R. Djordjevic, IEEE Trans. Microw. Theory Tech. **38**, 154–159 (1990); "Computation of cutoff wavenumbers of TE and TM Modes in waveguides of arbitrary cross sections using a surface integral formulation."

[26] M. Balagangadhar, T. K. Sarkar, J. Rejeb, and R. R. Boix, IEEE Trans. Microw. Theory Tech. **46**, 302–307 (1998); "Solution of the general Helmholtz equation in homogeneously filled waveguides using a static Green's function."

[27] B. S. Rodriguez, B. Sensale, and V. Leitao, J. Electron. Commun **62**, 576–581 (2008); "Determination of the TE and TM modes in arbitrarily shaped waveguides using a hypersingular boundary element formulation."

[28] J. P. Webb, IEEE Trans. Microw. Theory Tech. **36**, 1819–1824 (1988); "Finite element analysis of dispersion in waveguides with sharp metal edges."

[29] C. Shu, and Y. T. Chew, Prog. Electromagn. Res., PIER **21**, 1–19 (1999); "Application of multi-domain GDQ method to analysis of waveguides with rectangular boundaries."

[30] L. Proekt, S. Yuferev, I. Tsukerman, and N. Ida, IEEE Trans. Magn. **38**, 649–652 (2002). "Method of overlapping patches for electromagnetic computation near imperfectly conducting cusps and edges."

[31] M. Ikeuchi, K. Inoue, H. Sawami, and H. Niki, SIAM J. Appl. Math. **40**, 90–98 (1981). "Arbitrarily shaped hollow waveguide analysis by the α-interpolation method."

[32] Zehao Li, and L. R. Ram-Mohan, Phys. Rev. E **85**, 016706 (2012); "Taming hypersingular integrals using dimensional continuation."

[33] L. R. Ram-Mohan, *Finite Element and Boundary Element Applications in Quantum Mechanics* (Oxford University Press, Oxford, UK, 2002).

[34] R. Courant and D. Hilbert, *Methods of Mathematical Physics*, Vol I (Interscience, New York, 1953), p199.

[35] J. Moussa, L. R Ram-Mohan, J. Sullivan, T. Zhou, D. R. Hines, and S. A. Solin, Phys. Rev. B **64** (18), 184410 (2001). "Finite-element modeling of extraordinary m th metallic inclusions."

[36] Gouri Dhatt and Gilbert Touzot, *The Finite Element Method Displayed.* (John Wiley and Sons, New York, N. Y., 1984).

[37] P. G. Kassebaum, C. R. Boucher, and L. R. Ram-Mohan, J. Comput. Phys. **231** (17), 5747–5760 (2012). For other treatments of the Hermite interpolation on a triangle see: K. Bell, Int. J. Numer. Methods Eng. **1**, 101 (1969). *Finite Element Methods in Stress Analysis*, edited by I. Holand and K. Bell (Tapir, Trondheim, 1970), pp. 159, 213; J. H. Argyris, I. Fried, and D. W. Scharpf, "The TUBA family of plate elements for the matrix displacement method," Technical Report (Imperial College of Science and Technology, University of London, 1968), Technical Note 14.

[38] C. R. Boucher, Zehao Li, J. D. Albrecht, and L. R. Ram-Mohan, J. Appl. Phys. **115**, 1–10 (2014); "Hermite finite elements for high accuracy electromagnetic field calculations: A case study of homogeneous and inhomogeneous waveguides."

[39] C. R. Boucher, Zehao Li, C. I. Ahheng, J. D. Albrecht, and L. R. Ram-Mohan, J. Appl. Phys. **119**, 143106 (2016). "Hermite finite elements for high accuracy electromagnetic field calculations: A case study of homogeneous and inhomogeneous waveguides,"

[40] M. Dauge (2001), perso.univ-rennes1.fr/monique.dauge/core/index.html.

[41] S. Pandey, S. Bharadwaj, M. Santia, M. Hodek, J. D. Albrecht, and L. R. Ram-Mohan, J. Appl. Phys. **124**, 213106–1–13 (2018); "Cavity electrodynamics with Hermite interpolation: Role of symmetry and degeneracies."

[42] A. Dimitrov, H. Andrä, and E. Schnack, Int. J. Numer. Meth. Engng **52**, 805–827 (2001); (DOI: 10.1002/nme.230) "Efficient computation of order and mode of corner singularities in 3D-elasticity."

[43] L. Beilina, S. Korotov, and M. Krizek, Preprint 2003-02: Chalmers Finite Element Center, Chalmers University of Technology, Göteborg, Sweden (2003); "Application of the local nonobtuse tetrahedral refinement techniques near Fichera-like corners."

[44] Monique Dauge, Y. Lafranche, T. Ourmières-Bonafos, *Integral Equations and Operator Theory*, (Springer Nature Publishers, Switzerland, 2018) (https://doi.org/10.1007/s00020-018-2486-y) "Dirichlet spectrum of the Fichera layer."

[45] T. Apel, V. Mehrmann, and D. Watkins, *Comp. Methods Appl. Mech. Eng.*, **191**, 4459–4473 (2002); "Structured eigenvalue methods for the computation of corner singularities in 3D anisotropic elastic structures."

18
The gauge degree of freedom in electrodynamics

In this chapter:

- We consider the gauge degree of freedom inherent in electrodynamics. A systematic discussion of the implications of this freedom is presented. The magnetic and electric fields are expressed in terms of the magnetic vector potential $\mathbf{A}(\mathbf{r}, t)$ and the electric potential $\phi(\mathbf{r}, t)$. We study the implications of the gauge degree of freedom.
- As is usual, we use the Lorentz force $\mathbf{F} = q(\mathbf{E} + \mathbf{v} \times \mathbf{B})$ on a charged particle moving in electric and magnetic fields ($\mathbf{E}$, $\mathbf{B}$) to identify and define the Lagrangian density and the potential energy from which this force is derivable.
- The phase accumulation in the quantum mechanical wavefunction of a charged particle along its trajectory is evaluated. This leads us to consider the Aharonov-Bohm (AB) effect; the importance of the vector potential in quantum mechanics is brought out.
- Finally, we lay the groundwork for considering the AB effect in circular flat rings threaded by a magnetic field, and the same effect for Möbius rings.

18.1 Introduction

Maxwell's laws of electrodynamics are [1]

$$\nabla \cdot \mathbf{E}(\mathbf{r}, t) = \rho(\mathbf{r}, t)/\epsilon; \qquad \text{(Coulomb's law)} \tag{18.1}$$

$$\nabla \cdot \mathbf{B}(\mathbf{r}, t) = 0; \qquad \text{(No magnetic monopoles)} \tag{18.2}$$

$$\nabla \times \mathbf{E}(\mathbf{r}, t) = -\frac{\partial \mathbf{B}(\mathbf{r}, t)}{\partial t}; \qquad \text{(Faraday's law)} \tag{18.3}$$

$$\nabla \times \mathbf{H}(\mathbf{r}, t) = \mathbf{j}(\mathbf{r}, t) + \frac{\partial \mathbf{D}(\mathbf{r}, t)}{\partial t}. \qquad \text{(Ampere–Maxwell law)} \tag{18.4}$$

Finite Elements in Action. L. Ramdas Ram-Mohan, Oxford University Press. © L. Ramdas Ram-Mohan (2026).
DOI: 10.1093/oso/9780199563487.003.0018

Equations (18.1–18.4) together with the constitutive relations

$$\begin{aligned} \mathbf{D} &= \epsilon\mathbf{E} = \epsilon_0\mathbf{E} + \mathbf{P} \\ \mathbf{B} &= \mu\mathbf{H} = \mu_0\mathbf{H} + \mathbf{M}, \end{aligned} \tag{18.5}$$

and the equation of continuity,

$$\nabla\mathbf{j} + \partial\rho/\partial t = 0, \tag{18.6}$$

lead to an essentially complete description of physical phenomena related to electrodynamics at the classical level. [1]

In the absence of any charges in the region of interest, we have $\rho(\mathbf{r}, t) = 0$. Taking the curl of Eq. (18.3) and using Eq. (18.4), we arrive at the wave equation

$$\nabla^2\mathbf{E}(\mathbf{r}, t) - \mu\epsilon\,\partial_t^2\mathbf{E}(\mathbf{r}, t) = \mu\frac{\partial\mathbf{J}}{\partial t}. \tag{18.7}$$

Similarly, taking the curl of the Ampere-Maxwell equation, Eq. (18.4), and using $\nabla \cdot \mathbf{B} = 0$, we have

$$\nabla^2\mathbf{H}(\mathbf{r}, t) - \mu\epsilon\,\partial_t^2\mathbf{H}(\mathbf{r}, t) = -\nabla \times \mathbf{J}(\mathbf{r}, t). \tag{18.8}$$

The condition, $\nabla \cdot \mathbf{B}(\mathbf{r}, t) = 0$, suggests that we can write the magnetic field in terms of a vector potential $\mathbf{A}(\mathbf{r}, t)$,

$$\mathbf{B}(\mathbf{r}, t) = \nabla \times \mathbf{A}(\mathbf{r}, t), \tag{18.9}$$

so that the vector identity $\nabla \cdot (\nabla \times \mathbf{A}(\mathbf{r}, t)) = 0$ is automatically satisfied, and we are assured that Eq. (18.2) will hold if we represent the magnetic field in terms of the magnetic vector potential. This form allows us to represent the magnetic field in terms of the vector potential that is typically less singular. This is analogous to the case of electrostatics where the relation $\nabla \times \mathbf{E}(\mathbf{r}) = 0$ allows us to define an electrostatic scalar potential $\phi(\mathbf{r})$ such that $\mathbf{E}(\mathbf{r}) = -\nabla\phi(\mathbf{r})$, which then directly satisfies Gauss's law, Eq. (18.1), and $\nabla \times (\nabla\phi) = 0$. Here again, the field behaves as $\sim 1/r^2$ while the potential behaves as $\sim 1/r$.

In electrodynamics, with $\mathbf{B} = \nabla \times \mathbf{A}$, Faraday's law Eq. (18.3) takes the form

$$\nabla \times \left(\mathbf{E} + \frac{\partial\mathbf{A}}{\partial t}\right) = 0; \tag{18.10}$$

when the time-dependence is present, we can define

$$\mathbf{E}(\mathbf{r}, t) = -\nabla\phi(\mathbf{r}, t) - \frac{\partial\mathbf{A}(\mathbf{r},\, \mathbf{t})}{\partial t}. \tag{18.11}$$

Substituting for $\mathbf{E}$ and $\mathbf{B}$ in Maxwell's equations with $(\mathbf{A}, \phi)$ and taking the curl of the Faraday relation, Eq. (18.3), we obtain the wave equations†

$$\nabla^2 \mathbf{A}(\mathbf{r},t) - \mu\epsilon \frac{\partial^2}{\partial t^2}\mathbf{A}(\mathbf{r},t) = -\mu \mathbf{j}(\mathbf{r},t) + \nabla\left(\nabla\cdot\mathbf{A}(\mathbf{r},t) + \mu\epsilon\frac{\partial\phi(\mathbf{r},t)}{\partial t}\right),$$
$$\nabla^2\phi(\mathbf{r},t) + \frac{\partial}{\partial t}\left(\nabla\cdot\mathbf{A}(\mathbf{r},t)\right) = -\frac{\rho}{\epsilon}. \tag{18.12}$$

The four Maxwell equations, Eqs. (18.1–18.4), are written here as four equations corresponding to the three components of the vector potential and the scalar potential.

Suppose $\chi(\mathbf{r},t)$ is an arbitrary scalar field. Then the potentials

$$\mathbf{A}'(\mathbf{r},t) = \mathbf{A}(\mathbf{r},t) + \nabla\chi(\mathbf{r},t),$$
$$\phi'(\mathbf{r},t) = \phi(\mathbf{r},t) - \frac{\partial\chi(\mathbf{r},t)}{\partial t}, \tag{18.13}$$

yield the same fields $\mathbf{E}$ and $\mathbf{B}$, as can be verified directly from Eqs. (18.9, 18.11). This freedom to choose $\chi(\mathbf{r},t)$ is called the gauge degree of freedom, and $(\mathbf{A},\phi)$ and $(\mathbf{A}',\phi')$ are said to be gauge transforms of one another. Furthermore, $\mathbf{A}'$ and ϕ' satisfy the Maxwell equations,

$$\nabla\cdot\mathbf{B} = 0,$$
$$\nabla\times\mathbf{E} = -\frac{\partial\mathbf{B}}{\partial t},$$

as can be shown using vector identities. Since the definitions of the potentials needed to derive the physical electric and magnetic fields are arbitrary up to the choice of the function $\chi(\mathbf{r},t)$, called the gauge function, we find it convenient to choose the gauge such that $(\mathbf{A},\phi)$ satisfy the condition

$$\nabla\cdot\mathbf{A} + \mu\epsilon\frac{\partial\phi}{\partial t} = 0. \tag{18.14}$$

This is called the *Lorenz gauge* condition.‡ [2–5] This choice of the gauge directly decouples the Eqs. (18.12), so that we have

$$\nabla^2\mathbf{A}(\mathbf{r},t) - \mu\epsilon\frac{\partial^2}{\partial t^2}\mathbf{A}(\mathbf{r},t) = -\mu\mathbf{j}(\mathbf{r},t),$$
$$\nabla^2\phi(\mathbf{r},t) - \mu\epsilon\frac{\partial^2}{\partial t^2}\phi(\mathbf{r},t) = -\frac{\rho(\mathbf{r},t)}{\epsilon}. \tag{18.15}$$

†In a material medium with dielectric permittivity ϵ and magnetic permeability μ, we have the velocity of light given by $v^2 = 1/(\epsilon\mu)$. In vacuum, with $\epsilon \to \epsilon_0$ and $\mu \to \mu_0$, the velocity of light is given by $c^2 = 1/(\epsilon_0\mu_0)$=$2.997\,9245\,8\times10^8$ ms^{-1}.

‡The Lorenz gauge is frequently attributed in error to H. A. Lorentz.

This gauge corresponds to selecting $\mathbf{A}'$ and ϕ' such that Eq. (18.14) leads to

$$\nabla \cdot (\mathbf{A}' - \nabla\chi) + \mu\epsilon \frac{\partial}{\partial t}\left(\phi' + \frac{\partial \chi}{\partial t}\right) =$$
$$\left(\nabla \cdot \mathbf{A}' + \mu\epsilon \frac{\partial \phi'}{\partial t}\right) + \left(-\nabla^2\chi + \mu\epsilon \frac{\partial^2 \chi}{\partial t^2}\right) = 0.$$

In other words, the wave equations for $\mathbf{A}$ and ϕ would be invariant under the choice of the gauge function as long as the gauge function χ satisfies the wave equation.

Other choices of the gauge function are possible which provide some advantages in various circumstances. In particular, the Coulomb gauge is $\nabla \cdot \mathbf{A}(\mathbf{r}, t) = 0$, and it is frequently used in quantum mechanical calculations. In this gauge, the scalar potential ϕ satisfies the Poisson's equation, Eq. (18.12), and not the wave equation. The momentum operator commutes with the vector potential in this gauge, and this is one of the features in its favor.

For calculations of quantum mechanical energy levels in the presence of a static magnetic field, $\mathbf{B} = \hat{\mathbf{z}}B_0$, we can use the Landau gauge,

$$\mathbf{A}_{\text{Landau}} = B_0(0,\, x,\, 0), \tag{18.16}$$

or a more symmetrical form

$$\mathbf{A}_{\text{symmetric}} = B_0\left(-\frac{y}{2},\, \frac{x}{2},\, 0\right). \tag{18.17}$$

Both forms, Eq. (18.16) and Eq. (18.17), give the same magnetic field $\mathbf{B}$, and the two choices of the vector potential are related by a gauge transformation with the gauge function $\chi(\mathbf{r}) = -B_0\, xy/2$ such that $\mathbf{A}_{\text{symmetric}} = \mathbf{A}_{\text{Landau}} + \nabla\chi$.

The Landau gauge can be generalized to write

$$\begin{aligned} \mathbf{A} &= B_0(-ay,\, bx,\, 0), \\ a + b &= 1. \end{aligned} \tag{18.18}$$

18.2 The magnetic vector potential in classical dynamics

The following discussion leads to the identification of the kinematical momentum and the classical Hamiltonian expressed in terms of the electromagnetic potentials.

18.2.1 The Lorentz force

Following the treatment of Goldstein [6] for example, we show here that the Lorentz force on a point particle of mass m and charge q due to external electric and magnetic fields can be written in terms of the vector and scalar potentials. Since we have to manipulate vector components, we will use the usual Cartesian tensor notation.

The Lorentz force is given by

$$F_i = q\Big(E + (\mathbf{v}\times\mathbf{B})\Big)_i, \quad i = 1, 2, 3. \tag{18.19}$$

Since $E_i = -\partial_i\phi - \partial A_i/\partial t$ and $B_i = (\nabla\times\mathbf{A})_i$, we have

$$F_i = q\Big(-\partial_i\phi(\mathbf{r},t) - \partial A_i(\mathbf{r},t)/\partial t\Big) + q(\mathbf{v}\times(\nabla\times\mathbf{A}))_i.$$

We use

$$(\mathbf{v}\times(\nabla\times\mathbf{A}))_i = \epsilon_{ijk}v_j\epsilon_{klm}\partial_l A_m = v_j\partial_i A_j - (v_j\partial_j)A_i.$$

Also, with $\mathbf{A}(\mathbf{r},t)$ being an explicit function of $\mathbf{r}(t)$ and t, on using the chain rule of differentiation, we have

$$\frac{d}{dt}(A_i(\mathbf{r}(t),t)) = (v_j\partial_j)A_i + \frac{\partial A_i}{\partial t}. \tag{18.20}$$

We obtain

$$(\mathbf{v}\times(\nabla\times\mathbf{A}))_i = v_j\partial_i A_j + \left(\frac{\partial A_i}{\partial t} - \frac{d}{dt}(A_i(\mathbf{r},t))\right). \tag{18.21}$$

The Lorentz force is then given by

$$\begin{aligned} F_i &= q\Big(-\partial_i\phi(\mathbf{r},t) - \frac{\partial A_i}{\partial t}\Big) + q\Big((v_j\partial_i)A_j + \frac{\partial A_i}{\partial t} - \frac{d}{dt}A_i(\mathbf{r},t)\Big) \\ &= q\Big[-\partial_i\phi(\mathbf{r},t) + (\partial_i(v_jA_j) - A_j\partial_i v_j) - \frac{d}{dt}\left(\frac{\partial}{\partial v_i}(A_k v_k)\right)\Big]. \end{aligned} \tag{18.22}$$

Here, the second term in the parenthesis in the above expression, $-A_j\partial_i v_j$, is zero since $v_j = dx_j(t)/dt$, and $d/dt(\partial x_j/\partial x_i) = d/dt(\delta_{ij}) = 0$. This also follows from the fact that the coordinate and the velocity are independent variables.

The time-derivative term contains $\left(\frac{\partial}{\partial v_i}(A_k v_k)\right)$ which gives two terms $A_i\delta_{ik}$ and $v_k\partial A_k/\partial v_i$. The latter term is zero since the potentials are functions of the coordinate and time, not of the velocity variable. Note again that x_i and $v_i = \dot{x}_i(t)$ are independent variables.

We are then left with

$$F_i = q\Big[-\partial_i(\phi - v_jA_j) + \frac{d}{dt}\left(\frac{\partial}{\partial v_i}(\phi - v_jA_j)\right)\Big]. \tag{18.23}$$

Here, we have inserted the term $\partial\phi/\partial v_i = 0$ in the last parenthesis so as to write F in the above form. We now substitute

$$U = q\,\phi(\mathbf{r},t) - q\,\mathbf{v}\cdot\mathbf{A}(\mathbf{r},t), \tag{18.24}$$

to write

$$F_i = -\frac{\partial}{\partial x_i}U(\mathbf{r},t) + \frac{d}{dt}\left(\frac{\partial}{\partial v_i}U(\mathbf{r},t)\right). \tag{18.25}$$

The expression for the force contains a gradient of the potential and a time derivative of the velocity gradient of the potential. Since the potential energy U is dependent on the velocity, this term is not zero.

18.2.2 The classical Lagrangian with a velocity-dependent potential

The Lagrangian L for a charged particle in electromagnetic fields is now given by

$$\begin{aligned} L &= T - U, \\ &= \frac{1}{2} m\dot{x}_i^2 - q\,\phi + q\,v_i A_i. \end{aligned} \tag{18.26}$$

We recognize the potential energy terms as being the usual electrostatic contribution together with $\mathbf{j} \cdot \mathbf{A}$, which is associated with the motion of the charge in the external electromagnetic field.

Let us verify that the variation of the Lagrangian L leads to the equation of motion, the Lorentz force. We have, with $\dot{x}_i = v_i$,

$$\frac{\delta L}{\delta \dot{x}_i} = m\dot{x}_i + q\,A_i \equiv p_i, \tag{18.27}$$

and

$$\frac{\delta L}{\delta x_i} = -q\,\partial_i \phi + q\,\dot{x}_j \partial_i A_j.$$

Thus, the equation of motion is

$$\begin{aligned} \frac{d}{dt}\left(\frac{\delta L}{\delta \dot{x}_i}\right) - \frac{\delta L}{\delta x_i} &= 0, \\ \left(m\ddot{x}_i + q\,\frac{d}{dt} A_i\right) + q\,\partial_i \phi - q\,\dot{x}_j \partial_i A_j &= 0. \end{aligned}$$

Rearranging terms and noting that $q\,dA_i/dt = q\,\partial A_i/\partial t + q\,v_j \partial_j A_i$, we have

$$\begin{aligned} F_i &= q\left(-\partial_i \phi - \partial A_i/\partial t\right) + q\left(v_j \partial_i A_j - v_j \partial_j A_i\right) \\ &= q\mathbf{E}_i + q\left(\mathbf{v} \times (\nabla \times \mathbf{A})\right)_i \\ &= q\left(\mathbf{E} + \mathbf{v} \times \mathbf{B}\right)_i, \end{aligned}$$

as expected. Note that the canonical momentum $\mathbf{p}$, Eq. (18.27), is obtained by varying the Lagrangian with respect to the velocity, and with velocity-dependent potentials being present, we arrive at

$$\mathbf{p} = m\mathbf{v} + q\mathbf{A}. \tag{18.28}$$

The quantity mass$\times$velocity is referred to as the *kinematical momentum*, denoted by $\mathbf{\Pi}$, and is given by

$$m\mathbf{v} = \mathbf{\Pi} = \mathbf{p} - q\mathbf{A}. \tag{18.29}$$

The Hamiltonian is

$$H = \dot{x}_i p_i - L = \dot{x}_i(m\dot{x}_i + qA_i) - \left(\frac{1}{2}m\dot{x}_i^2 - q\,\phi + q\,v_i A_i\right)$$
$$= \frac{1}{2}m\dot{x}_i^2 + q\,\phi = \frac{1}{2m}(\mathbf{p} - q\mathbf{A})^2 + q\phi(\mathbf{r},t). \tag{18.30}$$

Hence, in the presence of electromagnetic fields, the Hamiltonian is altered such that $\mathbf{p} \to \mathbf{\Pi} = (\mathbf{p} - q\mathbf{A})$; the effect of the electric field is represented by the charge particle's potential energy $q\phi$.

18.3 Gauge freedom and quantum mechanics

When we transition to quantum mechanics from the classical considerations given above, we see that it is the potentials that play the important role in defining the dynamics of charged particles in external electromagnetic fields. The expectation values of physical observables must be the same under gauge transformations. We explore these issues in the following.

The Schrödinger Hamiltonian is given by

$$H = \frac{\Pi^2}{2m} + q\phi = \frac{(\mathbf{p} - q\,\mathbf{A})^2}{2m} + q\,\phi. \tag{18.31}$$

The canonical momenta commute, and we have $[p_i, p_j] = 0$, but the kinematical momenta do not. We can easily show that

$$[\Pi_i, \Pi_j] = i\hbar\, q\, \epsilon_{ijk} B_k. \tag{18.32}$$

It is very useful to identify the action integral and the Lagrangian that can be used for quantum mechanical considerations. We have $\mathcal{A} = \int dt L = \int dt \int d^3r \mathcal{L}$. Here, the Schrödinger Lagrangian density $\mathcal{L}$ for a time-dependent system is given by

$$\mathcal{L} = \psi^*(\mathbf{r},t)\left[\frac{1}{2m}\left(i\hbar \overleftarrow{\nabla} - q\mathbf{A}\right)\left(-i\hbar \overrightarrow{\nabla} - q\mathbf{A}\right) + q\phi - \frac{i\hbar}{2}\left(\overleftarrow{\partial_t} - \overrightarrow{\partial_t}\right)\right]\psi(\mathbf{r},t). \tag{18.33}$$

The notation allows us to write the Lagrangian density in a compact form, such that the derivative with a right- (left-) arrow acts only on the functions to the right (left) of it. The last two terms with the time derivative, $i\hbar\left(\overleftarrow{\partial_t} - \overrightarrow{\partial_t}\right)/2$ are replaced by E for a time-independent quantum mechanical problem. Frequently, the time-derivative terms are written in a more compact form $\overrightarrow{\partial_t} - \overleftarrow{\partial_t} = \overleftrightarrow{\partial_t}$.

18.4 The conserved current derived using gauge variation

From the Lagrangian density, we can derive the conserved current through a gauge variational method. [7–9] Consider the transformation

$$\psi(\mathbf{r},t) \to \psi'(\mathbf{r},t) = \exp(i\Lambda(\mathbf{r},t))\psi(\mathbf{r},t). \tag{18.34}$$

The new Lagrangian density $\mathcal{L}'$, in the absence of the electromagnetic field but under this phase transformation, is given by

$$\begin{aligned}\mathcal{L}' &= [-i\nabla\Lambda(\mathbf{r},t)\,\psi^*(\mathbf{r},t) + \nabla\psi^*(\mathbf{r},t)]\cdot\left(\frac{\hbar^2}{2m}\right)[i\nabla\Lambda(\mathbf{r},t)\psi(\mathbf{r},t) + \nabla\psi(\mathbf{r},t)]\\ &\quad - i\frac{\hbar}{2}\psi^*\,[i\partial_t\Lambda(\mathbf{r},t) + \partial_t\psi(\mathbf{r},t)] + i\frac{\hbar}{2}\,[-i\partial_t\Lambda(\mathbf{r},t) + \partial_t\psi(\mathbf{r},t)]\\ &= \mathcal{L} + \nabla\Lambda(\mathbf{r},t)\cdot\nabla\Lambda(\mathbf{r},t)\,\psi^*\frac{\hbar^2}{2m}\psi + \nabla\psi^*\frac{\hbar^2}{2m}\psi\,(i\nabla\Lambda)\\ &\quad - (i\nabla\Lambda)\,\psi^*\frac{\hbar^2}{2m}\nabla\psi - (\partial_t\Lambda)\,\hbar\psi^*\psi. \end{aligned} \tag{18.35}$$

A variation of the Lagrangian density with respect to Λ gives us

$$\begin{aligned}\frac{1}{\hbar}\lim_{\Lambda\to 0}\,\delta_\Lambda\mathcal{L}' &= \frac{1}{\hbar}\left[i\frac{\hbar^2}{2m}\nabla\Big(\psi^*(\mathbf{r},t)\nabla\psi(\mathbf{r},t) - (\nabla\psi^*)\psi\Big)\right.\\ &\qquad \left. - \hbar\partial_t\Big(\psi^*(\mathbf{r},t)\psi(\mathbf{r},t)\Big)\right]\delta\Lambda\\ &= 0, \end{aligned} \tag{18.36}$$

where the above variation is set to zero because of gauge invariance. The coefficient of $\delta\Lambda$ is zero for the arbitrary variations, leading to

$$\nabla\left[\frac{\hbar}{2mi}\left(\psi^*\overleftrightarrow{\nabla}\psi\right)\right] + \partial_t\left(\psi^*\psi\right) = 0,$$

which may be written as the equation of continuity for the probability density

$$\nabla\mathbf{j}(\mathbf{r},t) + \frac{\partial}{\partial t}\rho(\mathbf{r},t) = 0. \tag{18.37}$$

As can be anticipated, an integration of the above equation of continuity over a given volume Ω enclosed by a surface $\partial\Omega$ can be expressed using Gauss' theorem as

$$\oint d\vec{s}\cdot\mathbf{j} + \frac{d}{dt}\int d^3r\,\rho(\mathbf{r},t) = 0. \tag{18.38}$$

The usual arguments regarding localized probability density and currents may be followed to demand that the flux out of the surface as $r\to\infty$ is zero. We are then left with

$$\frac{d}{dt}Q = 0. \tag{18.39}$$

If we consider the lepton number gauge, we obtain that the electron number is conserved. Similarly, by starting from Eq. (18.33) and using the gauge function $\chi(\mathbf{r}.t)$, we can show that the electric charge is conserved. We see that the gauge invariance and charge conservation are intimately tied together.

The variation of the Lagrangian density with respect to the gradient of the gauge function $\chi(\mathbf{r}.t)$ can directly give us the current density $\mathbf{j}(\mathbf{r},t)$. We have

$$\begin{aligned}\mathbf{j}(\mathbf{r},t) &= \frac{1}{\hbar}\lim_{\nabla\chi\to 0}\frac{\delta\mathcal{L}}{\delta\nabla\chi} \\ &= \frac{1}{\hbar}\lim_{\nabla\chi\to 0}\frac{\delta}{\delta\nabla\chi}\psi^*(\mathbf{r},t)\left[\frac{1}{2m}\left(i\hbar\overleftarrow{\nabla} - q\mathbf{A} + \hbar\nabla\chi\right)\right. \\ &\qquad\qquad \left.\times\left(-i\hbar\overrightarrow{\nabla} - q\mathbf{A} + \hbar\nabla\chi\right) + q\phi\right]\psi(\mathbf{r},t).\end{aligned}$$

Performing the variation and the limiting procedure, we obtain

$$\mathbf{j}(\mathbf{r},t) = \frac{\hbar}{2mi}\left(\psi^*\overrightarrow{\nabla}\psi - \psi^*\overleftarrow{\nabla}\psi\right) - \frac{q}{m}\mathbf{A}\psi^*\psi, \tag{18.40}$$

where the first term in parentheses contains the usual probability current and the A-dependent term is given by the presence of the electromagnetic fields. This gauge variational approach to deriving the conserved current is very effective, and is generalizable to more complex systems of coupled differential equations with complex boundary conditions. [9]

18.5 The phase accumulation in the wavefunction along particle trajectories

In quantum mechanics, we require that the normalization of the wavefunction not change under the gauge transformations of Eq. (18.13); also, the expectation values of the coordinate $\mathbf{r}$ and the kinetic momentum $\mathbf{\Pi} = \mathbf{p} - q\mathbf{A}$ should not change.

If we label the wavefunctions for the $(\mathbf{A}, \phi)$ gauge by $\psi(\mathbf{r},t)$ and the wavefunctions after the gauge transformation to $(\mathbf{A}', \phi')$ by $\psi'(\mathbf{r},t)$, then we expect the two wavefunctions to be related by a unitary transformation U such that

$$\psi'(\mathbf{r},t) = U\psi(\mathbf{r},t). \tag{18.41}$$

The invariance properties of the matrix elements of the coordinate and the kinetic momentum under the gauge transformations are guaranteed, if we define

$$U = \exp[iq\,\chi(\mathbf{r},t)/\hbar]. \tag{18.42}$$

Here $\mathbf{r}$ commutes with any function of $\mathbf{r}$, including $\chi(\mathbf{r})$. We also have

$$\begin{aligned}U^\dagger(\mathbf{p} - q\mathbf{A}'(\mathbf{r},t))U &= U^\dagger(\mathbf{p} - q\mathbf{A}(\mathbf{r},t) - q\,\nabla\chi(\mathbf{r},t))\,U \\ &= \mathbf{p} - q\mathbf{A}(\mathbf{r},t).\end{aligned} \tag{18.43}$$

This follows from $\mathbf{p}U = U(\mathbf{p} + q\nabla\chi)$, where the gradient operator in the momentum $\mathbf{p}$ pulls down the gradient of the function in the exponent in U, as is seen using the chain rule of differentiation. From the above result, we can see that the Hamiltonian and the Lagrangian are gauge-invariant when the gauge-invariant kinetic momentum operator is used.

18.6 Absorbing the vector potential into a phase factor

The above considerations suggest that we can also pull out the presence of the $q\mathbf{A}(\mathbf{r},t)$ in the Lagrangian, or the Hamiltonian, and replace it with a phase factor. We showed in the above that the phase function is brought down by the gradient operator present in the momentum $\mathbf{p}$ as the gradient of the phase function; hence, we can use a line integral of the vector potential as the phase function to write

$$\psi(\mathbf{r},t) = \exp\left(i\frac{q}{\hbar}\int_{r_0}^{\mathbf{r}} d\boldsymbol{\ell}' \cdot \mathbf{A}(\mathbf{r}')\right)\psi_0(\mathbf{r},t), \tag{18.44}$$

with ψ_0 corresponding to the wavefunction in the absence of the magnetic vector potential. The kinetic momentum operator $\boldsymbol{\Pi}$ acting on ψ leads to

$$\begin{aligned}(\mathbf{p} - q\mathbf{A})\,\psi(\mathbf{r},t) &= \exp\left(i\frac{q}{\hbar}\int_{r_0}^{\mathbf{r}} d\boldsymbol{\ell}' \cdot \mathbf{A}(\mathbf{r}')\right)\left[(-i\hbar\boldsymbol{\nabla} - q\mathbf{A} + q\mathbf{A})\psi_0(\mathbf{r},t)\right] \\ &= \exp\left(i\frac{q}{\hbar}\int_{r_0}^{\mathbf{r}} d\boldsymbol{\ell}' \cdot \mathbf{A}(\mathbf{r}')\right)\mathbf{p}\,\psi_0(\mathbf{r},t).\end{aligned} \tag{18.45}$$

Hence, if we can solve for the wavefunction $\psi_0(\mathbf{r},t)$ in the absence of a magnetic field, we can obtain the correct wavefunction in the presence of the magnetic field by multiplying it with the phase factor of Eq. (18.44). We have here transferred the burden of solving for ψ onto determining the line integral of the vector potential. This idea has been very effectively exploited by Ueta [10] in actual quantum mechanical finite element calculations involving the presence of magnetic fields.

18.7 The Aharonov-Bohm effect

In Fig. 18.1, we consider two paths for a particle of charge q from a source located at $\mathbf{r}_0$ to a detector located at $\mathbf{r}$. As in the original Aharonov-Bohm (AB) thought experiment, suppose that a very long solenoid with a very small radius is present, with a magnetic field confined within it. The figure represents a cross-section across the solenoid and shows two paths P_1 and P_2 around the solenoid. The magnetic field lines returning to form closed loops are very far away from the paths of the electron from $\mathbf{r}_0$ to $\mathbf{r}$ and do not affect it directly. Aharonov and Bohm pointed out in 1959 that while the electron does not experience a magnetic field along its path, $\mathbf{B}$ being zero there, it is influenced by the vector potential which is not zero outside the solenoid (this is explicitly shown below). Since $\mathbf{B} = \nabla \times \mathbf{A} = 0$ along the two paths P_1 and P_2, we find that paths P_1' and P_2' in the neighborhood of the two paths are equivalent to the initial ones by Stokes' theorem. This is a consequence of the following:

$$\begin{aligned}\int_{\mathbf{r}_0,\mathbf{P}_1}^{\mathbf{r}} d\boldsymbol{\ell}' \cdot \mathbf{A} - \int_{\mathbf{r}_0,\mathbf{P}_{1'}}^{\mathbf{r}} d\boldsymbol{\ell}' \cdot \mathbf{A} &= \oint_{P_1 - P_1'} d\boldsymbol{\ell}' \cdot \mathbf{A} \\ &= \oint d\mathbf{s}' \cdot (\boldsymbol{\nabla}' \times \mathbf{A}(\mathbf{r}')) = \oint d\mathbf{s}' \cdot \mathbf{B}(\mathbf{r}') = \boldsymbol{\Phi}_B.\end{aligned} \tag{18.46}$$

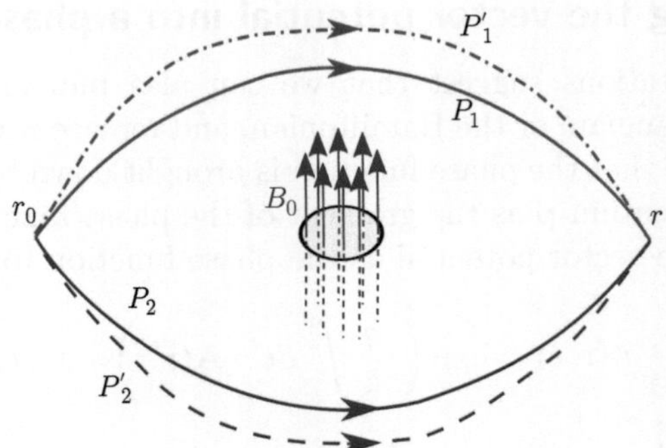

Figure 18.1 Two paths, one labeled by P_1 and the other labeled by P_2, of an electron from a source at $\mathbf{r}_0$ to a detector located at $\mathbf{r}$ are shown schematically. A magnetic field B_0 is present *in the circular region only*, with no direct influence along the electron paths. The vector potential outside the circular region, however, is not zero. Neighboring paths to P_1, P_2 are shown as dashed curves labeled by P_1' and P_2'. The paths are on a plane perpendicular to the magnetic field.

Here, the magnetic flux Φ_B through the area between the two paths P_1 and P_1' is zero since the field is zero, and the two paths are equivalent as far as the phase accumulation in going from $\mathbf{r}_0$ to $\mathbf{r}$. The same holds for the pair of paths P_2 and P_2'.

This equivalence of the paths is not true when we consider the two paths P_1 and P_2 together. The difference in the phase accumulation is given by

$$\begin{aligned}\Theta &= \left(\frac{q}{\hbar}\right)\int_{\mathbf{r}_0,\mathbf{P}_2}^{\mathbf{r}} d\boldsymbol{\ell}'\cdot\mathbf{A}(\mathbf{r}',t) - \left(\frac{q}{\hbar}\right)\int_{\mathbf{r}_0,\mathbf{P}_1}^{\mathbf{r}} d\boldsymbol{\ell}'\cdot\mathbf{A}(\mathbf{r}',t)\Big|_{P_1} \\ &= \frac{q}{\hbar}\oint d\boldsymbol{\ell}'\cdot\mathbf{A}(\mathbf{r}',t)\end{aligned} \tag{18.47}$$

The line integral around the closed path can be expressed as before using Stokes' theorem as the surface integral of the curl of $\mathbf{A}$ over the surface enclosed by the path. The phase is

$$\begin{aligned}\Theta &= \frac{q}{\hbar}\int d\mathbf{S}'\cdot(\boldsymbol{\nabla}\times\mathbf{A}) \\ &= \frac{q}{\hbar}\int_{\text{solenoid}} d\mathbf{S}'\cdot\mathbf{B}_0 = \frac{q}{\hbar}\Phi_B.\end{aligned} \tag{18.48}$$

In other words, the phase difference depends on the flux of the magnetic field through the solenoid, and this is not zero. As we see from Fig. 18.1, we no longer have a singly connected region for the paths of the charged particle. [11, 12]

18.8 Discussion of the Aharonov-Bohm effect

Within the context of classical physics, the replacement of the electric and magnetic fields by the potentials may be thought of as a matter of mathematical convenience and of no fundamental consequence. However, within the framework of quantum mechanics Aharonov and Bohm (AB), in 1959, clearly brought out the need to view the vector potential as a real physical field with observable effects that can be attributed directly to it. They pointed out, using essentially the arguments presented in Sec. 18.5, that for an electron of charge $q = -e$ emitted at $\mathbf{r}_0$ and detected at $\mathbf{r}$, the phase difference along the paths P_1 and P_2 of Fig. 18.1, the switching on of the magnetic field will lead to a shift in the interference fringes observed at the detector in the absence of any classical force on the electron. Furthermore, as the magnetic field is increased from zero, the phase difference of Eq. (18.47) shows a periodic behavior with period defined by

$$\Theta = \Theta_0 - 2\pi\, n\, \frac{\hbar}{e}, \quad n = \text{integer}. \tag{18.49}$$

Thus, with increasing magnetic field in the confined region, the interference pattern shifts in a periodic fashion with the period governed by the "quantized flux"[§]

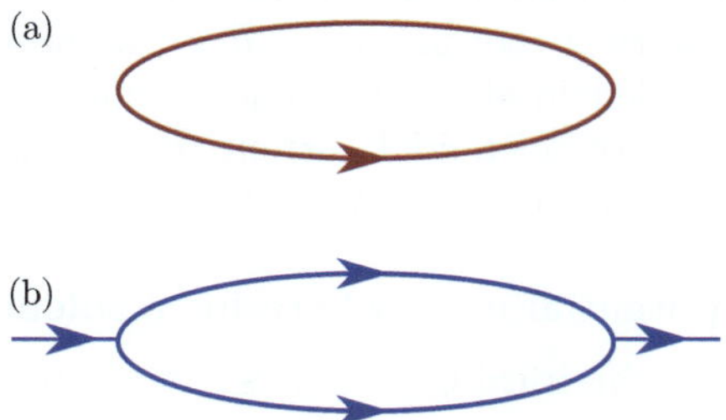

Figure 18.2 The eigenstates of a particle on a loop, shown in (a), are evaluated to determine the propagating states. In (b), the loop has two contacts attached to it at diametrically opposite points on the loop. The conditions for constructive and destructive interference in the transmission through the ring are determined in the text.

$$\Phi_0 = 2\pi\,\frac{\hbar}{e} = h/e = 4.135\,667\,514\,8 \times 10^{-15}\ \text{Weber}. \tag{18.50}$$

In a series of articles [13–15], the importance of the electromagnetic potentials was explored in detail by Aharonov and Bohm. The AB effect was soon confirmed

[§] *A check of units*: Energy of electromagnetic interaction $\int \mathbf{J}\cdot\mathbf{A}d^3r$ has units of

$$\{[j][A]\times[L^3]\} = \{([I]/[L^2])[A][L^3]\} = [I][A][L].$$

Now the units of the magnetic field are $[B] = [A]/[L]$ since the field $\mathbf{B}$ is the curl of $\mathbf{A}$. On the other hand, h/e = energy × (second/charge) = energy/current. Thus,

$$h/e = [A][I][L]/[I] = ([A]/[L])[L^2] = [B][L^2].$$

This is the magnetic flux, and it is given in units of Webers.

by experiments. [16] The review by Peshkin and Tonomura [17] provides an extensive theoretical as well as experimental status report as of 1989.

While others have recognized the importance of the vector potential in quantum mechanical systems, AB presented the issues in a most direct way. Ehrenberg and Siday [18] in 1949 noted as a curiosity, while analyzing the electron trajectories for designing electron optics, that the carriers will have their paths modified by potentials that exist even though the fields are absent along the trajectory. This observation was not recognized as a fundamental shift of the theoretical point of view in the merging of electrodynamics with quantum mechanics. In this connection, comments by Berry [19] are relevant.

18.8.1 A simple example

Consider an electron confined to the circumference of a circle of radius R_0, similar to a bead threaded by a circular wire. We suppose that a very long solenoidal wire of radius $a \ll R_0$ carrying a current I generates a magnetic field B_0 within it. We solve for the bound state energy levels of the electron, which does not experience a magnetic field in its orbit around the solenoid. The analysis is similar to Sec. 18.5 except that we have a circular path for the electron. The region inside the solenoid is excluded to the electrons. Using cylindrical coordinates (ρ, θ, z), we note that $\mathbf{B} = \nabla \times \mathbf{A}$ for the constant field inside the solenoid allows us to determine $\mathbf{A}$ everywhere. In cylindrical coordinates, the symmetry of solenoidal current requires that we have $A_\rho = 0 = A_z$ everywhere, while A_θ is not zero. Thus, $\mathbf{A} = (0, A_\theta, 0)$.

Deriving the vector potential everywhere for a solenoid

The curl operator in the cylindrical coordinates acting on the components of $\mathbf{A}$ with just one component $A_\theta(\rho)$ being present, leads to

$$\mathbf{B} = \nabla \times \mathbf{A} = \hat{z}\,\frac{1}{\rho}\left(\frac{\partial(\rho A_\theta)}{\partial \rho} - \frac{\partial A_\rho}{\partial \phi}\right). \qquad (18.51)$$

For a solenoid, we know that the magnetic field inside the solenoid of radius a oriented along the $\hat{z}$-axis is given by $\mathbf{B} = \hat{z}\mu_0\, n\, I$, where the number of turns of wire per unit length is n and the solenoidal current is I. If we take a circular path of radius $\rho < a$ around the axis of the solenoid, we have

$$\int \mathbf{B} \cdot d\mathbf{s} = (\pi\rho^2)\,[\mu_0 n\, I] = \int \nabla \times \mathbf{A} \cdot d\mathbf{s}.$$

Hence, by Stokes' theorem, we have

$$\int \mathbf{B} \cdot d\mathbf{s} = (\pi\,\rho^2)\,[\mu_0\, n\, I] = \oint \mathbf{A} \cdot d\boldsymbol{\ell} = 2\pi\rho\, A_\theta \qquad (18.52)$$

so that

$$A_\theta = \frac{\pi\,\rho^2 B}{2\pi\,\rho} = \frac{\rho B}{2}, \quad \text{for } \rho < a. \qquad (18.53)$$

In a similar manner, for a circular path of radius ρ outside the solenoid we have

$$\int \mathbf{B} \cdot d\mathbf{s} = \pi\, a^2\, B = A_\theta 2\pi\,\rho, \quad \text{for } \rho > a. \tag{18.54}$$

The magnetic flux is not zero only over the cross-section of the solenoid, and this has been taken into account in the above integral. We then have

$$A_\theta = \frac{B\,a^2}{2\rho}, \quad \text{for } \rho > a. \tag{18.55}$$

In summary, we have

$$\begin{aligned} \mathbf{A}(\rho) &= (0,\; B\rho/2,\; 0), \quad \text{for } \rho \leq a; \\ \mathbf{A}(\rho) &= \left(0, \frac{B\,a^2}{2\rho}, 0\right), \quad \text{for } \rho > a. \end{aligned} \tag{18.56}$$

The Hamiltonian for the circular bound state

In the absence of any electric potential, we note that the vector potential of Eq. (18.56) satisfies the Coulomb gauge $\boldsymbol{\nabla} \cdot \mathbf{A}(\mathbf{r}) = 0$. The gauge-invariant momentum for the electron in the circular orbit has no radial component and hence takes the form

$$\boldsymbol{\Pi} = \left[\hat{\theta}\left(-i\hbar\frac{1}{\rho}\frac{\partial}{\partial\theta} + \frac{e\Phi}{2\pi\,\rho}\right) + \hat{z}\frac{\partial}{\partial z}\right]. \tag{18.57}$$

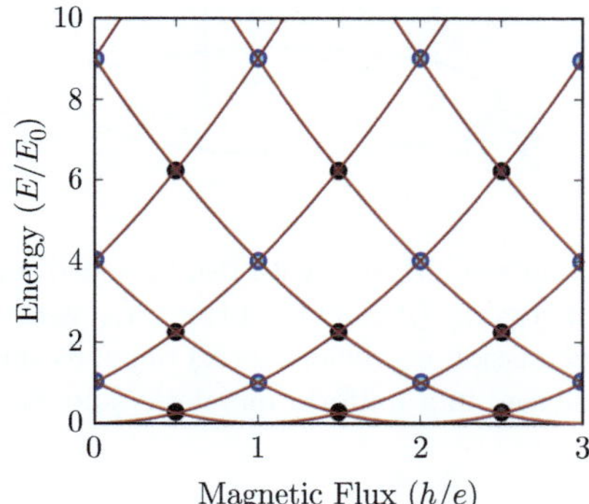

Figure 18.3 The states with eigenvalues n and $-$m can interfere for particular magnetic flux values. This figure shows the conditions for constructive interference (circles with light fill) and for destructive interference (circles with dark fill) as the magnetic field is varied. The same conditions hold for interference in the transmission current through a thin ring with external contacts. Here $E_0 = \hbar^2/2mr^2$, and the magnetic flux is in units of $\Phi_0 = h/e$.

The Hamiltonian is now given by

$$H = \frac{1}{2m\rho^2}\left(-i\hbar\frac{\partial}{\partial\theta} + \frac{e\Phi}{2\pi}\right)^2 = \frac{1}{2m\rho^2}\left(L_z + \frac{e\Phi}{2\pi}\right)^2,$$

$$= -\frac{\hbar^2}{2m\rho^2}\left(\frac{\partial}{\partial\theta} + i\frac{\Phi}{\Phi_o}\right)^2, \tag{18.58}$$

where the motion along the z-direction is dropped since it affects the motion of the electron in a very simple manner. From Eq. (18.58), we see that the bound state energies are dependent on the magnetic flux in the solenoid (the region excluded from the electron's path), and the eigenstates have quantized angular momentum. The eigenfunctions and the eigenvalues are given by

$$\psi_n(\theta) = \frac{1}{\sqrt{2\pi}}\, e^{in\theta}, \quad n = 0, \pm1, \pm2, \ldots$$

$$E_n(\nu) = \frac{\hbar^2}{2m\rho^2}(n + \Phi/\Phi_0)^2, \tag{18.59}$$

where the energy is dependent on n and on the magnetic flux with the period given by $\Phi_0 = 2\pi\,\hbar/e = h/e$. Note that when the flux increases by one unit of Φ_0, the index on the energy and on the wavefunction given by n is increased by one: $n \to n+1$ and $\psi_n \to \psi_{n+1}$, and the physical properties of the wavefunction are unchanged up to a phase. The symmetry implied here is the gauge symmetry in which the gauge transformation of the wavefunction is given by $U = \exp(i\theta)$. As pointed out by Peshkin and Tonomura, [17] the assumption that n are integers independently of the flux ensures that the wavefunctions are single-valued. This assumption is not needed for a simply connected region, whereas in a multiply connected region, as in our example, we do need this assumption.

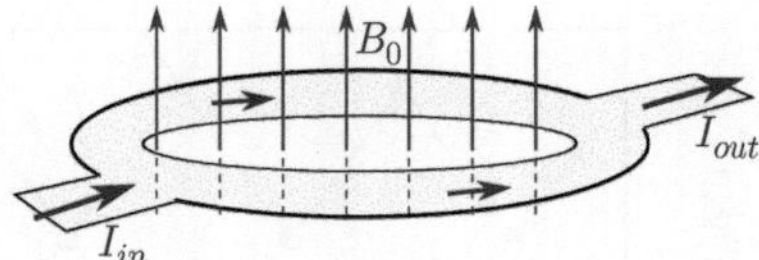

Figure 18.4 A mesoscopic finite-width ring with two attached contacts is shown in the presence of a magnetic field. The incoming (incident) current I_{in} gets transmitted and reflected by the structure, and the transmission coefficient $T(E)$ depends on the energy E of the incident particle. The system is immersed in a uniform magnetic field B_0 that is directed normal to the plane of the ring.

For the ring without any incoming and outgoing leads, as in Fig. 18.2(a), we see that the spectrum is periodic in Φ with period $\Phi_0 = h/e$. With $\Phi/\Phi_0 = \nu$, let us plot the energy levels as a function of the magnetic flux $\Phi = \nu\Phi_0$. With $E_n(\nu) \propto (n+\nu)^2$, we can have two energy curves crossing at some field such that $E_n(\nu) = E_m(\nu)$, implying that $(n+\nu)^2 = (m+\nu)^2$ for two states n and m. So we either have $n = m$ or $(n+\nu) = -(m+\nu)$. The two states then interfere when $\nu = -(n+m)/2$.

Now consider interference effects in the transmission through the thin ring when thin leads are attached diametrically opposite to one another, as in Fig. 18.2(b). Now the two states $\psi_n = e^{in\theta}$ and $\psi_m = e^{im\theta}$ both contribute to the transmission, and the total amplitude at the output port is $(e^{in\pi} + e^{im\pi})$. The quantity $e^{in\pi}/e^{im\pi} = e^{i(n-m)\pi}$ is 1 when $n - m$ is even, and -1 when $n - m$ is odd. Since n, m are integers and $n + m$ and $n - m$ have the same parity (both are even or both are odd), we see that the interference is constructive when ν is an integer and destructive when ν is a half-integer. This simple picture clearly explains the periodic behavior of the transmission as the magnetic flux is increased.

18.9 The AB effect in flat mesoscopic rings

Consider now two-dimensional rings of finite width with two contacts attached to them at diametrically opposite points, as in Fig. 18.4. When a current is passed through such a system, the current flows in and divides into two paths before coming together again at the exit to the ring, as shown schematically in Fig. 18.4. When the inner and outer radii of the ring, a and b respectively, are 20–70 nm and the contacts are of similar dimensions, quantum mechanical considerations become relevant.

The wavefunctions of the electrons in the two branches come together exhibiting interference effects. Thus, the transmission through such rings involves the waveguiding of the electron's wavefunctions. Since the paths are not simply connected, we can investigate the AB effect by switching on a magnetic field perpendicular to the plane of the ring. While the geometry is not exactly the same as in the scheme proposed by AB, we will once more have interference effects that depend on the magnetic flux through the ring. These oscillations in the transmission currents as the flux is increased have been observed experimentally. [20]

Tan and Inkson [21] have attempted to give an analytic solution to the wavefunctions in the annular ring of finite width for an applied bias, but have not included the effect of finite widths for the contacts. The injection of electrons into the ring depends on the geometry of the incoming contact, as does the outgoing current on the output contact geometry and the waveguiding of the carriers.

A complete calculation has now been performed, using the finite element method, [22, 23] to model the AB effect for the current propagation through the structure shown in Fig. 18.4. The eigenstates of a particle on a Möbius ring are first determined. [24] The transmission coefficient of the current through the structure when two thick leads are attached to it can be calculated. The current oscillations and the Landauer-Buttiker conductivity in flat rings and in Möbius rings have been calculated recently. [25]

References

[1] J. D. Jackson, *Classical Electrodynamics*, 3rd edition (J. Wiley, NY, 1999).

[2] J. D. Jackson and L. B. Okun, Rev. Mod. Phys. **73**, 663–680 (2001); "Historical roots of gauge invariance."

[3] L. Lorenz, Philos. Mag. **34**, 287–301 (1867); "On the identity of the vibrations of light with electrical currents," (translated from Annalen der Physics und Chemie, June 1867).

[4] R. Nevels and C.-S. Shin, IEEE Antennas Propag. Mag., **43**, 70–72 (2001); "Lorenz, Lorentz, and the Gauge."

[5] L. O'Raifeartaigh and N. Straumann Rev. Mod. Phys. **72**, 1–23 (2000); "Gauge theory: Historical origins and some modern developments."

[6] H. Goldstein, C. P. Poole, and J. L. Safko, *Classical Mechanics*, 3rd edition (Addison-Wesley, San Francisco, 2002).

[7] M. Gell-Mann and M. Levy, Il Nuovo Cimento **16** 53–73 (1960); "The axial vector current in beta-decay."

[8] J. J. Sakurai, *Invariance Principles and Elementary Particles* (Princeton University Press, Princeton, NJ, 1964); pp. 178–183.

[9] The gauge variational method for deriving the conserved current has proved to be very useful in the physics of semiconductor nanostructures and the determination of derivative boundary conditions at interfaces based on current continuity. See for example: L. R. Ram-Mohan, I. Vurgaftman, and J. R. Meyer, Microelectronics Journal **30**, 1031–1042 (1999); "Wave function engineering of antimonide quantum-well lasers." L. R. Ram-Mohan, Proc. 31st Int. Symp. on Compound Semiconductors (Seoul, Korea, Sept. 2004), "A Lagrangian approach to wavefunction engineering of layered quantum semiconductor structures," (Institute of Physics Conference Series vol 184) ed J-C Woo, H Hasegawa, Y-S Kwon, T Yao and K-H Yoo (Bristol: Institute of Physics Publishing) pp. 1–8 (2005). L R Ram-Mohan and K H Yoo, J. Phys.: Condens. Matter **18**, R901–R917 (2006); "Wavefunction engineering of layered semiconductors: theoretical foundations."

[10] Tsuyoshi Ueta and Y. Miyagawa, Phys. Rev. E **86**, 026707–1–8 (2012); "Local-gauge finite-element method for electron waves in magnetic fields."

[11] J. J. Sakurai, *Modern Quantum Mechanics*, revised edition, S. F. Tuan, editor (Addison-Wesley, NY, 1994).

[12] D. J. Griffiths, *Introduction to Quantum Mechanics*, 2nd edition (Pearson Prentice Hall, Upper Saddle River, NJ, 2005); p384.

[13] Y. Aharonov and D. Bohm, Phys. Rev. **115**, 485–491 (1959); "Significance of electromagnetic potentials in the quantum theory."

[14] Y. Aharonov and D. Bohm, Phys. Rev. **123**, 1511–1524 (1961); "Further considerations on electromagnetic potentials in the quantum theory."

[15] Y. Aharonov and D. Bohm, Phys. Rev. **130**, 1625–1632 (1963); "Further discussion of the role of electromagnetic potentials in the quantum theory."

[16] R. G. Chambers, Phys. Rev. Lett. **5**, 3–5 (1960); "Shift of an electron interference pattern by enclosed magnetic flux." Also see: G. Möllenstedt and W. Bayh, Naturwissenschaften **49**, 81–82 (1962); "Messung der kontinuierlichen Phasenschiebung von Elektronenwellen im kraftfeldfreien Raum durch das magnetische vektorpotential einer Luftspule" A. Tonomura et al., Phys. Rev. Lett. **56**, 792–795 (1986); "Evidence for Aharonov-Bohm effect with magnetic field completely shielded from electron wave."

[17] M. Peshkin and A. Tonomura, *The Aharonov-Bohm Effect*, Lecture Notes in Physics vol. 340 (Springer-Verlag, Berlin, Germany, 1989).
[18] W. Ehrenberg and R. E. Siday, Proc. Phys. Soc. B **62**, 8–21 (1949);"The refractive index in electron optics and the principles of dynamics."
[19] M. Berry, Physics Today, p8, Letters section, August 2010.
[20] R. A. Webb, S. Washburn, and C. P. Umbach, Phys. Rev. Lett. **54**, 2696–2699 (1985); "Observation of h/e Aharonov-Bohm oscillations in normal-metal rings."
[21] W. C. Tan and J. C. Inkson, Phys. Rev. B **53**, 6947–6050 (1996); "Landau quantization and the Aharonov-Bohm effect in a two-dimensional ring."
[22] L. R. Ram-Mohan, *Applications of the Finite Element and Boundary Element Methods in Quantum Mechanics* (Oxford University Press, 2002).
[23] L. R. Ram-Mohan, S. Saigal, D. Dossa, and J. Shertzer, Comput. Phys. **4**, 50–59 (1990); "The finite element method for energy eigenvalues of quantum mechanical systems."
[24] Zehao Li and L. R. Ram-Mohan, Phys. Rev. B **85**, 195438–1–9 (2012); "Quantum mechanics on a Mobius strip: energy levels, symmetries, optical transitions, and level splitting in a magnetic field."
[25] Zehao Li and L. R. Ram-Mohan, J. Appl. Phys. **114**, 164322–1–9 (2013); "The Aharonov-Bohm effect with a twist: Electron transport through finite-width Möbius rings."

Part VI

Further applications of FEM

A

Derivation of shape functions using group theory

In this chapter:

- We develop a technique for the derivation of higher-order shape functions having derivative continuity for finite element applications using the theory of group representations. A brief introduction to function transformations and group representation theory is included.
- In effect, the symmetry of the standard finite elements in 1D, 2D, and 3D can be used to obtain constraints on $\mathcal{C}_{(0)}$-, $\mathcal{C}_{(1)}$-, and $\mathcal{C}_{(2)}$-continuous shape functions of high order. We investigate the intra- and inter-element continuity properties of the polynomials that are derived by symmetry. The concept of equivalence representation is employed to generate shape functions associated with all element nodes using one of the set of shape functions connected with one node.
- The advantage of using $\mathcal{C}_{(n)}$-continuous polynomials is the faster convergence toward exact results, though at the price of a nominal bandwidth increase in matrix occupancy. There is a substantial gain in accuracy with even a sparse mesh.
- The additional derivative continuity gives smooth solutions across neighboring elements which is not afforded by Lagrange shape functions that are usually employed in engineering applications.

  The $\mathcal{C}_{(n)}$, $(n \geq 1)$, continuous interpolation polynomials are referred to as Hermite shape functions. Their use in solving Schrödinger's equation and in electrodynamics modeling is by now well established through our earlier works. The transformation from local standard elements to the global geometry involves Jacobian factors for the derivative degrees of freedom that are explicitly displayed. Alternately, we can use interface smoothing employing Fermi distribution functions from Statistical Physics or cubic Hermite functions, for example.

A.1 Introduction

Finite element analysis (FEA) has been shown to be a powerful method for the numerical solution of partial differential equations, and is known to be a versatile tool in applications to a wide variety of physical problems including structural mechanics, electromagnetic field modeling, and quantum mechanics. [1–3] The solution delivered by FEA depends critically on two essential steps. One is the discretization of the physical domain into 1D line elements, 2D rectangles and triangles, or 3D cubes and tetrahedra, that are faithful to the original physical domain. The other is the use of appropriate interpolation polynomials that will represent the solution on each of the discretized regions, or elements, with some level of accuracy over these finite elements. By a choice of using selectively finer and finer mesh refinements for the discretization where the solutions might vary considerably (h-refinement), and using higher-order polynomials (p-refinement) for the interpolation of the solution within each element, it is possible to obtain the desired accuracy for the solution. In addition, FEA may be thought of as the discretization of the action integral to directly evaluate the integral. Hence, it is very well founded in variational principles which justifies its wide applicability for physical problems.

It is usual to employ linear or quadratic Lagrange interpolation polynomials on the finite elements, so that we have inter-element continuity for the solutions, though they may display a "pixelated" form. This is known as $\mathcal{C}_{(0)}$-continuity. We can employ $\mathcal{C}_{(1)}$- or $\mathcal{C}_{(2)}$-continuous polynomials for a better representation of the physical solutions; these correspond to the first or second derivative continuity within and across elements, respectively. We have demonstrated the advantages of such polynomials which are called Hermite interpolation polynomials. [3] In general, Hermite interpolation polynomials can be devised to ensure $\mathcal{C}_{(n)}$-continuity across elements. Here $\mathcal{C}_{(n)}$-continuity means that the function and its derivatives up to n^{th} order are continuous within each element. The requirements on such polynomials depend on the type of derivative continuity needed in an element of a particular geometry, on the number of nodes per element, and the order of the polynomial derivatives desired at each node.

In the following, we show that group theory provides a beautiful approach for determining the interpolation polynomials on equilateral triangles and tetrahedra using their symmetry. We set the stage by perusing the Pascal triangle of Fig. A.1, which lists the terms in the complete polynomials of a given degree.

For complete 2D quintic polynomials, the Pascal triangle shows that 21 terms are present in the polynomials. We will obtain 18 of the parameters by having the nodal values of the function, its first, and second derivatives at the three vertices; this corresponds to the six values (degrees of freedom, or DoFs) at each vertex node. The remaining three DoFs are usually obtained by the normal derivative in the outward direction, with a node at each midpoint of the sides. Polynomials obtained in this manner are not in fashion, [4] mainly because the number of DoFs at the vertices of the triangle is different as compared with the DoFs at the mid-side nodes. Since much of FEA programming involves book-keeping issues anyway, this change in the DoFs from vertices to mid-side nodes should be considered more an irritation than

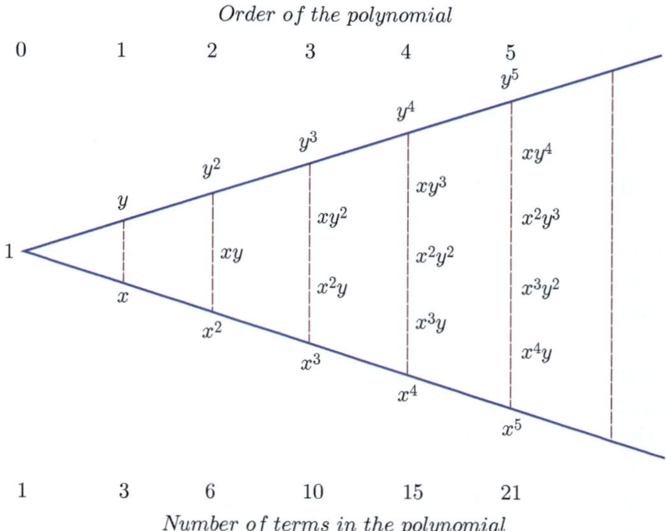

Figure A.1 The two-dimensional Pascal triangle is displayed.

as a limitation. However, there are other reasons to avoid using these polynomials. For instance, a side shared by two adjacent triangles will not have continuity of the normal derivative since the polynomials associated with the opposite vertex will influence the value of the function being represented. *Then adjacent triangles sharing a side, but having opposite nodes placed asymmetrically about the shared side will have different normal derivatives.* In addition, such a 21-DoF Hermite triangle will not deliver tangential continuity across adjacent elements since the mid-side nodes seek normal derivative continuity. Alternately, if the tangential derivative is specified at the mid-side nodes, the continuity of the normal derivatives across elements will degrade. A similar limitation arises for the case where all the three nodal DoFs, corresponding to the function value and its two first derivatives, are assigned to a single node at the centroid of the triangle.

A more practical way for finite element programming would be to remove the condition on normal derivatives at the mid-points, thereby making the triangle to be one with 18 DoFs. Now, the reduction of the number of parameters from 21 to 18 is achieved by imposing a condition that the normal derivative of the *across* the edges of the triangle vary as a cubic rather than a quartic polynomial. [5, 6] In other words, the cubic continuity of the normal derivative of the interpolated function across the edges of the triangle is *implicit* and is enabled without the use of a node on each edge. Meanwhile, tangential continuity across the three sides of the triangle depends on the function value, the first tangential derivative, and the tangential second derivatives, all of which are defined at the two nodes at the ends of the sides. Such methods of reducing the number of coefficients through combining terms in the 21-term polynomials are not readily generalized to other possible requirements. Thus, the choice of the type of data specified at each node, the number of nodes, and the continuity condition on the polynomial order of the nth derivative will require special schemes to determine the Hermite interpolation polynomials explicitly.

In the following, we present an alternate way in which *group representation theory* can be applied to derive the interpolation polynomials using the symmetry of the element. This method was first proposed by Kassebaum, Boucher, and Ram-Mohan (KBR). [7] Here we provide a detailed account for the derivation of polynomials presented in the KBR paper.

Sets of quintic polynomials on a triangular element obtained through the three methods mentioned above are compared below:

- Polynomials with 21 DoFs with explicit normal derivatives at the mid-side nodes are computationally efficient except that we need to match the normal derivatives at the common side for two adjacent triangles. This is not automatic since the normal derivatives across the shared edge from two adjacent triangles will not agree if these are arbitrary scalene triangles. The alternative choice of a node at the centroid of the triangle with 3 DoFs also has non-unique inter-element derivative continuity across the sides.
- Polynomials with 18 DoFs as presented first by Bell, [5] require that the normal derivatives vary as a cubic function across the sides of the triangle. We are thus reducing the number of DoFs by 3.
- In the KBR method, we obtain 19 conditions on the coefficients of the polynomials using the group representation theory. Additional conditions are obtained for the polynomial by demanding inter-element continuity and Taylor expansion compatibility. These conditions ensure a unique normal across the common side of two adjacent triangles, the behavior at and across the shared edge being determined only by the nodal DoFs at the two vertices. The method based on group representation theory can be generalized to obtain shape functions of any order in three, and hopefully, in higher dimensions.

This, Appendix A, is organized as follows. Since familiarity with function transformation and group theory is required to follow the arguments presented here, a primer on the active and passive transformations is provided in Sec. A.2. A brief introduction to group theory and a transformation by a symmetry operator is discussed in Sec. A.3. In Sec. A.4, the concept of function transformations is introduced with two illustrations. In Sec. A.5, as an example, we formally introduce the scheme based on group representation theory to derive the linear and Hermite interpolation polynomials in one dimension. We further extend this procedure in Sec. A.6 to develop $\mathcal{C}^{(2)}$-continuous quintic shape functions *in the equilateral triangular element* that can represent an arbitrary function with $\mathcal{C}_{(1)}$-continuity for the normal derivative across two triangular elements with a common edge. Transformation of these polynomials to an arbitrary triangular element is discussed in Sec. A.7, so that it is readily applicable in FEA.

In Appendix B, we list the shape functions on a line element, and the 18-DoF quintic polynomials on the standard equilateral and right triangular elements.

A.2 Coordinate transformation

A coordinate transformation is an operation which relates one set of basis vectors to another set of basis vectors. In the current context, we will assume it to be a linear transformation.

Given a set of basis vectors $\{\mathbf{e}_i \equiv (\mathbf{e}_1, \mathbf{e}_2, \mathbf{e}_3)\}$, every vector $\boldsymbol{V}$ can be written as a linear combination of these basis vectors

$$\boldsymbol{V} = v_1\mathbf{e}_1 + v_2\mathbf{e}_2 + v_3\mathbf{e}_3. \tag{A.1}$$

Let $\{\mathbf{e}'_i = (\mathbf{e}'_1, \mathbf{e}'_2, \mathbf{e}'_3)\}$ represent the new basis vectors. We can express them in terms of our original set as

$$\begin{aligned} \mathbf{e}'_1 &= b_{11}\mathbf{e}_1 + b_{12}\mathbf{e}_2 + b_{13}\mathbf{e}_3, \\ \mathbf{e}'_2 &= b_{21}\mathbf{e}_1 + b_{22}\mathbf{e}_2 + b_{23}\mathbf{e}_3, \\ \mathbf{e}'_3 &= b_{31}\mathbf{e}_1 + b_{32}\mathbf{e}_2 + b_{33}\mathbf{e}_3. \end{aligned} \tag{A.2}$$

And conversely,

$$\begin{aligned} \mathbf{e}_1 &= a_{11}\mathbf{e}'_1 + a_{12}\mathbf{e}'_2 + a_{13}\mathbf{e}'_3, \\ \mathbf{e}_2 &= a_{21}\mathbf{e}'_1 + a_{22}\mathbf{e}'_2 + a_{23}\mathbf{e}'_3, \\ \mathbf{e}_3 &= a_{31}\mathbf{e}'_1 + a_{32}\mathbf{e}'_2 + a_{33}\mathbf{e}'_3. \end{aligned} \tag{A.3}$$

Using Einstein's summation convention, we can write Eqs. (A.2, A.3) as

$$\mathbf{e}'_i = b_{ij}\mathbf{e}_j, \tag{A.4}$$

$$\mathbf{e}_j = a_{jk}\mathbf{e}'_k, \tag{A.5}$$

where $i, j, k = 1, 2, 3$. Let $\mathbf{e}'_i$ and $\mathbf{e}_j$ represent two sets of orthonormal bases so that

$$\mathbf{e}_i \cdot \mathbf{e}_j \quad = \quad \delta_{ij}; \quad \text{and} \quad \mathbf{e}'_i \cdot \mathbf{e}'_j = \delta_{ij}. \tag{A.6}$$

Substituting Eq. (A.5) in Eq. (A.4) we have,

$$\mathbf{e}'_i = b_{ij}a_{jk}\mathbf{e}'_k. \tag{A.7}$$

Taking a dot product with $\mathbf{e}'_n$ on both sides, we have

$$\begin{aligned} \mathbf{e}'_i \cdot \mathbf{e}'_n &= b_{ij}a_{jk} \ (\mathbf{e}'_k \cdot \mathbf{e}'_n), \\ \delta_{in} &= b_{ij}a_{jk} \ \delta_{kn}, \\ \delta_{in} &= b_{ij}a_{jn}. \end{aligned} \tag{A.8}$$

If we consider b_{ij} and a_{ij} to be the $\{i,j\}$-th elements of the two matrices $\boldsymbol{A}$ and $\boldsymbol{B}$, respectively, we have the relation

$$\boldsymbol{B}\cdot\boldsymbol{A}=\boldsymbol{I}, \tag{A.9}$$

where $\boldsymbol{I}$ is the identity matrix. Being orthogonal matrices, we have

$$\boldsymbol{A}=\boldsymbol{B}^{-1}=\boldsymbol{B}^{T}. \tag{A.10}$$

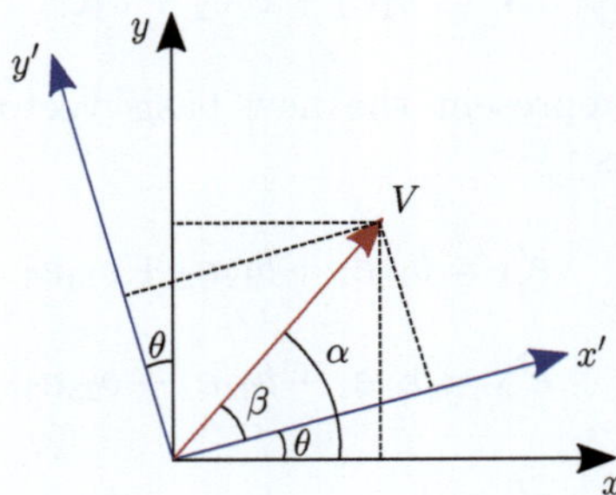

Figure A.2 The "passive" transformation of a vector is shown. The vector is held fixed while the coordinate system is rotated.

A.2.1 Rotation matrices in 2D

Let us consider a 2D coordinate transformation shown in Fig. A.2. It is straightforward to determine the matrix elements b_{ij} in this 2D rotation. Since the length of the vector $\boldsymbol{V}$ is the same in either of the coordinates (x,y) and (x',y'), $|\boldsymbol{V}|=|\boldsymbol{r}|=|\boldsymbol{r'}|$, we have

$$x=|\boldsymbol{r}|\cos\alpha;\quad y=|\boldsymbol{r}|\sin\alpha;$$
$$x'=|\boldsymbol{r}|\cos\beta;\quad y'=|\boldsymbol{r}|\sin\beta.$$

Now $\alpha=\beta+\theta$, so that

$$\begin{aligned} x' &= |\boldsymbol{r}|\cos(\alpha-\theta)=|\boldsymbol{r}|(\cos\alpha\cos\theta+\sin\alpha\sin\theta),\\ &= x\cos\theta+y\sin\theta. \end{aligned} \tag{A.11}$$

and, similarly

$$\begin{aligned} y' &= |\boldsymbol{r}|\sin(\alpha-\theta)=|\boldsymbol{r}|(-\cos\alpha\sin\theta+\sin\alpha\cos\theta),\\ &= -x\sin\theta+y\cos\theta. \end{aligned} \tag{A.12}$$

We see that the transformation matrix in 2D is given by

$$\begin{bmatrix} x'\\ y' \end{bmatrix}=\begin{bmatrix} \cos\theta & \sin\theta\\ -\sin\theta & \cos\theta \end{bmatrix}\cdot\begin{bmatrix} x\\ y \end{bmatrix}. \tag{A.13}$$

The generalization to 3D is performed using Euler angles defined for rotations about coordinate axes. [8, 9] The result of two transformation operations in 3D will, in general, depend on the order in which they are applied; i.e. the reversal of the order of two rotation operations can lead to a different orientation for the resulting axes depending on the order. The transformations are said to be non-commutative.

Note that the transformation matrix

$$\mathbf{A} = \begin{bmatrix} \cos\theta & \sin\theta \\ -\sin\theta & \cos\theta \end{bmatrix} = \begin{bmatrix} (\boldsymbol{e'}_1 \cdot \mathbf{e}_1) & (\boldsymbol{e'}_1 \cdot \mathbf{e}_2) \\ (\boldsymbol{e'}_2 \cdot \mathbf{e}_1) & (\boldsymbol{e'}_2 \cdot \mathbf{e}_2) \end{bmatrix}, \tag{A.14}$$

has matrix elements that are dot-products of the unit vectors; these entries in the matrix are seen to be the direction cosines of the angles between the coordinate vectors.

A.2.2 Passive transformation

The transformations discussed in the previous section are called passive transformations. We let the vector $\boldsymbol{V}$ be fixed in coordinate space, while the basis set transforms from $\{\boldsymbol{e}_i\}$ to $\{\boldsymbol{e}'_j\}$. Any vector can be represented in terms of components along either of the basis vectors as

$$\boldsymbol{V} = v_i \boldsymbol{e_i} = v'_j \boldsymbol{e}'_j. \tag{A.15}$$

(In matrix form, basis vectors $\{\boldsymbol{e}_i\}$ are represented by a column array and vector components $\{v_i\}$ by a row vector.) Substituting Eq. (A.5) into Eq. (A.15), we relate the vector components in the transformed basis to the components in the earlier basis as

$$\begin{aligned} v'_j \boldsymbol{e}'_j &= v_i (a_{ik} \boldsymbol{e}'_k), \\ v'_j \boldsymbol{e}'_j &= (v_i a_{ik}) \boldsymbol{e}'_k. \end{aligned} \tag{A.16}$$

Taking a dot product with $\boldsymbol{e}'_n$ on both sides, we get

$$\begin{aligned} v'_j \boldsymbol{e}'_j \cdot \boldsymbol{e}'_n &= v_i a_{ik} (\boldsymbol{e}'_k \cdot \boldsymbol{e}'_n), \\ v'_j \delta_{jn} &= (v_i a_{ik}) \delta_{kn}, \\ v'_n &= v_i a_{in}, \end{aligned} \tag{A.17}$$

which represents the rule for the transformation of the vector components under a passive transformation. Again, in Fig. A.2, notice that the vector $\boldsymbol{V}$ is fixed in space, but its components are different in the two coordinate systems designated by XY and $X'Y'$. Observe that a similar transformation of vector components is performed by the matrix $\boldsymbol{A}$ in Eq. (A.14).

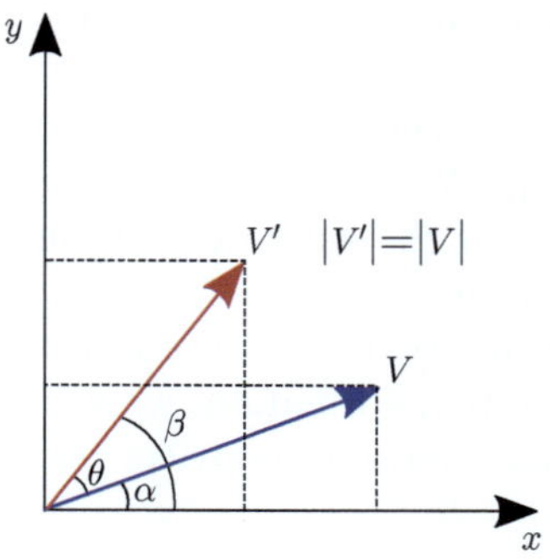

Figure A.3 The active transformation of a vector is shown.

A.2.3 Active transformation

In an active transformation, we let the coordinate basis be fixed but transform the physical vectors.[†] After an active transformation, we go to a new vector with transformed components with the basis vectors held fixed, in contrast with a passive transformation where we held the vector fixed. Figure A.3 shows the fixed coordinate system in which a vector $\boldsymbol{V}$ is rotated by an angle θ in the anti-clockwise sense to a new vector $\boldsymbol{V'}$. The components of $\boldsymbol{V} = \{v_x, v_y\}$ differ from those of $\boldsymbol{V'} = \{v'_x, v'_y\}$, while the magnitude of the two vectors are the same. With $\beta = \alpha + \theta$ and $|\boldsymbol{V}'| = |\boldsymbol{V}| = v$, we have

$$\begin{aligned} v'_x &= v\cos\beta = v\cos(\alpha+\theta), \\ &= (v\cos\alpha)\cos\theta - (v\sin\alpha)\sin\theta; \\ &= v_x\cos\theta - v_y\sin\theta. \end{aligned} \tag{A.18}$$

In a similar manner,

$$\begin{aligned} v'_y &= v\sin\beta = v\sin(\alpha+\theta), \\ &= (v\cos\alpha)\sin\theta + (v\sin\alpha)\cos\theta; \\ &= v_x\cos\theta + v_y\sin\theta; \end{aligned} \tag{A.19}$$

Expressing the above equations in matrix notation, we have

$$\begin{bmatrix} v'_x \\ v'_y \end{bmatrix} = \begin{bmatrix} \cos\theta & -\sin\theta \\ \sin\theta & \cos\theta \end{bmatrix} \cdot \begin{bmatrix} v_x \\ v_y \end{bmatrix}.$$

In general,

$$v'_i = b_{ij} v_j. \tag{A.20}$$

for $i, j = 1, 2, 3$. Here, b_{ij} is the $(i,\ j)$-th element of the matrix $\boldsymbol{B}$. The transformation matrix $\boldsymbol{B}$ is the transpose (inverse) of the orthogonal matrix $\boldsymbol{A}$ appearing in passive transformations.

Thus, we can go from the vector $\boldsymbol{V}$ to $\boldsymbol{V'}$ by doing the transformation of the components, as in above Eq. (A.20) and by keeping the basis vectors fixed. Notice that the transformation law for the active transformation, Eq. (A.20) is the inverse of the passive transformation Eq. (A.17).

In a passive transformation, we always stay in the transforming coordinate (body frame), whereas in the active transformation, we work with a fixed (lab frame) coordinate system. Hence, the transforming matrices for the two pictures are inverses of each other.

[†] Imagine that we are physically "taking hold" of the vector and rotating it in an "active transformation."

A.3 Transformation of coordinates by a symmetry operator

We now present a very brief introduction to *group representation theory* essentially to define the concepts and the notations. The reader is directed to standard books on group representation theory for a more complete treatment. [10–12]

A non-empty set G, with associated operators $(a, b, c \ldots)$, is said to be a "group," when the following four properties are satisfied:

1. *Closure*: The product of any two elements of the group is itself an element of the group.
2. *Associativity*: The operation is associative; that is, $(a \cdot b) \cdot c = a \cdot (b \cdot c)$ for all a, b, c in G.
3. *Identity*: There is an element ϵ in G such that $a \cdot \epsilon = \epsilon \cdot a = a$ for all a in G.
4. *Inverse*: For each element a in G, there is an element a^{-1} in G such that $a \cdot a^{-1} = a^{-1} \cdot a = \epsilon$.

For any given system, the set of all possible operations which leaves the system invariant forms a group known as its "symmetry group." Let R be an element of a symmetry group G. Once we know the way in which the corresponding operator R transforms the given geometry, we can deduce the rule with which the group element will transform the basis vectors (passive perspective) or the coordinates of a vector (active perspective) in space.

For example, the symmetry group of an equilateral triangle, C_{3v}, consists of all geometrical transformations that take the triangle to itself. The elements of the group C_{3v} are (see Fig. A.7)

1. the identity ϵ,
2. a rotation by $\pm 2\pi/3$ about the axis perpendicular to the plane of the triangle and passing through the centroid, denoted by C_3, C_3^{-1},
3. reflection about the planes bisecting the three vertex angles, denoted by $\sigma_1, \sigma_2, \sigma_3$.

Now consider the element C_3 in the symmetry group. Physically, C_3 is an anti-clockwise rotation by an angle $2\pi/3$ with respect to the z-axis. Therefore,

$$C_3 \cdot \begin{bmatrix} x \\ y \\ z \end{bmatrix} = \begin{bmatrix} \cos\left(\frac{2\pi}{3}\right) & \sin\left(\frac{2\pi}{3}\right) & 0 \\ -\sin\left(\frac{2\pi}{3}\right) & \cos\left(\frac{2\pi}{3}\right) & 0 \\ 0 & 0 & 1 \end{bmatrix} \cdot \begin{bmatrix} x \\ y \\ z \end{bmatrix}, \tag{A.21}$$

$$= \begin{bmatrix} -\frac{1}{2} & \frac{\sqrt{3}}{2} & 0 \\ -\frac{\sqrt{3}}{2} & -\frac{1}{2} & 0 \\ 0 & 0 & 1 \end{bmatrix} \cdot \begin{bmatrix} x \\ y \\ z \end{bmatrix}. \tag{A.22}$$

In general, we may write

$$R \cdot \begin{bmatrix} x \\ y \\ z \end{bmatrix} = \boldsymbol{D}(R) \cdot \begin{bmatrix} x \\ y \\ z \end{bmatrix}, \tag{A.23}$$

where R is an element in G and $\boldsymbol{D}(R)$ is the corresponding 3×3 matrix. For simplicity, the above expression can be written as

$$R \cdot \boldsymbol{r} = R_{ij}\, x_j. \tag{A.24}$$

Here, R_{ij} is the $(i,\, j)$-th element of the matrix $\boldsymbol{D}(R)$.

A.4 Function transformation

Let us consider a function $f(x_i)$ in coordinate space operated on by a group element R. In an active sense, the transformation takes the coordinates x_j to x'_i given by $x'_i = a_{ij}x_j$. Suppose that the function $f(x_i)$ is acted on by the operator P_R corresponding to the element R thereby yielding a new function $f'(x'_j)$. Here we write f' (not f), since the functional form will change due to the coordinate transformation. For example, $f(x) = x^2 + x$ under the reflection operation $x \to -x$ will go to a new function $f'(x) = x^2 - x$. Notice that due to the transformation, the functional form has changed. Just to be explicit about which transformation is being used, many texts will represent the function $f'(x)$ by $P_R f(x)$. Here $P_R f$ should be considered as a new designation for the transformed function.

Since the function value remains the same after the transformation, we have the relation

$$\begin{aligned} P_R f(x'_i) &= f(x_j), \\ P_R f(a_{ik}x_k) &= f(x_j). \end{aligned} \tag{A.25}$$

In vector notation, we write

$$P_R f(\boldsymbol{r}') = f(\boldsymbol{r}). \tag{A.26}$$

Here, $\boldsymbol{r}' = R \cdot \boldsymbol{r}$, which is an active transformation.

We can imagine a passive transformation as being on the rotated (transformed) coordinate system and tracing back to the original coordinate system. In the passive sense, we can write the above Eq. (A.26) as

$$P_R f(\boldsymbol{r}) = f(R^{-1} \cdot \boldsymbol{r}). \tag{A.27}$$

In Eq. (A.27), we have exchanged the labels for primed and unprimed coordinate systems. This is because we want to work always with the unprimed coordinate system.

Examples: One of the famous plane curves is the "clothoid" whose curvature is proportional to its arc length. The general form of the clothoid [13] is given in parametric form by

$$clothoid[n,a](t) = a\left(\int_0^t du \sin\left(\frac{u^{n+1}}{n+1}\right), \int_0^t du \cos\left(\frac{u^{n+1}}{n+1}\right)\right). \tag{A.28}$$

For $n = 2$, this is known as Euler's spiral or Cornu's spiral. Under an anti-clockwise rotation by $\pi/2$ about the z-axis in the passive sense, we obtain the new function

$$clothoid\,'[n,a](t) = a\left(-\int_0^t du \cos\left(\frac{u^{n+1}}{n+1}\right), \int_0^t du \sin\left(\frac{u^{n+1}}{n+1}\right)\right). \tag{A.29}$$

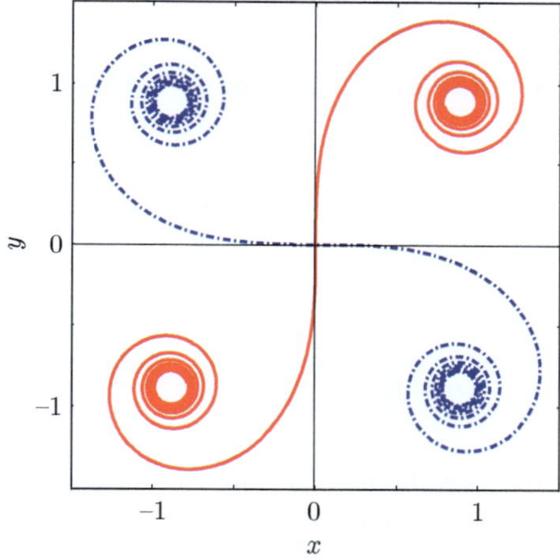

Figure A.4 The $clothoid[1,1]$ (continuous curve) and $clothoid\,'[1,1]$ (dotted curve) obtained through the function transformation of a rotation of $\pi/2$ in the anti-clockwise sense about the z-axis are shown.

As one can anticipate, the functional forms of the two clothoids are different. In Fig. A.4, we have shown the functions $clothoid[1,1]$ (continuous curve) and $clothoid\,'[1,1]$ (dotted curve).

As an example of rotation of 2D surfaces, consider a special hyperbolic paraboloid known as the *monkey's saddle.* [13] On a normal hyperbolic paraboloid a man can easily sit on the saddle as there are indentations for his legs, but not a monkey as there is no space to accommodate its tail. However, a monkey's saddle would be right for it as shown in the Fig. A.5. Its functional form is given by

$$monkeysaddle(x,y) = x^3 - 3xy^2. \tag{A.30}$$

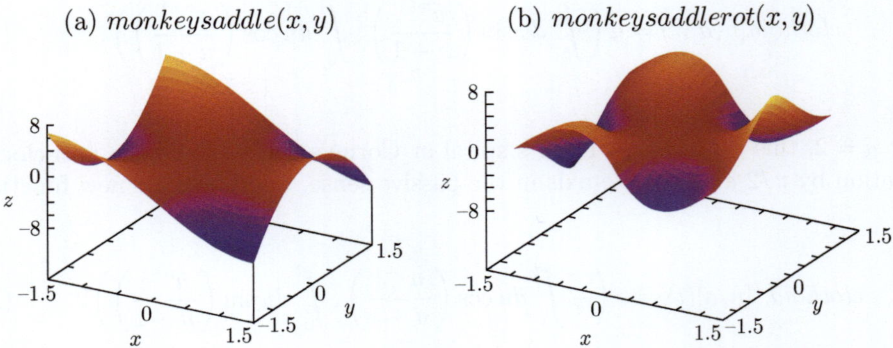

Figure A.5 The function $monkeysaddlerot(x, y)$ is obtained through an anticlockwise passive rotation by $\pi/4$ of the function $monkeysaddle(x, y)$.

Under the anti-clockwise rotation by an angle of $\pi/4$ about the z-axis, we get the new function

$$\begin{aligned} monkeysaddlerot(x, y) &= monkeysaddle(R^{-1} \cdot \boldsymbol{r}), \\ &= monkeysaddle\left(\frac{(x-y)}{\sqrt{2}}, \frac{(x+y)}{\sqrt{2}}\right), \\ &= \frac{(x-y)^3}{2\sqrt{2}} - \frac{3(x-y)(x+y)^2}{2\sqrt{2}}. \end{aligned} \tag{A.31}$$

These two examples illustrate the idea of function transformation.

As an example of function transformation with a group operator, we consider the symmetry group of an equilateral triangle C_{3v}; the group of rotation operators leaving the triangle invariant is

$$G = \{E, C_3, {C_3}^{-1}\}. \tag{A.32}$$

Clearly, G is a subgroup of C_{3v}. We would like to know the action of the elements of the group G on a function $F(x, y) = xy^2$.

Let $P_E, P_{C_3}, P_{C_3^{-1}}$ be the operators which transform a function under the rule given by the elements of the group G. Transformation of $F(x, y)$ under the identity element E is given by

$$\begin{aligned} P_E F(x, y) &= F(E^{-1}(x, y)), \\ &= F(x, y) = xy^2. \end{aligned} \tag{A.33}$$

To determine the transformation of $F(x, y)$ under the operation C_3, we write

$$P_{C_3} F(x, y) = F\left({C_3}^{-1}(x, y)\right).$$

We now need the transformation of the coordinate vector under C_3^{-1}. We have

$$C_3{}^{-1}\begin{bmatrix} x \\ y \\ z \end{bmatrix} = \begin{bmatrix} -\frac{1}{2} & -\frac{\sqrt{3}}{2} & 0 \\ \frac{\sqrt{3}}{2} & -\frac{1}{2} & 0 \\ 0 & 0 & 1 \end{bmatrix} \cdot \begin{bmatrix} x \\ y \\ z \end{bmatrix} = \begin{bmatrix} -\frac{x}{2} - \frac{\sqrt{3}y}{2} \\ \frac{\sqrt{3}x}{2} - \frac{y}{2} \\ z \end{bmatrix}. \tag{A.34}$$

Therefore,

$$\begin{aligned} P_{C_3}F(x,y) &= F\left(C_3{}^{-1}(x,y)\right), \\ &= F\left(-\frac{1}{2}x - \frac{\sqrt{3}}{2}y, \frac{\sqrt{3}}{2}x - \frac{1}{2}y\right), \\ &= \left(-\frac{x}{2} - \frac{\sqrt{3}y}{2}\right)\left(\frac{\sqrt{3}x}{2} - \frac{y}{2}\right)^2. \end{aligned} \tag{A.35}$$

In a similar manner, operation of $P_{C_3{}^{-1}}$ on $F(x,y)$ is given as

$$P_{C_3{}^{-1}}F(x,y) = F\left(C_3(x,y)\right) = \left(-\frac{x}{2} + \frac{\sqrt{3}y}{2}\right)\left(-\frac{\sqrt{3}x}{2} - \frac{y}{2}\right)^2. \tag{A.36}$$

A.5 Interpolation polynomials and symmetry in 1D

We now show how group representation theory can be used to determine the polynomials used for finite element interpolation in 1D.

A.5.1 Linear interpolation polynomials

Let us first derive the linear Lagrange interpolation polynomials in 1D through symmetry considerations. The simplest element in one dimension is a line with nodes located at each end at $x_1 = 1$ and $x_2 = -1$. Our "standard element" ranges over $[-1, 1]$.

It is evident that the standard element has a bilateral symmetry about the mid-point, denoted by S_2. The corresponding group elements are, the identity operation denoted by ϵ, and the mirror operation about the origin which is labeled by m. We construct the character table of S_2 in Table A.1 using standard procedures discussed

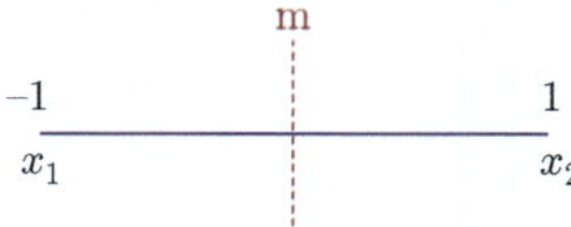

Figure A.6 The standard 1D line element. Nodes are located at $x = 1$ and $x = -1$ which are denoted as x_1 and x_2, respectively.

Table A.1 *Character table for the bilateral symmetry group, S_2. We have included the character for the equivalence representation Γ. The entries are the characters of different classes of the group. The last two columns contain the linear and quadratic functions corresponding to the representation in the first column.*

	ϵ	m	Linear	Quadratic
A_g	1	1	z	x^2, y^2, z^2, xy
A_u	1	-1	x, y	xz, yx
Γ	2	0	$\{x, y\}$	

in textbooks on group representation theory. [10, 12, 14] Here S_2 has two irreducible representations given by A_g, for the symmetric, and A_u for the antisymmetric representation, respectively. (The subscript g is for the German word "gerade" which means "even" and u is for "ungerade" which means "odd.")

An arbitrary scalar, s is invariant under any group operation. Under the mirror operation m, we have $m \cdot s = s$. Thus, s belongs to the representation A_g. The function value at a point, a numerical value, must be a scalar, so that it belongs to the representation A_g. Similarly, a vector $\boldsymbol{V}$ in 1D transforms as $m \cdot \boldsymbol{V} = -\boldsymbol{V}$ under the mirror operation. Hence, it belongs to the representation A_u.

Consider the linear interpolation polynomials with as-yet unknown coefficients in the interval $[-1,1]$. The polynomials must have the general form

$$\begin{aligned} N_1(x) &= a_1 + b_1 x, \\ N_2(x) &= a_2 + b_2 x. \end{aligned} \tag{A.37}$$

An arbitrary function $\psi(x)$ can be represented by $\psi(x) = \psi_1 N_1(x) + \psi_2 N_2(x)$, where ψ_1, ψ_2 are the values of the function at the node. With nodes located at ± 1, the nodal conditions on these shape functions at the two nodes are

$$N_1(-1) = 1; \quad N_1(1) = 0, \tag{A.38}$$

$$N_2(-1) = 0; \quad N_2(1) = 1. \tag{A.39}$$

Denoting the node at $x = -1$ by x_1 and at $x = +1$ by x_2 (see Fig. A.6) the nodes are seen to transform as

$$\begin{aligned} \epsilon \cdot \begin{bmatrix} x_1 \\ x_2 \end{bmatrix} &= \begin{bmatrix} x_1 \\ x_2 \end{bmatrix} = \begin{bmatrix} 1 & 0 \\ 0 & 1 \end{bmatrix} \cdot \begin{bmatrix} x_1 \\ x_2 \end{bmatrix}, \\ m \cdot \begin{bmatrix} x_1 \\ x_2 \end{bmatrix} &= \begin{bmatrix} x_2 \\ x_1 \end{bmatrix} = \begin{bmatrix} 0 & 1 \\ 1 & 0 \end{bmatrix} \cdot \begin{bmatrix} x_1 \\ x_2 \end{bmatrix}. \end{aligned} \tag{A.40}$$

These 2×2 coefficient matrices form a representation for the bilateral group which may be called the *nodal equivalence representation*, Γ. Hence, we have the representation

$$\Gamma(\epsilon) = \begin{bmatrix} 1 & 0 \\ 0 & 1 \end{bmatrix}, \quad \Gamma(m) = \begin{bmatrix} 0 & 1 \\ 1 & 0 \end{bmatrix}. \tag{A.41}$$

The traces of these matrices, viz. 2 and 0, are listed in the third row of Table A.1. Clearly, this is a reducible representation, as our group has only 1D representations. From Table A.1, we see that the equivalence representation can be decomposed as the sum of the two irreducible representations

$$\Gamma = A_g + A_u. \tag{A.42}$$

Nodal coordinates $\{x_1, x_2\}$ transform as per the representation Γ. Hence, the linear functions in this space will have components belonging to both the representations A_g and A_u, and these can be determined by operating with the corresponding projection operators, as discussed below.

Let G be a group of order h and Γ^i be an l_i-dimensional representation of G. For a group element R in G, its representation is given by a $l_i \times l_i$ square matrix $\Gamma^i(R)$. Then the projection operator [10], corresponding to the matrix element Γ^i_{mn} for $1 \leq m,\ n \leq l_i$ is given by

$$\mathcal{P}\left(\Gamma^i_{mn}\right) = \frac{l_i}{h} \sum_R \Gamma^i_{mn}(R) \cdot P_R, \tag{A.43}$$

where P_R is the operator corresponding to the element R. [10] From now onward, we denote any group element as R and the corresponding projection operators as P_R. The projection operator $\mathcal{P}\left(\Gamma^i_{mn}\right)$ projects out a function F on to a part f^i_{mn} which belongs to the mth row and n^{th} column of the representation Γ^i. Hence,

$$\mathcal{P}\left(\Gamma^i_{mn}\right) F = f^i_{mn}. \tag{A.44}$$

For the group S_2, we have two 1D irreducible representations. From Table A.1, the projection operator of the representation A_g is given by

$$\begin{aligned} \mathcal{P}(A_g) &= \frac{1}{2}\left[1.P_\epsilon + 1.P_m\right], \\ &= \frac{1}{2}\left[P_\epsilon + P_m\right]. \end{aligned} \tag{A.45}$$

Let us apply this procedure to find $N_2(x)$. We see that

$$\begin{aligned} \mathcal{P}(A_g)N_2(x) &= \frac{1}{2}\left[P_\epsilon + P_m\right][a_2 + b_2 x], \\ &= \frac{1}{2}\left[(a_2 + b_2 x) + (a_2 - b_2 x)\right], \\ &= a_2, \end{aligned} \tag{A.46}$$

and

$$\begin{aligned}\mathcal{P}(A_u)N_2(x) &= \frac{1}{2}\left[1.P_\epsilon - 1.P_m\right][a_2 + b_2 x],\\ &= \frac{1}{2}\left[(a_2 + b_2 x) - (a_2 - b_2 x)\right],\\ &= b_2 x.\end{aligned} \tag{A.47}$$

Hence,

$$N_2(x) = (\mathcal{P}(A_g)N_2(x) + \mathcal{P}(A_u)N_2(x)). \tag{A.48}$$

From Eq. (A.38)

$$N_2(1) = (\mathcal{P}(A_g)N_2(1) + \mathcal{P}(A_u)N_2(1)) = 1. \tag{A.49}$$

$\mathcal{P}(A_g)$ is symmetric and $\mathcal{P}(A_u)$ is antisymmetric under the mirror operation m. Since, the node x_2 can be obtained through the mirror operation on node x_1, we have

$$\begin{aligned}\mathcal{P}(A_u)N_2(-1) &= m \cdot \mathcal{P}(A_u)N_2(1),\\ &= -\mathcal{P}(A_u)N_2(1),\\ &= -b_2.\end{aligned} \tag{A.50}$$

Therefore, we have the relation

$$\mathcal{P}(A_u)N_2(-1) = -\mathcal{P}(A_u)N_2(1). \tag{A.51}$$

From Eqs. (A.48) and (A.50),

$$\begin{aligned}N_2(-1) &= m \cdot \left[\mathcal{P}(A_g)N_2(1) + \mathcal{P}(A_u)N_2(1)\right],\\ &= \left[\mathcal{P}(A_g)N_2(1) - \mathcal{P}(A_u)N_2(1)\right] = 0.\end{aligned} \tag{A.52}$$

We solve Eqs. (A.49) and (A.47) to obtain self-consistency conditions as

$$\begin{aligned}\mathcal{P}(A_g)N_2(1) &= \frac{1}{2},\\ \mathcal{P}(A_u)N_2(1) &= \frac{1}{2}.\end{aligned} \tag{A.53}$$

Using the above Eq. (A.53), we have

$$a_2 = \frac{1}{2},\ b_2 = \frac{1}{2}. \tag{A.54}$$

Hence, our shape function is given by

$$N_2(x) = \frac{1}{2}(1 + x). \tag{A.55}$$

To get $N_1(x)$, we can use the existence of mirror symmetry between x_1 and x_2.

$$N_1(x) = m \cdot N_2(x) = \frac{1}{2}(1 - x). \tag{A.56}$$

This example illustrates the method for the simplest of shape functions. The procedure comes into its own for 2D and 3D elements as we will see below.

A.5.2 1D Hermite interpolation polynomials: Group theory

The Hermite interpolation polynomials on a straight side element give us the freedom to impose derivative continuity at the nodes. As an example, let us determine the 1D cubic interpolation polynomials for an element having two nodes, with two DoFs at each node. Let, $N_i^{(0)}(x)$, $N_i^{(1)}(x)$ be the polynomials associated with node i, having nodal conditions

$$N_i^{(0)}(x_j) = \delta_{ij}; \quad \frac{d}{dx}N_i^{(0)}(x_j) = 0; \tag{A.57}$$

$$N_i^{(1)}(x_j) = 0; \quad \frac{d}{dx}N_i^{(1)}(x_j) = \delta_{ij}, \tag{A.58}$$

where $i, j = 1, 2$. The superscripts 0 and 1 represent the order of the derivative that is normalized to unity at one of the nodes, i. As before, we derive the interpolation polynomials defined over the range, $[-1, 1]$. We can represent the interpolated function $\psi_{i_{el}}(x)$, whose values and derivatives at nodes are known in the form

$$\psi_{i_{el}}(x) = \sum_{i=1}^{2} \left[N_i^{(0)}(x)\psi_i + N_i^{(1)}(x)\psi_i' \right]. \tag{A.59}$$

The subscript i_{el} provides the index for the element of interest. We have to assign values of two parameters ψ_i and ψ_i' at each node; hence, we have 2 DoFs at each node.

It is known that the symmetry of a function is specified by the symmetry of the boundary conditions that are imposed. [15] The boundary conditions must be satisfied by each of the components of the function that belong to the separate irreducible representations of the symmetry group of the boundary conditions. For the Lagrange interpolation polynomials in the above, decomposition into different irreducible representations at the boundary can be seen explicitly in the Eqs. (A.46, A.47).

A general polynomial over the range $[-1, 1]$, for the case of two nodes with 2 DoFs at each node, is a cubic function. (We have 2 DoF $\times$ 2 nodes $=$ 4 as the number of conditions needed to specify the polynomial. So we need to determine 4 coefficients; hence, we choose a cubic polynomial which has 4 terms.) As before, we denote the node at $x = -1$ as x_1 and the node at $x = 1$ as x_2.

Let us derive the shape function, $N_2^{(0)}(x)$ using group theory considerations. Let

$$N_2^{(0)}(x) = a_1 + b_1 x + c_1 x^2 + d_1 x^3. \tag{A.60}$$

Here, a_1, b_1, c_1, and d_1 are as-yet undetermined coefficients. Nodal conditions on this polynomial are

$$N_2^{(0)}(1) = 1, \; N_2^{(0)}(-1) = 0; \tag{A.61}$$

$$\frac{dN_2^{(0)}}{dx}(x=1) = \frac{dN_2^{(0)}}{dx}(x=-1) = 0. \tag{A.62}$$

We noted in the earlier section that the shape functions in this standard element will be in the equivalence representation Γ, which decomposes as $A_g + A_u$. We can write the A_g and A_u part of the shape function as

$$\mathcal{P}(A_g)N_2^{(0)}(x) = \frac{1}{2}\left[P_\epsilon N_2^{(0)}(x) + P_m N_2^{(0)}(x)\right] = \frac{1}{2}(a_1 + c_1 x^2), \tag{A.63}$$

$$\mathcal{P}(A_u)N_2^{(0)}(x) = \frac{1}{2}\left[P_\epsilon N_2^{(0)}(x) - P_m N_2^{(0)}(x)\right] = \frac{1}{2}(b_1 x + d_1 x^3). \tag{A.64}$$

Nodal conditions for the value of the shape function, $N_2^{(0)}(x)$, given in Eq. (A.61) are rewritten as

$$\mathcal{P}(A_g)N_2^{(0)}(1) + \mathcal{P}(A_u)N_2^{(0)}(1) = 1, \tag{A.65}$$

$$\mathcal{P}(A_g)N_2^{(0)}(-1) + \mathcal{P}(A_u)N_2^{(0)}(-1) = 0. \tag{A.66}$$

Since, $m \cdot x_2 = x_1$, we can simplify Eq. (A.66) as

$$\begin{aligned}\mathcal{P}(A_g)N_2^{(0)}(-1) + \mathcal{P}(A_u)N_2^{(0)}(-1) &= m \cdot \left[\mathcal{P}(A_g)N_2^{(0)}(1) + \mathcal{P}(A_u)N_2^{(0)}(1)\right], \\ &= \mathcal{P}(A_g)N_2^{(0)}(1) - \mathcal{P}(A_u)N_2^{(0)}(1) = 0. \end{aligned} \tag{A.67}$$

Here we have used the fact that $\mathcal{P}(A_g)$ is even and $\mathcal{P}(A_u)$ is odd under mirror operation. We solve the Eqs. (A.65) and (A.67) to get

$$\mathcal{P}(A_g)N_2^{(0)}(1) = \frac{1}{2}(a_1 + c_1) = \frac{1}{2}, \tag{A.68}$$

$$\mathcal{P}(A_u)N_2^{(0)}(1) = \frac{1}{2}(b_1 + d_1) = \frac{1}{2}. \tag{A.69}$$

We wish to obtain the representation for the derivative of the shape functions. For that purpose, we introduce the concept of *direct product* of two representations. Let Γ^1 and Γ^2 are m- and n-dimensional representations of the same group G. Then the direct product [10] of these two representations is defined as

$$\Gamma = \Gamma^1 \otimes \Gamma^2, \tag{A.70}$$

where Γ is an $m \times n$-dimensional representation of the group G.

Under the mirror operation, the derivative operator transforms as

$$m \cdot \frac{d}{dx} = -\frac{d}{dx}. \tag{A.71}$$

Hence, the derivative operator is in the representation A_u. Therefore, the derivatives of the shape functions belonging to the representation $A_g + A_u$ transform as

$$A_u \otimes (A_g + A_u) = A_u + A_g. \tag{A.72}$$

The derivative nodal conditions in Eq. (A.62) decompose at the node x_1 as

$$\mathcal{P}(A_g)\frac{dN_2^{(0)}}{dx}(1) + \mathcal{P}(A_u)\frac{dN_2^{(0)}}{dx}(1) = 0, \tag{A.73}$$

and at x_2 as

$$\begin{aligned}
\mathcal{P}(A_g)\frac{dN_2^{(0)}}{dx}(-1) + \mathcal{P}(A_u)\frac{dN_2^{(0)}}{dx}(-1) &= 0, \\
m \cdot \left[\mathcal{P}(A_g)\frac{dN_2^{(0)}}{dx}(1) + \mathcal{P}(A_u)\frac{dN_2^{(0)}}{dx}(1)\right] &= 0, \\
\mathcal{P}(A_g)\frac{dN_2^{(0)}}{dx}(1) - \mathcal{P}(A_u)\frac{dN_2^{(0)}}{dx}(1) &= 0.
\end{aligned} \tag{A.74}$$

Solving the Eqs. (A.73, A.74), we have

$$\mathcal{P}(A_g)\frac{dN_2^{(0)}}{dx}(1) = b_1 + 3d_1 = 0, \tag{A.75}$$

$$\mathcal{P}(A_u)\frac{dN_2^{(0)}}{dx}(1) = 2c_1 = 0. \tag{A.76}$$

The self-consistency conditions in Eqs. (A.68), (A.69), (A.75), and (A.76) can be cast into the matrix form

$$\begin{bmatrix} 1 & 1 & 0 & 0 \\ 0 & 2 & 0 & 0 \\ 0 & 0 & 1 & 1 \\ 0 & 0 & 1 & 3 \end{bmatrix} \cdot \begin{bmatrix} a_1 \\ c_1 \\ b_1 \\ d_1 \end{bmatrix} = \begin{bmatrix} 1 \\ 0 \\ 1 \\ 0 \end{bmatrix}. \tag{A.77}$$

Here the coefficients are rearranged in such a way that 4×4 matrix will take the block-diagonal form. We have two blocks each of dimension 2, which are consistent with the representations given in the Eq. (A.42, A.72). We invert the above matrix equation to determine the coefficient values,

$$\begin{bmatrix} a_1 \\ c_1 \\ b_1 \\ d_1 \end{bmatrix} = \begin{bmatrix} 1 & -\frac{1}{2} & 0 & 0 \\ 0 & \frac{1}{2} & 0 & 0 \\ 0 & 0 & \frac{3}{2} & -\frac{1}{2} \\ 0 & 0 & -\frac{1}{2} & \frac{1}{2} \end{bmatrix} \cdot \begin{bmatrix} 1 \\ 0 \\ 1 \\ 0 \end{bmatrix}, \tag{A.78}$$

and obtain the shape function

$$N_2^{(0)}(x) = \frac{1}{4}\left(2 + 3x - x^3\right). \tag{A.79}$$

Before we proceed to obtain other shape functions, we introduce a new classification for the shape function or the interpolation polynomials depending on the nodal conditions.

The functions whose values are set to 1 at a node and 0 at all the other nodes are called scalar shape functions. Hence, $N_1^{(0)}(x)$ and $N_2^{(0)}(x)$ are scalar shape functions, as their values are set to 1 at the node $x = -1$ and $x = 1$, respectively.

The functions whose first derivative value is set to 1 at a node and 0 at all the other nodes are called vector shape functions. Hence, $N_1^{(1)}(x)$ and $N_2^{(1)}(x)$ are vector shape functions as their x derivative values are set to 1 at the node $x = -1$ and at $x = 1$, respectively. This classification will be useful when we discuss the transformation properties of these shape functions from one node to another, as well as from the standard element to an arbitrary element.

Next, we derive $N_2^{(1)}(x)$ using a similar procedure. Let

$$N_2^{(1)}(x) = a_2 + b_2 x + c_2 x^2 + d_2 x^3. \tag{A.80}$$

The nodal conditions on this polynomial are

$$N_2^{(1)}(1) = N_2^{(1)}(-1) = 0.$$

$$\frac{dN_2^{(1)}}{dx}(x = 1) = 1, \quad \frac{dN_2^{(1)}}{dx}(x = -1) = 0.$$

Nodal conditions on the value of $N_2^{(1)}$ are invariant under the mirror operation. Hence, $N_2^{(1)}$ has components in irreducible representations of the group S_2. Hence, at $x = 1$, we write

$$\mathcal{P}(A_g)\frac{dN_2^{(1)}}{dx}(1) + \mathcal{P}(A_u)\frac{dN_2^{(1)}}{dx}(1) = 1, \tag{A.81}$$

and at $x = -1$, we have

$$\begin{aligned} \mathcal{P}(A_g)\frac{dN_2^{(1)}}{dx}(-1) + \mathcal{P}(A_u)\frac{dN_2^{(1)}}{dx}(-1) &= 0, \\ m \cdot \left[\mathcal{P}(A_g)\frac{dN_2^{(1)}}{dx}(1) + \mathcal{P}(A_u)\frac{dN_2^{(1)}}{dx}(1)\right] &= 0, \\ \mathcal{P}(A_g)\frac{dN_2^{(1)}}{dx}(1) - \mathcal{P}(A_u)\frac{dN_2^{(1)}}{dx}(1) &= 0. \end{aligned} \tag{A.82}$$

Solving Eqs. (A.81) and (A.82), we have

$$\begin{aligned} \mathcal{P}(A_g)\frac{dN_2^{(1)}}{dx}(1) &= b_2 + 3d_2 = \frac{1}{2}, \\ \mathcal{P}(A_u)\frac{dN_2^{(1)}}{dx}(1) &= 2c_2 = \frac{1}{2}. \end{aligned} \tag{A.83}$$

Similarly, for the shape function values of $N_2^{(1)}(x)$, we can deduce the conditions as

$$\begin{aligned} \mathcal{P}(A_g)\frac{dN_2^{(1)}}{dx}(1) &= a_2 + c_2 = 0, \\ \mathcal{P}(A_u)\frac{dN_2^{(1)}}{dx}(1) &= b_2 + d_2 = 0. \end{aligned} \tag{A.84}$$

We can cast the above four equations in matrix form as

$$\begin{bmatrix} 0 & 4 & 0 & 0 \\ 1 & 1 & 0 & 0 \\ 0 & 0 & 2 & 6 \\ 0 & 0 & 1 & 1 \end{bmatrix} \cdot \begin{bmatrix} a_2 \\ c_2 \\ b_2 \\ d_2 \end{bmatrix} = \begin{bmatrix} 1 \\ 0 \\ 1 \\ 0 \end{bmatrix}, \tag{A.85}$$

and invert the matrix to determine the coefficients a_2, b_2, c_2, and d_2. We then have

$$N_2^{(1)}(x) = \frac{1}{4}\left(-1 - x + x^2 + x^3\right). \tag{A.86}$$

Shape functions associated with node x_2 are now obtained through appropriate symmetry group transformations. We deduce $N_1^{(0)}(x)$ through the mirror operation on $N_2^{(0)}(x)$ to get

$$\begin{aligned} m \cdot N_2^{(0)}(x) &= N_2^{(0)}(\xi(x)), \\ &= N_2^{(0)}(-x), \\ m \cdot N_2^{(0)}(x) &= N_1^{(0)}(x), \\ &= \frac{1}{4}\left(2 - 3x + x^3\right). \end{aligned} \tag{A.87}$$

where $\xi(x) = m^{-1} \cdot x = -x$. Under the group transformation from the node x_2 to x_1, the shape functions $N_1^{(1)}$ and $N_2^{(1)}$ transform from one to the other as the component of a vector in 1D. Hence, we call $N_2^{(1)}$ and $N_1^{(1)}$ as vector shape functions.‡ While transforming to $N_1^{(1)}$ from $N_2^{(1)}$, we will have the *Jacobian* associated with the transformation as a coefficient. We write

$$N_1^{(1)}(x) = N_2^{(1)}(\xi(x))\,\frac{dx}{d\xi} = \frac{1}{4}(-1 + x + x^2 - x^3). \tag{A.88}$$

Let us verify that the above expression satisfies the required boundary conditions. We see that

$$\begin{aligned} N_1^{(1)}(1) &= N_2^{(1)}(-1)\frac{dx}{d\xi} = 0, \\ N_1^{(1)}(1) &= N_2^{(1)}(1)\frac{dx}{d\xi} = 0. \end{aligned} \tag{A.89}$$

Using the chain rule for differentiation, we have

$$\begin{aligned} \frac{dN_1^{(1)}(x)}{dx} &= \frac{dN_2^{(1)}(\xi(x))}{d\xi}\,\frac{d\xi}{dx}\,\frac{dx}{d\xi}, \\ &= \frac{dN_2^{(1)}(\xi(x))}{d\xi}. \end{aligned} \tag{A.90}$$

Hence,

$$\begin{aligned} \frac{dN_1^{(1)}}{dx}(1) &= \frac{dN_2^{(1)}}{d\xi}(\xi(1)) = \frac{dN_2^{(1)}}{d\xi}(-1) = 0. \\ \frac{dN_1^{(1)}}{dx}(-1) &= \frac{dN_2^{(1)}}{d\xi}(\xi(-1)) = \frac{dN_2^{(1)}}{d\xi}(1) = 1. \end{aligned} \tag{A.91}$$

First, we note that for vector shape functions transformation from one node to another will involve the *Jacobian* factor $\frac{dx}{d\xi}$ as the coefficient. Second, notice that scalar and vector shape functions have different transformation properties. A list of all 4-DoF cubic polynomials on a 1D line element that support $\mathcal{C}_{(1)}$-continuity across the element are listed in Table B.2. It is straightforward to derive the 6-DoF $\mathcal{C}_{(2)}$-continuous quintic Hermite polynomials in 1D in a similar manner. These are displayed in Table B.3. *One of the highlights of our formalism is that once we determine all the polynomials at a node, we can easily generate the polynomials associated with the other nodes in the element by group transformations.*

‡We will explain this concept in more detail in Sec. A.6.4.

A.6 Shape functions for a triangular element

Now consider the interpolation polynomials in 2D, specifically on an equilateral triangular element with nodes located at the vertices. The data at its vertices provide a total of 18 DoFs with which to describe an arbitrary function over the triangle, with 6 DoFs values at each of the three vertices. From Fig. A.1, we see that the next higher complete polynomial with more than 18 terms is the quintic polynomial which has 21 parameters.

Here the nodal conditions will have two types of symmetries. For example, let a function $f(x)$ be a constant f_1 at the vertex 1 and zero at vertices 2 and 3 of an equilateral triangle. A mirror operation which takes vertex 2 to 3 leaves $f(x)$ invariant. Hence, $f(x)$ has S_2 symmetry. Let $\partial_x f$ be set to 0 at all the three vertices. Any operation in the symmetry group C_{3v} leaves the $\partial_x f$ value invariant. Hence, $\partial_x f(x)$ is invariant under the operators of the group C_{3v}.

Let the desired shape function be written as an arbitrary complete 5th-order polynomial

$$N_i^{(m,n)}(x,y) = \sum_{j=1}^{21} c_j^{(i)} x^a y^b, (i = 1,2,3), \quad \text{such that } a + b \leq 5, \tag{A.92}$$

where $c_j^{(i)}$ are real coefficients to be determined.

The superscripts on N_i denote the derivative order necessary for the shape function to be equal to unity at its associated node, and they have the values $(m,n) = (0,0)$, $\{(0,1),(1,0)\}$, and $\{(2,0),(1,1),(0,2)\}$. The indices m and n are associated with the order of the x and y derivative of N_i which is set to 1 at the associated node.

For example, $N_1^{(0,0)}$ is the polynomial whose value is set 1 at the node $\boldsymbol{r}_1$ and 0 at all other nodes (see Fig. A.7). Similarly, $N_1^{(1,0)}$ and $N_1^{(0,1)}$ are the polynomials whose x and y derivative is set to 1 at the node $\boldsymbol{r}_1$ and zero at all other nodes, respectively. Again, $N_1^{(2,0)}$, $N_1^{(1,1)}$, and $N_1^{(0,2)}$ are the polynomials corresponding to the second derivatives ∂_{xx}, ∂_{xy}, and ∂_{yy}, respectively. Since we require values of the function, its two first derivatives and its three second derivatives to be defined at each node, we have 6 conditions, multiplied by 3 nodes $= 18$ Hermite interpolation polynomials on the equilateral triangle.

Scalar shape functions $N_i^{(0,0)}$ are normalized to unity at a specified vertex, i, and zero at the other two vertices. They have bilateral symmetry in the function value at the nodes. Since all of its derivatives are set to zero at all the 3 vertices, the derivatives of scalar shape functions will have the symmetry of an equilateral triangle.

Let us derive the expression for $N_1^{(0,0)}$ which is the shape function associated with the node at $\boldsymbol{r}_1$. The polynomial expansion is projected into the irreducible representations of the bilateral group S_2 and the group of the equilateral triangle C_{3v}. Character table for S_2 and for C_{3v} are given in Tables A.1 and A.2, respectively. The triangle can be constructed from one of its halves using the operations of S_2, so the nodal conditions can be applied to just two nodes that belong to one of the halves. Similarly, the nodal conditions that are symmetric with respect to C_{3v} can be applied at

Table A.2 Character table for the symmetry group of an equilateral triangle. We have included the character for the reducible representation Γ^{eq}, which is explained in Sec. A.6.1. The last two columns contain the linear and quadratic basis functions of the corresponding representation in the first column.

	ϵ	$2C_3$	$3m$	Linear	Quadratic
A_1	1	1	1	z	x^2+y^2, z^2
A_2	1	1	-1	–	–
E	2	-1	0	$\{x,y\}$	$\{x^2-y^2, xy\}, \{xz, yz\}$
Γ^{eq}	3	0	1	$\{x,y,z\}$	

just one vertex. The conditions on other two nodes are imposed implicitly through group theoretical operations. As noted before, irreducible representations of S_2 are A_g and A_u. The nodal conditions on the symmetry components are provided using the projection operator in Eq. (A.43) as

$$\mathcal{P}(A_g)N_1^{(0,0)}(\boldsymbol{r}_{11}) = 1; \qquad \mathcal{P}(A_u)N_1^{(0,0)}(\boldsymbol{r}_{11}) = 0, \tag{A.93}$$

$$\mathcal{P}(A_g)N_1^{(0,0)}(\boldsymbol{r}_2) = 0; \qquad \mathcal{P}(A_u)N_1^{(0,0)}(\boldsymbol{r}_2) = 0. \tag{A.94}$$

Note that we are putting these conditions manually in such a way that $N_1^{(0,0)}$ is normalized to unity at $\boldsymbol{r}_1$ and zero at $\boldsymbol{r}_2$ or $\boldsymbol{r}_3$, as per our original nodal conditions.

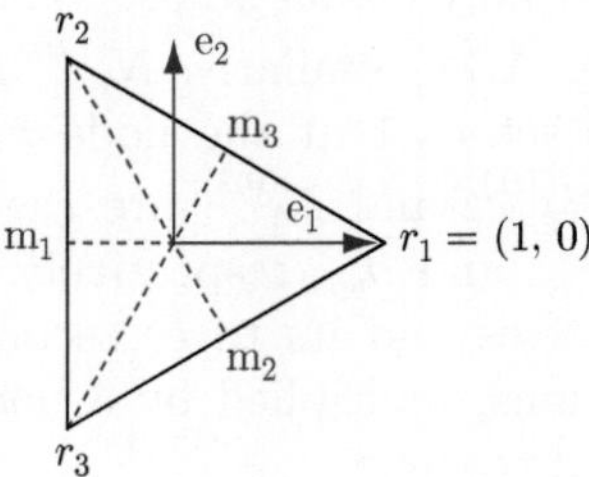

Figure A.7 The equilateral triangular element and the mirror operations of the group C_{3v} are shown.

$$\mathcal{P}(A_g)N_1^{(0,0)}(\boldsymbol{r}_{11}) = 1; \quad \mathcal{P}(A_u)N_1^{(0,0)}(\boldsymbol{r}_{11}) = 0, \tag{A.95}$$

$$\mathcal{P}(A_g)N_1^{(0,0)}(\boldsymbol{r}_2) = 0; \quad \mathcal{P}(A_u)N_1^{(0,0)}(\boldsymbol{r}_2) = 0. \tag{A.96}$$

From Fig. A.7, we note that

$$\mathcal{P}(A_g)N_1^{(0,0)}(x,y) = \frac{1}{2}\left[P_\epsilon N_1^{(0,0)}(x,y) + P_m N_1^{(0,0)}(x,y)\right],$$
$$= \frac{1}{2}[N_1^{(0,0)}(x,y) + N_1^{(0,0)}(x,-y)], \qquad \text{(A.97)}$$

and

$$\mathcal{P}(A_u)N_1^{(0,0)}(x,y) = \frac{1}{2}\left[P_\epsilon N_1^{(0,0)}(x,y) - P_m N_1^{(0,0)}(x,y)\right],$$
$$= \frac{1}{2}[N_1^{(0,0)}(x,y) - N_1^{(0,0)}(x,-y)]. \qquad \text{(A.98)}$$

We use the notation

$$N_1^{(0,0)} = \sum_{n=1}^{21} c_n p_n(x,y), \qquad \text{(A.99)}$$

where $p_n(x,y) = x^a y^b$, for all possible a, b such that $a + b \leq 5$. So Eq. (A.97) and Eq. (A.98) become

$$\mathcal{P}(A_g)N_1^{(0,0)}(x,y) = c_{21}x^5 + c_{19}x^4 + c_{18}x^3y^2 + c_{16}x^3 + c_{14}x^2y^2 + c_{12}x^2$$
$$+ c_{11}xy^4 + c_9xy^2 + c_7x + c_5y^4 + c_3y^2 + c_1, \qquad \text{(A.100)}$$

$$\mathcal{P}(A_u)N_1^{(0,0)}(x,y) = c_{20}x^4y + c_{17}x^3y + c_{15}x^2y^3 + c_{13}x^2y + c_{10}xy^3$$
$$+ c_8xy + c_6y^5 + c_4y^3 + c_2y. \qquad \text{(A.101)}$$

Now we can use Eqs. (A.95, A.96) to get 4 equations to determine the coefficients $\{c_i\}$.

The rest of the nodal conditions equate the derivatives of the shape function $N_i^{(0,0)}$ to zero at all the three nodes, which have the symmetry of the equilateral triangle. Therefore, these nodal conditions decompose as per the irreducible representations of the symmetry group C_{3v}. To get these conditions we need the concept of *"Equivalence Representation"* which is introduced in the next section.

A.6.1 Equivalence representation

We noted earlier that the group representation theory has reduced the problem of finding shape functions, associated with each node in an element, to that of finding shape functions that belong only to those nodes that cannot be transformed from other nodes under operations of the symmetry group of the element. If nodes can be transformed into each other under an element of the symmetry group, we refer to them as being *equivalent.* The notion is motivated from group theory applied to covalent molecules. [11, 12] Two bonds in a molecule are said to be *equivalent* if they transform into one another under an operation in the symmetry group of the molecule. For example, the shape of the ammonia molecule (NH_3) is a trigonal pyramid. Three hydrogen (H) atoms are at the vertices of the triangle and the nitrogen (N) atom is at the top vertex of the pyramid. Hence, the molecule has C_{3v} symmetry. All the three N-H bonds can be brought into each other under the symmetry operations of the group C_{3v}. Therefore, we say that all the three N-H bonds are equivalent.

In the case of the equilateral triangle element (see Fig. A.7), the nodes at $\boldsymbol{r}_2$ and $\boldsymbol{r}_3$ can be brought into the node at $\boldsymbol{r}_1$ by rotation and reflection operations of the symmetry group C_{3v}, respectively. Hence, all these three nodes are equivalent. Therefore, once we determine the shape function $N_1^{(0,0)}$, we can determine $N_2^{(0,0)}$ and $N_3^{(0,0)}$ by simply applying the symmetry operations on $N_1^{(0,0)}$. The shape functions associated with the derivative DoFs at $\boldsymbol{r}_1$ are treated in a similar manner once we account for the vector nature of $\{\partial_x, \partial_y\}$, and the tensor nature of the second derivatives $\{\partial_{xx}, \partial_{xy}, \partial_{yy}\}$ under a coordinate transformation.

Nodal equivalence representation

The transformation that takes one equivalent node into the other generates a representation for the symmetry group that we call as the *nodal equivalence representation.* [7] The matrix components, D_{ij}^{eq} of the nodal equivalence representation $\Gamma^{eq}(R)$ is given by

$$R \cdot x_i = \sum_{j=1}^{3} D_{ij}^{eq}\, x_j. \tag{A.102}$$

Let us obtain the nodal equivalence representation for the equilateral triangle element. From Fig. A.7, we see that under the mirror operation m_1, the node at $\boldsymbol{r}_1$ remain invariant. Whereas, the node at $\boldsymbol{r}_2$ and $\boldsymbol{r}_3$ transform into one another. Therefore,

$$m_1 \cdot \begin{bmatrix} x_1 \\ x_2 \\ x_3 \end{bmatrix} = \begin{bmatrix} x_1 \\ x_3 \\ x_2 \end{bmatrix} = \begin{bmatrix} 1 & 0 & 0 \\ 0 & 0 & 1 \\ 0 & 1 & 0 \end{bmatrix} \cdot \begin{bmatrix} x_1 \\ x_2 \\ x_3 \end{bmatrix}, \tag{A.103}$$

and the nodal equivalence representation of m_1 is given by

$$\Gamma^{eq}(m_1) \equiv \begin{bmatrix} 1 & 0 & 0 \\ 0 & 0 & 1 \\ 0 & 1 & 0 \end{bmatrix}. \tag{A.104}$$

We can use a similar procedure to obtain the representation Γ^{eq} for rest of the operators in the group C_{3v}. Notice that this representation will have only ones and zeros as its components. As in molecular symmetry and molecular bonds, we say that the trace or character of the nodal equivalence representation of a group operator R is equal to the number of nodes which remain invariant under R.

Now Γ^{eq} is a 3D representation of the group C_{3v}. For the group C_{3v}, we have only two 1D representations and one 2D representation as seen in Table A.2. Hence, the nodal equivalent representation is reducible. We can decompose Γ^{eq} in terms of the irreducible representations as

$$\Gamma^{eq} = A_1 + E. \tag{A.105}$$

Table A.3 Self-consistent nodal conditions for first and second derivatives of the polynomial $N_1^{(0,0)}$.

$\mathcal{P}(A_1)\left[\frac{\partial}{\partial x}N_1^{(0,0)}\right]_{\boldsymbol{r}_2} = 0;$	$\mathcal{P}(A_1)\left[\frac{\partial^2}{\partial x^2}N_1^{(0,0)}\right]_{\boldsymbol{r}_2} = 0;$
$\mathcal{P}(E_{11})\left[\frac{\partial}{\partial x}N_1^{(0,0)}\right]_{\boldsymbol{r}_2} = 0;$	$\mathcal{P}(E_{11})\left[\frac{\partial^2}{\partial x^2}N_1^{(0,0)}\right]_{\boldsymbol{r}_2} =;0$
$\mathcal{P}(E_{22})\left[\frac{\partial}{\partial x}N_1^{(0,0)}\right]_{\boldsymbol{r}_2} = 0;$	$\mathcal{P}(E_{22})\left[\frac{\partial^2}{\partial x^2}N_1^{(0,0)}\right]_{\boldsymbol{r}_2} = 0;$
$\mathcal{P}(A_1)\left[\frac{\partial}{\partial y}N_1^{(0,0)}\right]_{\boldsymbol{r}_2} = 0;$	$\mathcal{P}(A_1)\left[\frac{\partial^2}{\partial y^2}N_1^{(0,0)}\right]_{\boldsymbol{r}_2} = 0;$
$\mathcal{P}(E_{11})\left[\frac{\partial}{\partial y}N_1^{(0,0)}\right]_{\boldsymbol{r}_2} = 0;$	$\mathcal{P}(E_{11})\left[\frac{\partial^2}{\partial y^2}N_1^{(0,0)}\right]_{\boldsymbol{r}_2} = 0,$
$\mathcal{P}(E_{22})\left[\frac{\partial}{\partial y}N_1^{(0,0)}\right]_{\boldsymbol{r}_2} = 0;$	$\mathcal{P}(E_{22})\left[\frac{\partial^2}{\partial y^2}N_1^{(0,0)}\right]_{\boldsymbol{r}_2} = 0,$
	$\mathcal{P}(A_1)\left[\frac{\partial^2}{\partial x \partial y}N_1^{(0,0)}\right]_{\boldsymbol{r}_2} = 0,$
$\mathcal{P}(E_{22})\left[\frac{\partial^2}{\partial x \partial y}N_1^{(0,0)}\right]_{\boldsymbol{r}_2} = 0;$	$\mathcal{P}(E_{11})\left[\frac{\partial^2}{\partial x \partial y}N_1^{(0,0)}\right]_{\boldsymbol{r}_2} = 0.$

The nodal equivalence representation determines the projection operator components to which we need to apply the consistency conditions. From Eq. (A.105), we see that only A_1 and E contribute to the representation. Hence, we need to apply the self-consistency conditions only for the projection operators of A_1 and E.

We know that the nodal conditions on the first and second derivatives of $N_1^{(0,0)}$ have C_{3v} symmetry. Hence, we need to apply the self-consistency conditions for derivatives at any one of the nodes. We set each projected component of derivatives of $N_1^{(0,0)}$ to zero at $\boldsymbol{r}_2$. Thus, we get 15 remaining nodal conditions, 6 corresponding to the first and 9 to the second derivatives as shown in Table A.3. The components of $\mathcal{P}(A_2)$ for the first and second derivatives vanish as A_2 does not appear in Eq. (A.105).

The natural basis $\boldsymbol{e}_1, \boldsymbol{e}_2$ are chosen to represent the 2D E irreducible representation of C_{3v}, which comprises the column vector along x- and y-axis, respectively. We know that the character of a representation $\chi_i(R)$ for an element R, is equal to the sum of the diagonal elements of the matrix representation $\Gamma^{(i)}(R)$. *It is well known that for an arbitrary function, the sum of the diagonal projection operators*[§] *of a representation projects out the part belonging to that representation.* Hence, it is sufficient to apply the self-consistency conditions for diagonal projection operators $\mathcal{P}(E_{11}), \mathcal{P}(E_{22})$. Off-diagonal projection operators $\mathcal{P}(E_{12}), \mathcal{P}(E_{21})$ will not contribute any additional conditions.

[§]See p. 42 in Ref. [10].

Tensor equivalence representation

We stated earlier that the shape functions can be classified according to their nodal conditions. Shape functions whose value is set to unity at a node are called scalar shape functions. In the current case, $N_i^{(0,0)}$ for $i = 1, 2, 3$ are scalar shape functions. Similarly, $N_i^{(1,0)}$, $N_i^{(0,1)}$ are called as vector shape functions as one of their first derivatives is set to 1 at the ith node. The shape functions with one of the second derivatives is set to 1 at a node are called rank-2 tensor shape functions. Here, $\left\{N_i^{(2,0)},\ N_i^{(1,1)},\ N_i^{(0,2)}\right\}$ for $i = 1, 2, 3$ are rank-2 shape functions. Transformation rules from one node to another are different for the scalar, vector, and rank-2 tensor shape functions. We wish to obtain the representation for the group C_{3v} with scalar, vector, and rank-2 shape functions as its basis.

The scalar shape functions transform under a coordinate transformation individually as a scalar. Since scalar quantities belong to the representation A_1, the equivalence representation for scalar shape functions is given by

$$\begin{aligned} A_1 \otimes \Gamma^{eq} &= A_1 \otimes (A_1 + E), \\ &= A_1 + E. \end{aligned} \tag{A.106}$$

Note that the dimension of the representation is 3, same as the number of scalar shape functions in our set. The irreducible representation A_1 provides the scalar nature of the shape functions under transformation and the equivalent representation Γ^{eq} includes the contribution from all the three nodes. Hence, we need to take their *direct product*, which incorporates both the scalar nature of the shape functions and the nodal equivalence.

Vector shape functions transform pairwise. We can treat $\left\{N_i^{(1,0)}, N_i^{(0,1)}\right\}$ as a vector in a 2D vector space and deduce the appropriate transformation properties. Since $\{x, y\}$ form a basis for the representation E of the group C_{3v} (see Table A.2), and the pair $\{x, y\}$ transforms as a vector, the vector shape functions are in the representation

$$E \otimes \Gamma^{eq} = A_1 + A_2 + 2E, \tag{A.107}$$

a 6D representation, since we have 2 nodes $\times$ 3 vertices = 6 vector shape functions. Similarly, rank-2 tensor shape functions are in the representation

$$(E \otimes E)_{ord} \otimes \Gamma^{eq} = 2A_1 + A_2 + 3E, \tag{A.108}$$

where, we define $(E \otimes E)_{ord}$ to be a 3D representation of C_{3v} formed by the basis $\{x^2, xy, y^2\}$. (The direct product, $E \otimes E$ is a 4D reducible representation. Since, we need a 3D representation (as there are only 3 rank-2 shape functions at each node), we work with the representation $(E \otimes E)_{ord}$.)

We obtain another condition on the polynomial $N_1^{(0,0)}$ by taking the mirror symmetry of the nodal conditions with respect to the e_1-axis. The irreducible representation E has four projection operators corresponding to the four matrix element components,

E_{11}, E_{12}, E_{21}, and E_{22}. From Eq. (A.43), we see that

$$\mathcal{P}(E_{11}) = \left[\frac{1}{3}\left(2P_{\epsilon} - P_{c_3} - P_{c_3^2} + 2P_{m_1} - P_{m_2} - P_{m_3}\right)\right].$$

Under the mirror operation m_1, the $\mathcal{P}(E_{11})$ remain invariant. Hence,

$$\begin{aligned} m_1 \cdot \mathcal{P}(E_{11}) &= m_1 \cdot \left[\frac{1}{3}\left(2P_{\epsilon} - P_{c_3} - P_{c_3^2} + 2P_{m_1} - P_{m_2} - P_{m_3}\right)\right], \\ &= \frac{1}{3}\left[2P_{\epsilon} - P_{c_3} - P_{c_3^2} + 2P_{m_1} - P_{m_2} - P_{m_3}\right], \end{aligned} \tag{A.109}$$

implies that $\mathcal{P}(E_{11})$ is even. Similarly, the projection operator $\mathcal{P}(E_{12})$ is even, and the projection operators $\mathcal{P}(E_{21})$, $\mathcal{P}(E_{22})$ are odd with respect to the group operation m_1 (reflection through the $\boldsymbol{e}_1$-axis).

Since boundary conditions for $N_1^{(0,0)}$ are symmetric with respect to the mirror operation m_1, $N_1^{(0,0)}$ has to be an even function. Hence, it only comprises components in A_1 and E_{11}. From Eq. (A.107), $N_1^{(1,0)}$ contains components in A_1 and E_{11}, and $N_1^{(0,1)}$ contains components in A_2 and E_{22}. Similarly, from Eq. (A.108), $N_1^{(2,0)}$ has components in A_1 and E_{11}, $N_1^{(1,1)}$ has components in A_2 and E_{22}, and $N_1^{(0,2)}$ has components in A_1 and E_{11}.

We have 18 nodal conditions from Eqs. (A.95, A.96) and from Table A.3 to determine the coefficients of terms in the polynomial $N_1^{(0,0)}$. By applying these 18 *self-consistency conditions* for the projected components, we can express all the 21 coefficients in terms of the polynomial $N_1^{(0,0)}$ in terms of 3 as-yet unknown coefficients. These 3 coefficients are taken arbitrarily to be c_{14}, c_{15}, and c_{18}. For this choice of coefficients, the projected component for the element E_{22} is

$$\mathcal{P}(E_{22})N_1^{(0,0)}(x,y) = -\frac{1}{12}c_{15}(1+2x)^2 y(1-2x+x^2-3y^2). \tag{A.110}$$

Since $\mathcal{P}(E_{22})$ is odd with respect to the mirror operation m_1 and $N_1^{(0,0)}$ is an even function, the projected component $\mathcal{P}(E_{22})N_1^{(0,0)}$ should vanish. Hence, c_{15} must be set to zero. This is the 19th condition for the polynomial.

A.6.2 Inter-element continuity

Each side in the interior of the discretized physical region is shared by two triangular elements. We will face inconsistency, if the opposite nodes to the shared side from the two adjacent triangles influence the value of the shape functions along this side (see Fig. A.8). One way to remove this inconsistency is by requiring that the value of the shape functions associated with a node is zero along its opposite side. Since we desire $\mathcal{C}_{(1)}$-continuity across an element, we impose that even the normal derivative of the shape functions associated with a node is zero on the opposite side. This is the 20th condition. It so happens that for $N_1^{(0,0)}(x,y)$, with 19 conditions imposed through group theory are sufficient to make its value vanish at the opposite side. From Fig. A.7, we see that for the vertex $\boldsymbol{r}_1$, the normal at the opposite side is $\boldsymbol{e}_1$. Hence,

$$\frac{\partial}{\partial x} N_1^{(0,0)}(-\frac{1}{2}, y) = -\frac{1}{16} c_{14}(3 - 4y^2)^2 = 0. \tag{A.111}$$

This implies that the coefficient, $c_{14} = 0$. With this condition, in Fig. A.8, we see that along $\{23\}$ only the DoFs at 2 and 3 affect the inter-element properties.

We have fixed 20 of the 21 undetermined coefficients by the above mentioned conditions. As the shape function along the edge depends only on the vertices at the ends of the edge, we have built in the tangential derivative continuity along each edge. We deduce the last constraint by noting that the shape function must be compatible with the Taylor expansion of the normal derivative along an edge containing the vertex, as discussed in the following section.

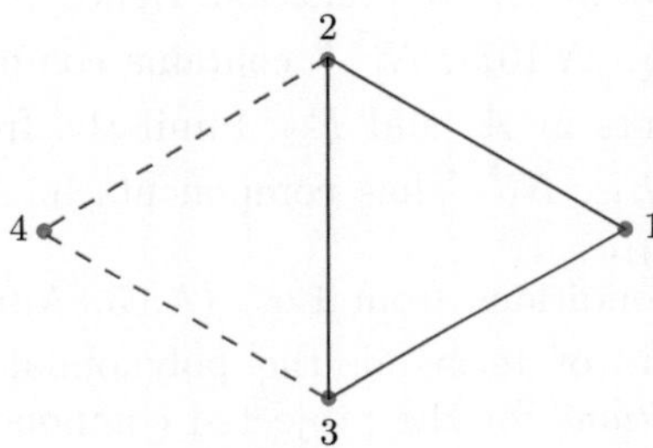

Figure A.8 Two adjacent triangles with nodes at $\{1, 2, 3\}$ and $\{2, 3, 4\}$ share a common side $\{23\}$. The polynomials associated with the node at 1 and 4 may have different values along the side $\{23\}$. In order to remove this inconsistency, we impose a condition that the polynomials associated with the node 1 and 4 and their corresponding normal derivatives are zero along the side $\{23\}$.

A.6.3 Taylor expansion compatibility

Let us suppose that the function which we express in terms of the shape functions is $f(x, y)$. A Taylor series expansion of the function away from a node $\{x_1, y_1\}$ and along one of the two edges is

$$\begin{aligned} f(x, y) &= f(x_1, y_1) + \left[\frac{\partial f}{\partial x}\right]_{(x_1, y_1)} (x - x_1) + \left[\frac{\partial f}{\partial y}\right]_{(x_1, y_1)} (y - y_1) \\ &+ \left[\frac{\partial^2 f}{\partial x^2}\right]_{(x_1, y_1)} \frac{1}{2}(x - x_1)^2 + \left[\frac{\partial^2 f}{\partial y^2}\right]_{(x_1, y_1)} \frac{1}{2}(y - y_1)^2 \\ &+ \left[\frac{\partial^2 f}{\partial x \partial y}\right]_{(x_1, y_1)} (x - x_1)(y - y_1) + \ldots . \end{aligned} \tag{A.112}$$

But according to the finite element Hermite interpolation scheme, we should be able to express the same function in the basis of interpolation polynomials given as

$$f(x,y) = f(x_1,y_1)N_1^{(0,0)} + \left[\frac{\partial f}{\partial x}\right]_{(x_1,y_1)} N_1^{(1,0)} + \left[\frac{\partial f}{\partial y}\right]_{(x_1,y_1)} N_1^{(0,1)} + \ldots. \qquad \text{(A.113)}$$

Let us take the normal derivative of the function $f(x, y)$. Using Eq. (A.112), we obtain

$$\begin{aligned}
\partial_{\boldsymbol{n}} f(x,y) &= \left[\frac{\partial f}{\partial x}\right]_{(x_1,y_1)} n_1 + \left[\frac{\partial f}{\partial y}\right]_{(x_1,y_1)} n_2 + \left[\frac{\partial^2 f}{\partial x^2}\right]_{(x_1,y_1)} (x-x_1)n_1 \\
&\quad + \left[\frac{\partial^2 f}{\partial y^2}\right]_{(x_1,y_1)} (y-y_1)n_2 + \left[\frac{\partial^2 f}{\partial x \partial y}\right]_{(x_1,y_1)} (x-x_1)n_2 \\
&\quad + \left[\frac{\partial f^2}{\partial x \partial y}\right]_{(x_1,y_1)} (y-y_1)n_1 + \ldots. \qquad \text{(A.114)}
\end{aligned}$$

where $\boldsymbol{n} = n_1\boldsymbol{e}_1 + n_2\boldsymbol{e}_2$ is the normal to an edge containing the vertex 1. Alternatively, using Eq. (A.113), we obtain

$$\begin{aligned}
\partial_{\boldsymbol{n}} f(x,y) &= f(x_1,y_1)\partial_{\boldsymbol{n}} N_1^{(0,0)} + \left[\frac{\partial f}{\partial x}\right]_{(x_1,y_1)} \partial_{\boldsymbol{n}} N_1^{(1,0)} \\
&\quad + \left[\frac{\partial f}{\partial y}\right]_{(x_1,y_1)} \partial_{\boldsymbol{n}} N_1^{(0,1)} + \ldots. \qquad \text{(A.115)}
\end{aligned}$$

Since function value at the node $f(x_1, y_1)$ does not appear in the above equation, by comparing with Eq. (A.113), we find the desired additional constraint in that the normal derivative of $N_1^{(0,0)}$ vanishes along the edge

$$\partial_{\boldsymbol{n}} N_1^{(0,0)}(\Sigma) = 0, \qquad \text{(A.116)}$$

where Σ is the set of points on the side connected to the vertex 1. From the Fig. A.7, we note that the normal vector for the edge containing vertices 1 and 2 is

$$\begin{aligned}
\boldsymbol{n} &= \cos(\frac{\pi}{6})\boldsymbol{e}_1 + \sin(\frac{\pi}{6})\boldsymbol{e}_2, \\
&= \frac{1}{2}\boldsymbol{e}_1 + \frac{\sqrt{3}}{2}\boldsymbol{e}_2.
\end{aligned} \qquad \text{(A.117)}$$

Equation (A.116) now takes the form

$$\begin{aligned}
\frac{1}{2}\left(\partial_x + \sqrt{3}\partial_y\right) N_1^{(0,0)}\left(x, \frac{1}{\sqrt{3}}(1-x)\right) &= \frac{27}{81}\left(c_{18} - \frac{40}{27}\right)(1-x)^2(1+2x)^2 \\
&= 0. \qquad \text{(A.118)}
\end{aligned}$$

This implies, $c_{18} = 40/27$. In Fig. A.8, we see that the polynomials associated with the node 1 are made quartic along the sides $\{12\}$ and $\{13\}$.

With all 21 conditions applied to the complete 5th-order polynomial in 2D, $N_1^{(0,0)}$ is fully specified to be

$$N_1^{(0,0)} = \frac{1}{27}\left(9 + 30x + 10x^2 - 30x^3 + 8x^5 - 10y^2 - 30xy^2 + 40x^3y^2\right). \qquad (A.119)$$

The scalar shape function at node $\boldsymbol{r}_2$ can be obtained by an anti-clockwise rotation by an angle $2\pi/3$ on $N_1^{(0,0)}$.

$$\begin{aligned} N_2^{(0,0)} &= C_3 \cdot N_1^{(0,0)}(x, y), \\ &= N_1^{(0,0)}\left(C_3^{-1}(x, y)\right), \\ &= N_1^{(0,0)}\left(\frac{1}{2}(-x + \sqrt{3}y), \frac{1}{2}(-\sqrt{3}x - y)\right). \end{aligned} \qquad (A.120)$$

On substitution into Eq. (A.119), we have

$$N_2^{(0,0)} = \frac{1}{54}(x - 1)(1 + 2x)^2(11 - x(2 + 3x) - 15y^2). \qquad (A.121)$$

Similarly, we obtain $N_3^{(0,0)}$ through an appropriate group operation in C_{3v}. As yet, we have determined the shape functions whose values are normalized at a particular node. They are classified as scalar shape functions. To determine the shape functions whose derivatives are normalized at a node, we need to take care of their tensor nature under a coordinate transformation. For the 18-DoF polynomials on an equilateral triangular element, we have rank-0 (scalar), rank-1 (vector), and rank-2 tensor shape functions. We would like to obtain rank-1 and rank-2 tensor shape functions associated with the node $\boldsymbol{r}_2$ and $\boldsymbol{r}_3$ (see Fig. A.7) by transforming the shape functions associated with the node $\boldsymbol{r}_1$, which we have already obtained through group theory.

Interpolation of a constant function

A constant function, $\psi(x, y) = c$ can be represented using Hermite interpolation polynomials as

$$\psi(x, y) = \sum_{i=1}^{4} \psi_i N_i^{(0,0)}(x, y) = c, \qquad (A.122)$$

since derivatives of ψ are zero throughout. Hence, vector and rank-2 shape functions will not contribute in the above interpolation. The function value at all the i nodes is $\psi_i = c$. *Therefore, we require a condition that the sum of the scalar interpolation polynomials add up to unity.* This condition is true in any dimension. For example, the sum of 1D linear interpolation polynomials on the standard line element (see Sec. A.5) is given by

$$N_1(x) + N_2(x) = \frac{1 + x}{2} + \frac{1 - x}{2} = 1. \qquad (A.123)$$

For the set of 18-DoF polynomials on a triangular element, we know that the scalar shape functions associated with node r_2 and r_3 are obtained through anti-clockwise and clockwise rotation by an angle $2\pi/3$ on $N_1^{(0,0)}$, respectively. Therefore,

$$
\begin{aligned}
N_1^{(0,0)}(\boldsymbol{r}) + N_2^{(0,0)}(\boldsymbol{r}) + N_3^{(0,0)}(\boldsymbol{r}) &= N_1^{(0,0)}(\boldsymbol{r}) + N_1^{(0,0)}(C_3^{-1} \cdot \boldsymbol{r}) \\
&\quad + N_1^{(0,0)}(C_3 \cdot \boldsymbol{r}), \\
&= 1. \qquad \text{(A.124)}
\end{aligned}
$$

If we use the above relation instead of the condition through the Taylor expansion (see Eq. (A.118)), we obtain the same expression for the shape function, $N_1^{(0,0)}(x, y)$ as before. In 3D, we will use a similar normalization condition for scalar shape functions.

A.6.4 Creating equivalent shape functions

We know that

$$\boldsymbol{r}_2 = C_3 \cdot \boldsymbol{r}_1, \qquad \text{(A.125)}$$

where, C_3 gives rise to an anti-clockwise rotation by an angle $2\pi/3$. The transformation rule for the scalar shape function is given by

$$N_2^{(0,0)}(\boldsymbol{r}_2) = N_1^{(0,0)}\left(C_3^{-1}\boldsymbol{r}_1\right), \qquad \text{(A.126)}$$

as explained in Sec. A.4. On the other hand, under group transformation, the shape functions $N_1^{(1,0)}$ and $N_1^{(0,1)}$ transform like the $x-$ and $y-$components of a vector in $x - y$ plane, respectively. The active transformation rule for coordinates of a vector $\boldsymbol{V}$ under C_3 rotation is given by

$$\begin{bmatrix} v'_x \\ v'_y \end{bmatrix} = \begin{bmatrix} a_{11} & a_{12} \\ a_{21} & a_{22} \end{bmatrix} \cdot \begin{bmatrix} v_x \\ v_y \end{bmatrix}, \qquad \text{(A.127)}$$

where a_{ij} are the matrix components of the group operation C_3. The transformation rule for the vector shape functions associated with vertex 2 is given by

$$\begin{bmatrix} N_2^{(1,0)}(\boldsymbol{r}) \\ N_2^{(0,1)}(\boldsymbol{r}) \end{bmatrix} = \begin{bmatrix} a_{11} & a_{12} \\ a_{21} & a_{22} \end{bmatrix} \cdot \begin{bmatrix} N_1^{(1,0)}\left(C_3^{-1}\boldsymbol{r}\right) \\ N_1^{(0,1)}\left(C_3^{-1}\boldsymbol{r}\right) \end{bmatrix}. \qquad \text{(A.128)}$$

In component form, we have

$$
\begin{aligned}
N_2^{(1,0)}(x, y) &= a_{11} N_1^{(1,0)}(b_{11}x + b_{12}y, b_{21}x + b_{22}y) \\
&\quad + a_{12} N_2^{(0,1)}(b_{11}x + b_{12}y, b_{21}x + b_{22}y); \qquad \text{(A.129)}
\end{aligned}
$$

$$
\begin{aligned}
N_2^{(0,1)}(x, y) &= a_{21} N_1^{(1,0)}(b_{11}x + b_{12}y, b_{21}x + b_{22}y) \\
&\quad + a_{22} N_2^{(0,1)}(b_{11}x + b_{12}y, b_{21}x + b_{22}y), \qquad \text{(A.130)}
\end{aligned}
$$

where b_{ij} are the components of the rotation matrix, C_3^{-1}.

The rank-2 shape functions are composed of 3 shape functions, $\left\{N_i^{(2,0)}, N_i^{(1,1)}, N_i^{(0,2)}\right\}$, for $i = 1, 2, 3$. A general rank-2 tensor $\boldsymbol{T}_1$ in 2D space has 3 parameters. $\boldsymbol{T}_1$ is represented by a 2×2 square matrix as

$$\boldsymbol{T}_1 = \begin{bmatrix} p_1 & p_3 \\ p_3 & p_2 \end{bmatrix}, \tag{A.131}$$

where, p_1, p_2, and p_3 are real numbers. Under a group transformation, the general rank-2 tensor transforms as

$$\boldsymbol{T}' = \boldsymbol{A} \cdot \boldsymbol{T} \cdot \boldsymbol{A}^{-1}. \tag{A.132}$$

Since C_3^{-1} is the inverse of rotation C_3, we have $a_{ik}b_{kj} = \delta_{ij}$. Hence, the transformation rule for rank-2 tensor shape functions associated with vertex 2 is given by

$$\begin{bmatrix} N_2^{(2,0)}(\boldsymbol{r}) & N_2^{(1,1)}(\boldsymbol{r}) \\ N_2^{(1,1)}(\boldsymbol{r}) & N_2^{(0,2)}(\boldsymbol{r}) \end{bmatrix} = \begin{bmatrix} a_{11} & a_{12} \\ a_{21} & a_{22} \end{bmatrix} \cdot$$

$$\begin{bmatrix} N_1^{(2,0)}(C_3^{-1}\boldsymbol{r}) & N_1^{(1,1)}(C_3^{-1}\boldsymbol{r}) \\ N_1^{(1,1)}(C_3^{-1}\boldsymbol{r}) & N_1^{(0,2)}(C_3^{-1}\boldsymbol{r}) \end{bmatrix} \cdot \begin{bmatrix} b_{11} & b_{12} \\ b_{21} & b_{22} \end{bmatrix}. \tag{A.133}$$

In a similar manner, we can derive the transformation rules for the shape functions associated with the vertex 3.

A.6.5 Vector shape functions

The shape functions whose first derivatives are set to unity at a vertex are called *vector shape functions.* The labeling is based on the fact that $\left\{N_1^{(1,0)}, N_1^{(0,1)}\right\}$ transforms like a vector under a symmetry operation of the group C_{3v}. As an exercise, let us apply the above-mentioned procedures to derive the polynomial $N_1^{(0,1)}(x, y)$.

At vertex 1, the partial derivative of $N_1^{(0,1)}$ with respect to y is set to unity, and its function value and the other first and all the three second derivatives are set to zero. Therefore, $\partial_y N_1^{(0,1)}$ will have bilateral symmetry. Hence,

$$\mathcal{P}(A_g)\frac{\partial N_1^{(0,1)}}{\partial y}(\boldsymbol{r}_1) = 1; \quad \mathcal{P}(A_u)\frac{\partial N_1^{(0,1)}}{\partial y}(\boldsymbol{r}_1) = 0,$$

$$\mathcal{P}(A_g)\frac{\partial N_1^{(0,1)}}{\partial y}(\boldsymbol{r}_2) = 0; \quad \mathcal{P}(A_u)\frac{\partial N_1^{(0,1)}}{\partial y}(\boldsymbol{r}_2) = 0.$$

Furthermore, $N_1^{(0,0)}$, $\partial_x N_1^{(0,0)}$, $\partial_{xx} N_1^{(0,0)}$, $\partial_{xy} N_1^{(0,0)}$, and $\partial_{yy} N_1^{(0,0)}$ are zero at all the three nodes. Hence, they have C_{3v} symmetry. We know that, nodal equivalence representation is $\Gamma^{eq} = A_1 + E$. Hence, we obtain 3 nodes $\times$ 5 DoF =15 self-consistency conditions as listed in Table A.4. We solve for these self-consistency conditions to obtain the polynomial in terms of 3 undetermined coefficients.

$$\begin{aligned} N_1^{(0,1)}(x,y) &= \frac{1}{108}(2x+1)\ [27c_{18}(2x+1)y^3 \\ &\quad + 27(x-1)y^2(c_{15}(2x+1)+2c_{14}) + 3(x-1)^3(-2c_{14}(x+2) \\ &\quad - 3c_{15}(2x+1)) + 54c_{14}y^4 \\ &\quad + (2x+1)y\left(4(3-4(x-1)x) - 9c_{18}(x-1)^2\right)]. \end{aligned} \tag{A.134}$$

From Eq. (A.107), we see that $N_1^{(1,0)}$ will have components in A_1 and E_{11}, and $N_1^{(0,1)}$ will have components in A_2 and E_{22}. Therefore, the component of $N_1^{(0,1)}$ in A_1 and E_{11} should vanish.

$$\begin{aligned} \mathcal{P}(A_1)N_1^{(0,1)} &= -\frac{1}{36}(4c_{14}+3c_{15})(2x+1)\left((x-1)^2-3y^2\right)\left(x^2+y^2-1\right), \\ &= 0. \end{aligned} \tag{A.135}$$

$$\begin{aligned} \mathcal{P}(E_{11})N_1^{(0,1)} &= \frac{1}{72}(2c_{14}-3c_{15})(2x+1)\left((x-1)^2-3y^2\right)\left((x-1)x-y^2\right), \\ &= 0. \end{aligned} \tag{A.136}$$

Solving these two simultaneous equations, we see that $c_{14} = c_{15} = 0$. So we are left with just one undetermined coefficient.

Table A.4 Self-consistent nodal conditions for the polynomial $N_1^{(0,1)}$ through C_{3v} symmetry.

$\mathcal{P}(A_1)N_1^{(0,1)}(\boldsymbol{r}_2) = 0,$	$\mathcal{P}(A_1)\frac{\partial}{\partial x}N_1^{(0,1)}(\boldsymbol{r}_2) = 0;$
$\mathcal{P}(E_{11})N_1^{(0,1)}(\boldsymbol{r}_2) = 0,$	$\mathcal{P}(E_{11})\frac{\partial}{\partial x}N_1^{(0,1)}(\boldsymbol{r}_2) = 0;$
$\mathcal{P}(E_{22})N_1^{(0,1)}(\boldsymbol{r}_2) = 0,$	$\mathcal{P}(E_{22})\frac{\partial}{\partial x}N_1^{(0,1)}(\boldsymbol{r}_2) = 0;$
$\mathcal{P}(A_1)\frac{\partial^2}{\partial x^2}N_1^{(0,1)}(\boldsymbol{r}_2) = 0;$	$\mathcal{P}(A_1)\frac{\partial^2}{\partial y^2}N_1^{(0,1)}(\boldsymbol{r}_2) = 0;$
$\mathcal{P}(E_{11})\frac{\partial^2}{\partial x^2}N_1^{(0,1)}(\boldsymbol{r}_2) = 0;$	$\mathcal{P}(E_{11})\frac{\partial^2}{\partial y^2}N_1^{(0,1)}(\boldsymbol{r}_2) = 0,$
$\mathcal{P}(E_{22})\frac{\partial^2}{\partial x^2}N_1^{(0,1)}(\boldsymbol{r}_2) = 0;$	$\mathcal{P}(E_{22})\frac{\partial^2}{\partial y^2}N_1^{(0,1)}(\boldsymbol{r}_2) = 0,$
$\mathcal{P}(A_1)\frac{\partial^2}{\partial x\partial y}N_1^{(0,1)}(\boldsymbol{r}_2) = 0;$	$\mathcal{P}(E_{11})\frac{\partial^2}{\partial x\partial y}N_1^{(0,1)}(\boldsymbol{r}_2) = 0.$
$\mathcal{P}(E_{22})\frac{\partial^2}{\partial x\partial y}N_1^{(0,1)}(\boldsymbol{r}_2) = 0;$	

To have the inter-element continuity as explained in Sec. A.6.2, the polynomial and its normal derivative along {23} has to be zero along the side opposite to the vertex 1 (see Fig. A.8). This is indeed true.

To obtain another condition on the polynomial, we impose the condition that the normal derivative along the sides containing the vertex 1 is a cubic function. We see that

$$\frac{1}{2}\left[\frac{\partial}{\partial x}+\sqrt{3}\frac{\partial}{\partial y}\right]N_1^{(0,1)}\left(x,\frac{1}{\sqrt{3}}(1-x)\right) = \frac{4c_{18}x^4}{3\sqrt{3}}-\frac{4c_{18}x^3}{3\sqrt{3}}-\frac{c_{18}x^2}{\sqrt{3}} + \frac{2c_{18}x}{3\sqrt{3}}+\frac{c_{18}}{3\sqrt{3}}+\frac{8x^4}{27\sqrt{3}} - \frac{32x^3}{27\sqrt{3}}+\frac{4x^2}{9\sqrt{3}}+\frac{40x}{27\sqrt{3}}+\frac{25}{54\sqrt{3}}. \tag{A.137}$$

After grouping terms, we set the coefficients of quartic terms to zero, leading to

$$c_{18}\frac{4}{3\sqrt{3}}+\frac{8}{27\sqrt{3}}=0. \tag{A.138}$$

Thus,

$$c_{18}=-\frac{2}{9}, \tag{A.139}$$

and the required interpolation polynomial is given by

$$N_1^{(0,1)}(x,y)=\frac{1}{54}(2x+1)^2y\left((6-7x)x-3y^2+7\right). \tag{A.140}$$

In a similar manner, we get the functional form for the polynomial $N_1^{(1,0)}$ as

$$N_1^{(1,0)}(x,y)=-\frac{1}{54}(x-1)(2x+1)^2\left(x(3x+2)+15y^2-11\right). \tag{A.141}$$

To determine $N_2^{(0,1)}$ and $N_2^{(0,1)}$, we need to consider the vectorial nature of the transformation. From Eq. (A.128), we have

$$\begin{bmatrix} N_2^{(0,1)}(x,y) \\ N_2^{(1,0)}(x,y) \end{bmatrix} = \begin{bmatrix} -\frac{1}{2} & -\frac{\sqrt{3}}{2} \\ \frac{\sqrt{3}}{2} & -\frac{1}{2} \end{bmatrix} \cdot \begin{bmatrix} N_1^{(1,0)}\left(\frac{\sqrt{3}y}{2}-\frac{x}{2},-\frac{\sqrt{3}x}{2}-\frac{y}{2}\right) \\ N_1^{(0,1)}\left(\frac{\sqrt{3}y}{2}-\frac{x}{2},-\frac{\sqrt{3}x}{2}-\frac{y}{2}\right) \end{bmatrix}. \tag{A.142}$$

Substituting appropriate functions, we get

$$N_2^{(0,1)}(x,y)=\frac{(2x+1)}{108}\left(-x+\sqrt{3}y+1\right)^2\left(11-6x^2-\sqrt{3}(3-2x)y-5x\right), \tag{A.143}$$

$$N_2^{(1,0)}(x,y)=\frac{1}{108}\left(-x+\sqrt{3}y+1\right)^2\left(4\sqrt{3}x^3+2x^2\left(5\sqrt{3}-4y\right) + x\left(-8y\left(\sqrt{3}y-3\right)-3\sqrt{3}\right) + y\left(6\left(\sqrt{3}-2y\right)y+23\right)-11\sqrt{3}\right). \tag{A.144}$$

From Fig. A.8, the polynomials associated with the node 1 and their normal derivatives are set to zero along the side {23}. But the second normal derivatives of such

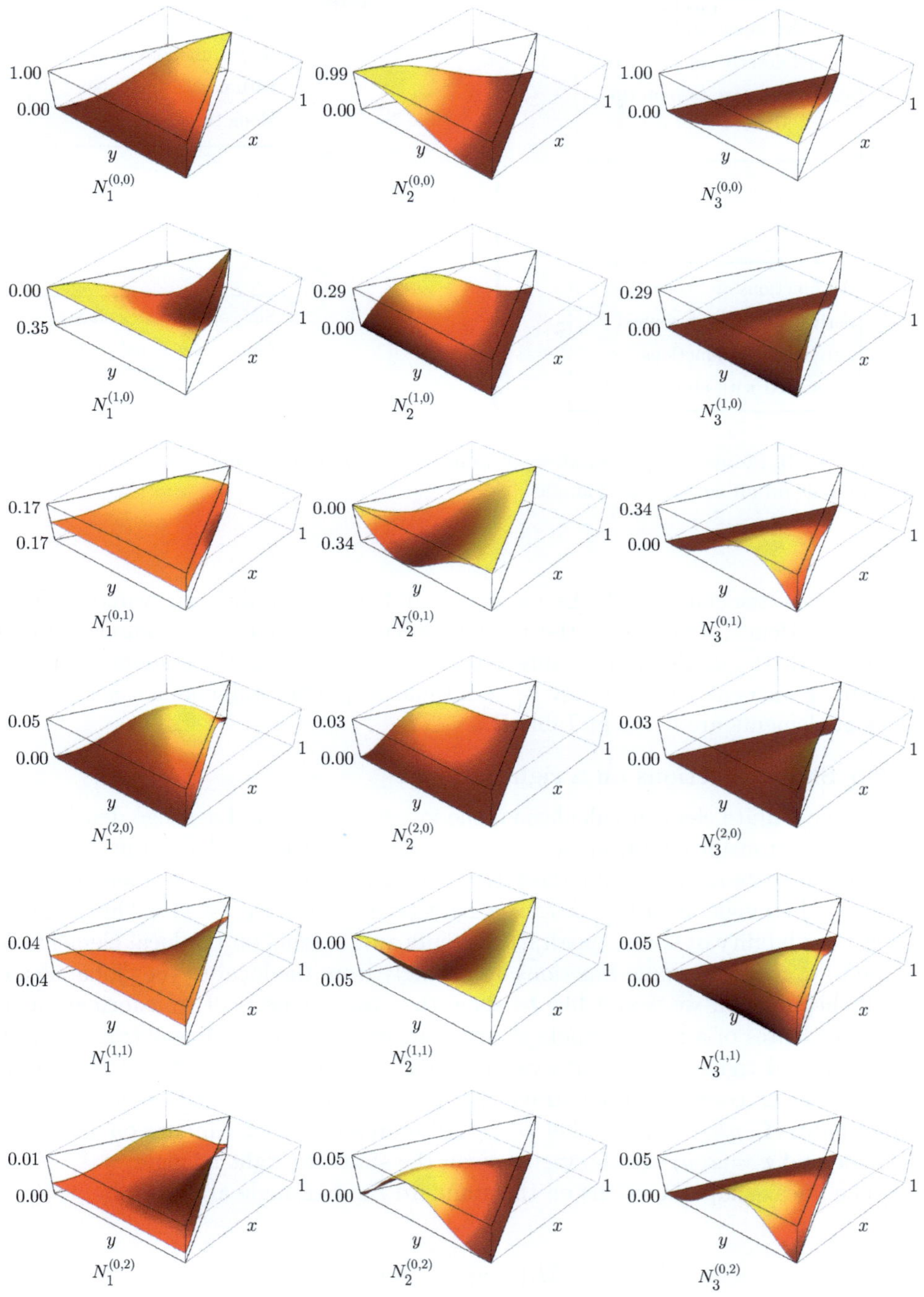

Figure A.9 The $\mathcal{C}_{(2)}$-continuous 18-DoF quintic Hermite interpolation polynomials that have $\mathcal{C}_{(1)}$-continuous normal derivatives across the element are plotted on the standard equilateral triangle.

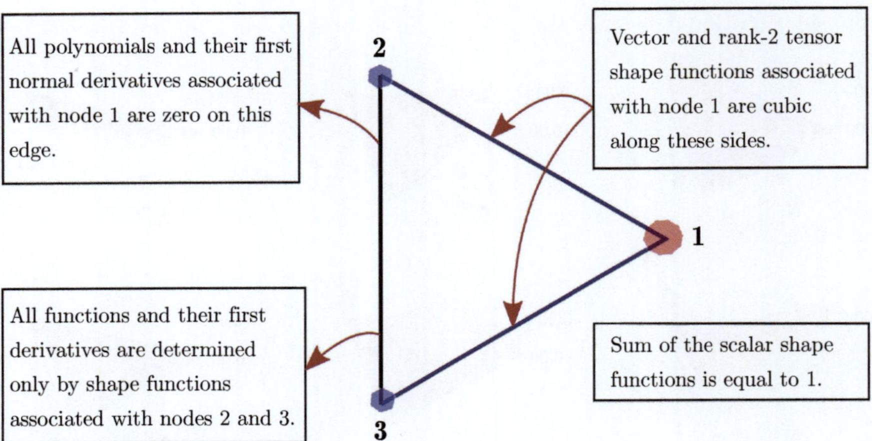

Figure A.10 Pictorial representation of additional conditions applied to determine 18-DoF polynomials on an equilateral triangle.

polynomials associated with the nodes 1 and 4 may have different values along the side $\{23\}$. Hence, we say that the 18-DoF polynomials are $\mathcal{C}_{(1)}$-continuous across the element. All 18 quintic Hermite interpolation polynomials in the standard equilateral triangular element that support $\mathcal{C}_{(1)}$-continuous quartic polynomials across shared sides of elements are listed in Table B.6.

A.6.6 Shape functions on a right triangle

In a typical finite element calculation, the integration of the Lagrange density is carried out by transforming from the general triangular element, defined in the "global" coordinate system, into a standard element. This "local" finite element can be an equilateral triangle as in Fig. A.7, or a right triangle (see Fig. A.11). This is dictated by the availability of Gauss quadrature points and weights on the local element. So it is useful to map the shape functions accordingly from the local to the global element.

To begin with, we would like to show that the geometrical transformation from the coordinates of a right triangle to an arbitrary triangle is a linear transformation. The standard right triangle with vertices at $P_1 = (0,0), P_2 = (1,0)$ and $P_3 = (0,1)$ is chosen. Let us consider an arbitrary triangular element with vertices located at the coordinates (ξ_i, η_i), $i = 1,2,3$. Let $\boldsymbol{\rho} = (\xi, \eta)$ represent the coordinates on an arbitrary triangle and $\boldsymbol{r} =(x,y)$ are the variables on the standard right triangle.

The linear shape functions on the standard right triangle are given by

$$\begin{aligned} M_1(x,y) &= 1 - x - y, \\ M_2(x,y) &= x, \\ M_3(x,y) &= y. \end{aligned} \tag{A.145}$$

A general function $\psi(\xi,\eta)$, with the given values ψ_i at the three vertices of the arbitrary triangle can be interpolated as

$$\psi(\xi,\eta) \approx \sum_{i=1}^{3} \psi_i M_i(x,y). \tag{A.146}$$

We can represent the coordinates ξ and η themselves in terms of linear shape functions as

$$\begin{aligned}\xi &= \xi_1 M_1(x,y) + \xi_2 M_2(x,y) + \xi_3 M_3(x,y),\\ &= \xi_1\ (1-x-y) + \xi_2\ x + \xi_3\ y,\\ &= (\xi_2-\xi_1)x + (\xi_3-\xi_2)y + \xi_1,\end{aligned} \tag{A.147}$$

and

$$\begin{aligned}\eta &= \eta_1 M_1(x,y) + \eta_2 M_2(x,y) + \eta_3 M_3(x,y),\\ &= \eta_1\ (1-x-y) + \eta_2\ x + \eta_3\ y,\\ &= (\eta_2-\eta_1)x + (\eta_3-\eta_2)y + \eta_1.\end{aligned} \tag{A.148}$$

This is an isoparametric mapping. Equations (A.147) and (A.148) can be written as a 2×2 matrix of the form

$$\begin{bmatrix}\xi-\xi_1\\ \eta-\eta_1\end{bmatrix} = \begin{bmatrix}\xi_2-\xi_1 & \xi_3-\xi_1\\ \eta_2-\eta_1 & \eta_3-\eta_1\end{bmatrix}\cdot\begin{bmatrix}x\\ y\end{bmatrix},$$

$$\boldsymbol{\rho}-\boldsymbol{\rho}_1 = \boldsymbol{J}\cdot\boldsymbol{r}. \tag{A.149}$$

This is a linear mapping between the variables (x,y) and (ξ,η). The determinant of the matrix $\boldsymbol{J}$ is the *Jacobian* of the transformation. For example, consider the transformation from the coordinates (x,y) in the standard triangular element (see Fig. A.11) to the coordinates (ξ,η) in our equilateral triangle (see Fig. A.7). From Eq. (A.149), the corresponding transformation is given by

$$\begin{bmatrix}\xi-1\\ \eta\end{bmatrix} = \begin{bmatrix}-\frac{3}{2} & -\frac{3}{2}\\ \frac{\sqrt{3}}{2} & -\frac{\sqrt{3}}{2}\end{bmatrix}\cdot\begin{bmatrix}x\\ y\end{bmatrix}, \tag{A.150}$$

where the Jacobian of the transformation is given by

$$\boldsymbol{J} = \begin{bmatrix}-\frac{3}{2} & -\frac{3}{2}\\ \frac{\sqrt{3}}{2} & -\frac{\sqrt{3}}{2}\end{bmatrix}. \tag{A.151}$$

Hence,

$$\begin{bmatrix}\xi\\ \eta\end{bmatrix} = \begin{bmatrix}-\frac{3}{2} & -\frac{3}{2}\\ \frac{\sqrt{3}}{2} & -\frac{\sqrt{3}}{2}\end{bmatrix}\cdot\begin{bmatrix}x\\ y\end{bmatrix} + \begin{bmatrix}1\\ 0\end{bmatrix}, \tag{A.152}$$

and the inverse transformation is given by

$$\begin{bmatrix} x \\ y \end{bmatrix} = \begin{bmatrix} -\frac{1}{3} & \frac{1}{\sqrt{3}} \\ -\frac{1}{3} & -\frac{1}{\sqrt{3}} \end{bmatrix} \cdot \begin{bmatrix} \xi \\ \eta \end{bmatrix} + \begin{bmatrix} \frac{1}{3} \\ \frac{1}{3} \end{bmatrix}. \tag{A.153}$$

Let us consider the transformation of the linear shape functions labeled by $\{M_i(x,y)\|i=1,2,3\}$ on the right triangle (see Eq. (A.145)) to obtain the linear shape functions $\{N_i(\xi,\eta)|\ i=1,2,3\}$ on our standard equilateral triangle. The corresponding passive coordinate transformation is obtained by inverting the Jacobian matrix in Eq. (A.151). Hence,

$$\begin{aligned} \begin{bmatrix} x \\ y \end{bmatrix} &= \begin{bmatrix} -\frac{3}{2} & -\frac{3}{2} \\ \frac{\sqrt{3}}{2} & -\frac{\sqrt{3}}{2} \end{bmatrix}^{-1} \cdot \begin{bmatrix} \xi - 1 \\ \eta \end{bmatrix}, \\ &= \begin{bmatrix} \dfrac{1-\xi}{3} + \dfrac{\eta}{\sqrt{3}} \\ \dfrac{1-\xi}{3} - \dfrac{\eta}{\sqrt{3}}. \end{bmatrix}. \end{aligned} \tag{A.154}$$

Since we have only scalar shape functions, transformation rule at the ith node is given by

$$\begin{aligned} N_i(\xi,\eta) &= M_i\left(x(\xi,\eta), y(\xi,\eta)\right), \\ &= M_i\left(\frac{1-\xi}{3} + \frac{\eta}{\sqrt{3}}, \frac{1-\xi}{3} - \frac{\eta}{\sqrt{3}}\right). \end{aligned} \tag{A.155}$$

Substituting Eq. (A.145) in Eq. (A.155), we get

$$\begin{aligned} N_1(\xi,\eta) &= \frac{2(\xi-1)}{3} + 1, \\ N_2(\xi,\eta) &= \frac{\eta}{\sqrt{3}} + \frac{1-\xi}{3}, \\ N_3(\xi,\eta) &= \frac{1-\xi}{3} - \frac{\eta}{\sqrt{3}}. \end{aligned} \tag{A.156}$$

Transformation rules for the Hermite shape functions

Let us consider the concrete case of mapping our Hermite shape functions from the equilateral triangle into a right angled triangle. Transformation properties closely follow our discussion in Sec. A.6.4. Let $M_i^{(m,n)}(x,y)$ be the shape functions on the right triangle, with $i=1,2,3$. The possible combinations of (m,n) are $\{(0,0),(1,0),(0,1),(2,0),(1,1),(0,2)\}$. Transformation properties for scalar shape functions follow simply as

$$M_i^{(0,0)}(x,y) = N_i^{(0,0)}\left(\xi(x,y), \eta(x,y)\right), \tag{A.157}$$

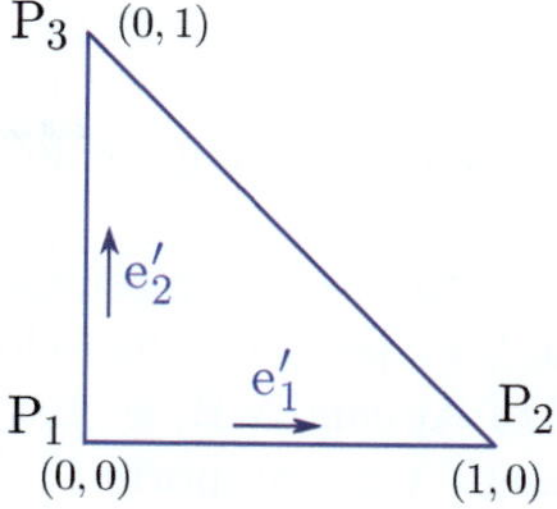

Figure A.11 The right triangular element typically used in Finite Element Analysis is shown.

given the shape functions $N_i^{(0,0)}$ on the equilateral triangle. For example,

$$\begin{aligned} M_1^{(0,0)}(x,y) &= N_1^{(0,0)}\left(\xi(x,y),\eta(x,y)\right), \\ &= N_1^{(0,0)}\left(\frac{1}{2}(-3x-3y+2),\frac{1}{2}\sqrt{3}(x-y)\right). \end{aligned} \tag{A.158}$$

Therefore,

$$\begin{aligned} M_1^{(0,0)}(x,y) = -(x+y-1)^2\Big(6x^3+3x^2(y-1)+x(3(y-2)y-2) \\ +(y-1)\left(6y^2+3y+1\right)\Big). \end{aligned} \tag{A.159}$$

We can derive $M_2^{(0,0)}(x,y)$ and $M_3^{(0,0)}(x,y)$ in a similar fashion.

In the case of vector shape functions $M_i^{(1,0)}, M_i^{(0,1)}$, the transformation properties are given by

$$\begin{bmatrix} M_i^{(1,0)}(x,y) \\ M_i^{(0,1)}(x,y) \end{bmatrix} = \begin{bmatrix} -\frac{1}{3} & \frac{1}{\sqrt{3}} \\ -\frac{1}{3} & -\frac{1}{\sqrt{3}} \end{bmatrix} \cdot \begin{bmatrix} N_i^{(1,0)}\left(\xi(x,y),\eta(x,y)\right) \\ N_i^{(0,1)}\left(\xi(x,y),\eta(x,y)\right) \end{bmatrix}. \tag{A.160}$$

For example,

$$\begin{aligned} M_1^{(1,0)}(x,y) = &-\frac{1}{3}N_1^{(1,0)}\left(\xi(x,y),\eta(x,y)\right) \\ &+\frac{1}{\sqrt{3}}N_1^{(0,1)}\left(\xi(x,y),\eta(x,y)\right). \end{aligned} \tag{A.161}$$

Substituting Eq. (A.152) into the above equation, we get

$$\begin{aligned} M_1^{(1,0)}(x,y) = &-\frac{1}{3}N_1^{(1,0)}\left(\frac{1}{2}(-3x-3y+2),\frac{1}{2}\sqrt{3}(x-y)\right) \\ &+\frac{1}{\sqrt{3}}N_1^{(0,1)}\left(\frac{1}{2}(-3x-3y+2),\frac{1}{2}\sqrt{3}(x-y)\right). \end{aligned} \tag{A.162}$$

Hence,

$$M_1^{(1,0)}(x,y) = -\frac{1}{2}x(x+y-1)^2\left[x(6x+3y-4)-4y-2\right]. \qquad \text{(A.163)}$$

Derivations of the rest of the vector shape functions follow the same steps. Transformation properties for the rank-2 shape functions follow from the discussion above and from Sec. A.6.4. In Table B.7 of Appendix B, we list a set of 18-DoF quintic Hermite polynomials on the right triangle that supports $\mathcal{C}_{(1)}$-continuous quartic polynomials across shared sides of elements. These interpolation polynomials were reported for the first time in the KBR paper, [7] where they also compared and contrasted these new polynomials with previously reported 18-DoF polynomials by Bell. [5]

A.7 Transformation to an arbitrary triangular element

In finite element analysis, the central idea is to express the solution as a linear combination of a set of shape functions multiplied by as-yet undetermined coefficients. We can always find such polynomials for an arbitrary element. However, having to find the explicit form in global coordinates in every element is very inefficient. Thus, we seek to develop a method to obtain the shape functions on an arbitrary element given a set of shape functions on a standard element.

A.7.1 Shape functions on the standard domain

On each element, we represent the solution ψ approximately in terms of the basis of shape functions $\{M_i\}$ as

$$\psi = \sum_i^n \psi_i M_i(x,y). \qquad \text{(A.164)}$$

We may choose the shape functions M_i such that the coefficients are always the value or derivative of the entire function ψ at one of the nodes. This has two advantages. First it enables us to extract the function values at points within the element directly from ψ_i through interpolation. This choice also makes it easy to implement boundary conditions by simply setting the nodal value of some components at the boundary. Each element will be determined by the vertices, $v_1 = (x_1, y_1), v_2 = (x_2, y_2), v_3 = (x_3, y_3)$, enumerated in an anti-clockwise order.

For our purposes, we need the quintic Hermite polynomials to represent the function values and first- and second-order derivatives at the three vertices, so that on each vertex, we have the direct product

$$\begin{bmatrix} 1 \\ \partial_x \\ \partial_y \\ \partial_{xx} \\ \partial_{xy} \\ \partial_{yy} \end{bmatrix} \otimes \begin{bmatrix} M_{v_i}^{(0,0)} & M_{v_i}^{(1,0)} & M_{v_i}^{(0,1)} & M_{v_i}^{(2,0)} & M_{v_i}^{(1,1)} & M_{v_i}^{(0,2)} \end{bmatrix} = \mathbf{I}_6. \qquad \text{(A.165)}$$

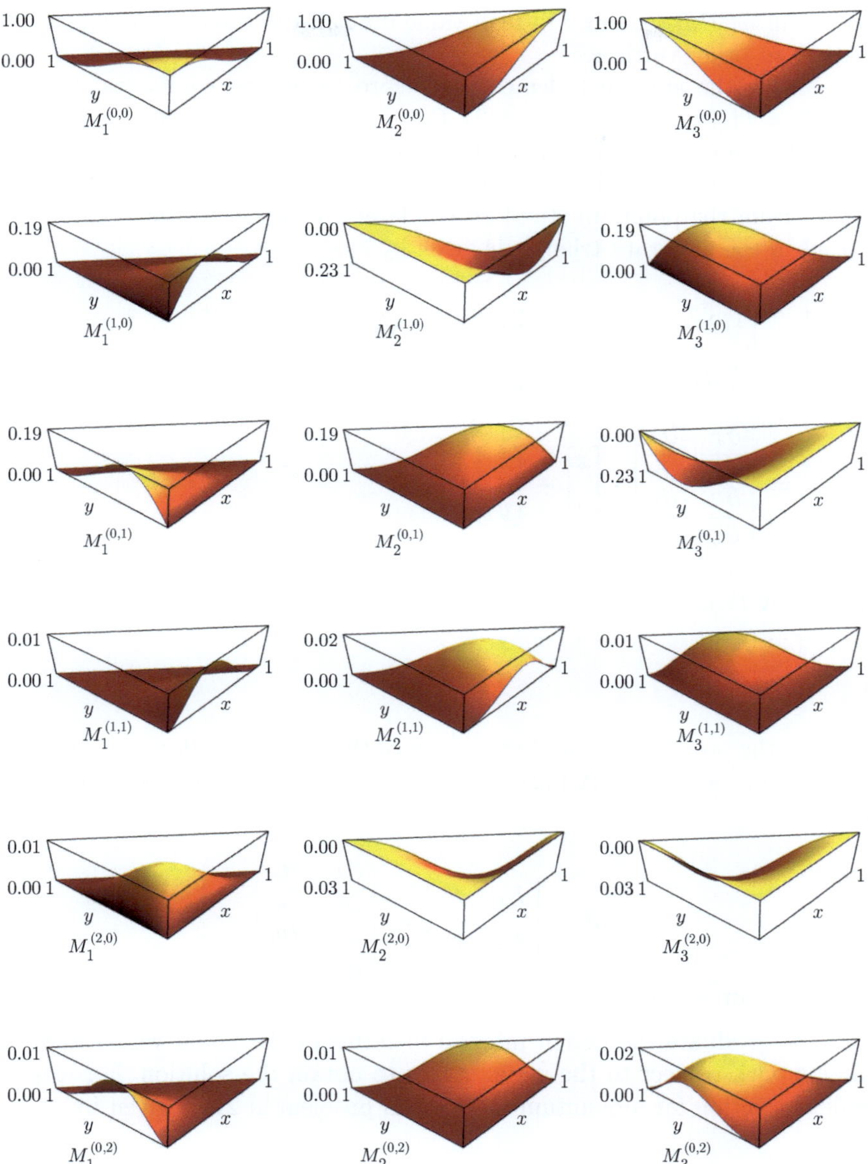

Figure A.12 The $\mathcal{C}_{(2)}$-continuous 18-DoF quintic Hermite interpolation polynomials that have $\mathcal{C}_{(1)}$-continuous normal derivatives across the element are plotted on the standard right triangle.

In other words, every shape function represents a "unit function" under these operations. For example, the first shape function will be 1 at v_1 and 0 at other vertices, and has zero derivatives at all vertices, and the x-derivative of the second shape function will be 1 at v_1, but will take value 0 at all the three vertices and have zero derivatives at other vertices.

We choose the standard element as before to be the equilateral triangle with vertices $P_1 = (1,0), P_2 = (-1/2, \sqrt{3}/2), P_3 = (-1/2, -\sqrt{3}/2)$. Let us suppose that an arbitrary triangular element has vertices located at $V_1 = (x_1, y_1), V_2 = (x_2, y_2)$, and $V_3 = (x_3, y_3)$ in the global coordinate system. The transformation of coordinates $\boldsymbol{\rho} = (\xi, \eta)$ from the points in standard equilateral triangle, to the global coordinates $\boldsymbol{r} = (x, y)$ in the arbitrary triangle is given by

$$\begin{bmatrix} x \\ y \end{bmatrix} = \begin{bmatrix} \frac{1}{3}(2x_1 - x_2 - x_3) & \frac{1}{3}\left(\sqrt{3}x_2 - \sqrt{3}x_3\right) \\ \frac{1}{3}(2y_1 - y_2 - y_3) & \frac{1}{3}\left(\sqrt{3}y_2 - \sqrt{3}y_3\right) \end{bmatrix} \cdot \begin{bmatrix} \xi \\ \eta \end{bmatrix} + \begin{bmatrix} \frac{1}{3}(x_1 + x_2 + x_3) \\ \frac{1}{3}(y_1 + y_2 + y_3) \end{bmatrix},$$

$$= \begin{bmatrix} \dfrac{\partial x}{\partial \xi} & \dfrac{\partial x}{\partial \eta} \\ \dfrac{\partial y}{\partial \xi} & \dfrac{\partial y}{\partial \eta} \end{bmatrix} \cdot \begin{bmatrix} x \\ y \end{bmatrix} + \begin{bmatrix} x_0 \\ y_0 \end{bmatrix}. \tag{A.166}$$

In vector notation

$$\boldsymbol{r} = \boldsymbol{J} \cdot \boldsymbol{\rho} + \boldsymbol{r}_0, \tag{A.167}$$

where $\boldsymbol{r}_0$ is the position vector of the center of the triangular element. This is a linear transformation, as in Eq. (A.149). The Jacobian matrix of the above transformation is given by

$$\boldsymbol{J} = \begin{bmatrix} J_{11} & J_{12} \\ J_{21} & J_{22} \end{bmatrix} = \begin{bmatrix} \dfrac{\partial x}{\partial \xi} & \dfrac{\partial x}{\partial \eta} \\ \dfrac{\partial y}{\partial \xi} & \dfrac{\partial y}{\partial \eta} \end{bmatrix}. \tag{A.168}$$

A.7.2 Element matrix

In FEA, we evaluate the action integral in terms of the nodal values and minimize the action with respect to the nodal values to obtain the solution. For example, the Schrödinger action for a quantum mechanical problem in 2D is given by

$$A = \int \left[\partial_\alpha \psi \dagger(\mathbf{r}) \cdot \partial_\alpha \psi(\mathbf{r}) + \psi^\dagger(\mathbf{r})(V - E)\psi(\mathbf{r})\right] d^2r, \text{ where, } \alpha = 1, 2. \tag{A.169}$$

The above integral is discretized over triangular elements. To evaluate the above integral within an arbitrary triangular element, we need to determine the shape functions corresponding to this arbitrary element. An efficient way is to derive a set of shape functions that satisfies Eq. (A.165) for the standard triangle and transform them into the given triangular element. This transformation is linear and the operations needed in this process are fairly rapidly evaluated.

Let us denote the coordinate system of standard element with $\boldsymbol{\rho} = (\xi, \eta)$ and an arbitrary global element with $\boldsymbol{r} = (x, y)$. Then

$$\mathbf{r} = \begin{bmatrix} x \\ y \end{bmatrix} = \begin{bmatrix} \frac{\partial x}{\partial \xi} & \frac{\partial x}{\partial \eta} \\ \frac{\partial y}{\partial \xi} & \frac{\partial y}{\partial \eta} \end{bmatrix} \begin{bmatrix} \xi \\ \eta \end{bmatrix} + \begin{bmatrix} x_0 \\ y_0 \end{bmatrix} = \mathbf{J}\boldsymbol{\rho} + \mathbf{r}_0, \qquad \text{(A.170)}$$

where $\boldsymbol{r}$ is a displacement vector.

Since the function ψ in Eq. (A.169) can be expanded in terms of shape functions, the action integral is given by

$$A = \int \left[(\partial_\alpha \sum_i \psi_i^* M_i) \cdot (\partial_\alpha \sum_j \psi_j M_j) + \sum_i \psi_i^* M_i (V - E) \sum_j \psi_j M_j \right] d^2 r.$$

Hence,

$$\begin{aligned} A &= \sum_{i,j} \psi_i^* \left(\int \left[\partial_\alpha M_j \cdot \partial_\alpha M_i + M_j (V - E) M_i \right] d^2 r \right) \psi_j \\ &= \sum_{i,j} \psi_i^* \, S_{ij} \psi_j, \end{aligned} \qquad \text{(A.171)}$$

where

$$A_{ij} = \int \left[\partial_\alpha M_j \cdot \partial_\alpha M_i + M_j (V - E) M_i \right] d^2 r.$$

Thus essentially, we need to evaluate every A_{ij}, which constitutes the entries of the element matrix.

We perform a coordinate transformation on the integral A_{ij} from (x, y) to (ξ, η).

$$\begin{aligned} S_{ij} &= \int \left[\nabla_{x,y} M_j \cdot \nabla_{x,y} M_i + M_j (V - E) M_i \right] d^2 x, \\ &= \int \left[\mathbf{J}^{-1} \nabla_{\xi,\eta} M_j \cdot \mathbf{J}^{-1} \nabla_{\xi,\eta} M_i + M_j (V - E) M_i \right] |\mathbf{J}| d^2 \xi. \end{aligned} \qquad \text{(A.172)}$$

For Gauss integration, the coordinate points and weights are readily available for the right triangular element. In other words, a transformation $\mathbf{T}$ such that $M_i = T_{ij} N_j$. Hence,

$$S_{ij} = \int \left[\mathbf{J}^{-1} \mathbf{T}_{ik} \nabla_{\xi,\eta} N_k \cdot \mathbf{J}^{-1} \mathbf{T}_{jl} \nabla_{\xi,\eta} N_l + \mathbf{T}_{ik} N_k (V - E) \mathbf{T}_{jl} N_l \right] |\mathbf{J}| d^2 \xi. \qquad \text{(A.173)}$$

A.7.3 Transformation to an arbitrary triangular domain

Let $\left\{ M_i^{(m,n)} \right\}$ be the set of shape functions on the triangular element in the global coordinates, which are to be determined given the shape function $\left\{ N_i^{(m,n)} \right\}$ in the standard equilateral element. Equation (A.167) transforms the vertex P_i of the standard equilateral triangle to the vertex V_i of the arbitrary triangle. Hence, the nodal conditions for the corresponding shape functions should remain the same. The nodal conditions for the scalar shape functions is written as

$$\begin{aligned} M_i^{(0,0)}(x_j, y_j) &= N_i^{(0,0)}\left(\xi_j, \eta_j\right), \\ &= N_i^{(0,0)}\left(\boldsymbol{J}^{-1}\cdot(\boldsymbol{r_j}-\boldsymbol{r}_0)\right), \\ &= \delta_{ij}, \end{aligned} \tag{A.174}$$

where, $i, j = 1, 2, 3$. Therefore, the transformation rule for the scalar shape function associated with the ith node is simply given by

$$M_i^{(0,0)}(x, y) = N_i^{(0,0)}\left(\boldsymbol{J}^{-1}\cdot(\boldsymbol{r}-\boldsymbol{r}_0)\right). \tag{A.175}$$

To obtain the transformation rules for the vector and rank-2 tensor shape functions, we need to take into account their tensorial nature under transformation (see Sec. A.6.4). Nodal conditions on the vector shape functions is written as¶

$$\begin{aligned} &\begin{bmatrix}\partial_x\\ \partial_y\end{bmatrix} \otimes \left[M_i^{(1,0)}(x_i, y_i) \quad M_i^{(0,1)}(x_i, y_i)\right] \\ &\qquad = \begin{bmatrix}\partial_\xi\\ \partial_\eta\end{bmatrix} \otimes \left[N_i^{(1,0)}(\xi_i, \eta_i) \quad N_i^{(0,1)}(\xi_i, \eta_i)\right] = \boldsymbol{I}_2. \end{aligned} \tag{A.176}$$

Let us consider

$$\begin{aligned} &\begin{bmatrix}\partial_\xi\\ \partial_\eta\end{bmatrix} \otimes \left[M_i^{(1,0)}\left(x_i(\xi,\eta), y_i(\xi,\eta)\right) \quad M_i^{(0,1)}\left(x_i(\xi,\eta), y_i(\xi,\eta)\right)\right] \\ &\qquad = \begin{bmatrix}\dfrac{\partial x}{\partial \xi} & \dfrac{\partial y}{\partial \xi}\\ \dfrac{\partial x}{\partial \eta} & \dfrac{\partial y}{\partial \eta}\end{bmatrix} \cdot \begin{bmatrix}\partial_x\\ \partial_y\end{bmatrix} \otimes \left[M_i^{(1,0)}(x_i, y_i) \quad M_i^{(0,1)}(x_i, y_i)\right], \end{aligned} \tag{A.177}$$

Substituting Eq. (A.176) into Eq. (A.177), we obtain

$$\begin{aligned} &\begin{bmatrix}\partial_\xi\\ \partial_\eta\end{bmatrix} \otimes \left[M_i^{(1,0)}(x_i, y_i), \quad M_i^{(0,1)}(x_i, y_i)\right] = \begin{bmatrix}\frac{\partial x}{\partial \xi} & \frac{\partial y}{\partial \xi}\\ \frac{\partial x}{\partial \eta} & \frac{\partial y}{\partial \eta}\end{bmatrix} \cdot \\ &\qquad \begin{bmatrix}\partial_\xi & \\ & \partial_\eta\end{bmatrix} \otimes \left[N_i^{(1,0)}(\xi_i, \eta_i), \quad N_i^{(0,1)}(\xi_i, \eta_i)\right] \\ &\qquad = \boldsymbol{J}^{\mathrm{T}} \begin{bmatrix}\partial_\xi\\ \partial_\eta\end{bmatrix} \otimes \left[N_i^{(1,0)}(\xi_i, \eta_i), \quad N_i^{(0,1)}(\xi_i, \eta_i)\right]. \end{aligned} \tag{A.178}$$

The operator $\left[\partial_\xi, \partial_\eta\right]^T$ is acting on both the left and right side of Eq. (A.178). If we remove this operator in Eq. (A.178), we get a transformation rule for the vector shape

¶Here, the direct product $\begin{bmatrix}\partial_x\\ \partial_y\end{bmatrix} \otimes \left[M_i^{(1,0)}(x_i, y_i) \quad M_i^{(0,1)}(x_i, y_i)\right]$ is defined to be $\begin{bmatrix}\partial_x M_i^{(1,0)} & \partial_x M_i^{(0,1)}\\ \partial_y M_i^{(1,0)} & \partial_y M_i^{(0,1)}\end{bmatrix}$, where all the derivatives are evaluated at the point (x_i, y_i).

functions on the arbitrary triangular element in terms of the vector shape functions on the standard equilateral triangle. Hence, we write

$$\begin{aligned}&\left[M_i^{(1,0)}(x,y) \quad M_i^{(0,1)}(x,y)\right] \\ &\qquad = \boldsymbol{J}^{\mathrm{T}} \cdot \left[N_i^{(1,0)}\left(\xi(x,y),\eta(x,y)\right) \quad N_i^{(0,1)}\left(\xi(x,y),\eta(x,y)\right)\right]. \end{aligned} \tag{A.179}$$

Transposing the above relation, we have

$$\begin{aligned}\begin{bmatrix} M_i^{(1,0)}(x,y) \\ M_i^{(0,1)}(x,y) \end{bmatrix} &= \boldsymbol{J} \cdot \begin{bmatrix} N_i^{(1,0)}\left(\xi(x,y),\eta\left(x,y\right)\right) \\ N_i^{(0,1)}\left(\xi(x,y),\eta(x,y)\right) \end{bmatrix} \\ &= \boldsymbol{J} \cdot \begin{bmatrix} N_i^{(1,0)}\left(\boldsymbol{J}^{-1}\cdot(\boldsymbol{r}-\boldsymbol{r}_0)\right) \\ N_i^{(0,1)}\left(\boldsymbol{J}^{-1}\cdot(\boldsymbol{r}-\boldsymbol{r}_0)\right) \end{bmatrix}. \end{aligned} \tag{A.180}$$

Now the nodal conditions on the second derivatives of the rank-2 tensor shape functions evaluated at the vertex (x_i, y_i) is written as

$$\begin{aligned}&\begin{bmatrix} \partial_{xx} \\ \partial_{xy} \\ \partial_{yy} \end{bmatrix} \otimes \left[M_i^{(2,0)} \quad M_i^{(1,1)} \quad M_i^{(0,2)}\right] = \\ &\quad \begin{bmatrix} \partial_{\xi\xi} \\ \partial_{\xi\eta} \\ \partial_{\eta\eta} \end{bmatrix} \otimes \left[N_i^{(2,0)} \quad N_i^{(1,1)} \quad N_i^{(0,2)}\right] = \mathbf{I}_3. \end{aligned} \tag{A.181}$$

Let us consider

$$\begin{aligned}&\begin{bmatrix} \partial_{\xi\xi} \\ \partial_{\xi\eta} \\ \partial_{\eta\eta} \end{bmatrix} \otimes \left[M_i^{(2,0)} \quad M_i^{(1,1)} \quad M_i^{(0,2)}\right] \\ &= \begin{bmatrix} \left(\frac{\partial x}{\partial \xi}\partial_x + \frac{\partial y}{\partial \xi}\partial_y\right) \cdot \left(\frac{\partial x}{\partial \xi}\partial_x + \frac{\partial y}{\partial \xi}\partial_y\right) \\ \left(\frac{\partial x}{\partial \xi}\partial_x + \frac{\partial y}{\partial \xi}\partial_y\right) \cdot \left(\frac{\partial x}{\partial \eta}\partial_x + \frac{\partial y}{\partial \eta}\partial_y\right) \\ \left(\frac{\partial x}{\partial \eta}\partial_x + \frac{\partial y}{\partial \eta}\partial_y\right) \cdot \left(\frac{\partial x}{\partial \eta}\partial_x + \frac{\partial y}{\partial \eta}\partial_y\right) \end{bmatrix} \otimes \left[M_i^{(2,0)} \quad M_i^{(1,1)} \quad M_i^{(0,2)}\right]. \end{aligned} \tag{A.182}$$

The above equation, Eq. (A.182), can be expanded to write

$$\begin{bmatrix} \partial_{\xi\xi} \\ \partial_{\xi\eta} \\ \partial_{\eta\eta} \end{bmatrix} \otimes \begin{bmatrix} M_i^{(2,0)} & M_i^{(1,1)} & M_i^{(0,2)} \end{bmatrix} \tag{A.183}$$

$$= \begin{bmatrix} \left(\frac{\partial x}{\partial \xi}\right)^2 & 2\frac{\partial x}{\partial \xi}\frac{\partial y}{\partial \xi} & \left(\frac{\partial y}{\partial \xi}\right)^2 \\ \frac{\partial x}{\partial \xi}\frac{\partial x}{\partial \eta} & \frac{\partial x}{\partial \xi}\frac{\partial y}{\partial \eta} + \frac{\partial x}{\partial \eta}\frac{\partial y}{\partial \xi} & \frac{\partial y}{\partial \eta}\frac{\partial y}{\partial \xi} \\ \left(\frac{\partial x}{\partial \eta}\right)^2 & 2\frac{\partial x}{\partial \eta}\frac{\partial y}{\partial \eta} & \left(\frac{\partial y}{\partial \eta}\right)^2 \end{bmatrix} \cdot \begin{bmatrix} \partial_{xx} \\ \partial_{xy} \\ \partial_{yy} \end{bmatrix} \otimes \begin{bmatrix} M_i^{(2,0)} & M_i^{(1,1)} & M_i^{(0,2)} \end{bmatrix}.$$

Using Eq. (A.181), we have

$$\begin{bmatrix} \partial_{\xi\xi} \\ \partial_{\xi\eta} \\ \partial_{\eta\eta} \end{bmatrix} \otimes \begin{bmatrix} M_i^{(2,0)} & M_i^{(1,1)} & M_i^{(0,2)} \end{bmatrix} \tag{A.184}$$

$$= \begin{bmatrix} \left(\frac{\partial x}{\partial \xi}\right)^2 & 2\frac{\partial x}{\partial \xi}\frac{\partial y}{\partial \xi} & \left(\frac{\partial y}{\partial \xi}\right)^2 \\ \frac{\partial x}{\partial \xi}\frac{\partial x}{\partial \eta} & \frac{\partial x}{\partial \xi}\frac{\partial y}{\partial \eta} + \frac{\partial x}{\partial \eta}\frac{\partial y}{\partial \xi} & \frac{\partial y}{\partial \eta}\frac{\partial y}{\partial \xi} \\ \left(\frac{\partial x}{\partial \eta}\right)^2 & 2\frac{\partial x}{\partial \eta}\frac{\partial y}{\partial \eta} & \left(\frac{\partial y}{\partial \eta}\right)^2 \end{bmatrix} \cdot \begin{bmatrix} \partial_{\xi\xi} \\ \partial_{\xi\eta} \\ \partial_{\eta\eta} \end{bmatrix} \otimes \begin{bmatrix} N_i^{(2,0)} & N_i^{(1,1)} & N_i^{(0,2)} \end{bmatrix}.$$

Hence, we can write down the transformation rule for the rank-2 tensor shape functions as

$$\begin{bmatrix} M_i^{(2,0)} & M_i^{(1,1)} & M_i^{(0,2)} \end{bmatrix} = \begin{bmatrix} \left(\frac{\partial x}{\partial \xi}\right)^2 & 2\frac{\partial x}{\partial \xi}\frac{\partial y}{\partial \xi} & \left(\frac{\partial y}{\partial \xi}\right)^2 \\ \frac{\partial x}{\partial \xi}\frac{\partial x}{\partial \eta} & \frac{\partial x}{\partial \xi}\frac{\partial y}{\partial \eta} + \frac{\partial x}{\partial \eta}\frac{\partial y}{\partial \xi} & \frac{\partial y}{\partial \eta}\frac{\partial y}{\partial \xi} \\ \left(\frac{\partial x}{\partial \eta}\right)^2 & 2\frac{\partial x}{\partial \eta}\frac{\partial y}{\partial \eta} & \left(\frac{\partial y}{\partial \eta}\right)^2 \end{bmatrix} \cdot$$

$$\begin{bmatrix} N_i^{(2,0)} & N_i^{(1,1)} & N_i^{(0,2)} \end{bmatrix}. \tag{A.185}$$

Transposing the above relation, we obtain

$$\begin{bmatrix} M_i^{(2,0)}(x,y) \\ M_i^{(1,1)}(x,y) \\ M_i^{(0,2)}(x,y) \end{bmatrix} = \begin{bmatrix} \left(\frac{\partial x}{\partial \xi}\right)^2 & 2\frac{\partial x}{\partial \xi}\frac{\partial y}{\partial \xi} & \left(\frac{\partial y}{\partial \xi}\right)^2 \\ \frac{\partial x}{\partial \xi}\frac{\partial x}{\partial \eta} & \frac{\partial x}{\partial \xi}\frac{\partial y}{\partial \eta} + \frac{\partial x}{\partial \eta}\frac{\partial y}{\partial \xi} & \frac{\partial y}{\partial \eta}\frac{\partial y}{\partial \xi} \\ \left(\frac{\partial x}{\partial \eta}\right)^2 & 2\frac{\partial x}{\partial \eta}\frac{\partial y}{\partial \eta} & \left(\frac{\partial y}{\partial \eta}\right)^2 \end{bmatrix}^{\boldsymbol{T}} \cdot \begin{bmatrix} N_i^{(2,0)}\left(\boldsymbol{J}^{-1}(\boldsymbol{r}-\boldsymbol{r}_0)\right) \\ N_i^{(1,1)}\left(\boldsymbol{J}^{-1}\cdot(\boldsymbol{r}-\boldsymbol{r}_0)\right) \\ N_i^{(0,2)}\left(\boldsymbol{J}^{-1}\cdot(\boldsymbol{r}-\boldsymbol{r}_0)\right) \end{bmatrix}. \tag{A.186}$$

Hence, the transformation rule for rank-1 shape functions in terms of the entries of the Jacobian matrix $\boldsymbol{J}$ is given by

$$\begin{bmatrix} M_i^{(2,0)} \\ M_i^{(1,1)} \\ M_i^{(0,2)} \end{bmatrix} = \begin{bmatrix} J_{11}^2 & J_{11}J_{12} & J_{12}^2 \\ 2J_{11}J_{21} & J_{12}J_{21} + J_{11}J_{22} & 2J_{12}J_{22} \\ J_{21}^2 & J_{21}J_{22} & J_{22}^2 \end{bmatrix} \cdot \begin{bmatrix} N_i^{(2,0)} \\ N_i^{(1,1)} \\ N_i^{(0,2)} \end{bmatrix}. \tag{A.187}$$

The set of 18-DoF polynomials contains scalar, vector, and rank-2 tensor shape functions. Hence, the transformation for the polynomials associated with i^{th} node is given by

$$\begin{bmatrix} M_i^{(0,0)} \\ M_i^{(1,0)} \\ M_i^{(0,1)} \\ M_i^{(2,0)} \\ M_i^{(1,1)} \\ M_i^{(0,2)} \end{bmatrix}(x,y,z) = \begin{bmatrix} 1 & & & & & \\ & J_{11} & J_{12} & & & \\ & J_{21} & J_{22} & & & \\ & & & J_{11}^2 & J_{11}J_{12} & J_{12}^2 \\ & & & 2J_{11}J_{21} & J_{12}J_{21} + J_{11}J_{22} & 2J_{12}J_{22} \\ & & & J_{21}^2 & 2J_{21}J_{22} & J_{22}^2 \end{bmatrix}_{6\times 6} \cdot \begin{bmatrix} N_i^{(0,0)} \\ N_i^{(1,0)} \\ N_i^{(0,1)} \\ N_i^{(2,0)} \\ N_i^{(1,1)} \\ N_i^{(0,2)} \end{bmatrix}\left(\boldsymbol{J}^{-1}(\boldsymbol{r}-\boldsymbol{r}_0)\right), \tag{A.188}$$

where $i = 1, 2, 3$.

A.8 Summary

A.8.1 Properties of different sets of polynomials on a triangle

It is very appropriate at this stage to summarize the properties of the different sets of Hermite interpolation polynomials we have considered here. This provides an overview of the properties of these polynomials in 2D triangles and 3D tetrahedra, allowing us to make choices as to which set would be applicable and natural for a given problem.

18 DoF

1. Complete quintic polynomials with 21 coefficients are treated with group representation theory.
2. Nodes are at the vertices of the triangle, and f, $\partial_x f$, $\partial_y f$, $\partial_{xx} f$, $\partial_{xy} f$, and $\partial_{yy} f$ are the DoF defined at each vertex.
3. Polynomials and their normal derivatives are set to zero at the opposite side of the given node.
4. Supports $\mathcal{C}_{(2)}$-continuous quintic polynomials in each element.
5. The set of 18-DoF polynomials first derived by KBR [7] supports tangential $\mathcal{C}_{(2)}$- and normal derivative $\mathcal{C}_{(1)}$-continuity across the element.

21 DoF

1. Coefficients of quintic polynomials with 21 terms are obtained through the matrix inversion method.
2. Nodes are located at the vertices and at the centroid of the triangle. We define $f, \partial_x f, \partial_y f, \partial_{xx} f, \partial_{xy} f, \partial_{yy} f$ as the DoFs at each vertex and f, $\partial_x f$, $\partial_y f$ at the centroid of the triangle.
3. Polynomials are zero at the side opposite to a given node. But their normal derivatives do not have unique values across the interface between two elements. Hence, the inter-element normal derivative continuity is not guaranteed.
4. 18-DoF polynomials are preferred over 21-DoF polynomials for any finite element calculations.

A.9 Concluding remarks

We have presented a new method based on the group representation theory to derive interpolation polynomials that satisfy the requirement of $\mathcal{C}_{(n)}$-continuity across elements using the symmetry of the finite element. We have applied this method to derive linear interpolation polynomials and cubic Hermite interpolation polynomials in 1D straight element. For an equilateral triangular finite element, the 18-DoF $\mathcal{C}_{(2)}$-continuous quintic Hermite interpolation polynomials with $\mathcal{C}_{(1)}$-continuous normal derivatives across the element were derived. Transformation of these polynomials to a right triangle element and to an arbitrary triangular element is discussed.

In three dimensions, we derived a set of 16-DoF $\mathcal{C}_{(1)}$-continuous cubic polynomials with $\mathcal{C}_{(0)}$-normal continuity across the element and a set of 40-DoF $\mathcal{C}_{(2)}$-continuous quintic Hermite polynomials with $\mathcal{C}_{(1)}$-normal continuity across the element on the standard regular tetrahedron. *These set of shape functions are reported here for the first time in the literature.* Transformation properties of these polynomials to the standard right and to an arbitrary tetrahedral element are discussed. The approach presented here provides clear guidelines for obtaining shape functions in three or perhaps higher dimensions of arbitrary straight-edged elements.

We have shown in our earlier work that the 18-DoF finite elements yields better accuracy by several orders of magnitude, with a smoother representation of fields than the vector finite element methods for electromagnetic field calculations in waveguides and photonic crystals. [16, 17] Hermite finite elements presented here do not generate any spurious solutions that afflict Lagrange finite elements, even though both are scalar in nature. These polynomials are shown to provide an efficient and accurate means of solving Maxwell's equations in a variety of systems, potentially offering a computationally inexpensive means of designing devices for optoelectronics and plasmonics of increasing complexity. The development of $\mathcal{C}_{(1)}$- and $\mathcal{C}_{(2)}$-continuous set of Hermite interpolating polynomials in 3D will make this method a feasible option for a large number of engineering applications.

The finite elements reported here provide a robust method for quantum mechanical calculations as well. Simulations based on these scalar finite elements are applicable to a broad class of physical systems, e.g. to semiconducting lasers which require simultaneous modeling of transitions in nanoscale quantum wells or dots together with EM cavity calculations, to modeling plasmonic structures in the presence of EM field emissions, and to on-chip propagation within monolithic integrated circuits in high-frequency electronics.

In conclusion, Hermite finite elements presented here provide significant advantages in both electromagnetic and quantum mechanical modeling of complex systems that should attract their more universal usage in physics and engineering applications.

References

[1] O. C. Zienkiewicz and Y. K. Cheung, *Finite Element Methods in Structural and Continuum Mechanics* (McGraw-Hill, New York, 1967); O. C. Zienkiewicz, *The Finite Element Method* (McGraw-Hill, New York, 1977).

[2] J. Jin, *The Finite Element Method in Electromagnetics*, 2nd edition (Wiley, New York, 2002).

[3] L. R. Ram-Mohan, *Finite and Boundary Element applications in Quantum Mechanics* (Oxford University Press Inc., New York, 2002).

[4] H. R. Schwarz, *Finite Element Methods* (Academic Press Inc., San Diego, 1988). pp. 134–137.

[5] K. Bell, Int. J. Numer. Methods Eng. **1**, 101–122 (1969); "A refined triangular plate bending finite element."

[6] J. H. Argyris, I. Fried, and D. W. Scharpf, Aeronaut. J. Royal Aeronaut. Soc. **72**, 701–709 (1968); "The TUBA family of plate elements for the matrix displacement method."

[7] P. G. Kassebaum, C. R. Boucher, and L. R. Ram-Mohan, J. Comp. Physics. **231**, 5747–5760 (2012); "Application of group representation theory to derive Hermite interpolation on a triangle."

[8] H. Goldstein, C. Poole, and J. Safko, *Classical Mechanics*, 3rd. edition (Addison-Wesley, New York, 2002).

[9] M. E. Rose, *Elementary Theory of Angular Momentum* (Dover Publications, New York, 1995).

[10] M. Tinkham, *Group Theory and Quantum Mechanics* (Dover Publications Inc., New York, 2003).

[11] M. Lax, *Symmetry Principles in Solid State and Molecular Physics* (J. Wiley and Sons, New York, 1974).

[12] M. S. Dresselhaus, G. Dresselhaus, and A. Jorio, *Group Theory: Application to the Physics of Condensed Matter* (Springer, Berlin, 2008).

[13] Alfred Gray, *Modern Differential Geometry of Curves and Surfaces* (CRC Press, Inc., 2000).

[14] R. McWeeny, *Symmetry, an Introduction to Group Theory and its Applications* (Pergamon Press, Oxford, 1963).

[15] A. Bossavit, SIAM J. Appl. Math. **53**, 1352–1380 (1993); "Boundary value problems with symmetry and their approximation by finite elements."

[16] C. R. Boucher, Z. Li, C. I. Ahheng, J. D. Albrecht, and L. R. Ram-Mohan, J. Appl. Phys. **119**, 143106 (2016); "Hermite finite elements for high accuracy electromagnetic field calculations: A case study of homogeneous and inhomogeneous waveguides."

[17] C. R Boucher, Z. Li, J. D. Albrecht, and L. R. Ram-Mohan, J. Appl. Phys. **115**, 1–10 (2014); "Efficient modeling of photonic crystals With local Hermite polynomials."

B

Shape functions for 1D, 2D, and 3D finite elements

In this chapter:

- We list for ready reference the typical interpolation polynomials that are used in FEM. Several of them have been derived in earlier chapters. We have shown how many of the shape functions are derivable through the application of group representation theory in Appendix A.
- The 1D interpolation polynomials include Lagrange and Hermite shape functions. The interpolation polynomials have powers of the coordinate variables multiplied by coefficients that have to be determined. They are readily obtained by inverting the matrix of values of the coordinate variables at the nodes, as has been shown in Ref. [1] and in Dhatt and Touzot [2].
- The 2D shape functions for triangular and square elements are given for Lagrange interpolation as well as for Hermite interpolation.
- The 3D shape functions for linear and quadratic Lagrange elements are given. These are followed by Hermite interpolation polynomials on a tetrahedron. All of these are easily derivable using symbolic software such as *MATHEMATICA*.

B.1 List of polynomials on a 1D line element

When the number of degrees of freedom (DoFs) in an element equals the number of terms in the interpolation polynomials, we can employ the matrix method delineated in Chap. 3 of Ref. [1]: the values of the coordinate terms in the interpolation polynomials at the nodes in the element are arranged in a matrix form with the as-yet unknown coefficients arranged in a second matrix multiplying the coefficient matrix. The right side of the equation consists of the values of the polynomials arranged in a matrix form (unit matrix). This is discussed in further detail in the reference mentioned above. The coefficients of the powers of the coordinates in the following tables, except where mentioned, are generated in this manner.

Table B.1 Linear interpolation polynomials on a 1D line element with nodes at $x = -1$ and $x = 1$.

$N_1(x) = \frac{1}{2}(1 - x);$	$N_2(x) = \frac{1}{2}(1 + x).$

Table B.2 Cubic Hermite interpolation polynomials on a 1D line element with nodes at $x = -1$ and $x = 1$ that support $\mathcal{C}_{(1)}$-continuity across the element. Each node has two degrees of freedom to represent the function $f(x)$ and $f'(x)$.

$N_1^{(0)}(x) = \frac{1}{4}\left(x^3 - 3x + 2\right);$	$N_1^{(1)}(x) = \frac{1}{4}(x-1)^2(x+1);$
$N_2^{(0)}(x) = \frac{1}{4}\left(-x^3 + 3x + 2\right);$	$N_2^{(1)}(x) = \frac{1}{4}(x-1)(x+1)^2.$

Table B.3 Quintic Hermite interpolation polynomial element with nodes at $x = -1$ and $x = 1$ that support $\mathcal{C}_{(2)}$-continuity across the element. Each node has three degrees of freedom to represent the function $f(x)$, $f'(x)$, and $f''(x)$.

$N_1^{(0)}(x) = \frac{1}{16}(1-x)^3(3x(x+3)+8),$	$N_1^{(1)}(x) = \frac{1}{16}(1-x)^3(x+1)(3x+5),$
$N_1^{(2)}(x) = \frac{1}{16}(1-x)^3(x+1)^2,$	$N_2^{(0)}(x) = \frac{1}{16}(x+1)^3(3(x-3)x+8),$
$N_2^{(1)}(x) = \frac{1}{16}(1-x)(x+1)^3(3x-5),$	$N_2^{(2)}(x) = \frac{1}{16}(x-1)^2(x+1)^3.$

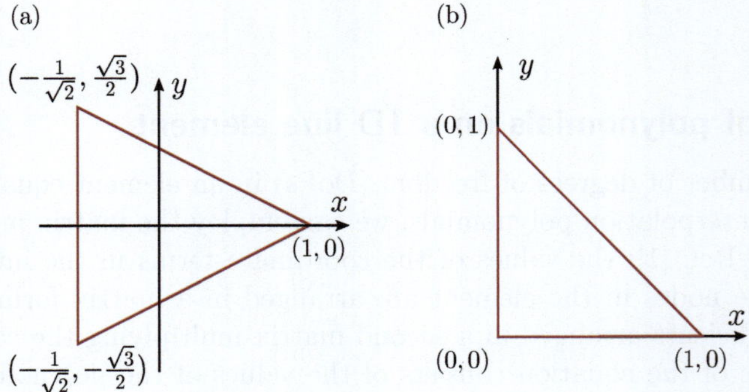

Figure B.1 (a) An equilateral triangular element and (b) a right triangular element with nodes at the vertices are shown.

B.2 List of polynomials on a triangular element

We list below the linear shape functions on an equilateral triangular element and on a right-triangular element. This is followed by the 18-DoF quintic Hermite polynomials on an equilateral triangular element.

Table B.4 Linear interpolation polynomials on the standard equilateral triangular element. Coordinates of the vertices are located at $(1, 0)$, $(-\frac{1}{2}, \frac{\sqrt{3}}{2})$, and $(-\frac{1}{2}, -\frac{\sqrt{3}}{2})$ which are labeled as vertex 1, 2, and 3, respectively.

$$N_1(x, y) = 1 + \tfrac{2}{3}(x - 1);$$
$$N_2(x, y) = \tfrac{1}{3}(1 - x) + \tfrac{y}{\sqrt{3}};$$
$$N_3(x, y) = \tfrac{1}{3}(1 - x) - \tfrac{y}{\sqrt{3}}.$$

Table B.5 Linear interpolation polynomials on the standard right triangular element. Coordinates of the vertices are $(0, 0)$, $(1, 0)$, and $(0, 1)$, which are labeled as vertex 1, 2, and 3, respectively.

$M_1(x, y) = 1 - x - y,$	$M_2(x, y) = x,$	$M_3(x, y) = y.$

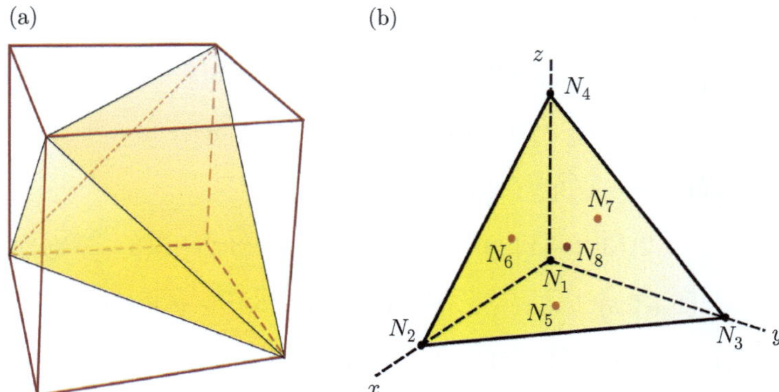

Figure B.2 (a) A regular symmetric tetrahedral element with vertices at $\{(1, 1, 1), (1, -1, -1), (-1, 1, -1), (-1, -1, 1)\}$ and (b) a right-angled tetrahedron with nodes 1–4 at the vertices and nodes 5–8 at the face centers are shown.

Table B.6 Quintic Hermite interpolation polynomials in the equilateral triangular reference element with vertices at $(1,0)$, $(-\frac{1}{2},\frac{\sqrt{3}}{2})$, and $(-\frac{1}{2},-\frac{\sqrt{3}}{2})$ that support $C^{(1)}$-continuous quartic polynomials across shared sides of elements.

$$N_1^{(0,0)}(x,y) = \tfrac{1}{27}(1+2x)^2(9-10y^2-2x(3+(1-x)x-5y^2)).$$

$$N_1^{(1,0)}(x,y) = \tfrac{1}{54}(x-1)(1+2x)^2(11-x(2+3x)-15y^2).$$

$$N_1^{(0,1)}(x,y) = \tfrac{1}{54}(1+2x)^2y(7+x(6-7x)-3y^2).$$

$$N_1^{(2,0)}(x,y) = \tfrac{1}{72}(1-x)(1+2x)^2(3-x(2+x)-5y^2).$$

$$N_1^{(1,1)}(x,y) = \tfrac{1}{108}(2x+1)^2y\left(x(7x-2)+3y^2-5\right).$$

$$N_1^{(0,2)}(x,y) = \tfrac{1}{216}(1+2x)^2((7+5x)y^2-(1-x)^3).$$

$$\begin{aligned}N_2^{(0,0)}(x,y) = \tfrac{1}{27}(&-4x^5+10\sqrt{3}x^4y+5x^3\left(3-4y^2\right)-5x^2\left(3\sqrt{3}y+1\right)\\&-5x\left(-3y^2+2\sqrt{3}y+3\right)+y(y\left(3\sqrt{3}y\left(2y^2-5\right)+5\right)\\&+15\sqrt{3})+9).\end{aligned}$$

$$N_2^{(1,0)}(x,y) = \tfrac{1}{108}(1+2x)(1-x+\sqrt{3}y)^2(11-5x-6x^2-\sqrt{3}(3-2x)y).$$

$$\begin{aligned}N_2^{(0,1)}(x,y) = \tfrac{1}{108}(1-x+\sqrt{3}y)^2(&4\sqrt{3}x^3+2x^2(5\sqrt{3}-4y)\\&+y(23+6(\sqrt{3}-2y)y)-x(3\sqrt{3}-8y(3-\sqrt{3}y))-11\sqrt{3}).\end{aligned}$$

$$N_2^{(2,0)}(x,y) = \tfrac{1}{432}(1+2x)^2(1-x+\sqrt{3}y)^2(3-3x+\sqrt{3}y).$$

$$N_2^{(1,1)}(x,y) = \tfrac{1}{216}(1+2x)(1-x+\sqrt{3}y)^2(11y+x(3\sqrt{3}+2\sqrt{3}x-2y)-5\sqrt{3}).$$

$$\begin{aligned}N_2^{(0,2)}(x,y) = -\tfrac{1}{432}\left(-x+\sqrt{3}y+1\right)^2(&4(1-4x)y^2-4\sqrt{3}(x-3)xy\\&+x(4x(x+3)-3)-8\sqrt{3}y^3+13\left(\sqrt{3}y-1\right)).\end{aligned}$$

$$\begin{aligned}N_3^{(0,0)}(x,y) = \tfrac{1}{27}(1-x-\sqrt{3}y)^2(&x(3-4x(2+x))\\&+2\sqrt{3}(2-x)xy-4(1-x)y^2-2\sqrt{3}y^3+3(3+\sqrt{3}y)).\end{aligned}$$

$$N_3^{(1,0)}(x,y) = \tfrac{1}{108}(1+2x)(1-x-\sqrt{3}y)^2(11-5x-6x^2+\sqrt{3}(3-2x)y).$$

$$\begin{aligned}N_3^{(0,1)}(x,y) = \tfrac{1}{108}(1-x-\sqrt{3}y)^2(&11\sqrt{3}-4\sqrt{3}x^3-2x^2(5\sqrt{3}+4y)\\&+y(23-6y(\sqrt{3}+2y))+x(3\sqrt{3}+8y(3+\sqrt{3}y))).\end{aligned}$$

$$N_3^{(2,0)}(x,y) = \tfrac{1}{432}(1+2x)^2(1-x-\sqrt{3}y)^2(3-3x-\sqrt{3}y).$$

$$N_3^{(1,1)}(x,y) = \tfrac{1}{216}(1+2x)(1-x-\sqrt{3}y)^2(\sqrt{3}(1-x)(5+2x)+(11-2x)y).$$

$$\begin{aligned}N_3^{(0,2)}(x,y) = \tfrac{1}{432}(1-x-\sqrt{3}y)^2(&13-4x^3+13\sqrt{3}y-4y^2\\&-8\sqrt{3}y^3-4x^2(3+\sqrt{3}y)+x(3+4y(3\sqrt{3}+4y))).\end{aligned}$$

Table B.7 Quintic Hermite interpolation polynomials derived for the equilateral triangular reference element that support $C^{(1)}$-continuous quartic polynomials across shared sides of elements are mapped into the right triangle with vertices at $(0,0)$, $(1,0)$, and $(0,1)$ which are labeled as the vertex 1, 2, and 3, respectively.

$$M_1^{(0,0)}(x,y) = (1-x-y)^2(3x^2(1-y) - 6x^3 + x(2+3(2-y)y) + (1-y)(1+3y+6y^2)).$$

$$M_1^{(1,0)}(x,y) = \tfrac{1}{2}x(1-x-y)^2(2+4y+x(4-6x-3y)).$$

$$M_1^{(0,1)}(x,y) = \tfrac{1}{2}y(1-x-y)^2(x(4-3y)+2(1-y)(1+3y)).$$

$$M_1^{(2,0)}(x,y) = \tfrac{1}{4}x^2(1-x-y)^2(2-2x-y).$$

$$M_1^{(1,1)}(x,y) = xy(1-x-y)^2.$$

$$M_1^{(0,2)}(x,y) = \tfrac{1}{4}y^2(1-x-y)^2(2-2y-x).$$

$$M_2^{(0,0)}(x,y) = x^2(x(10-3x(5-2x)) + 15(1-x)^2y - 15(1-x)y^2).$$

$$M_2^{(1,0)}(x,y) = \tfrac{1}{2}x^2(1-x)(6x^2 - 15(1-y)y - x(8-15y)).$$

$$M_2^{(0,1)}(x,y) = \tfrac{1}{2}x^2y(x(3y-4) + 3(2-y-y^2)).$$

$$M_2^{(2,0)}(x,y) = \tfrac{1}{4}x^2(x-1)(2(x-1)x - 5(1-x)y + 5y^2).$$

$$M_2^{(1,1)}(x,y) = \tfrac{1}{2}x^2y(y^2+y+x(2-y)-2).$$

$$M_2^{(0,2)}(x,y) = \tfrac{1}{4}x^2(1+x-y)y^2.$$

$$M_3^{(0,0)}(x,y) = y^2(15x^2(y-1) + 15x(1-y)^2 + y(10-3y(5-2y))).$$

$$M_3^{(1,0)}(x,y) = \tfrac{1}{2}xy^2(6-3x(1+x-y)-4y).$$

$$M_3^{(0,1)}(x,y) = \tfrac{1}{2}y^2(1-y)(15(x-1)x - (8-15x)y + 6y^2).$$

$$M_3^{(2,0)}(x,y) = \tfrac{1}{4}x^2y^2(1-x+y).$$

$$M_3^{(1,1)}(x,y) = \tfrac{1}{2}xy^2(x+x^2+2y-xy-2).$$

$$M_3^{(0,2)}(x,y) = \tfrac{1}{4}y^2(y-1)(5x^2 - 5x(1-y) - 2(1-y)y).$$

B.3 Hermite interpolation polynomials on a tetrahedron

We present the quintic Hermite interpolation polynomials obtained for tetrahedral elements. A sample right tetrahedral element is shown in Fig. B.2. The polynomials assure continuity for the function value f, the first derivatives f_x, f_y, f_z, and the second derivatives $f_{xx}, f_{xy}, f_{xz}, f_{yy}, f_{yz}, f_{zz}$ at each vertex (nodes 1, 2, 3, and 4 in Fig. B.2) of the tetrahedral element. At the face centers (nodes 4–8 in Fig. B.2), on the other hand, continuity in the function value and the first derivatives are guaranteed. These conditions add up to a set of **56** polynomials of 5th-order in x, y, z whose coefficients are obtained through the inversion of the coefficient matrix $\mathbf{M}$ in the equation

$$\mathbf{M} \cdot \mathbf{A} = \mathbf{I}, \tag{B.1}$$

where $\mathbf{A}$ is the matrix containing function and derivative values of the polynomial terms at each node, and $\mathbf{I}$ is the unit matrix. We can easily verify that the polynomials have unit value at their assigned nodes and derivative level, while having zero value at other nodes. The polynomials are listed in Table B.8.

Table B.8 Quintic Hermite polynomials on a tetrahedron that have $C^{(2)}$ continuity at the vertices and $C^{(1)}$ continuity at the face centers. Coordinates of the vertices are $(0,0,0), (1,0,0), (0,1,0), (0,0,1)$ which are labeled as 1, 2, 3, 4, respectively, while the coordinates of the face centers are $(\frac{1}{3},\frac{1}{3},0), (\frac{1}{3},0,\frac{1}{3}), (0,\frac{1}{3},\frac{1}{3}), (\frac{1}{3},\frac{1}{3},\frac{1}{3})$ and labeled as 5, 6, 7, 8, respectively.

$$\begin{aligned}
N_1^{(0,0,0)}(x,y,z) = 1 \; & -10x^3 + 15x^4 - 6x^5 - 87x^2y + 174x^3y - 87x^4y - 87xy^2 \\
& + 381x^2y^2 - 294x^3y^2 - 10y^3 + 174xy^3 - 294x^2y^3 + 15y^4 \\
& - 87xy^4 - 6y^5 - 87x^2z + 174x^3z - 87x^4z + 135xyz \\
& - 45x^2yz - 90x^3yz - 87y^2z - 45xy^2z - 69x^2y^2z + 174y^3z \\
& - 90xy^3z - 87y^4z - 87xz^2 + 381x^2z^2 - 294x^3z^2 - 87yz^2 \\
& - 45xyz^2 - 69x^2yz^2 + 381y^2z^2 - 69xy^2z^2 - 294y^3z^2 \\
& - 10z^3 + 174xz^3 - 294x^2z^3 + 174yz^3 - 90xyz^3 \\
& - 294y^2z^3 + 15z^4 - 87xz^4 - 87yz^4 - 6z^5,
\end{aligned}$$

$$\begin{aligned}
N_1^{(1,0,0)}(x,y,z) = x \; & -6x^3 + 8x^4 - 3x^5 - 26x^2y + 52x^3y - 26x^4y - 11xy^2 \\
& + 72x^2y^2 - 61x^3y^2 + 18xy^3 - 46x^2y^3 - 8xy^4 - 26x^2z \\
& + 52x^3z - 26x^4z + \tfrac{25xyz}{3} + \tfrac{188}{3}x^2yz - 71x^3yz - \tfrac{49}{3}xy^2z \\
& - 47x^2y^2z + 8xy^3z - 11xz^2 + 72x^2z^2 - 61x^3z^2 - \tfrac{49}{3}xyz^2 \\
& - 47x^2yz^2 + 32xy^2z^2 + 18xz^3 - 46x^2z^3 + 8xyz^3 - 8xz^4,
\end{aligned}$$

$$\begin{aligned}
N_1^{(0,1,0)}(x,y,z) = y \; & -11x^2y + 18x^3y - 8x^4y - 26xy^2 + 72x^2y^2 - 46x^3y^2 - 6y^3 \\
& + 52xy^3 - 61x^2y^3 + 8y^4 - 26xy^4 - 3y^5 + \tfrac{25xyz}{3} - \tfrac{49}{3}x^2yz \\
& + 8x^3yz - 26y^2z + \tfrac{188}{3}xy^2z - 47x^2y^2z + 52y^3z - 71xy^3z \\
& - 26y^4z - 11yz^2 - \tfrac{49}{3}xyz^2 + 32x^2yz^2 + 72y^2z^2 - 47xy^2z^2 \\
& - 61y^3z^2 + 18yz^3 + 8xyz^3 - 46y^2z^3 - 8yz^4,
\end{aligned}$$

$$\begin{aligned}
N_1^{(0,0,1)}(x,y,z) = z \; & -11x^2z + 18x^3z - 8x^4z + \tfrac{25xyz}{3} - \tfrac{49}{3}x^2yz + 8x^3yz \\
& - 11y^2z - \tfrac{49}{3}xy^2z + 32x^2y^2z + 18y^3z + 8xy^3z - 8y^4z \\
& - 26xz^2 + 72x^2z^2 - 46x^3z^2 - 26yz^2 + \tfrac{188}{3}xyz^2 - 47x^2yz^2 \\
& + 72y^2z^2 - 47xy^2z^2 - 46y^3z^2 - 6z^3 + 52xz^3 - 61x^2z^3 \\
& + 52yz^3 - 71xyz^3 - 61y^2z^3 + 8z^4 - 26xz^4 - 26yz^4 - 3z^5,
\end{aligned}$$

$$\begin{aligned}
N_1^{(2,0,0)}(x,y,z) = \tfrac{x^2}{2} \; & - \tfrac{3x^3}{2} + \tfrac{3x^4}{2} - \tfrac{x^5}{2} - \tfrac{5x^2y}{2} + 5x^3y - \tfrac{5x^4y}{2} + 4x^2y^2 \\
& - 4x^3y^2 - 2x^2y^3 - \tfrac{5x^2z}{2} + 5x^3z - \tfrac{5x^4z}{2} + \tfrac{4xyz}{3} \\
& + \tfrac{14}{3}x^2yz - 6x^3yz - \tfrac{10}{3}xy^2z - 2x^2y^2z + 2xy^3z + 4x^2z^2 \\
& - 4x^3z^2 - \tfrac{10}{3}xyz^2 - 2x^2yz^2 + 4xy^2z^2 - 2x^2z^3 + 2xyz^3,
\end{aligned}$$

$$\begin{aligned}
N_1^{(1,1,0)}(x,y,z) = xy \; & -4x^2y + 5x^3y - 2x^4y - 4xy^2 + 11x^2y^2 - 7x^3y^2 + 5xy^3 \\
& - 7x^2y^3 - 2xy^4 - \tfrac{7xyz}{3} + \tfrac{19}{3}x^2yz - 4x^3yz + \tfrac{19}{3}xy^2z \\
& - 9x^2y^2z - 4xy^3z + \tfrac{4}{3}xyz^2 - 2x^2yz^2 - 2xy^2z^2,
\end{aligned}$$

$$\begin{aligned}
N_1^{(1,0,1)}(x,y,z) = xz \; & -4x^2z + 5x^3z - 2x^4z - \tfrac{7xyz}{3} + \tfrac{19}{3}x^2yz - 4x^3yz \\
& + \tfrac{4}{3}xy^2z - 2x^2y^2z - 4xz^2 + 11x^2z^2 - 7x^3z^2 + \tfrac{19}{3}xyz^2 \\
& - 9x^2yz^2 - 2xy^2z^2 + 5xz^3 - 7x^2z^3 - 4xyz^3 - 2xz^4,
\end{aligned}$$

$$\begin{aligned}
N_1^{(0,2,0)}(x,y,z) = \tfrac{y^2}{2} \; & - \tfrac{5xy^2}{2} + 4x^2y^2 - 2x^3y^2 - \tfrac{3y^3}{2} + 5xy^3 - 4x^2y^3 + \tfrac{3y^4}{2} \\
& - \tfrac{5xy^4}{2} - \tfrac{y^5}{2} + \tfrac{4xyz}{3} - \tfrac{10}{3}x^2yz + 2x^3yz - \tfrac{5y^2z}{2} + \tfrac{14}{3}xy^2z \\
& - 2x^2y^2z + 5y^3z - 6xy^3z - \tfrac{5y^4z}{2} - \tfrac{10}{3}xyz^2 + 4x^2yz^2 \\
& + 4y^2z^2 - 2xy^2z^2 - 4y^3z^2 + 2xyz^3 - 2y^2z^3,
\end{aligned}$$

$$\begin{aligned}
N_1^{(0,1,1)}(x,y,z) = yz \; & - \tfrac{7xyz}{3} + \tfrac{4}{3}x^2yz - 4y^2z + \tfrac{19}{3}xy^2z - 2x^2y^2z + 5y^3z \\
& -4xy^3z - 2y^4z - 4yz^2 + \tfrac{19}{3}xyz^2 - 2x^2yz^2 + 11y^2z^2 \\
& -9xy^2z^2 - 7y^3z^2 + 5yz^3 - 4xyz^3 - 7y^2z^3 - 2yz^4,
\end{aligned}$$

Continued

Table B.8 *Continued*

$$
\begin{aligned}
N_1^{(0,0,2)}(x,y,z) = {} & \tfrac{4xyz}{3} - \tfrac{10}{3}x^2yz + 2x^3yz - \tfrac{10}{3}xy^2z + 4x^2y^2z + 2xy^3z + \tfrac{z^2}{2} \\
& - \tfrac{5xz^2}{2} + 4x^2z^2 - 2x^3z^2 - \tfrac{5yz^2}{2} + \tfrac{14}{3}xyz^2 - 2x^2yz^2 \\
& + 4y^2z^2 - 2xy^2z^2 - 2y^3z^2 - \tfrac{3z^3}{2} + 5xz^3 - 4x^2z^3 + 5yz^3 \\
& - 6xyz^3 - 4y^2z^3 + \tfrac{3z^4}{2} - \tfrac{5xz^4}{2} - \tfrac{5yz^4}{2} - \tfrac{z^5}{2},
\end{aligned}
$$

$$
\begin{aligned}
N_2^{(0,0,0)}(x,y,z) = {} & 10x^3 - 15x^4 + 6x^5 - 57x^2y + 114x^3y - 57x^4y + 63xy^2 - 69x^2y^2 \\
& + 6x^3y^2 - 126xy^3 + 126x^2y^3 + 63xy^4 - 57x^2z + 114x^3z \\
& - 57x^4z - 63xyz + 549x^2yz - 486x^3yz - 63xy^2z - 303x^2y^2z \\
& + 126xy^3z + 63xz^2 - 69x^2z^2 + 6x^3z^2 - 63xyz^2 - 303x^2yz^2 \\
& + 189xy^2z^2 - 126xz^3 + 126x^2z^3 + 126xyz^3 + 63xz^4,
\end{aligned}
$$

$$
\begin{aligned}
N_2^{(1,0,0)}(x,y,z) = {} & -4x^3 + 7x^4 - 3x^5 + 19x^2y - 38x^3y + 19x^4y - 20xy^2 + 21x^2y^2 \\
& - x^3y^2 + 40xy^3 - 40x^2y^3 - 20xy^4 + 19x^2z - 38x^3z + 19x^4z \\
& + 20xyz - 177x^2yz + 157x^3yz + 20xy^2z + 98x^2y^2z - 40xy^3z \\
& - 20xz^2 + 21x^2z^2 - x^3z^2 + 20xyz^2 + 98x^2yz^2 - 60xy^2z^2 \\
& + 40xz^3 - 40x^2z^3 - 40xyz^3 - 20xz^4,
\end{aligned}
$$

$$
\begin{aligned}
N_2^{(0,1,0)}(x,y,z) = {} & -5x^2y + 14x^3y - 8x^4y + 10xy^2 - 24x^2y^2 + 14x^3y^2 - 20xy^3 \\
& + 29x^2y^3 + 10xy^4 - \tfrac{29xyz}{3} + \tfrac{149}{3}x^2yz - 40x^3yz - \tfrac{4}{3}xy^2z \\
& - 29x^2y^2z + 11xy^3z + \tfrac{29}{3}xyz^2 - 40x^2yz^2 + 11xy^2z^2,
\end{aligned}
$$

$$
\begin{aligned}
N_2^{(0,0,1)}(x,y,z) = {} & -5x^2z + 14x^3z - 8x^4z - \tfrac{29xyz}{3} + \tfrac{149}{3}x^2yz - 40x^3yz \\
& + \tfrac{29}{3}xy^2z - 40x^2y^2z + 10xz^2 - 24x^2z^2 + 14x^3z^2 - \tfrac{4}{3}xyz^2 \\
& - 29x^2yz^2 + 11xy^2z^2 - 20xz^3 + 29x^2z^3 + 11xyz^3 + 10xz^4
\end{aligned}
$$

$$
\begin{aligned}
N_2^{(2,0,0)}(x,y,z) = {} & \tfrac{x^3}{2} - x^4 + \tfrac{x^5}{2} - 2x^2y + 4x^3y - 2x^4y + 2xy^2 - 2x^2y^2 - 4xy^3 \\
& + 4x^2y^3 + 2xy^4 - 2x^2z + 4x^3z - 2x^4z - 2xyz + 18x^2yz \\
& - 16x^3yz - 2xy^2z - 10x^2y^2z + 4xy^3z + 2xz^2 - 2x^2z^2 \\
& - 2xyz^2 - 10x^2yz^2 + 6xy^2z^2 - 4xz^3 + 4x^2z^3 + 4xyz^3 + 2xz^4
\end{aligned}
$$

$$
\begin{aligned}
N_2^{(1,1,0)}(x,y,z) = {} & x^2y - 3x^3y + 2x^4y - 2xy^2 + 5x^2y^2 - 3x^3y^2 + 4xy^3 - 6x^2y^3 \\
& - 2xy^4 + 2xyz - 10x^2yz + 8x^3yz + 6x^2y^2z - 2xy^3z - 2xyz^2 \\
& + 8x^2yz^2 - 2xy^2z^2,
\end{aligned}
$$

$$
\begin{aligned}
N_2^{(1,0,1)}(x,y,z) = {} & x^2z - 3x^3z + 2x^4z + 2xyz - 10x^2yz + 8x^3yz - 2xy^2z \\
& + 8x^2y^2z - 2xz^2 + 5x^2z^2 - 3x^3z^2 + 6x^2yz^2 - 2xy^2z^2 \\
& + 4xz^3 - 6x^2z^3 - 2xyz^3 - 2xz^4,
\end{aligned}
$$

$$N_2^{(0,2,0)}(x,y,z) = \tfrac{xy^2}{2} - 2x^2y^2 + 2x^3y^2 - xy^3 + 2x^2y^3 + \tfrac{xy^4}{2} - \tfrac{2xyz}{3} + \tfrac{8}{3}x^2yz - 2x^3yz + \tfrac{2}{3}xy^2z - 2x^2y^2z + \tfrac{2}{3}xyz^2 - 2x^2yz^2,$$

$$N_2^{(0,1,1)}(x,y,z) = -\tfrac{1}{3}xyz + \tfrac{4}{3}x^2yz + \tfrac{1}{3}xy^2z - 2x^2y^2z + \tfrac{1}{3}xyz^2 - 2x^2yz^2 + xy^2z^2,$$

$$N_2^{(0,0,2)}(x,y,z) = -\tfrac{2}{3}xyz + \tfrac{8}{3}x^2yz - 2x^3yz + \tfrac{2}{3}xy^2z - 2x^2y^2z + \tfrac{xz^2}{2} - 2x^2z^2 + 2x^3z^2 + \tfrac{2}{3}xyz^2 - 2x^2yz^2 - xz^3 + 2x^2z^3 + \tfrac{xz^4}{2},$$

$$N_3^{(0,0,0)}(x,y,z) = 63x^2y - 126x^3y + 63x^4y - 57xy^2 - 69x^2y^2 + 126x^3y^2 + 10y^3 + 114xy^3 + 6x^2y^3 - 15y^4 - 57xy^4 + 6y^5 - 63xyz - 63x^2yz + 126x^3yz - 57y^2z + 549xy^2z - 303x^2y^2z + 114y^3z - 486xy^3z - 57y^4z + 63yz^2 - 63xyz^2 + 189x^2yz^2 - 69y^2z^2 - 303xy^2z^2 + 6y^3z^2 - 126yz^3 + 126xyz^3 + 126y^2z^3 + 63yz^4,$$

$$N_3^{(1,0,0)}(x,y,z) = 10x^2y - 20x^3y + 10x^4y - 5xy^2 - 24x^2y^2 + 29x^3y^2 + 14xy^3 + 14x^2y^3 - 8xy^4 - \tfrac{29xyz}{3} - \tfrac{4}{3}x^2yz + 11x^3yz + \tfrac{149}{3}xy^2z - 29x^2y^2z - 40xy^3z + \tfrac{29}{3}xyz^2 + 11x^2yz^2 - 40xy^2z^2,$$

$$N_3^{(0,1,0)}(x,y,z) = -20x^2y + 40x^3y - 20x^4y + 19xy^2 + 21x^2y^2 - 40x^3y^2 - 4y^3 - 38xy^3 - x^2y^3 + 7y^4 + 19xy^4 - 3y^5 + 20xyz + 20x^2yz - 40x^3yz + 19y^2z - 177xy^2z + 98x^2y^2z - 38y^3z + 157xy^3z + 19y^4z - 20yz^2 + 20xyz^2 - 60x^2yz^2 + 21y^2z^2 + 98xy^2z^2 - y^3z^2 + 40yz^3 - 40xyz^3 - 40y^2z^3 - 20yz^4,$$

$$N_3^{(0,0,1)}(x,y,z) = -\tfrac{29}{3}xyz + \tfrac{29}{3}x^2yz - 5y^2z + \tfrac{149}{3}xy^2z - 40x^2y^2z + 14y^3z - 40xy^3z - 8y^4z + 10yz^2 - \tfrac{4}{3}xyz^2 + 11x^2yz^2 - 24y^2z^2 - 29xy^2z^2 + 14y^3z^2 - 20yz^3 + 11xyz^3 + 29y^2z^3 + 10yz^4,$$

$$N_3^{(2,0,0)}(x,y,z) = \tfrac{x^2y}{2} - x^3y + \tfrac{x^4y}{2} - 2x^2y^2 + 2x^3y^2 + 2x^2y^3 - \tfrac{2xyz}{3} + \tfrac{2}{3}x^2yz + \tfrac{8}{3}xy^2z - 2x^2y^2z - 2xy^3z + \tfrac{2}{3}xyz^2 - 2xy^2z^2,$$

$$N_3^{(1,1,0)}(x,y,z) = -2x^2y + 4x^3y - 2x^4y + xy^2 + 5x^2y^2 - 6x^3y^2 - 3xy^3 - 3x^2y^3 + 2xy^4 + 2xyz - 2x^3yz - 10xy^2z + 6x^2y^2z + 8xy^3z - 2xyz^2 - 2x^2yz^2 + 8xy^2z^2,$$

$$N_3^{(1,0,1)}(x,y,z) = -\tfrac{1}{3}xyz + \tfrac{1}{3}x^2yz + \tfrac{4}{3}xy^2z - 2x^2y^2z + \tfrac{1}{3}xyz^2 + x^2yz^2 - 2xy^2z^2,$$

Continued

Table B.8 *Continued*

$$\begin{aligned}
N_3^{(0,2,0)}(x,y,z) = {} & 2x^2y - 4x^3y + 2x^4y - 2xy^2 - 2x^2y^2 + 4x^3y^2 + \tfrac{y^3}{2} + 4xy^3 \\
& - y^4 - 2xy^4 + \tfrac{y^5}{2} - 2xyz - 2x^2yz + 4x^3yz - 2y^2z + 18xy^2z \\
& - 10x^2y^2z + 4y^3z - 16xy^3z - 2y^4z + 2yz^2 - 2xyz^2 + 6x^2yz^2 \\
& - 2y^2z^2 - 10xy^2z^2 - 4yz^3 + 4xyz^3 + 4y^2z^3 + 2yz^4,
\end{aligned}$$

$$\begin{aligned}
N_3^{(0,1,1)}(x,y,z) = {} & 2xyz - 2x^2yz + y^2z - 10xy^2z + 8x^2y^2z - 3y^3z + 8xy^3z + 2y^4z \\
& - 2yz^2 - 2x^2yz^2 + 5y^2z^2 + 6xy^2z^2 - 3y^3z^2 + 4yz^3 - 2xyz^3 \\
& - 6y^2z^3 - 2yz^4,
\end{aligned}$$

$$\begin{aligned}
N_3^{(0,0,2)}(x,y,z) = {} & -\tfrac{2}{3}xyz + \tfrac{2}{3}x^2yz + \tfrac{8}{3}xy^2z - 2x^2y^2z - 2xy^3z + \tfrac{yz^2}{2} + \tfrac{2}{3}xyz^2 \\
& - 2y^2z^2 - 2xy^2z^2 + 2y^3z^2 - yz^3 + 2y^2z^3 + \tfrac{yz^4}{2},
\end{aligned}$$

$$\begin{aligned}
N_4^{(0,0,0)}(x,y,z) = {} & 63x^2z - 126x^3z + 63x^4z - 63xyz - 63x^2yz + 126x^3yz + 63y^2z \\
& - 63xy^2z + 189x^2y^2z - 126y^3z + 126xy^3z + 63y^4z - 57xz^2 \\
& - 69x^2z^2 + 126x^3z^2 - 57yz^2 + 549xyz^2 - 303x^2yz^2 - 69y^2z^2 \\
& - 303xy^2z^2 + 126y^3z^2 + 10z^3 + 114xz^3 + 6x^2z^3 + 114yz^3 \\
& - 486xyz^3 + 6y^2z^3 - 15z^4 - 57xz^4 - 57yz^4 + 6z^5,
\end{aligned}$$

$$\begin{aligned}
N_4^{(1,0,0)}(x,y,z) = {} & 10x^2z - 20x^3z + 10x^4z - \tfrac{29xyz}{3} - \tfrac{4}{3}x^2yz \\
& + 11x^3yz + \tfrac{29}{3}xy^2z + 11x^2y^2z - 5xz^2 - 24x^2z^2 + 29x^3z^2 \\
& + \tfrac{149}{3}xyz^2 - 29x^2yz^2 - 40xy^2z^2 + 14xz^3 + 14x^2z^3 - 40xyz^3 \\
& - 8xz^4,
\end{aligned}$$

$$\begin{aligned}
N_4^{(0,1,0)}(x,y,z) = {} & -\tfrac{29}{3}xyz + \tfrac{29}{3}x^2yz + 10y^2z - \tfrac{4}{3}xy^2z + 11x^2y^2z - 20y^3z \\
& + 11xy^3z + 10y^4z - 5yz^2 + \tfrac{149}{3}xyz^2 - 40x^2yz^2 - 24y^2z^2 \\
& - 29xy^2z^2 + 29y^3z^2 + 14yz^3 - 40xyz^3 + 14y^2z^3 - 8yz^4,
\end{aligned}$$

$$\begin{aligned}
N_4^{(0,0,1)}(x,y,z) = {} & -20x^2z + 40x^3z - 20x^4z + 20xyz + 20x^2yz - 40x^3yz - 20y^2z \\
& + 20xy^2z - 60x^2y^2z + 40y^3z - 40xy^3z - 20y^4z + 19xz^2 \\
& + 21x^2z^2 - 40x^3z^2 + 19yz^2 - 177xyz^2 + 98x^2yz^2 + 21y^2z^2 \\
& + 98xy^2z^2 - 40y^3z^2 - 4z^3 - 38xz^3 - x^2z^3 - 38yz^3 + 157xyz^3 \\
& - y^2z^3 + 7z^4 + 19xz^4 + 19yz^4 - 3z^5,
\end{aligned}$$

$$\begin{aligned}
N_4^{(2,0,0)}(x,y,z) = {} & \tfrac{x^2z}{2} - x^3z + \tfrac{x^4z}{2} - \tfrac{2xyz}{3} + \tfrac{2}{3}x^2yz + \tfrac{2}{3}xy^2z - 2x^2z^2 + 2x^3z^2 \\
& + \tfrac{8}{3}xyz^2 - 2x^2yz^2 - 2xy^2z^2 + 2x^2z^3 - 2xyz^3,
\end{aligned}$$

$$\begin{aligned}
N_4^{(1,1,0)}(x,y,z) = {} & -\tfrac{1}{3}xyz + \tfrac{1}{3}x^2yz + \tfrac{1}{3}xy^2z + x^2y^2z + \tfrac{4}{3}xyz^2 - 2x^2yz^2 \\
& - 2xy^2z^2,
\end{aligned}$$

$$N_4^{(1,0,1)}(x,y,z) = -2x^2z + 4x^3z - 2x^4z + 2xyz - 2x^3yz - 2xy^2z - 2x^2y^2z + xz^2 + 5x^2z^2 - 6x^3z^2 - 10xyz^2 + 6x^2yz^2 + 8xy^2z^2 - 3xz^3 - 3x^2z^3 + 8xyz^3 + 2xz^4,$$

$$N_4^{(0,2,0)}(x,y,z) = -\tfrac{2}{3}xyz + \tfrac{2}{3}x^2yz + \tfrac{y^2z}{2} + \tfrac{2}{3}xy^2z - y^3z + \tfrac{y^4z}{2} + \tfrac{8}{3}xyz^2 - 2x^2yz^2 - 2y^2z^2 - 2xy^2z^2 + 2y^3z^2 - 2xyz^3 + 2y^2z^3,$$

$$N_4^{(0,1,1)}(x,y,z) = 2xyz - 2x^2yz - 2y^2z - 2x^2y^2z + 4y^3z - 2xy^3z - 2y^4z + yz^2 - 10xyz^2 + 8x^2yz^2 + 5y^2z^2 + 6xy^2z^2 - 6y^3z^2 - 3yz^3 + 8xyz^3 - 3y^2z^3 + 2yz^4,$$

$$N_4^{(0,0,2)}(x,y,z) = 2x^2z - 4x^3z + 2x^4z - 2xyz - 2x^2yz + 4x^3yz + 2y^2z - 2xy^2z + 6x^2y^2z - 4y^3z + 4xy^3z + 2y^4z - 2xz^2 - 2x^2z^2 + 4x^3z^2 - 2yz^2 + 18xyz^2 - 10x^2yz^2 - 2y^2z^2 - 10xy^2z^2 + 4y^3z^2 + \tfrac{z^3}{2} + 4xz^3 + 4yz^3 - 16xyz^3 - z^4 - 2xz^4 - 2yz^4 + \tfrac{z^5}{2},$$

$$N_5^{(0,0,0)}(x,y,z) = 81x^2y - 162x^3y + 81x^4y + 81xy^2 - 243x^2y^2 + 162x^3y^2 - 162xy^3 + 162x^2y^3 + 81xy^4 + 135xyz - 297x^2yz + 162x^3yz - 297xy^2z + 243x^2y^2z + 162xy^3z - 621xyz^2 + 567x^2yz^2 + 567xy^2z^2 + 486xyz^3,$$

$$N_5^{(1,0,0)}(x,y,z) = 27x^2y - 54x^3y + 27x^4y - 54xy^2 + 81x^2y^2 - 27x^3y^2 + 108xy^3 - 108x^2y^3 - 54xy^4 + 18xyz - 153x^2yz + 135x^3yz + 90xy^2z - 108xy^3z - 18xyz^2 + 108x^2yz^2 - 54xy^2z^2,$$

$$N_5^{(0,1,0)}(x,y,z) = -54x^2y + 108x^3y - 54x^4y + 27xy^2 + 81x^2y^2 - 108x^3y^2 - 54xy^3 - 27x^2y^3 + 27xy^4 + 18xyz + 90x^2yz - 108x^3yz - 153xy^2z + 135xy^3z - 18xyz^2 - 54x^2yz^2 + 108xy^2z^2,$$

$$N_5^{(0,0,1)}(x,y,z) = 27xyz - 27x^2yz - 27xy^2z - 108xyz^2 + 81x^2yz^2 + 81xy^2z^2 + 81xyz^3,$$

$$N_6^{(0,0,0)}(x,y,z) = 81x^2z - 162x^3z + 81x^4z + 135xyz - 297x^2yz + 162x^3yz - 621xy^2z + 567x^2y^2z + 486xy^3z + 81xz^2 - 243x^2z^2 + 162x^3z^2 - 297xyz^2 + 243x^2yz^2 + 567xy^2z^2 - 162xz^3 + 162x^2z^3 + 162xyz^3 + 81xz^4,$$

Continued

Table B.8 *Continued*

$$N_6^{(1,0,0)}(x,y,z) = 27x^2z - 54x^3z + 27x^4z + 18xyz - 153x^2yz + 135x^3yz - 18xy^2z + 108x^2y^2z - 54xz^2 + 81x^2z^2 - 27x^3z^2 + 90xyz^2 - 54xy^2z^2 + 108xz^3 - 108x^2z^3 - 108xyz^3 - 54xz^4,$$

$$N_6^{(0,1,0)}(x,y,z) = 27xyz - 27x^2yz - 108xy^2z + 81x^2y^2z + 81xy^3z - 27xyz^2 + 81xy^2z^2,$$

$$N_6^{(0,0,1)}(x,y,z) = -54x^2z + 108x^3z - 54x^4z + 18xyz + 90x^2yz - 108x^3yz - 18xy^2z - 54x^2y^2z + 27xz^2 + 81x^2z^2 - 108x^3z^2 - 153xyz^2 + 108xy^2z^2 - 54xz^3 - 27x^2z^3 + 135xyz^3 + 27xz^4,$$

$$N_7^{(0,0,0)}(x,y,z) = 135xyz - 621x^2yz + 486x^3yz + 81y^2z - 297xy^2z + 567x^2y^2z - 162y^3z + 162xy^3z + 81y^4z + 81yz^2 - 297xyz^2 + 567x^2yz^2 - 243y^2z^2 + 243xy^2z^2 + 162y^3z^2 - 162yz^3 + 162xyz^3 + 162y^2z^3 + 81yz^4,$$

$$N_7^{(1,0,0)}(x,y,z) = 27xyz - 108x^2yz + 81x^3yz - 27xy^2z + 81x^2y^2z - 27xyz^2 + 81x^2yz^2,$$

$$N_7^{(0,1,0)}(x,y,z) = 18xyz - 18x^2yz + 27y^2z - 153xy^2z + 108x^2y^2z - 54y^3z + 135xy^3z + 27y^4z - 54yz^2 + 90xyz^2 - 54x^2yz^2 + 81y^2z^2 - 27y^3z^2 + 108yz^3 - 108xyz^3 - 108y^2z^3 - 54yz^4,$$

$$N_7^{(0,0,1)}(x,y,z) = 18xyz - 18x^2yz - 54y^2z + 90xy^2z - 54x^2y^2z + 108y^3z - 108xy^3z - 54y^4z + 27yz^2 - 153xyz^2 + 108x^2yz^2 + 81y^2z^2 - 108y^3z^2 - 54yz^3 + 135xyz^3 - 27y^2z^3 + 27yz^4,$$

$$N_8^{(0,0,0)}(x,y,z) = -351xyz + 837x^2yz - 486x^3yz + 837xy^2z - 891x^2y^2z - 486xy^3z + 837xyz^2 - 891x^2yz^2 - 891xy^2z^2 - 486xyz^3,$$

$$N_8^{(1,0,0)}(x,y,z) = 18xyz - 99x^2yz + 81x^3yz - 18xy^2z + 108x^2y^2z - 18xyz^2 + 108x^2yz^2 - 54xy^2z^2,$$

$$N_8^{(0,1,0)}(x,y,z) = 18xyz - 18x^2yz - 99xy^2z + 108x^2y^2z + 81xy^3z - 18xyz^2 - 54x^2yz^2 + 108xy^2z^2,$$

$$N_8^{(0,0,1)}(x,y,z) = 18xyz - 18x^2yz - 18xy^2z - 54x^2y^2z - 99xyz^2 + 108x^2yz^2 + 108xy^2z^2 + 81xyz^3.$$

B.4 Concluding remarks

We have listed only a few shape functions here. Most of the other finite element shape functions are fairly straightforward to develop.

The shape functions on a square and a hexahedral (cube) element are obtained as a direct product of shape functions on 1D elements. For example, the cubic Hermite interpolation polynomials on a square lead to a finite element with 16 DoFs with each vertex requiring the four parameters f, f'_x, f'_y, and $f''_{x,y}$ as DoFs.

The use of group theory in determining the shape functions [3] in finite elements with high symmetry is described in Appendix A. This has the advantage of building in the symmetry into the shape functions for an optimal representation of a general function in terms of the interpolation polynomials. When employing interpolation polynomials we should keep in mind that we are representing the unknown solution to the global problem using piece-wise continuous polynomials in each element. The use of Hermite interpolation polynomials allows us to impose derivative continuity at the nodes common to neighboring elements. We may think of the interpolation internal to the element as our representing the solution function using polynomials that have derivative continuity also within each element. This representation allows us to approach the exact global solution faster with fewer elements and generates eigenvalues that are more accurate than Lagrange elements.

B.5 Problems

1. Using cubic Hermite interpolation on a 1D element with two nodes, employ the resulting cubic polynomials to obtain the cubic Hermite interpolation polynomials for
   (a) a square element, and
   (b) a cube element.
2. Obtain the interpolation polynomials for a cube element (a brick element, as it is also referred to) having 27 nodes corresponding to each coordinate having nodes at (–1, 0, 1).
   (a) How many DoFs does it have at each node?
   (b) Determine the number of terms in each 3D polynomial.
3. Derive the shape functions for interpolation within a symmetric standard tetrahedron having the four degrees of freedom $\{f(x,y,z), f'_x(x,y,z), f'_y(x,y,z), f'_z(x,y,z)\}$ at each node. This is a 16-DoF 3D element.

References

[1] L. R. Ram-Mohan, *Finite Element and Boundary Element Applications to Quantum Mechanics* (Oxford University Press, Oxford, UK, 2002).

[2] Gouri Dhatt and Gilbert Touzot, *The Finite Element Method Displayed*, (John Wiley & Sons, NY, 1984).

[3] P. G. Kassebaum, C. R. Boucher, and L. R. Ram-Mohan, J. Comp. Physics. **231**, 5747–5760 (2012); "Application of group representation theory to derive Hermite interpolation on a triangle."

C

Hermite least squares data fitting

In this chapter:

- We show that a least-squares fit to discrete data using piece-wise Hermite interpolation polynomials simultaneously gives the fitted polynomial function and the derivative of the function from the same data set. The physical range over which data is available is divided into finite regions suitably, and further subdivided into sub-regions called elements, and the function is represented in terms of the Hermite interpolation polynomials whose coefficients at special points (called nodes) in each element are function values and the derivatives of the function. The resulting fitted function has derivative continuity ($\mathcal{C}_{(1)}$-continuity) across all nodes and elements.
- The derivative of the function is directly obtained from the fitting parameters within the physical range. The break up of the range into elements leads to greater freedom in fitting the data locally with polynomials of low degree.
- We give examples of this method of data fitting and display its advantages. The data can be fitted without presupposing a specific theoretical form for the fitted function. With a large enough data set, it is possible to obtain not just $\mathcal{C}_{(1)}$-continuity but also piece-wise polynomial fits with $\mathcal{C}_{(2)}$ or higher continuity. Examples of the use of such higher-order Hermite interpolation polynomials in least squares data fitting, the Hermite Least Squares (HLS) fitting is presented.

C.1 Introduction

Very frequently, experimental data collected at discrete values of the abscissa are fitted to a *function*, while its *derivative* is essential in the theoretical analysis. For example, in metallic diffusion, the concentration profile can be determined by chemical or microprobe analysis. However, the flux is governed by the spatial derivative of the concentration of the metallic specie, and it is this quantity that is needed to determine diffusion coefficients. [1] Conversely, one measures the derivative spectrum in modulation spectroscopy, [2] and we then are interested in reconstructing the spectrum from the data. In the first case, the differentiation procedure could degrade the information obtained through data fitting. In the latter case, the derivative spectrum is usually fitted to a preassigned form of the line shape, and this is neither necessary nor essential in extracting out the line shape and the line width.

A third example is the determination of laser threshold from the output power versus input current (called the *L*-*I* graph) in a solid state laser. As the applied bias on the solid state structure is increased, the output power continues to be essentially zero till a threshold current is reached above which the light intensity is observed to increase sharply. The value of the threshold current is used as a characteristic of the quality of the laser, and it specifies the current required to "turn on" the laser. [3, 4] It is the value of the current at which the gain in photon emission balances the loss in photon absorption in the medium. [5] Above threshold, the gain overcomes the losses and the device emits coherent radiation. In actual measurements, this abrupt increase is considerably smoothed out. It then becomes difficult to specify the threshold value of the current precisely.

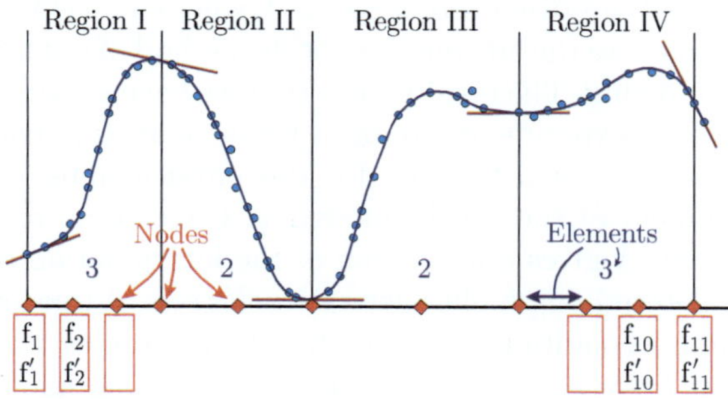

Figure C.1 A schematic diagram for the fitting of experimental data (filled circles) by local Hermite interpolation polynomials defined over each element. The full range of the data is split into regions, each of which is further divided into elements of uniform size in each region. The beginning and end nodes of the element are marked (◇). At each node, the parameters for the function f_i and for the derivative of the function, f_i' are used in the data fitting. The slopes at the ends of the regions are shown as tangent lines.

We show in this chapter that the use of Hermite interpolation polynomials [6] provides a very convenient framework for data fitting in all the above cases, and gives an effective way of obtaining a fit to the data. Simultaneously, the derivative of the fitted function (or *vice versa*) in terms of parameters that are determined on an equal footing can be used to reconstruct the function and the derivative function directly. We are then able to extract the quantities of physical interest much more conveniently and reliably. We note that this method is very general in its applicability, and should find extensive use in data analysis. The examples we have cited are not unique in that every discipline in science has similar situations of requiring the derivative function from the data, or the original function from the derivative data.

In Sec. C.2, we review briefly the usual data fitting procedure with "global" basis sets. [7, 8] We next show that a break-up of the physical range over which the discretized data is available into smaller regions, and the regions further divided into sub-regions, provides an alternative approach. Now we can locally fit the data with low-order polynomials using the method of least squares, thereby allowing for greater flexibility in the data fitting. A simple choice for the basis set is to use piece-wise polynomials that are specific to each element. In particular, we can use local Lagrange polynomials. These polynomials will give inter-element continuity ($\mathcal{C}_{(0)}$-continuity) across elements. This idea is extended to the use of Hermite interpolation polynomials which allows first-derivative continuity ($\mathcal{C}_{(1)}$-continuity) across elements, thereby yielding a smooth function given in terms of low-order polynomials. At first sight, this method may seem to be similar to spline smoothing (see Ref. [7] pp. 113–116). However, here we obtain derivative continuity over the range of an element through minimization in the fitting, with the advantage of obtaining the fitted function and the derivative function immediately over each element. Also, if there is noise in the data, spline smoothing may not work at times.

Examples of Hermite interpolation polynomials with still higher-order continuity are given in the applications discussed in Sec. C.3, where we treat the above-mentioned three examples to illustrate the advantages of this method of data fitting – the Hermite Least Squares (HLS) fitting. The results show that we have at hand a reliable approach to the representation of experimental data in terms of local polynomials with derivative continuity which can then be used for the physical interpretation of the data and for the development of theoretical modeling. Concluding remarks are presented in Sec. C.4, and Chap. A gives a discussion of the interpolation polynomials used. A simple method for determining higher derivative degrees of freedom from the discrete data for the boundary conditions used in the fitting process would be to employ a matrix derivation of *finite difference* formulae for obtaining higher-order derivatives; this leads to reliable inputs in the Hermite fitting with $\mathcal{C}_2, \mathcal{C}_3$, or higher-order derivative continuity.

C.2 Data fitting by the least squares method

C.2.1 Using basis functions for fitting a function

We review very briefly the standard approach to data fitting. [7, 8] The p data points for the function $f(x)$ are given as $\{x_i, f_{exp}(x_i) \equiv f_{exp,i}; i = 1, \ldots, p\}$ pairs over the range $[x_{min}, x_{max}]$. The experimentally measured values have been designated by $f_{exp,i}$.

To initiate the discussion, we may choose a polynomial of some "reasonable" degree, $n(< p)$, to represent the function and write

$$f(x) = \sum_{k=0}^{n} a_k x^k, \tag{C.1}$$

with our basis functions for representing the function to be the powers of x. We can generalize this choice of the basis functions to expand $f(x)$ in terms of n arbitrary fixed functions $X_k(x)$. The basis functions can be nonlinear, but the expansion is linear in the parameters a_k. The least-squares fit to the function is performed by finding the minimum of the *merit function*

$$\chi^2 = \sum_{i=1}^{p} \left[\frac{f_{exp,i} - \sum_{k=1}^{n} a_k X_k(x_i)}{\sigma_i} \right]^2, \tag{C.2}$$

with respect to the parameters a_k. Here, σ_i is the measurement error or the standard deviation of the ith data point. The σ_i are set to a constant value of unity initially, and are determined through the fitting procedure.

It is usual [7, 8] to construct what is called a *design matrix* $\mathbf{A}$ whose $p \times n$ elements are

$$A_{ip,jn} = \frac{X_{jn}(x_{ip})}{\sigma_{jn}}, \quad (ip = 1, \ldots, p; jn = 1, \ldots, n) \tag{C.3}$$

with $p \geq n$ (in order to fit the data with fewer parameters than the number of data points). Thus, the matrix has p rows, one for each data point, and n columns corresponding to each parameter. We use extended variable names such as ip, jp, etc., as is typical in programming languages in order to identify the indices. We also define, for convenience, a p-component vector $\mathbf{b} = \{f_{exp,i}/\sigma_i\}$ and the n-component vector $\boldsymbol{\alpha} = \{a_i\}$ for the fitting parameters. We may write the merit function in the form

$$\chi^2 = (\mathbf{b} - \mathbf{A} \cdot \boldsymbol{\alpha})^T \cdot (\mathbf{b} - \mathbf{A} \cdot \boldsymbol{\alpha}). \tag{C.4}$$

A variation of χ^2 with respect to the individual parameters a_k leads to the n equations, called the *normal equations*,

$$\mathbf{A}^T \cdot \mathbf{A} \cdot \boldsymbol{\alpha} = \mathbf{A}^T \mathbf{b}. \tag{C.5}$$

Here, the $n \times n$ matrix ($\mathcal{D} = \mathbf{A}^T \cdot \mathbf{A}$) is inverted to determine the parameters $\boldsymbol{\alpha}$. It is shown by Press et al., [7] that the uncertainties in the estimated parameters a_i are given by the variance

$$\sigma^2(a_{jj}) = (\mathcal{D}^{-1})_{jj} = (A^T \cdot A)^{-1}_{jj}. \tag{C.6}$$

Also, the off-diagonal elements of the matrix on the right represent the covariances between the parameters a_k and a_{k+1}. We refer the reader to the *Numerical Recipes* [7] for an excellent discussion of the variances. It is implicitly understood in the above approach that the basis functions are globally defined over the entire region $[x_{min}, x_{max}]$.

C.2.2 Use of Lagrange polynomials over smaller regions

A more flexible and effective approach would be to breakup the region $[x_{min}, x_{max}]$ into smaller regions, and to subdivide these further into still smaller sub-regions which we will call elements. This breakup of a physical domain is typically used in finite element analysis which has found extensive use in engineering applications such as structural mechanics, [9, 10] fluid dynamics, [11–13] in modeling electromagnetic fields in complex geometries, [14] and most recently in quantum mechanical modeling of nanoscale systems. [6] The concept of "divide and conquer" can be extended to data fitting as well!

This breakup of the region of interest is illustrated in Fig. C.1, where four regions are used. Regions I and IV are further divided into three elements with the end points of the elements being designated as nodes (shown as $\diamond$ in the figure). Similarly, Regions II and III are divided into two elements each. Fig. C.1 displays the nodes at the beginning and end of each element. In Fig. C.1, the nodes are labeled by the function values f_{in}, which are accompanied with f'_{in}. (The figure uses the index i for convenience. Also, we will consider the derivative values in the next section.) We can also invoke additional nodes internal to each of the elements. We then represent the function separately in each element in terms of low-order polynomials such that the function takes on the value f_{in} at the nodes; this is easily done using Lagrange interpolation polynomials within each element. Note that here the f_{in} are not necessarily the experimental values, but are values at nodes; we will be performing a least-squares fit to the data given at $f_{exp,i}$ with parameters f_{in} that have a direct relation to the fitted function values at nodes.

Suppose we consider quadratic interpolation and employ one additional interior node (typically at the mid-point of the element), we can then write for the element with index iel and nodes at $x_{in}^{(iel)}$ (with $in = 1,2,3$),

$$f(x) = \sum_{in=1}^{3} f_{in} \prod_{\substack{jn=1 \\ jn \neq in}}^{3} \frac{(x_{jn} - x)}{(x_{jn} - x_{in})},$$

$$= \sum_{in=1}^{3} f_{in} L_{in}(x), \quad x_1 \leq x \leq x_3. \tag{C.7}$$

Here, $L_{in}(x)$ are quadratic Lagrange interpolation polynomials, with the property that

$$L_{in}(x_{jn}) = \delta_{in,jn}. \tag{C.8}$$

With the polynomial taking on the value unity at the node, we can employ the value of the fitted function itself as the parameter for the fitting. In other words, the fitting parameters a_i used earlier in Eq. (C.1) in this case directly give the fitted value of the function at the corresponding node.

In using the method of least squares for the data fitting, it is assumed that in each element, we have more data points than parameters f_i associated with the basis function representation. In the present case, with quadratic interpolation within each element, we should attempt to have more than three data points in the range $[x_1^{iel}, x_3^{iel}]$; the size of the elements is after all under our control, and we can usually conform to this rule. The function continuity across the element is assured by employing the same parameter $f_{in=3}$ of element iel as $f_{in=1}$ for element $(iel+1)$. The Lagrange interpolation polynomials are derived over a standard element $[-1, 1]$ in Appendix A and Appendix B, and a simple linear mapping from the standard element to the actual (or global) element readily gives the Lagrange polynomials.

In the present example, let us suppose there are three elements in a single region, then the total number of nodes is $(3 * 2 + 1 = 7)$. Suppose there are 4, 4, and 5 data points in these elements. With $p = 11$ data points at hand, the 11×7 design matrix takes the form:

$$\underline{data} \quad \underline{\#nodes\,(or\; parameters\, f_i)} \longrightarrow$$

$$\underline{point} \downarrow$$

$$\mathbf{A} = \begin{pmatrix} \tilde{L}_1^1(x_1) & \tilde{L}_2^1(x_1) & \tilde{L}_3^1(x_1) & 0 & 0 & 0 & 0 \\ \tilde{L}_1^1(x_2) & \tilde{L}_2^1(x_2) & \tilde{L}_3^1(x_2) & 0 & 0 & 0 & 0 \\ \tilde{L}_1^1(x_3) & \tilde{L}_2^1(x_3) & \tilde{L}_3^1(x_3) & 0 & 0 & 0 & 0 \\ \tilde{L}_1^1(x_4) & \tilde{L}_2^1(x_4) & \tilde{L}_3^1(x_4) & 0 & 0 & 0 & 0 \\ 0 & 0 & \tilde{L}_1^2(x_5) & \tilde{L}_2^2(x_5) & \tilde{L}_3^2(x_5) & 0 & 0 \\ 0 & 0 & \tilde{L}_1^2(x_6) & \tilde{L}_2^2(x_6) & \tilde{L}_3^2(x_6) & 0 & 0 \\ 0 & 0 & \tilde{L}_1^2(x_7) & \tilde{L}_2^2(x_7) & \tilde{L}_3^2(x_7) & 0 & 0 \\ 0 & 0 & \tilde{L}_1^2(x_8) & \tilde{L}_2^2(x_8) & \tilde{L}_3^2(x_8) & 0 & 0 \\ 0 & 0 & 0 & 0 & \tilde{L}_1^3(x_9) & \tilde{L}_2^3(x_9) & \tilde{L}_3^3(x_9) \\ 0 & 0 & 0 & 0 & \tilde{L}_1^3(x_{10}) & \tilde{L}_2^3(x_{10}) & \tilde{L}_3^3(x_{10}) \\ 0 & 0 & 0 & 0 & \tilde{L}_1^3(x_{11}) & \tilde{L}_2^3(x_{11}) & \tilde{L}_3^3(x_{11}) \\ 0 & 0 & 0 & 0 & \tilde{L}_1^3(x_{12}) & \tilde{L}_2^3(x_{12}) & \tilde{L}_3^3(x_{12}) \\ 0 & 0 & 0 & 0 & \tilde{L}_1^3(x_{13}) & \tilde{L}_2^3(x_{13}) & \tilde{L}_3^3(x_{13}) \end{pmatrix}$$

The superscripts on $\tilde{L}_{in}(x_j) = L_{in}(x_j)/\sigma_j$ denote the element number iel, while the subscripts denote the node number within each element. Each row corresponds to a data point, and each entry in the row includes the factor of the measurement error σ_j in the denominator. Here, the three Lagrange polynomials $L_{in}^{iel}(x)$ depend on location of the three nodes in each element. The first block of entries in the above matrix are for the first element, the next block of entries for the second element, and so on. The number of rows for each block depends on the number of data points falling within the element, and we have chosen to display the design matrix with 4, 4, and 5 data points in the elements, respectively. Note that with three elements, we are using the same nodal parameter for node 3 of element 1 and node 1 of element 2, thereby assuring continuity of the function across the first two elements. The same applies across all the elements.

The design matrix is enlarged to include the contributions from all elements over the entire physical domain. The merit function in this case is minimized with respect to the nodal values f_{in} to obtain

$$\mathcal{D} \cdot \boldsymbol{\alpha} = \mathbf{A}^T \cdot \mathbf{A} \cdot \boldsymbol{\alpha} = \mathbf{A}^T \mathbf{b}, \tag{C.9}$$

where $b_{ip} = f_{exp,ip}/\sigma_{ip}$ are given from experimental measurement, and $\alpha_{in} = f_{in}$, the nodal parameters in our element-by-element interpolation with Lagrange polynomials. An inversion of the matrix $\mathcal{D}$ using standard matrix computation libraries such as LAPACK [15] then allows us to determine the parameters f_{in} for the function interpolations within each element. As mentioned above, the diagonal elements of $\mathcal{D}^{-1}$ give the variance for each of the parameters, so that we can verify whether the fit to the data has acceptable accuracy. An iteration should be performed by reinserting the determined variances into the above considerations in order to obtain the actual values of the variances, since we initially assumed all of these to be unity.

If we desire to fix the value of the function at the first and the last node, say, or for that matter at any node, we implement this in a manner analogous to applying boundary conditions in the finite element method. [6] The terms in the normal equations, Eq. (C.5), with the variables that are fixed by our requirement are moved to the right side of the normal equations since they are known quantities. We now have fewer equations to solve. The reader is referred to Ref. [6] (see Chap. 5, pp. 120–121) for a simple approach, which has been called giving benediction to the matrix, for retaining the same matrix size while applying boundary conditions and solving the equations.

It is invariably tempting to use high-order polynomials to do the data fitting, but this is usually fraught with issues in obtaining reliable derivatives for the fitted function. In the following, we use higher-order polynomials with derivative continuity which allows us to obtain the derivatives directly. Other methods of determining the parameters can also be problematic. For example, if one were to use the simplex method to determine the parameters through the least squares data fitting, each attempt at the fit could give parameters for the polynomial that can thread through the data points very well but have a bulge upward or downward between the points with every starting choice or seed value of the parameters in the method. This makes the evaluation of the derivatives by such methods totally unreliable.

We should note that the Lagrange interpolation may be smooth (quadratically smooth, in our illustration above) within each element, but this cannot guarantee derivative continuity across elements. With the nodal values being shared across elements at the ends, we have assured only $\mathcal{C}_{(0)}$-continuity for the fitted function. We remedy this limitation in the following section.

C.2.3 Least squares fitting with Hermite polynomials

We can improve the continuity properties for the fitted function, within the elements and across them, by allowing for additional parameters in defining the function at the nodes. Let us suppose initially that the discrete values of the function at the nodes are known. If the first derivatives of the function are also known at the nodes, then we can represent the function in terms of two new sets of polynomials $N_i, \overline{N}_i$. It is usual to define these polynomials, called shape functions, over the region $-1 \leq \xi \leq 1$ and then use a linear mapping from the local element $[-1, 1]$ to the global element $[x_a, x_b]$. This is discussed in Chap. A. Here, we are using the parlance of the finite element method in identifying the elements as local versus global.

We write

$$f(x) = f_i N_i(\xi) + \left(\frac{df(x)}{dx}\right)_i \left(\frac{dx}{d\xi}\right) \overline{N}_i(\xi). \tag{C.10}$$

The $N_i, \overline{N_i}$ are functions of ξ, and the derivative $df(x_i)/dx$ with respect to x then requires a scale factor $dx/d\xi$ dictated by the chain rule of differentiation. As shown in Chap. A, the interpolation polynomials are designed so that $N_i(\xi_j) = \delta_{ij}$, analogous to the Lagrange polynomials, while $\overline{N}_i(\xi_j) = 0$, at the nodes located at ξ_j. This ensures that the function $f(x)$ takes its discrete values f_i at the nodes while the $\overline{N}$-polynomials do not affect these values. At the same time, we choose the coefficients of polynomials such that $dN_i(\xi_j)/d\xi = 0$ and $d\overline{N}_i(\xi_j)/d\xi = \delta_{ij}$. This ensures that the derivative of the function with respect to x takes on its appropriate value at the nodes. (We have simplified the notation here and in the following by referring to nodal values by i, j rather than in, in.)

By writing Eq. (C.10) in a compact form

$$f(x) = f_\alpha N_\alpha(x), \tag{C.11}$$

with the parameters ordered so that

$$f_\alpha = [(f_1, f'_1), (f_2, f'_2), \ldots], \tag{C.12}$$

we can relate the present interpolation to the notation we introduced for the Lagrange interpolation. In other words, the merit function can still be written using $\mathbf{b}_i = \{f_{exp,i}/\sigma_i\}$ and $\boldsymbol{\alpha}_i = \{(f_i, f'_{,x,i})\}$ by extending the array of the parameters to allow for all the derivative parameters at the node i. We have

$$\chi^2 = (\mathbf{b} - \mathbf{A} \cdot \boldsymbol{\alpha})^T \cdot (\mathbf{b} - \mathbf{A} \cdot \boldsymbol{\alpha}), \tag{C.13}$$

again, except that the design matrix $\mathbf{A}$ includes the values of all the shape functions at the data points. The parameters and the corresponding shape functions for each node are grouped together sequentially. For example, the design matrix for the curve in Fig. C.1 entails arranging the columns to correspond to $(f_1, f'_1), (f_2, f'_2), \ldots, (f_{11}, f'_{11})$ with as many rows as there are experimentally measured points. This arrangement is said to group together the "degrees of freedom" for the function at each node. The values of f_1 and f'_1 can be used as input, with the derivative at x_1 determined by a 7- or 9-point forward difference method as described in Chap. A. In Fig. C.1, we have shown the tangents to the curve at the beginning and end of regions. For $\mathcal{C}_{(1)}$-continuity we employ the shape functions of Eq. (C.10).

The rest of the procedure is by now straightforward: we set up the normal equations using the design matrix and solve for the unknown parameters (f_i, f'_i). The variances are also obtained in a manner similar to the Lagrange interpolation approach. Now the inter-element behavior is guaranteed to have derivative continuity because of our using the same nodal parameters for the nodes common to adjacent elements.

Several comments are in order:

(i) With the HLS fit, we obtain nodal parameters that at once allow us to obtain the derivative of the fitted function using Eq. (C.10). Thus, the same data fitting gives us the derivative function that is as reliable as the original function. The same variances apply to both the function and the derivative.

(ii) If we are given derivative data, from modulation spectroscopy for example, we can construct the design matrix and the normal equations with the expression for the derivative function, Eq. (C.10), with values of the derivatives of the shape functions evaluated at data points appearing in the design matrix. This allows us to obtain the HLS fitting parameters from which we can directly obtain the function representing the spectrum.

(iii) Frequently, the slope of the fitted function can be expected to be a fixed value at the ends of the physical region, say. In this case, we again view the normal equations as simultaneous equations in which the known terms are to be taken to the right side of the equation before inversion of the coefficient matrix. If the measured value of the function is reliable at the end points, we can use them as input values. If the derivatives have to be determined at the end points, we can make use of n-point finite difference methods with forward/backward differencing; here n can be chosen to be 5, 7, or 9 points for higher-order differencing formulae for greater reliability. For convenience, a short derivation of difference formulae using a matrix inversion approach is included in Appendices A and B. We discuss these aspects of data fitting in the following section with examples. Note that if there is noise in the data at the ends, the determination of derivatives at the end points will become an issue. However, the set of simultaneous equations requires only some convenient input values in order to invert it, if it is useful to pin down some of the parameters elsewhere, based on physical considerations.

(iv) Once the variances are determined, they can be used as input and the procedure iterated once to determine improved parameters and new variances.

C.3 Application of HLS data fitting to physical problems

In this section, we consider the three examples mentioned in the Introduction (Sec. C.1).

C.3.1 Fitting data from modulation spectroscopy

Modulation spectroscopy is now a well-established method [2] for improving the signal-to-noise ratio in spectroscopy through the periodic variation of a physical parameter through an external perturbation that affects the observed spectrum. Many types of modulation have been explored including electromodulation, magnetomodulation, frequency modulation, etc. For example in piezomodulation, a piezoelectric material is attached to the sample to constrict it periodically. The applied stress leads to a periodic variation of the dielectric constant, so that the derivative of the usually expected optical spectrum is obtained. In investigating inter-band optical transitions in semiconductor quantum wells and other quantum heterostructures, the piezomodulation method was used to obtain spectra of unprecedented quality. [16–22]

To illustrate the issue of data fitting and modulation spectroscopy, we consider experiments using reflectivity R to obtain the signature of the electron-hole excitonic recombination in semiconductors. This was done by employing electric field modulation during reflectivity measurements on CdMnTe quantum wells. [23] We show the modulated reflectivity $((\Delta R/\Delta E) \times (1/R))$ as a function of photon energy E in Fig. C.2, where the displayed data are obtained by digitizing the reported derivative

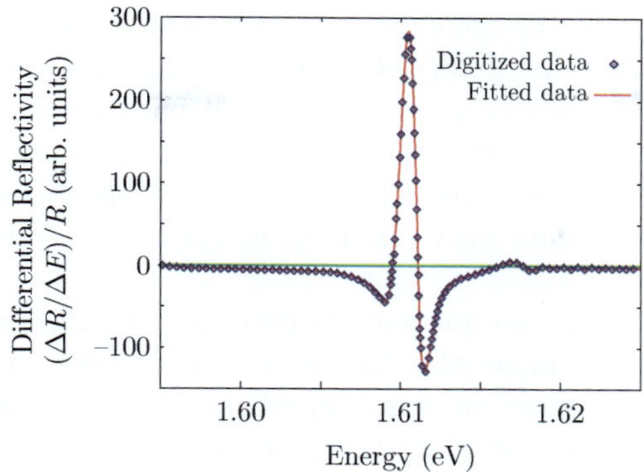

Figure C.2 The wavelength-modulated reflectivity spectrum of CdMnTe is the measured derivative spectrum reported in Ref. [23]. The digitized data (◇) and a fit to it (continuous curve) using $\mathcal{C}_{(3)}$-continuous Hermite polynomials are shown. Four regions were used for the data fitting, as described in the text.

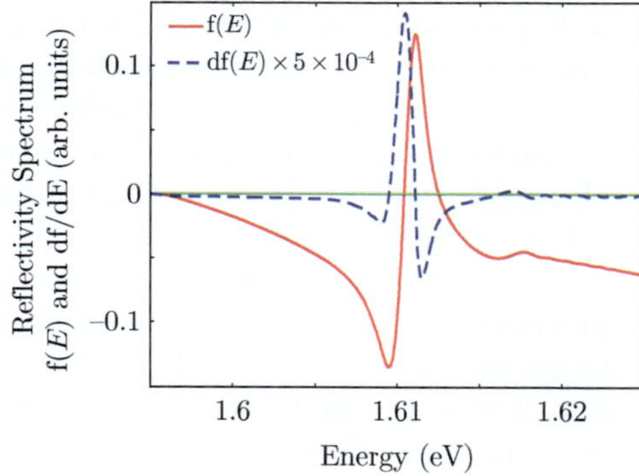

Figure C.3 The electromodulated reflectivity spectrum of CdMnTe (the measured derivative spectrum – dashed curve – scaled by a factor 5.0×10^{-4}) is shown with the integrated spectrum (continuous curve) as obtained from a $\mathcal{C}_{(1)}$- Hermite interpolation polynomial fit to the derivative data.

spectrum. [23] The digitization was performed using the software *Engauge* [24] that is freely available over the Internet.

The data was fitted using the HLS method with $\mathcal{C}_{(3)}$-continuous Hermite shape functions since the graph is oscillatory. The result is shown in Fig. C.2 as the continuous curve. Four regions were used for fitting the derivative data. In region I, 10 elements were employed over the range 1.5946–1.61 eV; in region II, three elements were used over the range 1.61–1.611 eV; region III was fitted with three elements over 1.611–1.612 eV; and in region IV over the range 1.612–1.6261 eV, 10 elements were used. The function value was set to zero at the initial node, while the derivative degrees of freedom f', $f^{(ii)}$, and $f^{(iii)}$ were determined by a forward-differencing scheme at that node. The four parameters at all other nodes were determined by solving the normal equations, with inter-element continuity assured by using values of the function and its three derivatives at the common node for adjacent elements. Once the four parameters at each of the nodes are determined, we can construct the actual spectrum using the $\mathcal{C}_{(3)}$ Hermite shape functions and the nodal parameters. In Fig. C.3, the electromodulated reflectivity spectrum of CdMnTe (the measured derivative spectrum – dashed curve – is scaled by a factor of 5.0×10^{-4}) is shown, with the integrated (or actual) spectrum (continuous curve) as obtained from a $\mathcal{C}_{(3)}$-HLS fit to the derivative data.

It is interesting that we do not have to make any assumptions about the shape of the spectrum. In this example, the derivative of the dielectric function with respect to the photon energy is being measured in reflectivity which depends on the complex refractive index of the material. The dielectric function is a complicated one which we will not reproduce here, but direct the reader to Ref. [25]. Suffice it to mention that the differential reflectance is given by

$$\frac{\Delta R}{R} = \frac{1}{R}\frac{\partial R}{\partial \epsilon_r}\Delta\epsilon_r + \frac{1}{R}\frac{\partial R}{\partial \epsilon_i}\Delta\epsilon_i, \tag{C.14}$$

where the partial derivative factors are functions of both the real and imaginary parts of the dielectric function. In piezomodulation the energy band gap in the semiconductor is altered by the applied piezo-stress, and the functional form of the modulated reflectivity has a resonant form as seen in Fig. C.3. Usually, the function Eq. (C.14) is fitted to the data and the line-width parameter, and the resonant energy is deduced from the fit. The HLS fitting is model-independent. Expressions for the derivative terms when the dielectric function is approximated by $\epsilon = 1 + A/[(E - E_g) + i\Gamma]$, with E_g being the energy bandgap in the semiconductor are given by Pollak. [26]

Note that once the procedure of $\mathcal{C}_{(1)}$-HLS fitting is set up, it is easy to extend the idea to allow $\mathcal{C}_{(3)}$-continuity, or for that matter, derivative continuity of still higher order. We see from Fig. C.3 that the fit to the data is very good, and the actual spectrum is obtained with very little effort once the nodal parameters are obtained.

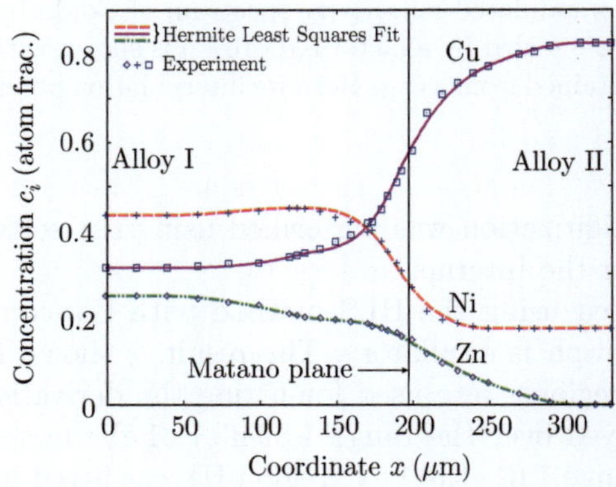

Figure C.4 A $\mathcal{C}_{(1)}$-Hermite polynomial fit to concentration data in a 3-component metallic diffusion between two ternary alloys of Zn, Ni, and Cu are shown. Five regions with beginning/end coordinates [0, 40, 125, 250, 310, 330] μm. One element was used in each region except for the middle region 125–250 μm, where two elements were employed to capture the behavior of the concentration curves around the planar Matano interface. (The data was reported earlier. [27])

C.3.2 Using HLS fitting for multi-component metallic diffusion

Consider metallic bars of two different alloys of n metal component species with different concentrations of metals. When the bars are put end-to-end and are heated in a furnace for a period of time, the atoms of the metals inter-diffuse from one region to the other across the interface. With conservation of mass as a constraint on the sum of the concentrations, we have $n-1$ independent concentrations $C_i(x)$. The variation of the concentrations as a function of the lateral coordinate x is determined experimentally by slicing the sample and using electron microprobe analysis or by chemical means. The central issue in multi-component metallic diffusion is to be able to predict the concentration profiles. This requires the determination of the diffusion coefficients.

However, there are several hurdles to overcome before this can be done: (*i*) The data usually has considerable noise so that the data fitting is nontrivial. (*ii*) The relation between the flux components and the concentration gradient requires an extension of the usual Fick's law to a multi-component relation [28]

$$J_i(x) = D_{ij}\frac{\partial C_j}{\partial x}, \qquad (i, j = 1, 2, \ldots, n-1). \tag{C.15}$$

Note that the concentrations are measured, but their gradients are needed in the above relation. Moreover, we have $(n-1)^2$ diffusion coefficients, but only $n-1$ relations in Eq. (C.15), and so there are more unknowns than we can solve for. (*iii*) Finally, the diffusion coefficients are themselves functions of the compositions, making diffusion a nonlinear problem. Here our focus is on the data fitting. However, we note that by approximating the diffusion coefficients to be constants over suitably small ranges of the coordinate, we can linearize the problem; [29] and taking moments of Eq. (C.15) over these small intervals, it is possible to generate additional constraint equations that allow us to obtain the diffusion coefficients with some degree of accuracy. [30] A transfer matrix approach developed for the diffusion equation [27, 31, 32] then provides an elegant solution to the coupled concentration profiles once the diffusion coefficients are estimated.

In this chapter, we consider the use of HLS fitting to obtain reliable derivatives of the concentration profiles. An example of concentration profiles is shown in Fig. C.4, in which the data is fitted using a $\mathcal{C}_1$-Hermite polynomial fit for diffusion between two ternary alloys of Zn, Ni, and Cu. Five regions with beginning/end coordinates [0, 40, 125, 250, 310, 330]μm are used. One element was used in each region except for the middle region 125–250μm, where two elements were employed to capture the behavior of the concentration curves around the planar interface known as the Boltzmann-Matano plane. [33] It is seen from Fig. C.4 that the fitting is excellent. The concentration gradients obtained from the HLS fitting are then used in further calculations as discussed elsewhere. [27]

C.3.3 Determining the threshold current in LI-graphs

As a final example of the use of HLS fitting, consider the case of determining the laser threshold from experimental data. A typical plot of output power $P(i)$ as a function of the input current through a semiconductor laser is shown in Fig. C.5. The rise of

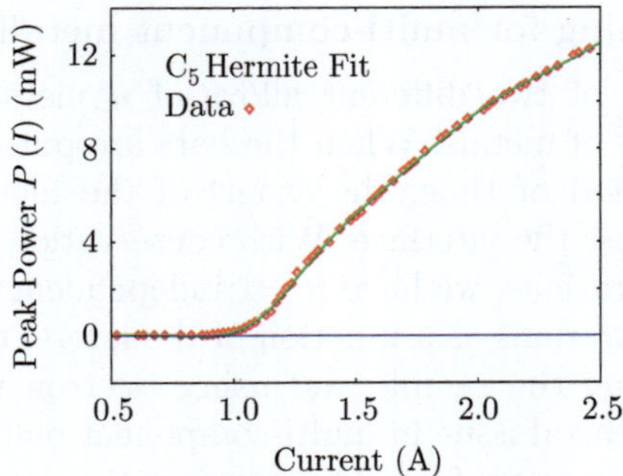

Figure C.5 A typical curve of the peak power of a semiconductor quantum cascade laser versus the input current, the so-called LI-curve. The digitized data (◇) is fitted with $\mathcal{C}_{(5)}$-continuous Hermite interpolation polynomials which provide 4th-derivative continuity across elements. The data was fitted using one region ranging over 0.52–2.69 A with one element.

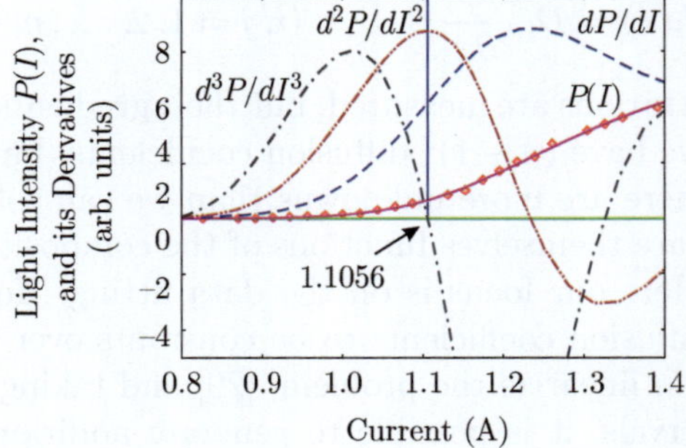

Figure C.6 A plot of the peak power $P(I)$ and its first three derivatives is shown. $P(I)$ and its derivatives are obtained from a $\mathcal{C}_{(5)}$-Hermite fit of the data shown in Fig. C.5. The curves are displayed here in the region around the laser threshold. The function dP/dI is scaled by a factor 0.5, d^2P/dI^2 is scaled by a factor of 0.1, and the third derivative d^3P/dI^3 is multiplied by a scaling factor of 0.015.

the power at the threshold is not quite abrupt, and the experimental data usually has noise that does not allow a good determination of the threshold conditions. The basics of the onset of lasing are presented in Refs. [3, 4]. In the present example, we have used $\mathcal{C}_{(5)}$-Hermite shape functions with just one element over the full region from 0.52–2.69 A. At low current, the power and its first five derivatives are set to zero, and the power alone is set to its observed value for large current. Other Hermite parameters are determined by solving the normal equations. We can now pin down the value of the current at threshold by plotting the power and its first three derivatives. As we see from Fig. C.6, the second derivative peaks at the threshold at 1.106 A, and the third derivative becomes zero there. We are thus able to identify the threshold quite well by inspecting the higher derivatives of the output power.

With the noise in the data, the higher derivatives used here to pin down the threshold behavior would not be reliably obtained by other methods, whereas the HLS fitting yields the function and its derivatives determined in terms of interpolation polynomial fitting, with the parameters associated with the degrees of freedom (i.e. f_i, f'_i, etc.) all determined directly from the data by a least square fitting procedure. In other words, the function and its derivatives are determined to the same level of reliability.

C.4 Concluding remarks

We have shown that the combination of finite element concepts with $\mathcal{C}_{(n)}$-continuous Hermite interpolation polynomials leads to a data fitting procedure that is locally adjustable through the insertion of more or fewer elements in each sub-region. The ubiquitous problem of requiring derivatives of fitted functions is resolved in a novel way in this chapter. The method is also applicable for obtaining functions when their derivative data is available, as in modulation spectroscopy. The approach presented here, thus, addresses a fairly general problem faced in the interpretation of physical data. It is hoped that this chapter will be useful to researchers and to students in undergraduate physics courses.

C.5 Problems

1. Construct a 100-point data set for $\sin(x)$ discretized over the range $[0, \pi]$ and add noise to the data by using a random number generator. Consider the fitting of this data set using cubic-Hermite polynomials defined in each of 10 elements over the range of values of x.
   (a) What happens to the variances when the number of elements is reduced to 5?
   (b) What happens to the variances when the number of elements is increased to 20?
   (c) Obtain the derivative of the interpolating function and compare it with the known derivative $\cos(x)$, in this case.
2. Obtain an expression for the differential reflectance when the dielectric function is

$$\epsilon = 1 + A/[(E - E_g) + i\Gamma]. \tag{C.16}$$

   Note that the dielectric function can be restated in terms of the complex refractive index as $\epsilon = (n + i\kappa)^2$ so that the reflection coefficient of light at a planar surface of the material can be calculated. This then allows the determination of the variation $\Delta R/R$.
3. (a) Show that the quadratic Lagrange interpolation polynomials of Eq. (C.7) agree with the usual expression

$$L_i(\xi) = \prod_{\substack{j=1 \\ j \neq i}}^{3} \frac{(\xi_j - \xi)}{(\xi_j - \xi_i)}, \tag{C.17}$$

   with $\xi_i = -1, 0, 1$, respectively.

(b) Compare the expressions for cubic Lagrange polynomials.

4. (a) Verify that the derivative factor $dx/d\xi$ is needed with the terms containing $\overline{N}_i(\xi)$ in Eq. (C.10) in order to faithfully reproduce f' at the nodes.
   (b) Show that the linear mapping Eq. (2.3) from $-1 \leq \xi \leq +1$ to $x_a \leq x \leq x_b$ leads to the function values $f(\xi)$ and $f(x)$ to be the same for ξ_0 in the standard element and the corresponding point x_0 in the global element.
5. (a) Working in double precision arithmetic, develop a 7-point finite difference algorithm to obtain the first six derivatives of any function.
   (b) Apply the 7-point formula on a test function, such as $\sin(x)$, whose derivatives are known. For the range of values from $[0, \pi/2]$ with 100 intervals, tabulate the calculated values of the first four derivatives and compare them with the analytical values. Notice the rapid loss of accuracy as we go to higher derivatives. Show that 7-point central differencing formulae lead to $\sim$8-digit accuracy for the third and the fourth derivatives while the second derivative in this example is accurate to 12 digits.
   (c) The above method can be extended to forward-differencing and to backward-differencing relations for the derivatives at a given data point. Are these forward/backward differencing results as accurate as the central differencing scheme?

References

[1] J. S. Kirkaldy and D. J. Young, *Diffusion in the Condensed State* (The Institute of Metals, London, 1987), pp. 226–272.

[2] M. Cardona, *Modulation Spectroscopy* in *Solid State Physics, Supplement* **11**, (Academic, New York, 1969), edited by M. Cardona, F. Seitz, D. Turnbell and H. Ehrenreich.

[3] M. Shur, *Physics of Semiconductor Devices*(Prentice-Hall, Upper Saddle River, NJ, 1990).

[4] S. L. Chuang, *Physics of Optoelectronic Devices* (J. Wiley, New York, NY, 1995) p. 309.

[5] M. G. A. Bernard and G. Duraffourg, Phys. Status Solidi **1** 699–703 (1961); "Laser conditions in semiconductors."

[6] L. R. Ram-Mohan, *Finite Element and Boundary Element Applications in Quantum Mechanics*, (Oxford University Press, Oxford, UK, 2002), Ch. 3 pp. 64–77.

[7] W. H. Press, S. A. Teukolsky, W. T. Vetterling, and B. P. Flannery, *Numerical Recipes in C: The Art of Scientific Computing* (Cambridge University Press, Cambridge, UK, 1992), Ch. 15.

[8] P. R. Bevington and D. K. Robinson, *Data Reduction and Error Analysis for the Physical Sciences*, 2nd edition (McGraw-Hill, New York, NY, 1992), Ch. 7.

[9] O. C. Zienkiewicz, R. L. Taylor, and J. Z. Zhu, *The Finite Element Method*, 6th edition (Elsevier Butterworth-Heinemann, Burlington, MA, 2005).

[10] T. J. R. Hughes, *The Finite Element Method*, (Prentice-Hall, Englewood Cliffs, NJ, 1987).

[11] O. C. Zienkiewicz, R. L. Taylor, and P. Nithiarasu, *The Finite element Method for Fluid dynamics*, 6th edition (Elsevier Butterworth-Heinemann, Burlington, MA, USA 2005).

[12] T. J. Chung, *Finite Element Analysis in Fluid Dynamics*, (McGraw-Hill, New York, 1978).

[13] J. N. Reddy and D. K. Gartling, *The Finite Element Method in Heat Transfer and Fluid Dynamics* (CRC Press, Boca Raton, FL, 1994).

[14] J.-M. Jin, *The Finite Element Method in Electromagnetics*, 2nd edition (Wiley-Interscience, New York, NY, 2002).

[15] E. Anderson, Z. Bai. C. Bischof, S. Blackford, J. Demmel, J. Dongarra, J. Du Croz, A. Greenbaum, S. Hammarling, A. McKenney, and D. Sorensen, *LAPACK Users' Guide* (Society for Industrial and Applied Mathematics, Philadelphia, PA,1999) 3rd edition. The software is available at the website: http://www.netlib.org/lapack/.

[16] Y. R. Lee, A. K. Ramdas, F. A. Chambers, J. M. Meese, and L. R. Ram-Mohan, Appl. Phys. Lett. **50**, 600–602 (1987); "Piezomodulated electronic spectra of semiconductor heterostructures: GaAs/AlGaAs quantum well structures."

[17] D. Dossa, Lok C. Lew Yan Voon, L. R. Ram-Mohan, C. Parks, R. G. Alonso, A. K. Ramdas, and M. R. Melloch, Appl. Phys. Lett. **59**, 2706–2708 (1991); "Observation of above-barrier quasi-bound states in asymmetric single quantum wells by piezomodulated reflectivity."

[18] R. G. Alonso, C. Parks, A. K. Ramdas, H. Luo, N. Samarth, J. K. Furdyna, and L. R. Ram-Mohan, Phys. Rev. B **45**, 1181–1186 (1992); "Modulated reflectivity spectrum of ZnSe/$Zn_{1-x}Cd_xSe$/ZnSe strained single quantum wells."

[19] C. Parks, R. G. Alonso, A. K. Ramdas, L. R. Ram-Mohan, D. Dossa, and M. R. Melloch, Phys. Rev. B **45**, 14215–14224 (1992); "Piezomodulated reflectivity of $Al_{x1}Ga_{1-x1}As$/GaAs/$Al_{x2}Ga_{1-x2}As$ asymmetric and symmetric single quantum wells."

[20] C. Parks, A. K. Ramdas, L. R. Ram-Mohan, and M. R. Melloch, Phys. Rev. B **48**, 5413–5421 (1993); "Piezomodulated reflectivity study of minibands in $Al_xGa_{1-x}As$/GaAs superlattices."

[21] C. Parks, A. K. Ramdas, M. R. Melloch, G. Steblovsky, L. R. Ram-Mohan, and H. Luo, Solid State Commun. **92**, 563–567 (1994); "Observation of electronic states confined in surface quantum wells and above quantum barriers in modulated reflectivity."

[22] C. Parks, A. K. Ramdas, M. R. Melloch, G. Steblovsky, L. R. Ram-Mohan, and H. Luo, J. Vac. Sci. Technol. B **13**, 657–659 (1995); "Modulated reflectivity spectroscopy of electronic states confined in surface quantum wells and above quantum barriers."

[23] H. Alawadhi, I. Miotkowski, V. Souw, M. McElfresh, A. K. Ramdas, and S. Miotkowska, Phys. Rev. B **63**, 155201–155210 (2001); "Excitonic Zeeman effect in the zinc-blende II-VI diluted magnetic semiconductors $Cd_{1-x}Y_xTe$

(Y=Mn, Co, and Fe)." We have digitized Fig. 3(a) of this paper for data analysis. We thank Professor Ramdas for suggesting his article for our data analysis.

[24] The software Engauge for digitizing graphs obtained from a scanner is available at http://digitizer.sourceforge.net/.

[25] P. Y. Yu and M. Cardona, *Fundamentals of Semiconductors*, Third edition (Springer, Berlin, Germany, 2001) pp. 315–333.

[26] F. H. Pollak, *Modulation Spectroscopy and Surface Plotovoltage Spectroscopy of Semiconductor Quantum Wires and Quantum Dots*, in *Nano-Optoelectronics - Concepts, Physics and Devices*, M. Grundmann, Editor (Springer, Berlin, 2002).

[27] L. R. Ram-Mohan, and M. A. Dayananda, Acta Materialia **54**, 2325–2334 (2006); "A transfer matrix method for the calculation of concentrations and fluxes in multicomponent diffusion couples."

[28] L. Onsager, Phys. Rev. **37**, 405–426 (1931); "Reciprocal relations in irreversible processes. I." L. Onsager, Ann. N. Y. Acad. Sci. **46**, 241–265 (1945); "Theories and problems of liquid diffusion."

[29] M. A. Dayananda, Metallur. Mater. Trans. **30A**, 535–543 (1999); "A new analysis for the determination of ternary interdiffusion coefficients from a single diffusion couple."

[30] K. M. Day, L. R. Ram-Mohan, and M. A. Dayananda, J. Phase Equilib. Diffus. **26**, 579–590 (2005); "Determination and assessment of ternary interdiffusion coefficients from individual diffusion couples."

[31] K. Kulkarni, A. M. Girgis, L. R. Ram-Mohan, and M. A. Dayananda, Philos. Mag. **87**, 853–872 (2007); "A transfer matrix analysis of quaternary diffusion."

[32] K. Kulkarni, L. R. Ram-Mohan, and M. A. Dayananda, J. Appl. Phys. **102**, 064908–064912 (2007); "A matrix Green's function analysis of multicomponent diffusion in multilayered assemblies."

[33] L. Boltzmann, Ann. der Phys. **53**, 959–964 (1894); "Zur Integration der Diffusiongleichung bei Variablen Diffusions-coefficienten." C. Matano, Jap. J. Phys. (Trans.) **8**, 109–113 (1933); "On the relation between the diffusion coefficients and concentrations of solid metals (The Nickel-Copper System), Jpn. J. Phys. (Trans.)."

About the Author

L. Ramdas Ram-Mohan received his BSc Honors degree in physics from the University of Delhi in 1964, and his PhD at Purdue University in 1971. He is presently Professor of Physics, Electrical & Computer Engineering, and Mechanical Engineering at Worcester Polytechnic Institute (WPI). Ram-Mohan's interests have ranged over elementary particle theory, nuclear matter and phase transitions, theory of metals, the linear and nonlinear optical properties of quantum semiconductor heterostructures, and the modeling of quantum semiconductor structures using the finite element method. He has worked as a consultant on problems in theoretical physics at several universities and research laboratories. He has over 250 journal articles to his credit. He has had a continuing interest in computational techniques since 1977 as an adjunct to his theoretical work. He is the President of Quantum Semiconductor Algorithms, a software and consultancy company started by him in 1993. He works closely with experimentalists while maintaining his independent theoretical program of research.

Ram-Mohan is a Fellow of the American Physical Society (APS), the Optical Society of America (OSA), the American Vacuum Society (AVS), the American Association for the Advancement of Science (AAAS), the Institute of Physics (UK), and the Australian Institute of Physics. He received the national Engineering Excellence Award of the Optical Society of America in 2003 for his development of the paradigm of wavefunction engineering and its use in designing nanoscale semiconductor structures for optoelectronic applications. In 1996, he was recognized with the Distinguished Alumnus Award by the School of Science at Purdue University. He is the recipient of the Exemplary Faculty Prize, the Trustees' Award for Creative Scholarship, and the Outstanding Teaching Award at WPI.

Author Index

Subject Index